MONGOLIA

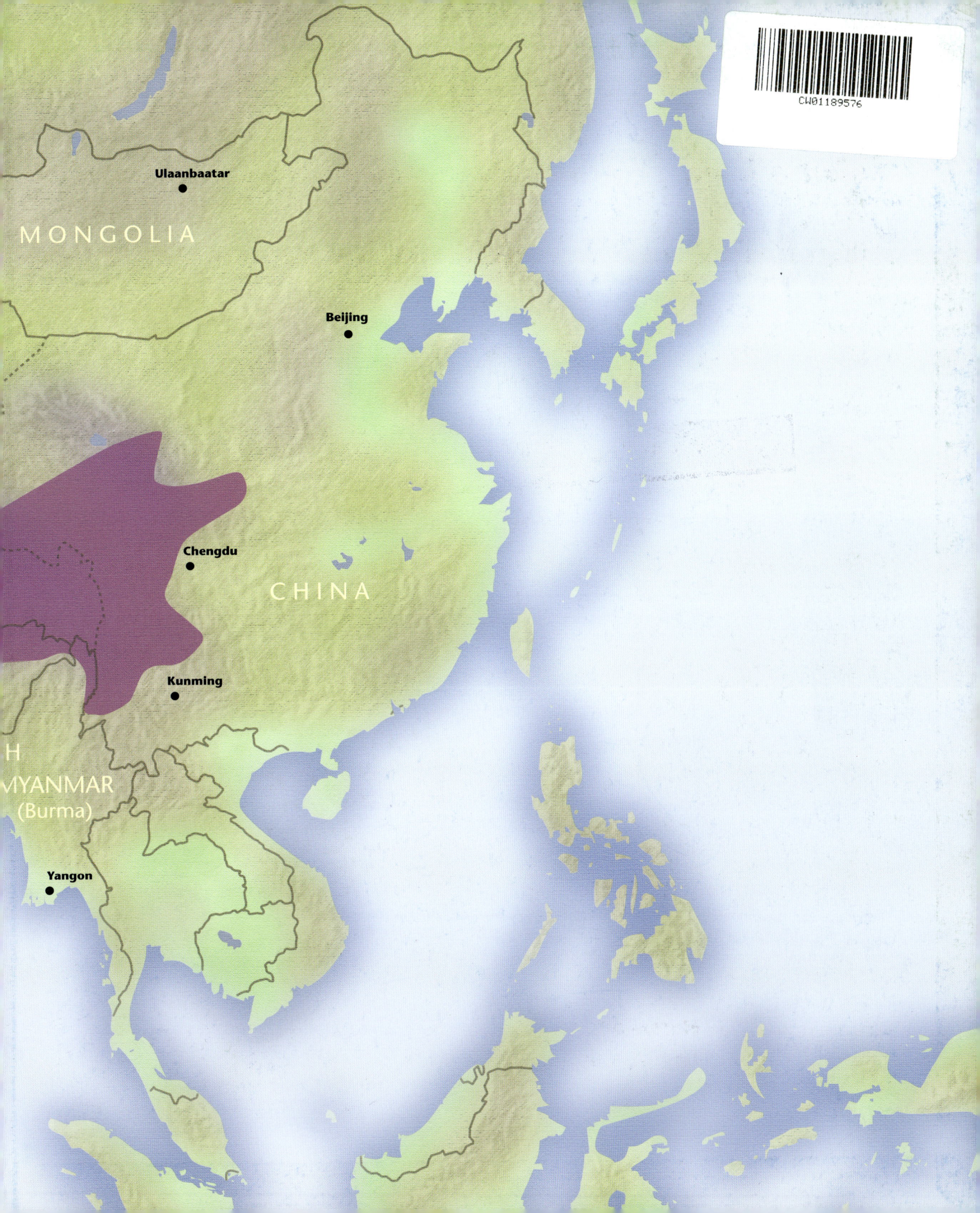

Ulaanbaatar

Beijing

Chengdu

CHINA

Kunming

MYANMAR
(Burma)

Yangon

THE GENUS
MECONOPSIS
Blue poppies and their relatives

THE GENUS MECONOPSIS

Blue poppies and their relatives

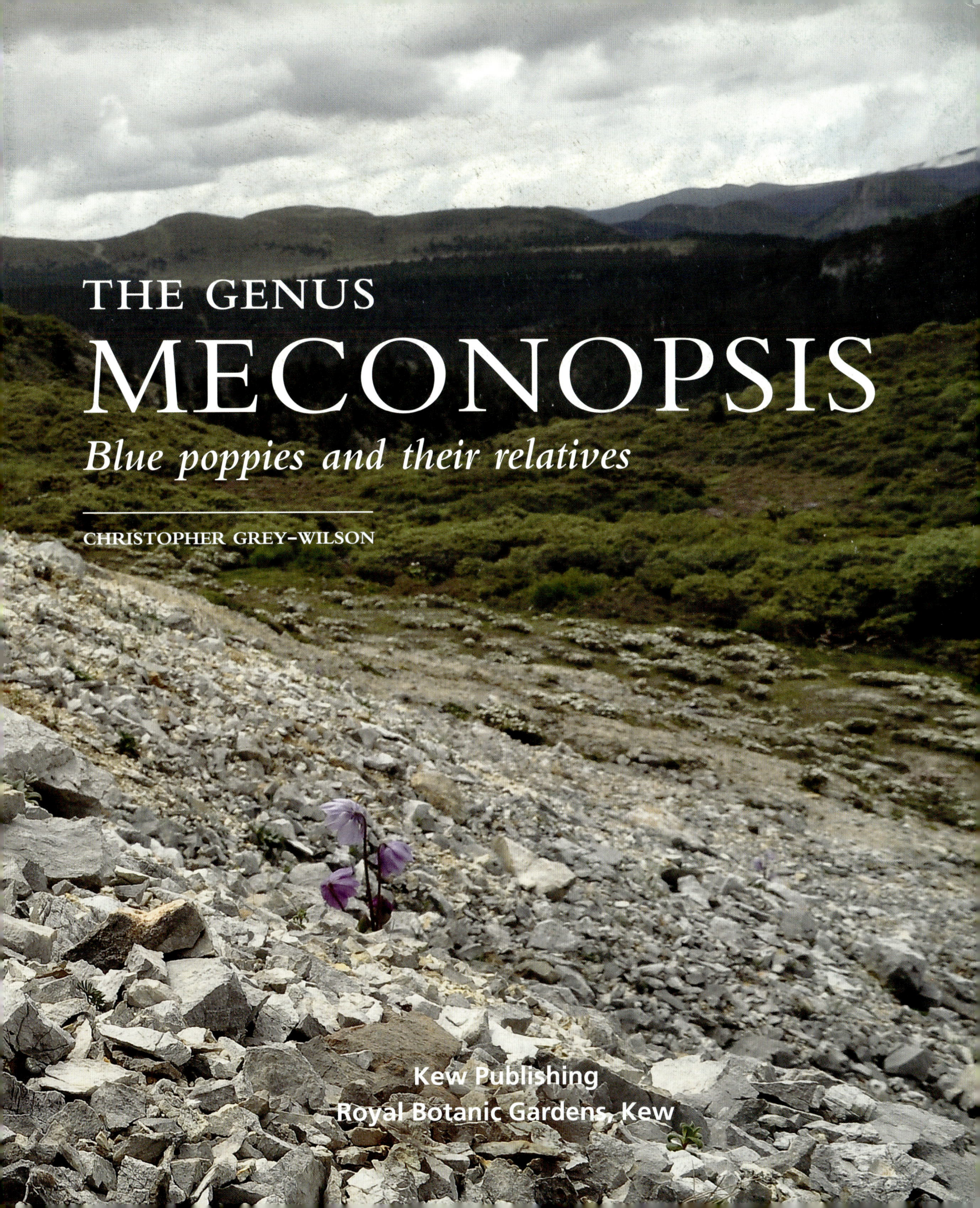

CHRISTOPHER GREY-WILSON

Kew Publishing
Royal Botanic Gardens, Kew

Kew
ROYAL BOTANIC GARDENS

© The Board of Trustees of the Royal Botanic Gardens, Kew 2014

Text © Christopher Grey-Wilson

Illustrations © the artists, as stated in captions

Photographs © the photographers, as stated in captions

The author has asserted his right to be identified as the author of this work in accordance with the Copyright, Designs and Patents Act 1988

All rights reserved. No part of this publication may be reproduced, stored in a retrieval system, or transmitted, in any form, or by any means, electronic, mechanical, photocopying, recording or otherwise, without written permission of the publisher unless in accordance with the provisions of the Copyright Designs and Patents Act 1988.

Great care has been taken to maintain the accuracy of the information contained in this work. However, neither the publisher, the editors nor author can be held responsible for any consequences arising from use of the information contained herein. The views expressed in this work are those of the individual author and do not necessarily reflect those of the publisher or of the Board of Trustees of the Royal Botanic Gardens, Kew.

First published in 2014 by
Royal Botanic Gardens, Kew
Richmond, Surrey, TW9 3AB, UK
www.kew.org

ISBN 978-1-84246-369-7

Distributed on behalf of the Royal Botanic Gardens, Kew in North America by the University of Chicago Press, 1427 East 60th Street, Chicago, IL 60637, USA

British Library Cataloguing in Publication Data
A catalogue record for this book is available from the British Library

Jacket illustrations
Front: *Meconopsis horridula* by Harry Jans; Back: clockwise from top left, *M. tibetica* by Harry Jans; *M. integrifolia* subsp. *souliei* by Harry Jans; *M. sherriffii* by Toshio Yoshida; *M. superba* by James Cobb; *M. bella* subsp. *subintegrifolia* by Tim Lever.
Frontispiece: *Meconopsis pseudovenusta* by Harry Jans

Project editor: Ruth Linklater
Design, typesetting and page layout: Christine Beard
Production manager: Georgina Smith

Printed in Italy by Printer Trento s.r.l.

FSC MIX Paper from responsible sources FSC® C015829

For information or to purchase all Kew titles please visit shop.kew.org/kewbooksonline or email publishing@kew.org

Kew's mission is to inspire and deliver science-based plant conservation worldwide, enhancing the quality of life.

Kew receives half of its running costs from Government through the Department for Environment, Food and Rural Affairs (Defra). All other funding needed to support Kew's vital work comes from members, foundations, donors and commercial activities including book sales.

CONTENTS

NEW TAXA AND COMBINATIONS	vii
1. INTRODUCTION	1
2. BRIEF HISTORY OF THE GENUS MECONOPSIS	5
3. MORPHOLOGY OF THE GENUS MECONOPSIS	11
4. CULTIVATION OF MECONOPSIS SPECIES	27
5. DISTRIBUTION AND HABITAT	35
6. TAXONOMIC TREATMENT	41
Species delineation	41
Meconopsis Vig. emend. Grey-Wilson	42
Infrageneric classification	43
Key to the subdivisions of the genus Meconopsis	48
Key to the species of Meconopsis	49
Subgenus Meconopsis	56
section Meconopsis	56
section Polychaetia	70
series Robustae	70
series Polychaetia	80
Subgenus Discogyne	118
Subgenus Grandes	136
section Grandes	136
series Grandes	137
series Integrifoliae	172
section Simplicifoliae	193
series Simplicifoliae	193
series Puniceae	204
section Racemosae	232
series Racemosae	232

6. TAXONOMIC TREATMENT (contd)

series Heterandrae	275
section Impeditae	281
series Impeditae	281
series Henricanae	304
series Delavayanae	315
section Forrestianae	318
section Cumminsia	329
series Primulinae	329
series Cumminsia	337
section Bellae	358
Excluded Species	365

7. THE GENUS PARAMECONOPSIS — 367

8. THE GENUS CATHCARTIA — 369

section Villosae	370
section Chelidonifolia	374

GLOSSARY — 379

REFERENCES — 382

APPENDIX 1
(2061) Proposal to conserve the name Meconopsis (Papaveraceae) with a conserved type — 386

APPENDIX 2
Collectors and Researchers' Biographies — 388

APPENDIX 3
Chinese Place Names — 392

INDEX TO SCIENTIFIC NAMES — 394

INDEX TO COMMON NAMES — 400

NEW TAXA AND COMBINATIONS

Cathcartia section **Chelidonifoliae** (Prain) Grey-Wilson **comb. nov.**	374
section **Villosae** (G. Taylor) Grey-Wilson **comb. nov.**	370
chelidonifolia (Bureau & Franch.) Grey-Wilson **comb. nov.**	374
oliveriana (Franch. ex Prain) Grey-Wilson **comb. nov.**	376
Meconopsis section **Cumminsia** (Prain) Grey-Wilson **comb. nov.**	47
section **Impeditae** Grey-Wilson, **section nov.**	46
series **Heterandrae** Grey-Wilson, **series nov.**	46
series **Integrifoliae** Grey-Wilson, **series nov.**	44
series **Puniceae** *Grey-Wilson* **series nov.**	45
subgenus **Cumminsia** (Prain) Grey-Wilson, **subgenus nov.**	45
subgenus **Grandes** (Prain) Grey-Wilson, **subgenus nov.**	44
baileyi subsp. **multidentata** Grey-Wilson **subsp. nov.**	150
baileyi subsp. **pratensis** Kingdon Ward ex Grey-Wilson **stat. nov.**	148
bella subsp. **grandifolia** Grey-Wilson **subsp. nov.**	362
bella subsp. **subintegrifolia** Grey-Wilson **subsp. nov.**	362
bijiangensis subsp. **chimiliensis** Grey-Wilson **subsp. nov.**	271
georgei f. **castanea** (H. Ohba, Tosh. Yoshida & H. Sun) Grey-Wilson **comb. & stat. nov.**	274
horridula subsp. **drukyulensis** Grey-Wilson **subsp. nov.**	245
impedita subsp. **rubra** (Kingdon Ward) Grey-Wilson **stat. nov.**	286
integrifolia subsp. **souliei** (Fedde) Grey-Wilson **stat. nov.**	180
×**kongboensis** Grey-Wilson non Hort. ex R. L. Harley **hybr. nov.**	203
lancifolia subsp. **eximia** (Prain) Grey-Wilson **comb. & stat. nov.**	322
lancifolia subsp. **lepida** (Prain) Grey-Wilson **comb. & stat. nov.**	324
lhasaensis Grey-Wilson **sp. nov.**	248
lijiangensis (Grey-Wilson) Grey-Wilson **stat. nov.**	182
ludlowii Grey-Wilson **sp. nov.**	350
paniculata var. **pseudoregia** Grey-Wilson **subsp. nov.**	112
paniculata var. **rubra** Grey-Wilson **var. nov.**	110
prainiana var. **lutea** (G. Taylor) Grey-Wilson **comb. nov.**, descr. emend.	263
simplicifolia subsp. **grandiflora** Grey-Wilson **subsp. nov.**	197
speciosa subsp. **cawdoriana** (Kingdon Ward) Grey-Wilson **comb. & stat. nov.**	231
speciosa subsp. **yulongxueshanensis** Grey-Wilson **subsp. nov.**	231
sulphurea Grey-Wilson **sp. nov.**	188
sulphurea subsp. **gracilifolia** Grey-Wilson **subsp. nov.**	192
zhongdianensis Grey-Wilson **sp. nov.**	258
Parameconopsis Grey-Wilson **genus nova**	367
cambrica (L.) Grey-Wilson **comb. nov.**	367

Meconopsis pseudovenusta; Shikashan, Xianggelila County, NW Yunnan, 4400 m.

PHOTO: TOSHIO YOSHIDA

1. INTRODUCTION

1934 seems a long time ago yet, surprisingly, that was the year in which the last detailed monograph of the genus *Meconopsis* was published. George Taylor, the author, was a botanist at the British Museum (Natural History) at the time and was later to become the Director of the Royal Botanic Gardens, Kew. When I first started at Kew in early 1968 as a junior botanist, Sir George, as he had become, was still the Director and he still maintained a keen fascination for the genus, although he never attempted to update his monograph. This was perhaps a pity because a great deal more information on the genus in the wild had come to light in the intervening years and it would have been fascinating to know how he would have interpreted all the additional information. Unfortunately, I was never able to discuss *Meconopsis* with George Taylor, although it is clear that his interpretation of the genus is radically different to my own, yet he was working at a very different time when resources and information on the genus were far scarcer and analysis was based almost exclusively on a detailed study of all the available herbarium material. Despite this, the 1934 monograph remained the benchmark for the genus right up until the 1990s when others started to take an interest. When I was first at Kew I was assigned to the "early families" in the Herbarium; that is all the plant families from number 1 to 33 (Ranunculaceae to Tiliaceae) according to the Bentham & Hooker (1862–83) sequence favoured at the time. This run of families included the poppy family, Papaveraceae, which along with the Ranunculaceae held most fascination for me.

Apart from the many colourful species of *Papaver*, *Meconopsis* holds the greatest allure in the poppy family. This is not surprising as the genus contains the fabled blue poppies which have captivated the imagination of botanist and gardeners since their discovery in the early years of the twentieth century and they still do to this day. There is something quite irresistible about blue poppies: go to any garden when they are in bloom and watch the reaction of visitors when they come across them and it is easy to understand that few flowers in our gardens can elicit such admiration. However, this thrill is nothing compared with that of seeing a blue poppy growing on a distant mountain in the wild.

The genus *Meconopsis*, as currently circumscribed, contains 79 species which are all native to the Himalaya and western China, including the Tibetan Plateau. The species are polycarpic perennials, or they are monocarpic, and range from dainty little plants of high mountainous habitats to substantial plants of woodland, woodland margins and open shrubberies at mid-altitudes (mainly 2500–4200 m). The largest grouping by far within the genus is to be found in subgenus *Cumminsia* which contains more than half of all the species. While it is fair to say that many of these are little known, several have had a long association with cultivation, these including *M. aculeata* Royle, *M. prattii* Prain and *M. racemosa* Maxim. (the latter two *M. horridula* of horticulture), as well as the little alpine gem *M. delavayi* Franch. ex Prain. Most of the plants in this subgenus are monocarpic plants, being almost exclusively herbaceous, with the leaves dying away in the autumn and the plants over-wintering as resting buds close to the soil surface. However, it is in two of the other subgenera that the greatest horticultural merit has been sought. Subgenus *Grandes* contains familiar garden plants such as the beautiful blue poppies, *M. baileyi* Prain and *M. grandis* Prain, as well as *M. punicea* Maxim. with its startlingly red, pendent blooms, the perennial harebell poppy, *M. quintuplinervia* Regel, the yellow monocarpic *M. integrifolia* (Maxim.) Franch. and *M. sulphurea* Grey-Wilson (*M. pseudointegrifolia* hort.). These are boldly perennial or monocarpic herbaceous species, overwintering as large resting buds at ground level or just below. The boldest of all meconopsis belong to subgenus *Meconopsis* which contains species that have attractive evergreen leaf-rosettes and large racemes or panicles that can attain 2.5 m in height, occasionally more, and often carrying many flowers. Subgenus *Meconopsis* contains many species of great horticultural appeal including *M. dhwojii* G. Taylor, *M. paniculata* (D. Don) Prain, *M. regia* G. Taylor, *M. staintonii* Grey-Wilson, *M. superba* King ex Prain and *M. wallichii* Hook. All the species in subgenus *Meconopsis* have handsome over-wintering leaf rosettes and these are part of their attraction in cultivation. They spend two, sometimes as many as five, years in the rosette stage before developing their large and impressive, multi-flowered inflorescences.

The taxonomy of subgenus *Meconopsis* (formerly included in section *Polychaetia*), and indeed the whole genus, has vexed botanists over the years, including eminent taxonomists such as William Bentham, David Prain and George Taylor. Although I have made various changes to the infrageneric and specific classification of the genus I do so in the knowledge that further research is required, especially into the phylogenetics of the genus. In recent years field observations have added greatly to our knowledge of the genus and, while new species have been discovered in the past 20 years or so, I have little doubt that others will come to light as botanists and explorers are able to reach the remoter and lesser known regions of the Sino-Himalayan mountains. I have been very fortunate in being able to access the photographic collections of a number of fine photographers and their pictures have really brought this exciting genus to life. Unfortunately, I have far too many photographs at my disposal so have had to impose a severe selection process. While undertaking this limitation exercise I have endeavoured to show various stages of the plants' natural cycle but, at the same time, have tried to give readers an insight into the variation found within the various taxa, the photographs perhaps conveying this far more vividly than a detailed botanical description ever could.

The first blue poppy that I saw in the wild was in July 1973 in the high remote mountains of Dolpo in western Nepal, an expedition led by the late Dr George Smith whose fascination was not *Meconopsis* but the genus *Androsace*. High on the screes of the Khung Khola valley, within a stone's throw of the Tibetan border, *M. horridula* was in full flower, the plants scattered across the rocks, their flowers like the finest stained glass in shades of blue, sometimes with a hint of mauve or purple. On later trips to Nepal I was to see this charming, yet prickly, little species, and others, on many occasions. From that first moment I was hooked on the genus, yet the chance to study them in detail really did not arise until some years later in the 1990s after I had left Kew to work for the Alpine Garden Society.

It is perhaps fitting that the Royal Botanic Gardens, Kew, are publishing this monograph, not only because of the association with George Taylor as monographer and Director, but also the historical link to David Prain, also one-time director at Kew, who did much of the early research and taxonomy of the genus. I trust that had either of these stalwart characters been alive today they would be astonished at the amount of additional information on *Meconopsis* that has come to light in recent times. Perhaps, like me, they would be wholly entranced by the glorious colour photographs of the species taken in the wild that now abound and which have been such an important part of this present work.

In this monograph I have quoted widely from the works of several of the great plant hunters, Reginald Farrer and Frank Kingdon Ward in particular, and make no apologies for this. Their comments on *Meconopsis* in the wild are invaluable, giving important details of the plants and their habitats, details that are often hard to ascertain from herbarium specimens and field notes alone. Their first-hand eloquence cannot be surpassed and I have made no attempt to re-interpret their words.

Undoubtedly the greatest benefit of the resurgence of interest in Chinese plants has been the fine opportunities now available to see and study the plants in their native habitats. Such studies may often reveal features and trends which are difficult or impossible to observe from herbarium material alone. Furthermore, studies in the wild often make it possible to understand and interpret herbarium material more readily. Herbarium specimens provide extremely valuable and often historical information about species, their habitats and distribution, besides being a fine record of the activities of botanists and plant hunters from the past. Perhaps most valuable of all, the examination of a wide range of herbarium specimens linked with detailed observation in the wild allows the botanist to interpret variation, the key to any taxonomic study, particularly a monographic work. Such variation may at times be difficult for the casual observer, gardener and horticulturist to appreciate because often they are looking at a very small sample, with some species being represented in cultivation by a single or just a few gatherings from the wild.

Today the gathering of herbarium material is just as important as ever, especially in those regions that have only recently become accessible for one reason or another. Just as important is revisiting locations explored by plant hunters many years ago. Well-gathered and prepared specimens can prove valuable sources of information to the researcher; dried herbarium specimens may seem dull to many but they are an essential element to any serious study, especially when used in conjunction with field observations and other data (DNA, and pollen studies for instance). In addition, in recent years the advent of the digital camera has made it possible to record plants in ever greater detail. No tool has been so beneficial in scientific field work, for the digital camera has allowed one to take unlimited numbers of photographs illustrating all aspects of a particular plant and its habitat in the field. Previously the number of photographs taken was severely limited by cost and the bulk of film, while at the same time, there was always that nervous wait at the conclusion of a field trip or expedition while the films were processed, wondering all the time whether the photos had come out properly. The digital camera presents no such problems, allowing the user instant results. What would those great collectors of the past, the George Forrests and Frank Kingdon Wards, have made of such an asset, for even in today's accessible world their explorations reached places that we still have not revisited in the modern era? A digital camera fitted with a GPS device allows details of location, latitude and longitude co-ordinates, altitude and date to be recorded with each photo taken. This tremendous facility allows the field worker to record plants and habitats in ever greater detail and ever more accurately, something that plant collectors of old (even as little as 30 years ago) could only have dreamt of.

Many countries today (for instance Bhutan and China, but many others) while allowing access to areas of interest yet restrict the collecting of biological material, whether this be dried specimens, living plants or seed in the case of plants. Such restrictions often extend outside the biological field to minerals, fossils and so on. These measures have severely limited the amount of material reaching botanic gardens, especially in Europe and North America. In the absence of herbarium material, good digital images of plants in the wild are proving invaluable. In this respect I am deeply grateful to all those who have allowed me access to their digital images. They are the modern plant collectors using camera rather than trowel in their quest; their contributions and observations have been profound in the construction of this account. Most prominent is my good friend Harry Jans who has taken a marvellous series of *Meconopsis* photos during his many visits to western China and Tibet in recent years, and continues to do so. In addition must be added Joe Atkin, Bill Baker (RBG Kew), John & Hilary Birks, Anne Chambers, Eiko Chiba, James Cobb, Phillip Cribb, Alan Dunkley, Paul Egan, Pam Eveleigh, Ron McBeath, John Mitchell, Martin Walsh, Tim Lever, Rosie Steele, Bill Terry, David & Margaret Thorne and Dieter Zschummel.

Finally, my Japanese colleague Toshio Yoshida, with whom I have collaborated on several articles on *Meconopsis*, has very generously allowed me access to his remarkable photographic collection: many of his photos are included in this work.

ACKNOWLEDGEMENTS

I am extremely grateful to the following who have been a great help in the preparation of this work, especially by supplying field data or giving me access to photographs taken in the wild: Makoto Amano, Chris Brickell, the late Dick Brummitt (RBG Kew), Graham Catlow, Ian Christie, James Cobb, Alan Coombes, Kenneth Cox, Peter Cox, the late James Cullen, Tom Gregory, Charlotte (Lottie) Grey-Wilson, Christine Grey-Wilson, Sun Hang, Finn Haughli, Jim Jermyn, Tony Kirkham (RBG Kew), Erica Larcom, Rong Li, John McNeill (RBG Edinburgh), Margaret North, Ian McNaughton, Alan Oatway, Colin Pendry (RBG Edinburgh), David Rae (RBG Edinburgh), David Rankin, Dieter Schacht, Evelyn Stevens, Henry & Margaret Taylor, Melanie Thomas (RBG Kew), Minoru Tomiyama, Hirokazu Tsukaya, Mark Watson (RBG Edinburgh), Toshio Yoshida, Wu Zhikun, Zhiling Dao. I am also deeply grateful to the Directors of both the British Museum (Natural History) and the Royal Botanic Gardens Kew, as well as the Regius Keeper of the Royal Botanic Garden Edinburgh for allowing me access to their important and irreplaceable collections. Finally, I must thank the significant contribution of the Meconopsis Group based in Scotland, a number of whose members are included in the preceding list, whose encouragement, enthusiasm and support have been paramount in the production of this monograph. After years of partial neglect *Meconopsis* has become a focus of attention in recent years, especially as parts of the Himalaya and western and south-western China have become revisited and more widely explored and the feed-back from expeditions and tours have greatly added to our knowledge of the distribution and habitat preferences of the different species but, above all, to the fascinating inherent variability found in nature.

Overleaf:
Meconopsis sherriffii, flower details; near Danji, Lunana, C Bhutan, 4700 m.
PHOTO: TOSHIO YOSHIDA

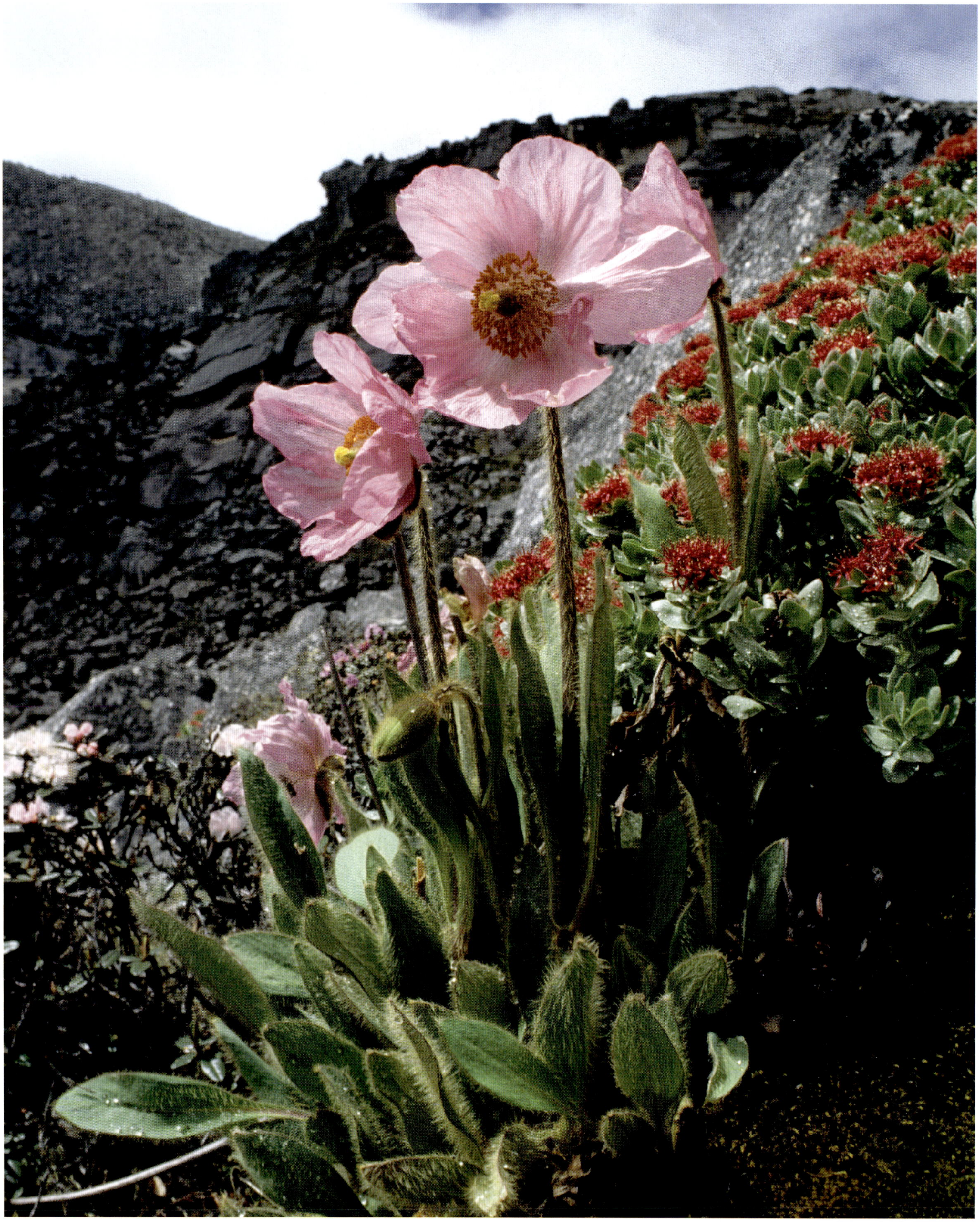

2. BRIEF HISTORY OF THE GENUS MECONOPSIS

The genus *Meconopsis* was established by L. A. G. Viguier, Montpellier, in 1814 based on a single species *M. cambrica*, the so-called Welsh poppy, a species first described and placed in the genus *Papaver* by Linnaeus in 1753, although the plant was well known before Linnaeus's time[1]. *Meconopsis* was distinguished from the closely similar *Papaver* primarily by the presence of a short style and the absence of a sessile stigmatic disk on top of the ovary. *Meconopsis cambrica* is restricted to Western Europe including Ireland, Wales (the Welsh poppy), western and south-western France and north-west Spain, although it is widely naturalised in other parts of western Europe, primarily as a garden escape.

Viguier's proposal that the Welsh poppy be called *Meconopsis cambrica* was taken up in 1815 by De Candolle in the 'Flora de France'; however, this view took quite a while to find acceptance. For instance in 1816 Desportes, while agreeing that the plant in question should be removed from *Papaver*, assigned it to *Argemone* as *A. cambrica*. The situation became more complicated when in 1821 A. Gray placed the Welsh poppy in the genus *Cerastites*, while four years later, in 1825, Sprengel linked the plant to the American celandine and placed it in the genus *Stylophorum* Nutt. Although Bernhardi accepted the genus *Meconopsis* for the Welsh poppy as early as 1833, it wasn't until some years later in 1848 (*Gen*. 1: 1140) that A. Gray was able to show convincingly that Nuttall's genus *Stylophorum* was distinct and not congeneric with Viguier's *Meconopsis*; thereafter the latter name was finally accepted.

In 1824 the first Asiatic species, *Meconopsis napaulensis* DC., was described based on Nathaniel Wallich collections from Nepal that had been sent to De Candolle in Geneva. De Candolle in fact was by no means convinced, treating the plant as a doubtful species and placing it in his section *Stylophorum* which, as had previously been established, had no place in the genus in any case. The following year two further Himalayan species of *Papaver*, *P. paniculatum* and *P. simplicifolium*, both based on other Wallich collections, were described by David Don (1825), these later being transferred to *Meconopsis* (*M. paniculata* (D. Don) Prain and *M. simplicifolia* (D. Don.) Walp. respectively). Bentham (1835) greatly extended the range of the genus by adding two additional species, *M. crassifolia* and *M. heterophylla*, both Californian natives; however, there was no general agreement on these which in the following years found themselves transferred in turn to the genus *Stylophorum* by Steudel in 1841 and to *Papaver* by Greene in 1888 and 1894 respectively. In 1930, prior to the publication of his *Meconopsis* monograph, George Taylor created the genus *Stylomecon* for the Californian taxa, recognising but a single species, *S. heterophylla*, and this situation has remained stable until recently.

Additional Himalayan species were added to *Meconopsis* in the mid-nineteenth century: *M. aculeata* Royle from Kashmir in 1839; *M. wallichii* in 1852, which W. J. Hooker described in *Curtis's Botanical Magazine* based on seed from Sikkim collected by his son J. D. Hooker; *M. robusta*, based on another Nathaniel Wallich collection, and *M. horridula* based upon specimens collected by J. D. Hooker in Sikkim, were described by Hooker & Thomson in 'Flora Indica' thus bringing the total number for the Himalaya to seven.

The first Chinese species to be added to the genus was *Meconopsis quintuplinervia* (Regel, *Gartenflora* 25) in 1876, followed the next year by *M. racemosa* and *M. integrifolia* (the latter first described as a species of *Cathcartia* and transferred in 1886 to *Meconopsis* by Franchet). At the same time, Franchet described two additional species of *Cathcartia*, *C. delavayi* and *C. lancifolia*, but changed his mind in 1896 by transferring them to *Meconopsis* where they have remained ever since. Additional species were added in the interim, the startling red-flowered *M. punicea* by Maximowicz in 1889, the glorious blue *M. betonicifolia* by Franchet in the same year, and *M. chelidonifolia* [now *Cathcartia chelidonifolia*] and *M. henrici* by Bureau and Franchet in 1891.

---

[1] Early botanists treated the Welsh Poppy either as a species of *Papaver* or a species of *Argemone*. Pre Linnaeus, in fact in the early part of the seventeenth century, *Argemone* was considered in a rather arbitrary manner and it was not until the time of Tournefort in 1694 that the matter was resolved. Tournefort in fact established the use of the name *Argemone* which Linnaeus later adopted in his 'Species Plantarum' of 1753, outlining the circumscription of both *Argemone* and *Papaver*. At the same time, those pre-Tournefort argemones were all relegated to the genus *Papaver*. This is well outlined in the *Journal of Botany* 33: 130–2 (1694).

Thereafter a number of additional Asiatic species were added, particularly by David Prain: *Meconopsis bella* in 1894 and four additional species *M. grandis*, *M. primulina*, *M. sinuata* and *M. superba*, in 1895. However, it was not until the following year (1896) that Prain established the first detailed study of the genus published in *The Journal of the Asiatic Society of Bengal* which contained a preliminary classification of the species, some 23, known at the time. In this paper Prain divided the species into seven prime groups. This was updated in 1906 by him when he published a full enumeration of the known species, 'A review of the genera *Meconopsis* and *Cathcartia*' published in volume 20 of the *Annals of Botany*. Although much modified, Prain retained much of these groupings in the 1906 account but, at the same time, recognising two major sections, *Eumeconopsis* and *Polychaetia*, the first based on all those species with simple hairs, the latter those with barbellate hairs. Three years later, in 1909, Fedde published his account of the Papaveraceae in *Das Pflanzenreich*: in this Prain's two sections are raised to subgeneric level, while his groupings (nine in all) are given the rank of section.

This was followed in 1915 when Prain published another revision of the genus, recognising this time some 43 species, but based essentially on his 1906 account, although taking into account the additional species described in the intervening years. At the same time he added two new series (*Cumminsia* and *Decorae*) whilst amending and adding to the existing ones.

Frank Kingdon Ward, who had some firm views on the subgeneric classification of *Meconopsis*, did not altogether agree with Prain's revisions and, in a series of articles published in the *Gardener's Chronicle* in 1926, he outlined his thoughts on the genus, with copious notes on the species recognised at the time, adding as he did so two further series, *Integrifoliae* [in obs.] and *Superbae*.

In 1934 George Taylor published his substantial monograph *An Account of the Genus Meconopsis* (often cited as '*The Genus Meconopsis*') which recognised some 41 species. Taylor expressly explored the genus to find other criteria upon which to base his primary classification. The result was a substantially different account: "The classification of *Meconopsis* proposed in the present work differs fundamentally from that hitherto accepted. The two major groups now defined do not correspond in any way to Prain's sections, and are differentiated on entirely different characters. While the subordinate groupings approximate closely in both systems, their status and arrangement are here altered very materially".

Taylor recognised two subgenera *Eumeconopsis* containing all but two of the species, and *Discogyne*. Within the former, three sections were recognised *Cambricae*, *Eucathcartia* and *Polychaeatia*. A summary of the classification as outlined by Taylor is presented in Table 1.

In addition, a further subgenus *Eumeconopsis* species (*M. argemonantha*) was listed as 'Insufficiently known species'.

Taylor's monograph became the benchmark for *Meconopsis* devotees for many years and very little was done to challenge his accepted version and the classification remained more or less intact right up until the 1980s, any new species being described in the interim period being slotted into the existing 1934 classification. Perhaps the most notable additions were *M. sherriffii* (1937) and *M. taylorii* (1972). As a result of further exploration in the Himalaya and adjacent regions of Tibet and western China other species were being discovered; however, little more was added to the classification of the genus, that is until 1980 when two Chinese botanists, C. Y. Wu and H. Chuang, published a paper in *Acta Botanica Yunnanica* describing three new sections (*Forrestii*, *Racemosae* and *Simplicifoliae*, the last upgraded from George Taylor's series *Simplicifoliae*) and two new series within *Meconopsis*. This rather insular Chinese-species based overview of the genus has had to be taken into consideration when preparing this monograph and while the basic elements of Wu &

Table 1

SUMMARY OF CLASSIFICATION OF MECONOPSIS ACCORDING TO TAYLOR (1934)

SUBGENUS **Eumeconopsis**

  SECTION Cambricae (*M. cambrica*)

  SECTION Eucathcartia
    SERIES Chelidonifoliae (*M. chelidonifolia*, *M. oliveriana*)
    SERIES Villosae (*M. smithiana*, *M. villosa*)

  SECTION Polychaetia
    SUBSECTION Eupolychaetia
      SERIES Superbae (*M. regia*, *M. superba*)
      SERIES Robustae (*M. dhwojii*, *M. gracilipes*, *M. longipetiolata*, *M. napaulensis*, *M. paniculata*, *M. robusta*, *M. violacea*)
    SUBSECTION Cumminsia
      SERIES Simplicifoliae (*M. punicea*, *M. quintuplinervia*, *M. simplicifolia*)
      SERIES Grandes (*M. betonicifolia*, *M. grandis*, *M. integrifolia*)
      SERIES Primulinae (*M. florindae*, *M. lyrata*, *M. primulina*)
      SERIES Delavayanae (*M. delavayi*)
      SERIES Aculeatae (*M. aculeata*, *M. forrestii*, *M. georgei*, *M. henrici*, *M. horridula*, *M. impedita*, *M. lancifolia*, *M. latifolia*, *M. neglecta*, *M. pseudovenusta*, *M. sinuata*, *M. speciosa*, *M. venusta*)
      SERIES Bellae (*M. bella*)

SUBGENUS **Discogyne** (*M. discigera*, *M. torquata*)

Chuang's new sections have been accepted, they have had to be considerably emended in the process.

In the same year two further sections were proposed by different authors: section *Nyingchienses* by Z. H. Zhou and section *Pinnatifoliae* by Wu & Zhuang. Neither of these sections can be upheld in the current revision as the characters that distinguish them overlap profoundly with existing divisions (*Nyingchienses* with section *Simplicifoliae*, and *Pinnatifoliae* with subgenus *Discogyne*).

In the modern era, as the remoter mountains of the Sino-Himalaya have become more readily accessible, there has been a surge of new information on *Meconopsis*, a wealth of new data (particularly photographs of the species in the wild) and new species discovered, greatly adding to our knowledge. For this, praise must be given to tireless field work undertaken by various professional and amateur botanists and plant explorers, notably John & Hilary Birks, Harry Jans, Tim Lever, David & Margaret Thorne, Martin Walsh and, perhaps above all, Toshio Yoshida.

Along with this there have been extensive molecular studies of the Papaveraceae, especially by Carolan *et al.* and Kadereit *et al.*[2] which have thrown new insights into the relationship between the genera of the Papaveraceae and, while a full detailed and widely sampled analysis of the species of *Meconopsis* has yet to be carried out, their researches included in the various papers have, nonetheless, covered some 34 species of *Meconopsis* (the genus now stands at 79!). In simple and generalised terms the result of these researches has shown clearly that, first, the western European *M. cambrica* is not related to the many Asiatic representatives of the genus and that, second, the same applies to the small hub of species around *M. villosa* (including *M. chelidonifolia*, *M. oliveriana* and *M. smithiana*) which are best relegated to the genus *Cathcartia*. This in turn has exposed problems with the typification of the genus *Meconopsis* as outlined by L. A. Viguier in 1814, based as it was on a single species, *M. cambrica* (see p. 367 for the resolution of this dilemma). This means that five of the species in Taylor's monograph have had to be removed from *Meconopsis* leaving his account with 36 species in all. With 79 species in the current account it can be seen that the genus has more than doubled in size since it was monographed in 1934.

With all these revelations it has proved a considerable task to take into account all the recent findings (both molecular and morphological) in redefining the genus and to establish a workable and accessible classification. While I can be accused of being an old-fashioned botanist there is every merit in producing an account that is unashamedly morphologically biased. Like many other field and herbarium botanists, the distribution pattern of species is of great interest to me, as are the habitat requirements of the various taxa but, above all, it is the variation of each taxon within its prescribed distribution in the wild that is the most fascinating, and sometimes the most difficult to explain in relatively simple terms. While many molecular studies have given us great insight into the relationship of one family or genus to another, and one species to another, the great weakness has been a detailed molecular study of variation within species. While I, in looking at material in herbaria and in the wild (specimens and photographs) have sometimes trawled through dozens of specimens of a particular and established taxon, molecular analysis has generally relied heavily on a single sampling in most instances, if the taxon has been looked at all. While it can be appreciated that sampling of material for molecular studies is very time-consuming, it is also expensive; however, there is at the same time, a strong feeling that research has tilted significantly towards molecular studies to the detriment of gross morphological taxonomy.

In the present classification (outlined in detail on p. 43) I have proposed four subgenera, *Meconopsis*, *Cumminsia*, *Discogyne* and *Grandes* and these are compared to Taylor's monograph in Table 2.

In addition, there has been a considerable realignment of sections and series to accommodate the many new taxa that have been described after 1934, including series *Heterandrae* which includes two species described in 2010 and 2011. Despite these changes, much of George Taylor's framework for the genus remains intact, although some of his series have been upgraded to sections and additional series have been created.

---

[2] In 2006 James Carolan *et al.* published a paper entitled "Phylogenetics of Papaver and Related Genera Based on DNA Sequences from ITS Nuclear Ribosomal DNA and Plastid trnL Intron and trnL-F Intergenic Spacers", which clearly showed that *Papaver* is not monophyletic unless *Meconopsis cambrica* and two closely related genera, *Roemeria* and *Stylomecon*, are included. The molecular analysis which the authors present "support the view of Kadereit et al. (1997) that *Stylomecon heterophylla* arose from within *Papaver* and should not be considered a separate genus". The distinguishing stylar capsule which persuaded early authors to include this taxon within *Meconopsis* is a bogus one. Kadereit *et al.* (2006) reveal the close morphological similarities (including leaf shape, globose flower buds, petal colour and pale anthers with filiform filaments) between *Papaver californicum* and *Stylomecon heterophylla*; however, they differ in the morphology of the fruit capsules. Concluding from the evidence presented they stress that the two species have diverged fairly recently and that "The separation of *S. heterophylla* from *Papaver* therefore is not justified based solely on differences in capsule characteristics". In short, there is a good argument in favour of the view that the stylar character prevalent in the majority of *Meconopsis* species, in *Papaver* (*Meconopsis*) *cambrica* and *Stylomecon* (*Papaver*) *heterophylla* has arisen independently on several occasions within the Papaveraceae, thus supporting the theory of parallel evolution in this instance.

MECONOPSIS, CONSERVATION OF THE NAME

As has been stated, the genus *Meconopsis* was established by L. A. G. Viguier, Montpellier, in 1814 based on a single species *M. cambrica*, a species first described and placed in the genus *Papaver* by Linnaeus in 1753, although the plant was well known before Linnaeus's time. *Meconopsis* was distinguished from the closely similar *Papaver* primarily by the presence of a short style and the absence of a sessile stigmatic disk on top of the ovary. *Meconopsis cambrica* is restricted to Western Europe including Ireland, Wales (the Welsh poppy), western and south-western France and north-west Spain, although it is widely naturalised in other parts of western Europe, primarily as a garden escape.

While the typification of *Meconopsis* is straightforward in taxonomic terms (the genus is based upon the European *M. cambrica*) it is clear from the facts presented below that the Asian species, representing as they do the very large bulk of the genus, occupy a different lineage and that the two should be separated. The means to do this presents a taxonomic dilemma. Under the widely recognised 'International Rules of Botanical Nomenclature' the European *M. cambrica*, being the type of the genus, would represent a monotypic genus, while the Asian species (more than 50 in total) would need to have a new generic name. At first it would seem that the genus *Cathcartia* might conceivably provide an easy solution, simply transferring all the Asian *Meconopsis* species to that genus; however, modern research shows quite clearly that *Cathcartia* itself also represents a distinct genus based upon *M. villosa* (Hook. f.) G. Taylor and containing at the most just three or four species: modern floras such as *The Flora of Bhutan*, have re-instated the genus *Cathcartia*. There is in fact no generic name extant for the bulk of the species and a new name would have to be devised.[3] *Meconopsis* is a well-known and popular genus, especially in the horticultural world, with quite a few species well-established in cultivation, several for more than a hundred years. To invent a new genus would undoubtedly upset and prove an annoyance to horticulturists and gardeners as well as some botanists. With this in mind I proposed that the name *Meconopsis* be preserved for the large majority of species, while the present *M. cambrica* be assigned a new generic name (*Parameconopsis*, see p. 367). This proposal, was I believe, the simplest solution involving a single taxon; the alternative would have been a large number of unnecessary and irritating new combinations under a new generic heading.

As a genus, *Meconopsis* has countless listings on the Internet, the vast majority of these referable to the Asian species. At the same time, the genus *Meconopsis* has a high profile in horticultural trade, particularly the

---

[3] The genus *Cerastites* Gray of 1821 (*Nat. Arrang. Brit. Pl.* ii. 703) was derived by transferring a number of Linnaeus' *Papaver* species into a new genus, namely *P. dubium*, *P. hybridum* and *P. cambricum* based on rather doubtful characters that have not since been upheld. To this Gray added two additional species *Cerastites laciniatus* and *C. macrocephalus*.

Table 2

REALLOCATION OF SUBGENERIC, SECTIONAL AND SERIES DIVISIONS RECOGNISED BY TAYLOR (1934) UNDER THE FOUR SUBGENERA OF THE NEW CLASSIFICATION (GREY-WILSON 2014). For the full new classification see p. 43.

Grey-Wilson (2014)	Taylor (1934)		
Subgenus *Meconopsis*	= subgenus *Eumeconopsis*		
	section *Polychaetia*	subsection *Eupolychaetia*	series *Robustae* series *Superbae*
	section *Cambricae*, *Meconopsis cambrica* is transferred to the genus *Parameconopsis* section *Eucathcartia* now the genus *Cathcartia*		
Subgenus *Cumminsia*	= subgenus *Eumeconopsis*		
	section *Polychaetia*	subsection *Cumminsia*	series *Aculeatae* series *Bellae* series *Delavayanae* series *Primulinae*
Subgenus *Discogyne*	remains the same		
Subgenus *Grandes*	= subgenus *Eumeconopsis*		
	section *Polychaetia*	subsection *Cumminsia*	series *Grandes* series *Simplicifoliae*

famous Himalayan blue poppies (*M. bailey* Prain and *M. grandis* Prain and their numerous spectacular hybrids and cultivars). In addition, the Meconopsis Group, based in Scotland, which is one of the prime areas for *Meconopsis* cultivation, has an international following. Any change in the generic name for the Asian species would have profound consequences and there would be a great reluctance to follow any suggested change however correct this is under the International Rules.

George Taylor in his *An Account of the Genus Meconopsis* (1934) stated that "…….Viguier proposed his new genus, regarding it as intermediate between *Papaver* and *Argemone*. It was some time before his view found acceptance, and for a number of years botanists regarded the genus *Meconopsis* as unnecessary, or, if admitting it, generally obscured its identity by the inclusion of species which really did not belong to it."

In 1824 the first Asiatic representative, *Meconopsis napaulensis*, was added by De Candolle, while a year later two further species of *Papaver*, *P. paniculatum* and *P. simplicifolium* were described by D. Don and later transferred to *Meconopsis*. Material of all three taxa had been sent from Nepal by Nathaniel Wallich to De Candolle in Geneva (representatives of all three reside in the Prodromus Herbarium in Geneva, with duplicates at Kew, primarily in the Wallich Collection). However, in 1835 Bentham greatly extended both the taxonomic and geographical boundary of *Meconopsis* by adding two species from California, *M. crassifolia* and *M. heterophylla*, but these latter two species were correctly removed to a new genus, *Stylomecon*, by George Taylor in 1930.

Over the intervening years further Asiatic species were added to *Meconopsis*. By 1896 David Prain (*J. Asiat. Soc. Bengal* 64 (2): 316) was able to recognise 23 species. Today the genus contains in excess of 50 species, all but *M. cambrica* (L.) Vig. Asian in origin, being confined primarily to the Himalaya, the mountains of western China and southern Tibet (Xizang). The genus is predominantly one that has evolved in the Sino-Himalayan monsoon belt with the widespread and polymorphic *M. horridula* Hook. f. & Thomson reaching out across the drier reaches of the Tibetan Plateau away from the influence of the monsoon. In addition one localised species, *M. torquata* Prain is restricted to a limited region in the dry hinterland around Lhasa in Tibet.

Ernst (1962) in an extensive review entitled 'A comparative view of the morphology of Papaveraceae' has clearly demonstrated in his comparative morphological studies in the Papaveraceae sensu lato that *Meconopsis cambrica* is quite distinct from the Asiatic members of the genus *Meconopsis*. I cannot do better than quote directly from this important paper (see pp. 62–155):

"There are no consistent morphological characters by which *Meconopsis* can be distinguished from all other taxa of the subfamily Papaveroideae. In Europe and the Mediterranean region, *Meconopsis cambrica* is distinct from the other taxa of the subfamily Papaveroideae, i.e. *Papaver* and *Roemeria*, by the single striking feature of a distinct style between the ovary and the stigmatic lobes. But *M. cambrica* cannot be distinguished from the North American taxa of the Papaveroideae by this one character because *Argemone*, *Romneya* and *Stylomecon* also have more or less distinct styles; consequently, other morphological characters must be used. The distinction can be facilitated by adding that *Meconopsis* not only has a definite style and lacks a cartilaginous disk-like roof on the ovary, but also lacks true dorsal traces to the gynoecia and does have pseudo-dorsal traces originating from the placental bundles. Thus *M. cambrica* is separable from both the North Amercian and the European-Mediterranean Papaveroideae."

"Four facts seem important when *M. cambrica* is compared with the Himalayan taxa. First, several of the Himalayan taxa do not have distinct styles, e.g. *M. villosa* [sic *Cathcartia villosa*], *M. punicea*, some forms of *M. integrifolia* [*M. integrifolia* sensu stricto], and *M. torquata*. In these taxa, a description would have to read "style obsolete or absent". Second, *M. discigera* and *M. torquata* both have very conspicuous cartilaginous disc-like roofs to the ovary. Third, some taxa of *Meconopsis* have distinct dorsal traces to the gynoecia, e.g. *M. villosa* [*Cathcartia villosa*], *M. chelidonifolia*, *M. oliveriana*, *M. delavayi*, and *M. punicea*; several other taxa have more or less vestigial dorsal traces, e.g. *M. georgei*, *M. integrifolia*, *M. lyrata*, *M. lancifolia*, *M. quintuplinervia*, and *M. regia*. Fourth, none of the Asiatic taxa of *Meconopsis* have distinct pseudo-dorsal traces to the gynoecia. Thus, the Asiatic taxa of *Meconopsis* do not fit the specifications of the genus; furthermore, there are no other morphological criteria by which *M. cambrica* can be reconciled with the Himalayan taxa. It seems probable that, had one of the Himalayan taxa been the type species for *Meconopsis*, *M. cambrica* would not have been included."

"Stated another way, *Meconopsis cambrica* more closely resembles the European-Mediterranean taxa of *Papaver* than the Asiatic taxa of *Meconopsis* with which it is supposed to be congeneric. Also, *M. cambrica* has more in common structurally with the two North American taxa *Papaver californicum* and *Stylomecon heterophylla* than with the Asiatic *Meconopsis*. Furthermore, some of the commonly accepted taxa of *Papaver* lack a truly distinct cartilaginous disc, for instance, *Papaver* section *Scapiflora*. The genus *Meconopsis* cannot be defined in such a way that includes both *M. cambrica* and the Himalayan taxa, and at the same time is distinct *from Papaver*, *Stylomecon* and *Roemeria*."

"Thus, three courses of procedure are possible: (1) *either Papaver, Stylomecon, Meconopsis* (including *Cathcartia*), and *Roemeria* must be given a single all-inclusive generic epithet, or (2) these taxa must be defined only arbitrarily on the grounds of geographical distribution without regard to structure, or (3) next taxonomic divisions of more uniform character must be made."

W. R. Ernst's morphological study of the Papaveraceae has been backed by more recent investigations, particularly that of James C. Carolan *et al.* in their detailed paper published in 2006 and entitled 'Phylogenetics of *Papaver* and related Genera Based on DNA Sequences from ITS Nuclear Ribosomal DNA and Plastid *trnL* Intron and *trnL-F* Intergenic Spacers':

"The only European species of *Meconopsis, M. cambrica,* is well separated from the representatives of Asian *Meconopsis* in the molecular analysis here. *Meconopsis cambrica* occupies a well-supported (99BP, fig. 3) sister-group position to the remaining sections of *Papaver* (excluding *Argemonidium, Californicum, Horrida* and *Meconella*). This supports the view (Kadereit *et al.* 1997) that two distinct lineages within *Meconopsis* s.l exist and that *Meconopsis* in its current circumscription is neither monophyletic nor distinct from *Papaver*. *Meconopsis cambrica* shares diagnostic *trnL-F* indels with the majority of *Papaver* (excluding *Argemonidium, Californicum, Horrida* and *Meconella*). *Meconopsis cambrica* could have arisen either in parallel with the Asian representatives of *Meconopsis*, clade 2, i.e. core *Papaver* (sects. *Carinatae, Papaver, Pilosa, Pseudopilosa, Oxytona, Meconidium* and *Rhoeadium*), or from within a lineage best recognised as members of an expanded *Meconopsis*. Both these views were proposed by Kadereit *et al.* (1997), who favoured the latter view based on geographical, phytochemical and morphological considerations. Topological considerations alone favour parallel evolution as *M. cambrica* is embedded in clade 2 in our *ITS/trnL-F* trees."

"It is evident from the results (Carolan *et al.*, 2006) of this analysis that incongruence exists with previous taxonomic classifications regarding the positioning of *M. cambrica*. If *M. cambrica* is recognised as *Meconopsis, Papaver* s.s. (i.e. after the exclusion of the groups discussed above) is not monophyletic. It is suggested to include *M. cambrica* (as *Papaver cambricum* L.) in *Papaver*. However, an appropriate treatment of this species is difficult owing to a lack of apparent morphological similarities with extant *Papaver* species. There is no obvious section or group of species with which to place *Papaver cambrica*. Although unsatisfactory from a taxonomic perspective it may be necessary to describe a new monotypic section for this species within *Papaver*. The alternative is to leave it as incertae sedis until further evidence is found regarding its placement."

The typification of *Meconopsis* after the removal of *M. cambrica*, the type of the genus, requires careful consideration. The second species to be described within the genus as now defined was *M. napaulensis* DC., described in 1824. However, this species has been long misinterpreted due primarily to the poor and rather fragmentary nature of the type material (based on Nathaniel Wallich collections from central Nepal). This I outlined in a paper entitled 'The true identity of *Meconopsis napaulensis* DC.' in 2006. Subsequently *M. paniculata* (D. Don) Prain appreared in 1925 and *M. simplicifolia* (D. Don) Walp. in 1842. Both the latter were in fact first described as species of *Papaver* by Don in his 1825 '*Prodromus Fl. Nepal*' also based on Wallich collections from central Nepal. Taking this into account I proposed that *M. regia*, based on *Lall Dhwoj* 18 (BM, holotype; E, K isotypes) from Barpak, W Nepal at 12–15,000 ft, should represent the type of the reconstituted genus *Meconopsis*; neither *M. paniculata* nor *M. simplicfolia* were chosen because of the poor and incomplete nature of the type collections.

In summation, both comparative morphological and DNA research strongly indicates that, as presently circumscribed, the Asiatic species of *Meconopsis* do not share the same lineage as the sole European representative *M. cambrica*. This required separating the two geographical entities; however, the genus *Meconopsis* was originally based on a single species, the western European *M. cambrica* and under the 'International Rules of Botanical Nomenclature' this would have necessitated placing the Asiatic species in a new genus. This was deemed undesirable because of the long association of the name *Meconopsis* with the Asiatic species both in the botanical as well as the horticultural world. Therefore it seemed desirable to find a way to preserve the name *Meconopsis* for the Asiatic members and to place *M. cambrica* under a new generic name for which *Parameconopsis* is now proposed. While some may argue that *M. cambrica* could be placed in the genus *Papaver* (as it was originally described) it would require its own subfamily and would sit uneasily within that genus. While it is certain that the various genera in the Papaveraceae require realignment in the light of recent research, there is a strong possibility that *Papaver* itself may need to be divided. However, none of this in fact detracts from the purpose of conserving the name *Meconopsis* for the Asiatic representatives of the genus.

The actual proposal to conserve the name *Meconopsis* (*Papaveraceae*) with a conserved type which was published in *Taxon* 61 (2): 473–474 in April 2012 can be found in Appendix 1 at the end of this work.

3. MORPHOLOGY OF THE GENUS MECONOPSIS

Meconopsis are perennial or monocarpic herbs that are rather brittle with stems and leaves that are easily damaged or bruised; as with the majority of members of the Papaveraceae, cut surfaces generally ooze a thin yellowish or orange, occasionally whitish, latex that soon dries blackish or brownish. Most parts of the plant are usually thinly to densely adorned with soft hairs or stout subpungent bristles (often wrongly referred to in the literature as spines). The perennial species (for instance *M. grandis* and *M. quintuplinervia*) are tufted plants forming a dense 'crown' of shoots but occasionally, as in these two species and in *M. betonicifolia*, they may be stoloniferous with thin underground stolons sending up leafy shoots some distance from the parent plant. The perennial species die down to resting buds during the autumn, these being located at or just below the soil surface. The monocarpic species can on occasion behave as biennials but generally speaking they take three or more years to reach flowering size and having set seed, the plants die. Only very rarely do monocarpic species produce a lateral shoot and continue on for an additional flowering: in cultivation, at least, this is generally due to damage to the plant as it comes into the flowering phase. A few species, *M. punicea* and *M. baileyi* in particular, can behave as monocarpic or perennial (polycarpic) plants. This is especially apparent in cultivated plants and there is also some evidence from the wild that this may also occur there; however, on the whole, species are either monocarpic or polycarpic. The monocarpic species behave in one of two ways, either as evergreen leaf-rosettes that enlarge annually until the flowering stage is reached or, alternatively, the leaves die back at the end of the growing season to an over-wintering resting bud located at or close to the soil surface. One species, *M. bulbilifera*, produces tiny bulbils in the axils of the cauline leaves: these can start to grow out while still attached to the plant but generally they separate from the parent plant as it dies down in the autumn, overwintering on the ground amongst the surrounding vegetation, then growing away in the spring to form new plants. *M. chelidonifolia* (now located in the genus *Cathcartia*) behaves in a similar manner.

ROOT

Meconopsis generally, but not always, bear a swollen, fleshy taproot. This varies from species to species. In some (e.g. *M. concinna*, *M. henrici* and *M. lancifolia*) the rootstock is short, napiform, short-dauciform or sub-fusiform, seldom exceeding 10 cm in length, while in others (e.g. *M. impedita*, *M. pseudovenusta* and *M. speciosa*) the taproot is long and carrot-like, gradually tapered to the tip, sometimes reaching as much as 40 cm in length. While most taproots are unbranched, branching does occur but is generally non-species specific. *M. bulbilifera* tends to have a rather diffuse, branching rootstock of thin strands, while *M. quintuplinervia* produces a weak taproot initially which is soon replaced by a branched fibrous rootstock and this also tends to happen in other polycarpic species such as *M. grandis* and *M. simplicifolia*. The monocarpic evergreen species of subgenus *Meconopsis* such as *M. regia* and *M. paniculata* generally produce a stout dauciform taproot. Taproots are generally adorned with a sparse to dense amount of fibrous roots, especially towards the tip. At the top end where the root transfers into the stem there are usually the fibrous remains of old leaf bases or petioles, these often accompanied by sparse or dense bristles: as the plants age this can build up into a substantial wad as it does in the yellow-flowered *M. discigera*.

STEM

Meconopsis species can be roughly divided into two groups, those with a prominent stem and those without (the so-called acaulescent species). In all species no obvious stem is apparent until plants reach their flowering phase. In reality all the leaves (few to many depending on the species) are crowded together on a very reduced stem and the development of a stem or stems is directly related to whether or not the species bears an inflorescence or if all the flowers are borne on basal scapes. Except in a few species, *M. forrestii* and *M. lancifolia* in particular, the stems bear leaves, from few and well-spaced to numerous and dense. Stems are terete longitudinally ridged or unridged, glabrous or variously adorned with generally patent hairs or bristles, sometimes densely so.

INDUMENTUM

Most species of *Meconopsis* bear some sort of pubescence but individuals within populations can be found that are entirely glabrous (for instance in *M. bella*, *M. concinna* and *M. pseudovenusta*). Hairs can be thin and flexible, brittle or not, or stouter and firmer, sometimes stiff and pungent or bristle-like. Hairs of two types are found in *Meconopsis*. In many species the hairs are apparently simple and this applies to all the members of the large subgenus *Cumminsia* which contains well over half of all species in the genus. Simple hairs range from thin and brittle to thick, stiff and spine-like; *M. horridula* sensu stricto is the most prickly species in the genus and the pungent bristles on the leaves, but more especially on the pedicels and fruit capsules can be painful to the touch, both in the living and in the dried state. Although classified often as simple, the hairs are in fact minutely barbellate: this can be observed with an ordinary microscope, although scanning electron microscope (SEM) images reveal the details far more clearly. (Fig. 1A–G).

The other subgenera, subgenus *Meconopsis*, subgenus *Discogyne* and subgenus *Grandes*, all possess barbellate (sometimes referred to as dendritic) hairs, that is hairs that are ridged and with short barbs at intervals, these alternate or sometimes subopposite. Barbellate hairs vary greatly in length and are particularly noticeable on the leaf-lamina. In some species, especially those of section *Robustae*, the indumentum of long barbellate hairs is underlain by a shorter, often denser indumentum of similar hairs. In his 1934 monograph of the genus, George Taylor refers to an underlay of 'substellate' hairs; however, this is misleading as the hairs are not stellate but simple short barbellate hairs. In some members of subgenus *Grandes* (*M. baileyi* for instance), while the hairs of leaves and stems are barbellate, those of the fruit capsules may be simple, or mostly simple with a few barbellate hairs present; however, in the *Robustae* (e.g. *M. paniculata* and *M. regia*) all the hairs, including those of the fruit capsules are barbellate. The leaves of *M. regia* bear only very long and thin, dense, barbellate hairs that give the surface a silky appearance, due in part to their density but also because they all closely parallel one another and are directed towards the tip of the lamina. In contrast, the hairs of *M. paniculata* are of two types, long and short, and this gives the leaves a more rugged, less sleek appearance.

Pubescence bestows a certain look on different species and is most developed in the handsome rosettes of those found in section *Robustae*, the monocarpic evergreen species. In *Meconopsis regia*, for instance, the leaves are covered in a thick appressed felt of golden hairs giving the rosettes an attractive and unmistakeable velvety sheen. In the closely related *M. superba* the dense hairs are silvery white, giving the rosettes a sericeous appearance. *M. paniculata* has a rougher indumentum of fawn or rufous hairs. The leaf-rosettes of the monocarpic evergreen species are extremely attractive and take on their boldest colours during the autumn and winter, gold and russet hues predominating. In cultivation this feature adds greatly to the garden value of these imposing plants. The indumentum in the species of subgenus *Discogyne* can also greatly enhance the developing leaf-rosettes.

In some species the stout bristles on the leaves are distinctly expanded and somewhat raised and wart-like at the base and this area may be the same colour as the leaf, paler or, more often, stained with violet-brown or black, giving the leaves a polka-dot appearance. This particular feature is common in some forms of *Meconopsis horridula* but is probably most marked in two Chinese species, *M. balangensis* and *M. rudis*. The swelling and darkening of the bristle bases often extends to other parts of the plant such as stems, sepals and even the fruit capsule.

LEAVES

Meconopsis reveal a diverse range of leaf forms. The leaf-lamina can be papery or sometimes somewhat fleshy (as in *M. pseudovenusta*) but they are often quite brittle at the same time and easily bruised. Apart from the upper cauline leaves of some species all leaves are petiolate, the petioles generally linear, sometimes narrowly winged, generally expanded and more-or-less sheath-like at the base. In acaulous species the petioles may be part buried in the ground or in mossy litter (e.g. *M. impedita*, *M. muscicola*) but generally the leaves are all above ground.

The leaf lamina varies greatly from species to species, sometimes within a species, but it can be a very useful diagnostic character. The lamina ranges from simple and undivided (lanceolate, elliptic to cordate, lyre, fiddle or paddle-shaped) to those that are variously lobed or segmented. In section *Robustae* (e.g. *Meconopsis napaulensis*, *M. paniculata* and *M. staintonii*) the leaves are often elliptic to lanceolate or oblanceolate in outline with the lower half or two-thirds pinnatisect (pinnately lobed), with the lowermost lobes sometimes bipinnately lobed, while the upper part of the lamina is pinnatifid or, alternatively, the entire lamina may be shallowly to deeply pinnatifid. Other members of the closely related section *Superbae* such as *M. regia* and *M. superba*, on the other hand, have unlobed leaves.

Leaf-lamina shape becomes more complex in some of the smaller species where the full range from simple undivided leaves to pinnate or bipinnate can be found.

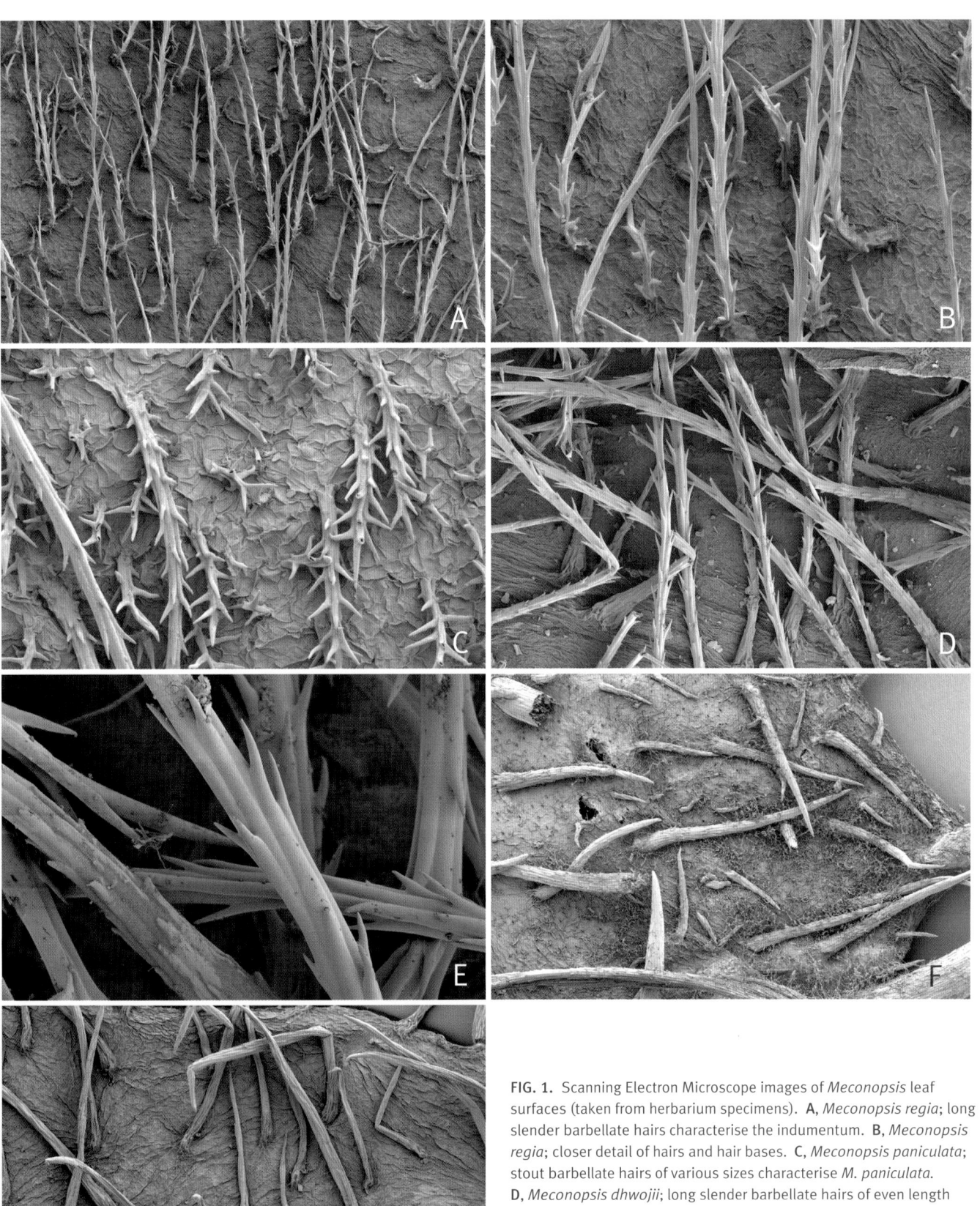

THE GENUS MECONOPSIS

MORPHOLOGY OF THE GENUS MECONOPSIS

FIG. 1. Scanning Electron Microscope images of *Meconopsis* leaf surfaces (taken from herbarium specimens). **A,** *Meconopsis regia*; long slender barbellate hairs characterise the indumentum. **B,** *Meconopsis regia*; closer detail of hairs and hair bases. **C,** *Meconopsis paniculata*; stout barbellate hairs of various sizes characterise *M. paniculata*. **D,** *Meconopsis dhwojii*; long slender barbellate hairs of even length spring from a slightly expanded base. **E,** *Meconopsis dhwojii*; closer detail of **D**. **F,** *Meconopsis horridula*; the stout 'simple' hairs of *M. horridula* and other members of subgenus *Cumminsia* are in fact minutely barbellate as seen through the scanning electron microscope (SEM). **G,** *Meconopsis lancifolia* subsp. *eximia*; weak slender 'simple' hairs are typical of *M. lancifolia* sensu lato.

As an added complication some species sometimes bear dimorphic leaves, especially some of the acaulous ones such as *Meconopsis bella*, *M. impedita* and *M. pseudovenusta*. In these, the lowermost leaves in the rosette, which are often smaller, are entire while the ones above are progressively more divided; however, in all three species individuals can be found which have all the leaves undivided or all divided so that the picture becomes quite complex.

Caulous species bear a few to many spirally arranged leaves which gradually merge into the bracts. In the species of subgenus *Grandes*, the uppermost leaves or bracts (generally 3–7) form a pseudowhorl. Cauline leaves are generally petiolate, with the uppermost leaves with the shortest petioles, or they may be sessile or subsessile, sometimes with an auriculate base.

BRACTS

The majority of species that bear an inflorescence also bear bracts but some species such as *Meconopsis forrestii* and *M. lancifolia*, which have a simple raceme, are bractless. Many of the racemose members of subgenus *Aculeatae* bear bracts subtending only the lowermost flowers in the inflorescence. At the opposite extreme, members of subgenus *Meconopsis* such as *M. autumnalis*, *M. napaulensis*, *M. paniculata* and *M. regia* bear substantial inflorescences in which all the flowers are bracteate and in those with well-developed paniculate inflorescences, bracteoles are also present.

Bracts are not clearly defined from leaves and there is generally a transition from one to the other with both become increasingly, and generally gradually, smaller up the inflorescence. While in some, the basal leaves may be markedly pinnatifid or pinnatisect the uppermost bracts may be only shallowly lobed, subentire or even entire. The indumentum of leaves and bracts is similar. Bracts may be ascending, horizontally aligned or deflexed, depending on the species.

As has been noted, in subgenus *Grandes* section *Grandes* the upper stem leaves or bracts are aggregated into a false whorl immediately beneath the flowers, although there may on occasion be subordinate flowers borne below the whorl from the middle or lower leaf-axils.

INFLORESCENCE

Generally speaking, *Meconopsis* can be divided into two main groups according to the way in which they bear their flowers. About one-third of all species are acaulescent monocarpic or occasionally polycarpic herbs (e.g. *M. bella*, *M. concinna*, *M. heterandra*, *M. impedita*, *M. pseudovenusta*, *M. quintuplinervia*, *M. simplicifolia* and *M. venusta*). In these the flowers are produced on basal scapes, sometimes just one per leaf-rosette but in others a few to many are produced in succession from a single rosette. In the latter the central (or terminal) flower of the rosette is the first to open, followed successively by those closest to the terminal one (i.e. from the uppermost leaf-axils in the rosette). The first flowers to open from a single leaf-rosette are often rather larger than those that follow.

On occasions some of the flowers in acaulous species (those closest to the centre) become partly agglutinosed by fusion of the lower part of the scapes or pedicels, thus forming a false inflorescence: this is referred to by David Prain (1915) as a 'compound scape'. This is most marked in some forms of the polymorphic *Meconopsis horridula*, but it can also be observed in other species (*M. pseudovenusta* for instance). It is important to note that in such false inflorescences, although part of the scapes become fused, the internal traces travel separately from the base of the plant to the flower. In a true inflorescence the traces from the different flowers in the inflorescence interconnect within the rachis in an ordered manner.

There are basically three types of inflorescence found in *Meconopsis*:

1. A simple raceme with 2–3 or numerous flowers. e.g. *M. balangensis*, *M. forrestii*, *M. rudis* and *M. speciosa*.

2. A paniculate inflorescence: in the majority of these the uppermost flowers are solitary and racemose, while the lower flowers are borne in pedunculate cymules (often just 2–4 but up to 11, occasionally more). On occasions inflorescences of this type can support 200 flowers or more. e.g. *M. paniculata*, *M. staintonii*, *M. wallichii* and *M. wilsonii*.

3. A pseudoumbellate inflorescence in which all or most of the flowers arise from congested leaf-like bracts at the top of the stem. e.g. *M. baileyi*, *M. grandis* and *M. integrifolia*. On occasions such inflorescences can be reduced to a single flower as in *M. sherriffii* and some forms of *M. grandis*.

In all *Meconopis* it is the uppermost terminal flower that is the first to open, followed in succession by those further down the stem. In lateral cymules it is again the terminal flower that is the first to open. One feature that is often seen in many species in the genus is the fact that the first flower or flowers to open are the largest, the later ones often successively smaller. In addition, the first flower to open may often have rather more than the normal compliment of petals. This can be observed in both wild and in cultivated plants.

PEDUNCLES AND PEDICELS

Both peduncles and pedicels are terete but while the former can at times be ridged or part-ridged, pedicels are never so: ridges on both stems and peduncles are generally the result of leaf or bract bases being decurrent, the thin extensions often running for some distance and, on occasions, overlapping with that of adjacent leaves or bracts. While they may be occasionally glabrous or subglabrous, in most species they are pubescent, the pubescence taking the form of soft or stiff, sometimes rather brittle, hairs, to more substantial, sometimes pungent, bristles. Although mostly patent, hairs can sometimes be ascending or even subappressed or occasionally somewhat retrorse. Pubescence is often densest just beneath the flowers. The pedicels often elongate significantly as the fruits develop; this is most marked in species like *Meconopsis grandis* and its allies and many of the species which bear solitary basal scapes.

FLOWERS

The *Meconopsis* flower, like those of other members of the Papaveraceae, is a relatively simple structure, with four whorls of organs: sepals, petals, stamens and with a number of fused carpels comprising the ovary.

Flowers can be variously shaped from campanulate to bowl- or saucer-shaped, occasionally more or less flat when fully opened or with the petals somewhat deflexed. Their posture can be anything from fully- to half-nodding, to lateral-facing, ascending or upright, depending on the species. Posture can sometimes lead to confusion, especially when examining herbarium specimens for, after pollination in those species with nodding or half-nodding blooms, the flowers gradually assume an upright position as they fade and the fruit capsules begin to develop. (Fig. 2A–G).

It is not without significance that the majority of species bear nodding or half-nodding flowers. Flowering as they do during the monsoon season, the large bright petals, apart from attracting pollinators, act as an umbrella keeping the important male and female sex organs dry, otherwise they would be swamped and this would certainly affect the production of seeds. It is interesting to note that in *Meconopsis horridula*, which encompasses various forms over its range, those growing in the heavier rainfall areas such as Nepal and Bhutan bear nodding flowers, while those to the north of the Himalaya have a tendency to produce lateral-facing or slightly upwardly inclined flowers. Some found in the drier north-eastern part of the range in Qinghai province in particular, can even have upright flowers.

The majority of *Meconopsis* species are unscented but scent has been noted in *M. bella*, *M. prainiana*, some forms of *M. pseudointegrifolia* and *M. speciosa*, while some of the members of the section *Primulinae* are said to have faintly scented flowers. Of these, only *M. speciosa* is recorded as possessing a strong fragrance that has been likened to that of the hyacinth (*Hyacinthus orientalis*).

BUDS

The majority of species bear pendent or semi-pendent buds whether or not they have pendent flowers. In those that do not, the flowers gradually take on a horizontal, ascending or upright position as the flowers open. An exception is *Meconopsis simplicifolia* whose buds are generally upright from the start.

The buds are usually ovoid or ellipsoid and often vary in size even on the same plant, with the first produced being the largest.

SEPALS

As in all members of the Papaveraceae, the flowers are heavily enveloped by the sepals. In *Meconopsis*, as in *Cathcartia* and *Papaver*, there are just two, very rarely three, imbricate, often boat-shaped sepals that completely conceal the petals and other floral organs within. Each sepal bears a narrow, sometimes rather broad, hyaline margin that is obscured in bud. As the buds swell the sepals are eventually forced apart by the expanding petals and are soon shed. Sepals, like the leaves and stems can be smooth and glabrous, but in the vast majority of species the sepals are adorned with patent to ascending bristles, sometimes very densely so. Because the sepals are readily shed and lost, they are generally omitted from the species descriptions. Sepals are generally green but can be variously flushed with blackish brown, purple or reddish maroon, sometimes heavily so, especially towards the base.

PETALS

By far the showiest part of the *Meconopsis* flower, petal numbers vary from species to species. While 4 is the basic number in the majority of species (e.g. *M. betonicifolia*, *M. compta*, *M. concinna*, *M. lyrata*, *M. paniculata*, *M. punicea*, *M. speciosa* and *M. wallichii*) others can have more. For instance, both *M. integrifolia* and *M. pseudointegrifolia* normally bear 6–8 petals, *M. simplicifolia* 6–9, while the flowers of *M. impedita* can have 4–10 petals. Having said this it is not uncommon for the first flower to open of normally 4-petalled species to have 5 or 6 petals and this can be observed in both wild and cultivated specimens.

THE GENUS MECONOPSIS
MORPHOLOGY OF THE GENUS MECONOPSIS

Petal colour varies enormously in the genus, although greens and browns are unknown. The three commonest colours are blue, red and yellow, but all shades of blue, purple-blue, mauve and lavender are seen in the wild. Only *Meconopsis superba* has white flowers while one or two species can have different coloured variants: for instance *M. georgei* has both red- and yellow-flowered variants, *M. prainiana* blue and yellow forms, while *M. wallichii* exists in both blue (or lavender) and red forms. Albinos have been recorded in the wild relatively rarely but they have been noted in *M. horridula* sensu lato, *M. prainiana*, *M. punicea*, *M. simplicifolia*, *M. sinomaculata*, *M. speciosa* and *M. staintonii*. However they are rarely common. *M. prainiana* is particularly interesting as the colour variants appear to have a geographical basis with the eastern populations (SE Tibet) being blue, purple or wine-coloured, the southern populations (E Bhutan and NE India) being white, while populations to the west of the former and north of the latter are yellow. Two colour forms are recognised in *M. wallichii*, one, the commonest, with blue-lavender flowers and the other with deep red flowers, yet the two have not been recorded growing together. In recent times pink- and red-flowered forms of *M. paniculata* have been found in Bhutan and in adjacent Arunachal Pradesh (NE India) growing in otherwise normally yellow-flowered colonies.

FIG. 2. Flowers, showing various shapes and disposition as well as the relative position of the stamens and style.

A *Meconopsis superba*. PHOTO: JAMES COBB
B *Meconopsis betonicifolia*. PHOTO: HARRY JANS
C *Meconopsis tibetica*. PHOTO: DAVID & MARGARET THORNE
D *Meconopsis integrifolia* subsp. *souliei*. PHOTO: HARRY JANS
E *Meconopsis horridula* subsp. *horridula*. PHOTO: HARRY JANS
F *Meconopsis sulphurea* subsp. *sulphurea*. PHOTO: HARRY JANS
G *Meconopsis punicea*. PHOTO: HARRY JANS

In this work, although I have in several instances recognised colour variants of species in the wild, I have not made any attempt to distinguish every colour variant within a species. To do so would be foolish and serve no purpose; however, where colour difference is very marked (as with yellow or red forms), or where habitat or geographical distribution play a significant role then I have made the distinction.

Petals are thin and papery, often semi-translucent, and seemingly delicate for plants that often grow in very exposed habitats. They are predominantly glabrous, but the Tibetan *M. torquata* has petals which bear some bristles on the exterior, towards the base, these being similar to those found on the pedicels. *M. horridula* var. *spinulifera* (see under *M. racemosa*), also from Tibet, is reported to be the same, although this requires confirmation. Petals on herbarium specimens can sometimes appear to bear bristles, but this is usually because the bristles have pushed through the petals from other parts of the plant during the pressing process, more especially from the ovary.

Petal margins can be entire to slightly to markedly erose, flat or somewhat undulate, while the petals themselves can be smooth, flat, undulate or somewhat fluted.

In bud the petals are tightly crumpled, expanding like the wings of a butterfly once released by the sepals. While this is apparent in all the species it is most striking in *Meconopsis punicea* whose long red petals can expand to more than four times from bud to mature flower.

STAMENS

Most *Meconopsis* flowers bear numerous stamens, but they can be as few as 12 in members of series *Lyratae*. The stamens are arranged in a number of tightly packed series, with the outermost maturing first.

Filaments

All species possess long filaments and in the vast majority they are linear and thread-like, sometimes finely tapered at the top. In *Meconopsis henrici*, *M. psilonomma* and *M. sinomaculata* the filaments are slightly expanded in the centre or towards the base and are rather flattish, especially those of the outermost stamens. In contrast, the stamens of *M. balangensis* and the related *M. heterandra* are dimorphic: the outer stamens spreading with linear filaments, while the inner are narrow-oblong and inflated (like a slender sausage balloon), congested and incurving around the ovary like a protective jacket. The function of this is not understood but it is an intriguing and distinctive feature. Filaments can be pale, often white, but in many species they are the same colour as the petals, although generally darker, that is with the exception of yellow-flowered species and variants that always have white filaments.

Anthers

These are basifixed and are generally narrow oblong to linear-oblong in outline, occasionally oval or ovate. Whatever their shape, they dehisce by longitudinal slits to release the pollen. Anthers vary in colour from white to blue, violet, purple or reddish but this colour is usually heavily masked in the mature stamen by the pollen which is most frequently yellow or orange, but it can be whitish, cream or even grey. Unfortunately, many records of anther colour on herbarium field notes actually refer to the pollen colour so that records of anther colour on many species is confused or at the very least ambiguous. However, properly recorded anther colour can be a useful diagnostic factor. For instance *Meconopsis henrici* and *M. psilonomma* look superficially very similar in flower, yet the former has consistently golden anthers while those of *M. psilonomma* are unusually (for the genus) buff.

POLLEN

Douglas Henderson (Royal Botanic Garden Edinburgh) published a paper entitled *The Pollen Morphology of Meconopsis* in *Grana Palynologia* in 1965. This was the first serious attempt to examine in detail the pollen of the genus and in the paper he used George Taylor's 1934 monograph *The Genus Meconopsis* as a template.

As a result of his studies Henderson was able to identify a number of morphological types of pollen (quoted here verbatim), leaving aside, that is, *Meconopsis cambrica* (now *Parameconopsis cambrica*) and the species transferred to *Cathcartia* (*Meconopsis villosa* and its allies):

1. Discigera-type

Pollen grains with 7–8 indistinct, scattered, pore-like areas, spheroidal (23–28 μ). Distance between apertures about 8–10 μ. Pores circular, 2.5–4 μ diameter, with indistinct margins. Exine about 1–1.3 μ thick but slightly thinner at the margins of the pores, with minute spinules. Spinules about 1.2 μ apart, less than 0.1 μ long. No differentiation into sexine and nexine, infra-tectal bacula absent. *Meconopsis discigera* [*M. bhutanica*]★ and *M. torquata*.

Henderson remarks that the large ill-defined, subcircular pores in the Discigera-type are very characteristic and only found elsewhere in the Impedita-type, but in this infra-tectal bacula are present, but lacking in both *Meconopsis discigera* and the closely related *M. torquata*.

2. Impedita-type

Pollen grains 6-aperaturate, spheroidal (18.5–23 μ diameter). Apertures scattered, about 5 μ diameter, usually circular but occasionally slightly elliptic, about 8 μ apart. Exine about 1 μ thick. Tectum slightly rough with minute spinules about 0.5 μ apart. Endosexine distinctly baculate. *Meconopsis impedita* and including *M. florindae*.

See under Discigera-type for general remarks. In addition, the Impedita-type has obvious affinities with the Primulina-type as regards tectal ornamentation.

3. Horridula-type

(a) Pollen grains 3-colpate, spheroidal (15–20 μ diameter). Apocolpium about 5 μ diameter. Exine about 0.8–1 μ thick. Sexine about 0.6 μ thick. Tectum with minute spinules. Spinules less than 0.1 μ long, about 0.5 μ apart. Endosexine baculate. Nexine about 0.3 μ thick. *Meconopsis horridula* [probably *M. racemosa* or *M. zhongdianensis*].

(b) Pollen grains 3-colpate, spheroidal (22–25 μ) diameter). Apocolpium 4–5 μ diameter. Exine about 1 μ thick. Tectum with minute spinules. Spinules less than 0.2 μ long, about 0.7 μ apart. Endosexine baculate. *Meconopsis aculeata, M. argemonontha, M. bella, M. delavayi, M. forrestii, M. georgei, M. henrici, M. horridula* [probably *M. prattii*], *M. lancifolia, M. pseudovenusta, M. quintuplinervia, M. speciosa, M. venusta*).

The most frequent pollen grain type found in *Meconopsis* and Henderson points out that it is "very similar to that of a number of genera of Papaveraceae including the genus *Papaver* itself".

4. Betonicifolia-type

Pollen grains without apertures, spheroidal (25–27 μ diameter). Exine about 1 μ thick with conspicuous spinules, 1–1.5 μ long, 1–1.2 μ broad at base, about 2.5 μ apart; between the large spinules very small, rather closely arranged spinules about 0.2 μ long and less than 0.1 μ broad. *Meconopsis betonicifolia* [*M. baileyi*], *M. integrifolia, M. punicea, M. simplicifolia, M. taylorii*.

The lack of apertures along with the conspicuous spinules and very thin exine are the most obvious features of the Betonicifolia-type.

6. Sherriffii-type

Pollen grains in rather loose, tetrahedral and rhomboidal tetrads, without apertures, spheroidal (28–32 μ diameter). Exine about 0.5 μ thick, spinulose. Spinules 1–1.5 μ long, 1–1.2 μ broad at base, about 4–5 μ apart. *Meconopsis grandis* (probably subspecies *orientalis*), *M. sherriffii*.

Henderson remarks that the "individual grains of the tetrads are exactly similar to those in the Betonicifolia-type". This is interesting in itself as the eastern populations of *Meconopsis grandis*, representing subsp. *grandis* and subsp. *orientalis*, have their pollen in tetrads, while the western population, subsp. *jumlaensis*, has solitary pollen grains or monads. The significance of this both evolutionary and taxonomically is unclear. It is certainly true to say that the Betonicifolia-type and the Sherriffii-type are very closely allied: both types bear pollen without apertures.

7. Primulina-type

Pollen grains 3- and 6-colpate, spheroidal (19–30 μ diameter). Acocolpium 4.5–5.5 μ diameter. Exine about 1 μ thick with minute spinules. Spinules less than 0.1 μ long, 0.5 μ apart. Infra-tectal bacula absent. *Meconopsis dhwojii, M. latifolia, M. lyrata, M. napaulensis* [*M. wilsonii*], *M. paniculata, M. primulina, M. regia, M. robusta, M. sinuata, M. superba*.

The Primulina-type is characterised by relatively large pollen grains, a very thin exine with tiny spinules and a tendency to have 6 apertures. Henderson points out that this type is "thus intermediate between the rather regular, 3-colpate, small grains of the cambrica and horridula types and the large, porate or inaperturate spinulose grains of the impedita and betonicifolia-types".

8. Gracilipes-type

Pollen grains without apertures, spheroidal (17.5–22 μ diameter). Exine about 1.5 μ thick with fine spinules. Spinules less than 0.2 μ long, 0.5 μ apart. No differentiation into sexine and nexine; infratectal bacula absent. *Meconopsis gracilipes*.

Very close to the pollen type found in the Primulina-type but the absence of any discernible apertures sets this lone species apart.

Henderson's paper closely follows George Taylor's (1934) classification of the genus. In the intervening years adjustments have been made to the taxonomy, species have been divided and new species described. Fortunately, Henderson lists the source material for his pollen studies so that in a few instances his names can be readjusted according to this present monograph. Those names in [] are those being used in this work.

Henderson includes a table which shows the species used in the study arranged in a series of pollen morphological affinity, pointing out at the same time that "….. the groups are not absolutely limited, in fact the "series shows an almost continuous gradation".

The study reveals that the pollen type reasonably

conforms with Taylor's classification, although there are a few anomalies. Of these the following stand out:

Meconopsis gracilipes which is, morphologically, very closely related to *M. dhwojii* is separated from it in the pollen classification.

Meconopsis quintuplinervia, which has its closest morphological ally in *M. punicea*, with which it freely hybridises, also finds itself in a different pollen type, being divorced from its other allies in the Betonicifolia-type grouping.

Other similar anomalies include *Meconopsis lyrata*, *M. primulina* and *M. sinuata*.

In taxonomic terms it would be difficult to explain such anomalies. However, pollen studies have greatly helped in the overall understanding of the genus and its relationship within the Papaveraceae as a whole. James Cobb (1989) points out that that "..... it seems to me that if very closely related species that are interfertile have radically different pollen then the characteristics of the pollen are a later and secondary development and tell nothing of relatedness". While I would agree with this in many ways, it is perhaps worth pointing out that many non-purely morphological studies (pollen and DNA analyses for instance) are based on a very small sampling, often only a single individual representing each species. Upon such studies (and *Meconopsis* is one such of many in recent years), sweeping reclassifications and assumptions have been made, often greatly disturbing the pre-existing morphological evidence. In the case of *Meconopsis* and pollen we simply do not have enough samples to make such judgements or to evaluate the implications of the data extracted. We do not know, for instance, how much pollen variation is found within a species, particularly those widespread ones such as *M. horridula* sensu stricto and *M. paniculata* to give two examples. This is very apparent in the blanket treatment proposed by George Taylor (1934) of *M. napaulensis*, a species which he considered ranged from west Nepal to western China; however, the material upon which he based his monograph has subsequently been divided into 6 species (Grey-Wilson 2006a). In Henderson's (1965) paper, two sources of material are indicated under the name *M. napaulensis*, one based on a George Forrest collection (no. 15883) now identified as *M. wilsonii* subsp. *australis* from north-western Yunnan, and the second on a plant of dubious origin cultivated at the Royal Botanic Garden Edinburgh. It is clear that a large and more thorough sampling is required in order to provide a firmer basis to the significance of pollen in the overall taxonomic studies of the genus. Regretfully, DNA studies fair no better in this respect, especially when one considers that in studying the general morphology of the various species, a number of different sheets, often measuring in the tens, sometimes in the hundreds, of herbarium specimens are examined in order to determine the variation found within a given taxon. It is the inherent variation found within species that I and many botanists find most fascinating and indeed most challenging. At the same time, it is the relationship between species and, at a higher level, genera, that is the core of any monographic treatment and, while it would be foolish to disregard the findings of pollen, DNA and other studies, they need to be taken into consideration in the overall study. After all, in the end a clearer understanding of species and their relationships is sought while, at the same time, most readers will need to be able to identify species by sight not via some intricate chemical analysis.

OVARY

The ovary is generally ovoid or ellipsoid, occasionally ellipsoid-cylindric, sometimes very narrowly so, consisting of 3–17 fused carpels, depending on the species. The zone marking the carpels is marked by a longitudinal groove or ridge which may be prominent or rather obscure. Ovaries may be smooth but more often they are adorned with appressed to ascending bristles, these sometimes dense and pungent, as in *Meconopsis horridula*. In pubescent kinds, the suture lines along ovaries and fruit capsules may be pubescent or glabrous, this often being specific to a particular taxon.

Style

Most species of *Meconopsis* bear a distinct style, generally several mm long. In a number (*M. integrifolia*, *M. punicea* and *M. torquata* for example) the style is obsolete or subobsolete, a character generally considered to be secondary in evolutionary terms. The style can be linear or somewhat conical and then, in that instance, expanded at the base onto the top of the ovary. The style persists into the fruiting stage, generally expanding and often thickening as the fruits develop so that the style in fruit may be twice the length it was in flower. In *M. simplicifolia* the style may reach 14 mm long, but this is an exception.

The members of subgenus *Discogyne* are very distinctive. The species bear a short linear style that expands greatly at the base to cover the whole top of the ovary like a cap or disc. The disc is often brightly coloured, deep red, crimson or purple, occasionally whitish, flat or somewhat undulate with a fringed, shallowly or deeply lobed margin, the lobes corresponding to the number of compartments of the fruit capsule. The stylar disk of the *Discogyne* should not be confused with the stigmatic disc of many species of *Papaver*. While the former is evolved from the basal development of the style, the latter is developed from the stigmatic lobes or rays. Evolutionarily, it is likely that these two features developed quite independently even though the two genera clearly shared the same progenitors in their evolutionary history.

Stigma

The stigma varies from species to species and can be capitate or club-shaped with as many lobes as the ovary compartments. In some, the lobes spread apart like fingers prior to pollination, but in the majority they are held closely together. In many species observed the outer anthers ripen before the stigmas are receptive but as the flowers mature so both are receptive at the same time. This ensures at least some cross-pollination in wild populations.

Occasionally the base of the stigmatic lobes are decurrent onto the style, or when this is absent, onto the top of the ovary, appearing there as thin flanges. This is most manifest in *M. integrifolia*, especially in the forms in which the style is obsolete or almost so, but it is not confined to this species.

FRUIT CAPSULE

The shape and details of the fruit capsules (especially the ornamentation, number of carpels and the details of dehiscence) are important diagnostic characters in the genus. In all species, whatever the orientation of the flowers, the capsules are upright, that is with the exception often of *Meconopsis bella* in which the fruiting pedicels generally head away tangentially as the fruits develop, often becoming somewhat contorted in the process: this process brings the fruit capsule close to the 'ground' (often a ledge or rock face).

The many-seeded capsules are derived from the fusion of a number of carpels (3–) 4–11 normally in number (but as many as 17 in *Meconopsis taylorii*), with free parietal placentation, and dehisce by the same number of valves as there are carpels. Thus fruit capsules are said to be 4–5-valvate and so on. Dehiscence is achieved when the top of each carpel peels back producing a pore or valve through which the seeds can escape, this process being aided by agitation, primarily by the action of the wind. The size of the pores varies from species to species but generally occupies no more than the top third of the mature capsule length. (Figs 3 & 4).

Most species bear a persistent style on top of the capsule, although in a few the style is very short or more or less obsolete, while in subgenus *Discogyne*, as we have seen, the base of the style is expanded into a cap that covers the top of the ovary and later the fruit capsule itself (see under style, above).

Seeds

The number of seeds per capsule varies enormously from species to species and often within species, the first fruits in an inflorescence often being larger than subsequent ones and containing a greater number of seeds. In small species like *Meconopsis compta* and *M. primulina* there may be relatively few seeds (between 12 and 30) but in some of the more robust species like *M. grandis*, *M. paniculata* and *M. regia*, large fruit capsules may contain well in excess of 150 seeds.

Seeds vary greatly in size from species to species, most filling the range from about 0.5 mm to 3 mm in length: of those measured to date both *Meconopsis dhwojii* and the closely related *M. gracilipes* have the smallest seeds, 0.5–0.8 mm long, those of *M. discigera* are 0.8–1 mm long, *M. bella* 1.2–1.4 mm long and the recently described *M. yaoshanensis* 2.3–2.9 mm long.

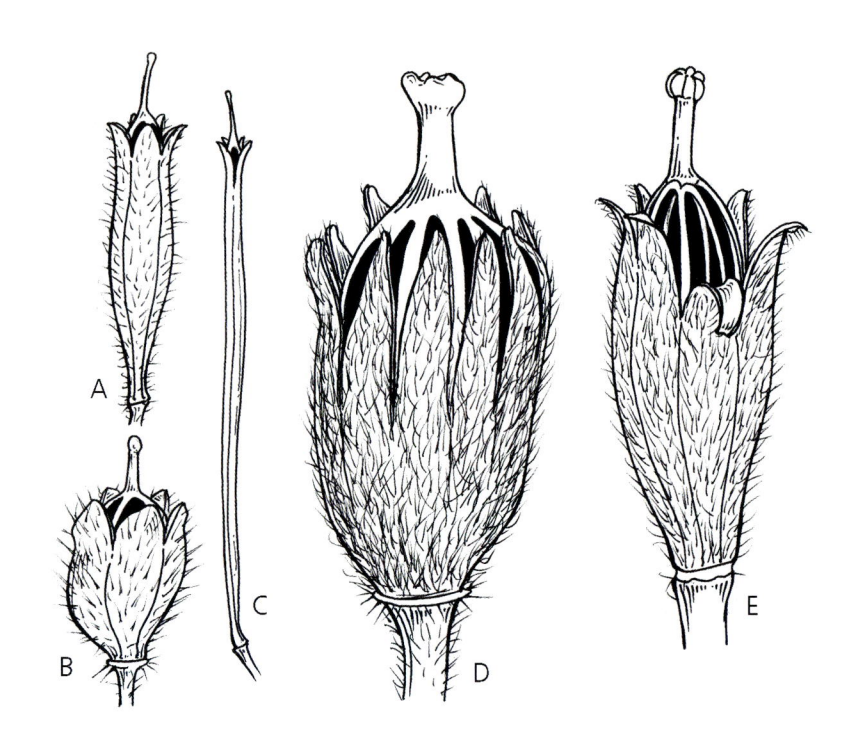

FIG. 3. Dehisced fruit capsules: **A**, *Meconopsis impedita* subsp. *impedita*; **B**, *M. rudis*; **C**, *M. delavayi*; **D**, *M. superba*; **E**, *M. sulphurea* subsp. *sulphurea*. All × 1½.
DRAWN BY CHRISTOPHER GREY-WILSON

THE GENUS MECONOPSIS
MORPHOLOGY OF THE GENUS MECONOPSIS

FIG. 4. Fruit capsules (photographed from herbarium collections; numbers in brackets the same as the species numbers in the text). PHOTOS: CHRISTOPHER GREY-WILSON

A *Meconopsis wallichii* (11); *Grey-Wilson & Phillips* 4527
B *Meconopsis paniculata* (13); *Stainton, Sykes & Williams* 8648
C *Meconopsis paniculata* (13); *Curzon* 52
D *Meconopsis violacea* (15); *Kingdon Ward* 6905
E *Meconopsis betonicifolia* (23); *Forrest* 21573
F *Meconopsis speciosa* subsp. *speciosa* (41); *Rock* 23288
G *Meconopsis speciosa* subsp. *cawdoriana* (41); *Ludlow, Sherriff & Elliot* 12257
H *Meconopsis horridula* subsp. *horridula* (42); *Halliwell* 175
J *Meconopsis prainiana* (47); *Ludlow & Sherriff* 6236
K *Meconopsis georgei* (50); *Rock* 23287
L *Meconopsis impedita* subsp. *impedita* (53); *Forrest* 18969
M *Meconopsis venusta* (58); *Schneider* 3128
N *Meconopsis forrestii* (66)

In shape, seeds tend to be like a half monkey-nut (roughly kidney-shaped) or, so I have heard it observed, like tiny armadillos. The surface of the testa is variously ornamented but basically consists of rows of honeycomb-like, more or less 4–5-sided, cells which collapse in the dried state to leave a rim. The overall impression given by these cells is a rather reticulate pattern to the surface of the seed. While some species like *Meconopsis baileyi*, *M. dhwojii* and *M. horridula* have relatively small seeds and a fine reticulation, those of *M. grandis*, *M. sherriffii* and *M. quintuplinervia* tend to be much coarser, although they also have larger seeds on average.

There are some interesting electron micrograph scans of various *Meconopsis* seeds and a couple of hybrids in James Cobb's book *Meconopsis* (1989).

The taxonomic implications of seed anatomy would make a very interesting study, especially if it were based on authenticated wild-source material. Such a study would be an important addition to our knowledge of the genus as a whole.

CHROMOSOMES

While several detailed cytological studies have been carried out on individual species, chromosome investigation in the genus cannot be stated to more than still in its infancy. Of the 79 species currently recognised, just 20 have had their chromosomes counted. There are big gaps in the study thus far: for instance none of the 7 species in subgenus *Discogyne* has been examined and in subgenus *Cumminsia* (the largest in the genus) no species in sections *Bellae*, *Cumminsia* and *Forrestianae*, comprising 16 species in total, have been examined. While it is highly desirable that wild material forms the basis of any detailed examination, many of the earlier records were based on cultivated material. The results presented below from various sources have been realigned to conform with the species as defined in this monograph.

Chromosome numbers recorded as follows:

M. aculeata 2n=28 (Sugiura 1937); 2n=28 (Jee *et al.* 1989); 2n=56 (Ratter 1968)★★★

M. baileyi, as *M. betonicifolia* 2n=82 (Ratter 1968)★★

M. × *cookei* 2n=84★★

M. delavayi 2n=56 (Xie 1999)★★★

M. dhwojii 2n=56 (Ratter 1968)★

M. gracilipes 2n=56 (Ratter 1968)★

M. grandis 2n=118, 120 (Ratter 1968); 2n=164 (McAllister 1999)★★

M. horridula 2n=56 (Ratter 1968), probably *M. racemosa*★★★

M. integrifolia 2n=74 (Ratter 1968)★★

M. latifolia 2n=56 (Ratter 1968)★★★

M. napaulensis, as *M. longipetiolata* 2n=56 (Ratter 1968)★

M. napaulensis 2n=56 (Ratter 1968), probably *M. staintonii*★

M. paniculata 2n=28 (Sugiura 1940); 56 (Ratter 1968)★

M. pseudovenusta 2n = 56 (Ying *et al.* 2006)★★★

M. quintuplinervia 2n=84 (Ratter 1968), 2n=76 or 84 (Huang *et al.* 1996)★★

M. regia 2n=56 (Ratter 1968)★

M. rudis, as *M. horridula* 2n=56★★★

M. simplicifolia 2n=82, 84 (Ratter 1968)★★

M. speciosa 2n=6 (Xie 1999)★★★

M. sulphurea, as *M. integrifolia* 2n=76 (Ying *et al.* 2006)★★

M. zhongdianensis, as *M. racemosa* 2n=56 (Ying *et al.* 2006)★★★

In addition several species formerly included in *Meconopsis*:

Cathcartia chelidonifolia 2n=28 (Ratter 1968)

Cathcartia villosa 2n=32 (Ratter 1968)

Parameconopsis cambrica 2n=28 (Sugiura 1940 & Ratter 1968), 2n=28 (Safonova 1991)

From these results several generalisations can be deduced. First, that all the members of Subgenus *Meconopsis* (★ above) analysed to date have 2n=56. Second, that members of subgenus *Grandes* (★★ above) consistently bear higher numbers, 2n= 74, 76, 82, 84, 118, 120 or 164. Third, species of subgenus *Cumminsia* (★★★ above) bear 2n=56, that is with the exception of *M. aculeata*, 2n=28 or 56.

Ying *et al.* (2006) point out that "Previous cytological studies indicated some confusion concerning the basic chromosome number of the genus, and some different chromosome numbers were observed, such as n=14, 16, 28, 37, 41 or 42, and 59 or 60. For example, the chromosome number found in *M. integrifolia* [now divided into several allied species!] (2n=76) differ substantially from the n=37 determined by Ratter (1968). Although a possible explanation for this disagreement could be the occurrence of aneuploidy reduction from a duodecaploid condition, this type of high haploid chromosome number has not been reported previously within *Meconopsis* species. But, the chromosome numbers of 2n=28, 56 and/or n=14, 28 are most prevalent in *Meconopsis*."

The question of the main base number remains speculative. Ernst (1965) and later Ratter (1968) suggested that x=7 and 8 are the basic chromosome numbers in the genus. However, according to Sugiura (1940) and Ratter (1968), the behaviour of chromosomes at meiosis shows that the basic number of *Meconopsis* is also possibly x=14. Both Ian McNaughton and John Richards (recorded from talks given to the *Meconopsis* Group in

March 2013 and October 2013 respectively) state that it seems likely that the base number in *Meconopsis* as a whole is x=14, and the diploid count 2n=28. This means that species with a chromsome count of 2n=56 are tetraploids, the commonest level found amongst the species thus far recorded. *M. baileyi* and *M. simplicifolia* would be hexaploids, while the high count for *M. grandis* of 2n=164 can be classed as a dodecaploid. Whatever the base number, it is clear that polyploidy, however it arose, has played an important part in the history and evolution of the genus with tetraploids, hexaploids and octoploids present amongst the species, while even higher ploidy levels are also present, as in *M. grandis*.

Two questions arise from the results presented above. First, variation at hexaploid level (e.g. 2n=76, 82, 84) and at even higher ploidy levels is frequent amongst plants and may be accounted for by technical difficulties (preparation and counting samples, for instance) but it could be genuine as chromosome loss is buffered by lower gene dosage effects at higher ploidy levels. Second, some other discrepancies are evident and this is most marked in *Meconopsis grandis* (2n=118, 120, 164). Ratter took his samples from living material and at that time in the late 1960s many plants now considered to be of hybrid origin were thought to be variations of *M. grandis* and it may be that he did not have genuine authenticated *M. grandis* to investigate. While this is not the place to investigate the origins of the 'big perennial blue poppies' in cultivation (of which there are now many named and unamed cultivars), investigations do throw some light on the species themselves.

With a chromosome count of 2n=118 or 120, and assuming a base number of x=14, then the '*M. grandis*' recorded by Ratter would probably prove to be a sterile nonaploids (9x). This means that the plants were almost certainly of hybrid origin. Work by the *Meconopsis* Group and others has investigated the origins of some of the 'big perennial blue poppies' and, although there is still much to do, results so far have established that the sterile cultivars with chromosome counts of 110–130 are probably all nonaploids (in essence triploids) of hybrid origin and as a result they are sterile. The well known and most widely cultivated 'Slieve Donard' (2n=110–114) is one such plant. The origin of this 'triploid' cultivar has not been proven beyond doubt but it is highly suspected that it arose spontaneously in cultivation in the mid 1930s between *M. baileyi* (at that time considered to be *M. betonicifolia*), 2n=82 and *M. grandis* subsp. *grandis*, 2n=164. In more recent times the cultivar 'Lingholm' has become widely available to gardeners. Being fertile it sets abundant seed and is readily raised by this method and, as might be expected, delivers variable offspring. It is generally believed that 'Lingholm' arose directly from 'Slieve Donard' by simple doubling of the chromosomes but, taking into account the characters of the two, especially the details of leaves and bracts, this hypothesis does not all together add up. Whatever its origin, 'Lingholm' has evidently arisen from a nonaploid plant.

Summing up, two species, *Meconopsis baileyi* (BB) and *M. grandis* (SSBB), probably gave rise in gardens to the plethora of 'big perennial blue poppies'. Many of these are sterile nonaploids (SSB). Another species, *M. simplicifolia* (SS) is sometimes thought to have been involved in the 'hybrid pool' but it has always been quite scarce in cultivation and, as a result, it is generally assumed that the primary nonaploid hybrids involved only *M. baileyi* and *M. grandis* (either subsp. *grandis* or subsp. *orientalis*).

This may or may not throw some light on the origin of *Meconopsis grandis*. As has been noted, *M. baileyi* and *M. simplicifolia* are known hexaploids (2n=76, 82, 84), while *M. grandis* is a dodecaploid (2n=164). It has also been suggested that *M. grandis* bears features of both *M. baileyi* (BB) and *M. simplicifolia* (SS) and that it might have arisen in south-eastern Tibet (Xizang) as a hybrid where both these species are sympatric. To become fertile the putative hybrid (SB) would have had to double chromosomes to become what we now know as *M. grandis* (SSBB). However, there are problems with this hypothesis. For a start, as far as it is known, *M. grandis* does not occur in the wild with *M. baileyi*, although it certainly overlaps in eastern Nepal and Bhutan with *M. simplicifolia*. There is some evidence today that *M. grandis* and *M. simplicifolia* may have hybridised in eastern Nepal (see p. 000). *M. grandis* is divided into three subspecies, one, subsp. *jumlaensis* is restricted to western Nepal, a great distance west of any known location for either *M. baileyi* or *M. simplicifolia*, a distance compounded by the very rugged nature of the Himalaya. While it can be argued that *M. grandis* subsp. *orientalis* shows some intermediate characters between the two putative parents the same cannot be said of subsp. *grandis*. Another explanation needs to be sought. It may be that *M. grandis* did not arise as a simple cross between two parent species but involved subsequent back-crossing with one or other parent, but the disparity in distribution of the three taxa as we know them today defies simple logic. The most likely explanation, to me at least, is that *M. grandis* arose from the progenitor that gave rise to all three species, which then radiated out as evolution and the complex glacial history of the Himalaya and Tibetan Plateau determined their subsequent development.

The cultivation of *Meconopsis* (species, hybrids and cultivars) and the development of the 'big perennial blue poppies' in gardens is a large subject and will be the focus of a separate book at present in preparation by myself in close collaboration with Dr Evelyn Stevens and the *Meconopsis* Group.

POLLINATION

Little is known about the pollination of *Meconopsis*. However, species of bumblebee are almost certainly the prime pollinators, both in the wild and in cultivation. Bumblebees can live at very high altitudes and can be frequently observed around *Meconopsis* in flower. *Meconopsis* do not secrete nectar, but the large number of stamens produce copious pollen which the bumblebees collect by rummaging around the flowers. The pendent or semi-pendent flowers of the majority of species are ideal for bumblebees which fly into the flowers often alighting on the stamens before accessing the anthers, or alternatively they land on the central core of stigmas. The forest zone species such as *M. wallichii* and *M. wilsonii* are also visited by honey bees which are wild or, in some areas (Bhutan and western China for instance) from hive bees kept close to mountain villages, and these can sometimes be seen set out on terraces or on the mountain slopes. Butterflies have been noted sitting on the flowers of a number of species, attracted no doubt by the bright blue or yellow flowers of species like *M. baileyi*, *M. paniculata* or *M. wallichii*. Butterflies require nectar so that the blooms are no use to them and, although they may occasionally effect pollination, they are unlikely to move from flower to flower or plant to plant as do bumblebees. Species of hoverfly (Syrphidae) are certainly attracted to the flowers and are often seen on the stamens and may be important pollinators. Flies of various sorts, including hoverflies, can frequently be seen sheltering within the blooms in the wild, especially during periods of inclement weather or overnight. Apart from shielding flies from the rain, the blooms may also give insects added protection and warmth, especially during cold montane nights. Many of the Sino-Himalayan species of *Meconopsis* have nodding to half-nodding flowers and these act as an umbrella to sheltering insects, besides protecting the sexual organs in the centre of the bloom from excessive wet.

In truth the only reward for visiting insects is pollen and this may also account for the fact that the majority of *Meconopsis* species are not sweetly scented. Most species bear numerous stamens and produce copious pollen for visiting insects. It is fascinating watching a bumblebee alight on a *Meconopsis* flower. As described above they either land directly on the anthers by clutching the filaments, or they land directly on the stigmas. In the latter case the bee will deposit some of the pollen from previously visited flowers onto the receptive stigmas (it is probable that in most species both anthers and stigmas ripen together); however, it is the outer anthers that mature first, followed progressively by those nearer the centre of the flower and by the time they are ripe the stigmas may have already received pollen from another flower or have lost their receptivity. However they land, bumblebees scramble around the stamens on their sides as they collect the pollen and transfer it to their pollen baskets or sacs. This system is common to many members of the Papaveraceae and can be readily observed in the common or field poppy, *Papaver rhoeas*. It will be observed that some pollen is not transferred to the pollen sacs but merely brushes off onto the body of the insect and it is this pollen that is likely to be brushed off onto the stigmas when the insect visits another flower.

When the *Meconopsis* flower first opens, the stamens are bunched closely around the ovary and stigma; however, they very quickly splay outwards away from the centre of the flower. This separation of anthers and stigmas is known technically as 'herkogamy' and its prime effect is to prevent automatic self-pollination.

COMPATABILITY

This is also dealt with under the next chapter, 'Cultivation' as it affects *Meconopsis* species and hybrids in cultivation, while this passage makes a brief examination of species in the wild. Little is known about compatibility of the wild species, although it can be gleaned from cultivation that species are more likely to set abundant seed when grown in groups. This would appear to indicate that in some *Meconopsis* species at least, there is a partial self-incompatibility system in place that favours outcrossing. In many plants the anthers mature in advance of the stigmas allowing cross-pollination to take place; this system is known as protandry. This does not appear to be the case in *Meconopsis*, and as in *Papaver*, anthers and stigmas mature more or less at the same time, although the first anthers to mature in a given flower may do so in advance of stigma receptivity.

While the breeding system in *Meconopsis* has not been studied to any degree, it has in the closely allied genus *Papaver* which exhibits a very similar flower structure. Proctor, Yeo and Lack have written extensively on breeding systems in their book *The Natural History of Pollination* produced in the *New Naturalist* series in 1996 and from which the following quotes have been elicited:

"In some species with a bowl or disc-shaped flower such as the greater celandine (*Chelidonium majus*) and perhaps, the poppies (*Papaver* spp.) and buttercups (*Ranunculus* spp.), the stigmas are in the centre of the flower and the anthers are splayed apart. Visiting insects home in on the centre of the flower and land on the stigma, depositing pollen if they have some, and then move to the edge to take off, walking over the anthers and collecting pollen in the process (Webb & Lloyd 1986). This seems an effective and simple method of promoting cross-pollination."

"Some bee-pollinated flowers have become specialised pollen-flowers, offering visitors abundant pollen but very little nectar. Examples are the poppies (*Papaver* spp.), peonies (*Paeonia* spp.) and the sunroses and rockroses (*Cistus* and *Helianthemum*; Cistaceae), whose big colourful flowers with masses of stamens yield profuse quantities of pollen ......".

"Typical bee flowers provide both nectar and pollen as rewards for the visiting insect. However, there are two important groups of bee-pollinated flowers in which the primary reward is pollen, and there is little or no nectar. The poppies (*Papaver* spp., Papaveraceae) are a familiar example, the numerous stamens providing abundant pollen avidly collected by visiting bees."

The usual way in which plants avoid self-fertilisation is by a system of self-incompatibility. This can be effected by a physical barrier or, more, usually, by a physiological one. In the latter, although copious pollen from the same plant alights on the stigma fertilisation does not happen. However, pollen from a separate individual (I am referring in this instance to the same species) will in all probability result in fertilisation. This means that for some reason or another the plant is able to discriminate between its own pollen and that of another individual. While several self-incompatibility systems are recognised (see Proctor *et al.* 1996) only one need concern us here and that is 'gametophytic self-incompatibility' and I can do no better here than quote directly from this important reference:

"This is a broad heading covering a number of systems, in all of which the growing pollen tube is recognised and rejected. The best-studied form involves the pollen grain germinating on the stigma and the tube growing down the style but being stopped before it reaches the ovules. The tube may be blocked or it may burst in the style in a way that is similar to what happens when it reaches the ovule in a successful fertilisation; it is as if the bursting is triggered too early. The nuclei produced by the pollen grain, one of which forms the pollen tube nucleus, are regarded as the 'gametophyte', i.e. that part of the plant that produces gametes or fertile cells, from its presumed origin in ancestral gymnosperms ...... This is why the system is called gametophytic. It is known from a diverse range of plant families such as the legumes (Fabaceae), poppies (Papaveraceae), nightshades (Solanaceae) and lilies (Liliaceae) .... A few of the species with gametophytic self-incompatibility have been studied genetically. In poppies (*Papaver*), clovers (*Trifolium*), evening primroses and probably many others, the incompatibility system appears to be controlled by a single gene locus that has many forms (alleles); the presence of an allele identical to either of the parents' alleles will stop the pollen tube growing. The corn poppy (*Papaver rhoeas*) is the best studied of these, and O'Donnell & Lawrence (1984) estimated that there were between 25 and 45 alleles in natural populations. They reported similar numbers from study on evening primroses and unconfirmed reports of higher numbers in clovers. They showed, in the poppy, that any one plant was fully capable of breeding with over 80% of the others in the population and totally incompatible with less than 5% mainly close relatives."

Unfortunately, at present it is only possible to conjecture on the system found in *Meconopsis* but it seems highly likely that it will closely match that found in *Papaver*. For instance it has often been reported that *Meconopsis* are self-sterile or more properly self-incompatible (see Cobb 1989). While this is certainly true of the majority of species it is by no means certain that it applies to all. Much of the evidence, it has to be said, comes from cultivation. Gardeners have noted that isolated plants generally produce little or no seed while those grown in groups often produce abundant seed provided, that is, that there were plenty of pollinators around at flowering time, as well as ample moisture during fruit development. Even so, isolated plants will sometimes produce a little seed and this may be nature's way of ensuring that at least the next generation is assured, although very much reduced: this is especially important with monocarpic species which complete their life-cycle once they have flowered. Self-incompatibility also means that, within certain groups of *Meconopsis* in the garden, hybridisation occurs frequently unless species are well isolated from one another. For example the plethora of big blue perennial poppies have all arisen in cultivation primarily by accident and between *M. baileyi* and *M. grandis*. The same applies to members of subgenus *Meconopsis*, *M. regia*, *M. paniculata* and *M. staintonii* in particular giving rise in gardens to a medley of different colours and forms often with the exclusion of the original parent species which have, in everyday jargon, 'been hybridised out'. The fact that some of these hybrids are fertile and produce plenty of seed means that, in the garden setting backcrossing between hybrids and parent species, or between the hybrids themselves, greatly complicates the picture and, unless the crosses have been made deliberately and the results documented, then their history becomes quickly obscured and their origins speculative.

Pollen transferred to a flower of a different individual of the same species will be compatible and lead to fertilisation and the development of fruit and seeds. Some pollen in such a simple pollination system is bound to end up on the stigmas of the same flower or another flower on the same plant and in the majority of instances this will probably prove to be incompatible for one reason or another. Thus the plant is able to discriminate between its own pollen and that of another individual. While this is important in constantly mixing the genes in wild populations where species are often isolated or able to keep their identity, the reverse is true in cultivation, especially where a number of species are grown in close proximity. In such instances hybridisation is all too common.

The frequently asked question is "why doesn't this happen in nature?". In most instances the species are well isolated from one another so that cross-pollination cannot occur. In evolutionary terms the species do not need, or have not evolved, an incompatibility barrier between species to prevent hybridisation, so that the species identity can remain pure. There are a very few examples of hybrids being observed in the wild. The best known is *M.* × *cookei* which can often be seen where the parent species (*M. punicea* and *M. quintuplinervia*) grow in close proximity in northern China. This hybrid is sterile and does not set viable seed and is thus unlikely to contaminate the gene pool in the wild. Another hybrid, *M.* × *kongboensis* (*M. sulphurea* × *M. simplicifolia*), the 'Ivory Poppy' has been found in south-eastern Tibet and would appear to be sterile or practically so (see p. 203). Interestingly, both these hybrids can be polycarpic, although most appear to be monocarpic plants; however, polycarpic plants are likely to persist in the wild for a number of years.

4. CULTIVATION OF MECONOPSIS SPECIES

GENERAL REQUIREMENTS

To understand the cultivation of *Meconopsis* species it is necessary to understand the conditions under which they grow in the wild. Not all species share the same type of habitat: while some grow in open woodland or in scrub, others grow out on high alpine pastures, while still others inhabit screes and moraines or rock crevices, sometimes at considerable altitudes and in very exposed positions. The majority of species are found in the montane monsoon belt, especially in the Himalaya and south-western China, although a few (*M. horridula* sensu lato and *M. torquata* in particular) colonise the far drier rain shadow areas of the Tibetan Plateau. In their more northerly Chinese haunts (Shaanxi, northern Sichuan, southern Gansu and south-eastern Qinghai) species like *M. integrifolia*, *M. psilonomma*, *M. punicea*, *M. quintuplinervia* and *M. sinomaculata* grow on the fringes of the monsoon belt, where the annual summer rainfall is considerably lower than it is further south or in the Himalaya. Despite this, most species have a regime of cold winters, often blanketed under snow, dry springs and autumns, with most of the precipitation occurring during the summer months, primarily July, August and September. In these regions one might reasonably expect the highest temperatures to be recorded during these summer months and indeed they do in much of the Indian subcontinent and China at low altitudes; however, in the higher mountains summer temperatures are depressed by the heavy and often persistent monsoon cloud and saturating mists.

I have been fortunate to travel in the Himalaya (Nepal in particular) and western and south-western China on many occasions. In the spring the air is often crisp and dry and shortly before the monsoon arrives it becomes hot and stifling. Once the rains break out the temperatures are reined back and the atmosphere becomes very humid. I can remember several occasions in July in both Nepal and south-western China, spending days on end in very humid, leech-ridden conditions with dense mists swirling around and obscuring the mountains from view. In fact I have had friends who have taken a three-week summer trek in Nepal and never once caught a glimpse of snowy Himalayan peaks. However, even in the middle of the monsoon there can be clear periods with glorious views of the mountains, especially at altitude and at sunrise when the cloud and mist has sunk into the valleys and lowlands way below; however, in the morning, once the sun has warmed the slopes, the clouds well up in the valleys, soon rising up the slopes and the shroud returns. These are the conditions in which *Meconopsis* has evolved.

Despite the heavy summer rains, species never endure boggy, waterlogged conditions around their roots during the summer months. During the winter months plants are relatively dry, often protected beneath the snow, generally as an overwintering bud or crown of dormant buds. The large monocarpic species of subgenus *Meconopsis* are an exception, producing an increasingly large evergreen rosette over several years until they reach flowering size, but the majority of these are to be found close to or below the tree-line.

How does this relate to their cultivation in gardens? In simple terms most species require moist, humid, cool summers and relatively dry winters. In most parts of Europe and North America such conditions are totally the opposite, where drier, hotter summers and wet winters prevail. It is for this reason that *Meconopsis* thrive best under the favourable conditions of more northerly latitudes. In Britain, northern England, parts of Wales, Scotland and Northern Ireland are the most favoured places, as they are for other predominantly monsoon genera such as *Nomocharis*, *Primula* and *Rhododendron*. In other parts of Europe, Denmark, Norway and Sweden seem best suited. Species and hybrids grow particularly well, for instance, at Tromsø Botanic Garden in northern Norway, lying at a latitude of almost 70°N, well north of anywhere in the British Isles and inside the Arctic Circle. In North America they thrive in some parts of Canada, being especially good in the vicinity of Vancouver, and in the USA. Bill Terry, who gardens in British Columbia, has the most comprehensive collection of *Meconopsis* in North America, while Vancouver Botanic Gardens also boasts some fine stands of *Meconopsis*, as well as an excellent collection of Himalayan plants in general. Much further east, towards the Atlantic seaboard, Les Jardins de Métis (Reford Gardens) close to the St Lawrence River is renowned for its huge drifts of blue poppies, in this instance *M. baileyi*.

These comments might seem to be a warning to those living in more southerly locations (southern Britain for instance) that it is not worth attempting growing these delightful plants but you can still find some fine stands of the easier species, *Meconopsis baileyi* and some of the *baileyi-grandis* derived hybrids such as 'Slieve Donard' and 'Lingholm', *M. punicea* and *M. integrifolia*. I can remember some fine stands of *M. baileyi*, *M. dhwojii*, *M. integrifolia*, *M. regia* and *M. wallichii* in the Woodland Garden at the Royal Botanic Gardens, Kew, in the 1970s. At the same time, it is undoubtedly true to say that, with the general increase in average summer temperatures and unpredictable rainfall (in Britain at least), it has become increasingly difficult to maintain *Meconopsis* in southern gardens with any reliability. One of their chief enemies is high summer temperatures to which they succumb all too readily, even with the careful application of water to prevent desiccation: they simply hate high temperatures. However, despite this, fine stands can still be seen in valley gardens in southern Britain where conditions are relatively less extreme and the atmosphere more humid in summer: the Saville Gardens in Windsor Great Park, Berkshire, is a good example. Some growers devise misting units over their *Meconopsis* beds to help keep up the humidity during the late spring and summer months: this not only keeps up the humidity levels but also helps alleviate high temperatures to some extent. Ironically, as I write these words in the late autumn of 2012, the country has been subjected to one of the coolest and wettest years on record and this has been of considerable benefit to monsoon-loving Sino-Himalayan plants in particular, especially *Meconospsis*, but also *Lilium*, *Primula*, *Rhododendron* and autumn-flowering *Gentiana*.

SOILS AND COMPOSTS

It has long been debated whether or not *Meconopsis* prefer acid or neutral soils rather than alkaline ones. The fact is that given the right conditions and plenty of humus (organic matter) in the compost they will thrive in all equally well, that is with the exception of very acid or very alkaline composts, a pH between say 5.5 and 8.5 being ideal.

Flower colour, especially blue can be easily affected by the soil pH. The best blues are produced on neutral to acid soils, while plants on alkaline soils, even mildly alkaline soils have a tendency to transform the blues to purplish, pinkish or muddy hues. Many growers find that divisions of the same plant placed in different parts of the same garden can produce purer or less pure blues and this almost certainly is reflected by the soil conditions.

The blue colour is explained well in Bill Terry's book *Blue Heaven* (2009), quoting from Jennifer Schultz Nelson (University of Illinois): "The Himlayan Blue Poppy is one of the few true blue flowers in the world ……. The blue colour is provided by the pigment delphinidin, named for being originally isolated from delphinium. For the delphinidin in the flower to appear blue, the environment inside the plant's cells must be acidic. The soil provides this acid, otherwise the flowers appear pinkish purple. This 'acid factor' is what makes blue such a rare find in the plant kingdom. Not only does a plant have to have the gene to make delphinidin in its flower cells, it must also be able to maintain a level of acidity within the cell to make the pigment appear blue. Few plants can accomplish this."

The large perennial (*Meconopsis baileyi* and *M. grandis* in particular) and leafy monocarpic species (*M. integrifolia* and *M. sulphurea*) and the evergreen monocarpic species (especially *M. paniculata*, *M. staintonii*, *M. wallichii* and *M. wilsonii*) all require a deep well-worked soil which has had ample humus dug in. Organic matter can include very well-rotted manure, leaf mould or peat, if from a reliable and sustainable source.

A suitable seed compost should contain as much expanded organic matter as possible, sieved peat, well-rotted garden compost or leaf mould (also sieved) are all ideal and will help retain water. A proprietary peat-based, soil-less compost is excellent for most species, depending rather on whether or not you have scruples about using peat in any form in the garden. The addition of coarse grit or Perlite will increase aeration in an otherwise moisture-retentive compost and will certainly benefit the young seedlings. If the garden compost or leaf mould can be sterilised in some way then potentially harmful pathogens can be eliminated before they harm the newly germinated seed or indeed young plants. Some growers advocate the addition of finely sieved, dried sphagnum moss to the compost, particularly the upper layers. This material has considerable moisture holding capacity and keeps up the humidity around germinating seeds and young seedlings: any drying out at this stage would otherwise prove fatal. The recommended compost would then consist of 50% soil-less compost, 30% sieved sphagnum moss and the remaining 20% coarse grit or Perlite or a mixture of both. A small amount of John Innes base fertiliser can be added but only at about a half the recommended rate.

PROPAGATION

Growing from seed

Seed of *Meconopsis* species is best sown the moment it is ripe unless it is stored very carefully and under optimum conditions which will be discussed shortly. This means that seed is available for sowing from midsummer onwards. Sown in the summer, germination for most will take place in 2–3 weeks and the resultant seedlings can be pricked out into pots and established before the winter sets in. Sown fresh, the seed will give a high germination rate. For many reasons it is not always practical to sow fresh seed, especially if it has been obtained from a seed firm or via a seed exchange or collected from the wild. In these instances seed is best sown in the early spring but the percentage germination will be lower than if the same seeds had been sown in the summer the moment ripe. This is because *Meconopsis* seed is not long-lived and quickly loses its viability unless stored under optimum controlled conditions. Of course the optimum conditions for storage, where temperature and humidity can be precisely regulated can only really be achieved by a specialised seed bank, like the one at the Royal Botanic Gardens Kew's Wakehurst Place. For the average grower the seed should be stored in dry and cool conditions. Even so, the viability of some species (for example *M. punicea* and *M. quintuplinervia*) is rapidly lost and, in this instance, autumn sowing is advisable.

Thin sowing is essential in order to avoid overcrowding and the associated problems of fungal infections i.e. damping off. Once sown, the pans, pots or trays, should be part immersed in a container of water in order to soak up plenty of moisture, but overwatering should be avoided as it will certainly hamper germination and cause rotting of the seeds. Watering overhead at this stage is not recommended as it can easily wash the seeds to one side of the container or even out of it altogether and the careful thin sowing will have been wasted. Sown pots can be placed in a glasshouse or cold frame in partial shade; however, all *Meconopsis* require some light to germinate, so pots should not be covered over. Although not essential, there is little doubt that some bottom heat, as can be provided by a heated frame, is beneficial and will speed up germination. Ideally the temperature should be about 18–20°C by day and half that at night. Germination normally takes place in two or three weeks and, generally speaking, if it does not do so in this time then it probably will not germinate at all.

Gardeners have always argued the pros and cons of sowing time. As stated, sown the minute it is ripe, seed will often germinate within a couple of weeks or so and the seedlings can be pricked out at the first true leaf stage. Left too long in their sowing pots the seedlings soon become starved and tangled and they are then likely to be damaged when separated out and potted on. Seedlings then have to be over-wintered in a glasshouse or frame and some losses can be expected, but the survivors will be ready for planting out in the late spring once the worst frosts have ceased. As noted, seed sown in the late winter or early spring will in all probability have lost some viability and will produce fewer seedlings per hundred seeds sown; however, once pricked out they will grow away rapidly and will certainly produce plants large enough to plant out in the same year: planting out by mid or late summer gives young plants time to establish before winter arrives. All pot-grown seedlings benefit from ample moisture and regular weak liquid feeds in order to keep them growing strongly.

Growing on

Seedlings are best pricked out into a nutrient-rich, humus-based compost at the first or second true leaf stage. Newly pricked out seedlings are delicate and should be shielded from strong sunlight and winds and must be kept moist at all times, as any desiccation will prove fatal. Under the right conditions the seedlings will put on rapid root growth and more leaves will expand. Regular weak liquid feeds will keep young plants vigorous and healthy, so that by mid-summer they should be large enough to plant out in the garden in most instances. Small seedlings or late sown ones are probably best overwintered in pots in a frame or glasshouse for planting out the following year, although this is less satisfactory and some fatalities are likely. For very rare or difficult species like *Meconopsis horridula* sensu stricto it is worthwhile considering just sowing half a dozen seeds in a pot, avoid pricking out and simply potting on the seedlings into increasingly larger pots until they reach maturity. Such plants will need regular feeding to keep them growing strongly.

Division of perennial species

In practical terms few *Meconopsis* species can be divided as the majority are taprooted monocarpic species. However, there are exceptions and these include some of the polycarpic species. Of these *M. quintuplinervia*, which forms spreading fibrous-rooted clumps, is by far the best example. Plants can either be dug up and divided carefully by teasing several-shooted portions apart or, alternatively, small portions can be removed from the periphery of an established plant. *M. grandis* can be divided like any herbaceous perennial: some forms tend to sucker and these suckers can be removed carefully without the need to lift the mother plant; a sharp knife will suffice to slice away the chosen portions. The big perennial blue poppies (primarily hybrids between *M. baileyi* and *M. grandis*) which are widely grown in gardens are readily lifted and divided in order to increase stock and maintain vigour. This is especially important for those that are sterile and do not set seed, then division is the only means of increase. The best time to divide is after flowering, although some growers find success by dividing just before growth commences in the early spring. Whenever it is done, the divisions must be replanted as soon as possible in order to protect them from desiccation, and kept well-watered thereafter until re-established. Divisions can be planted back in the garden or established in pots which makes them easier to hand on to other gardeners or collectors. It goes without saying that it takes some courage to lift and divide rare species or cultivars, or indeed if you only have a single plant to experiment with. Undoubtedly, if you want to increase species then seed raising is by far the best method, although a certain amount of patience is required.

Root cuttings

Root cuttings are a tried and tested method of propagation in some members of the allied genus *Papaver*, notably *P. orientalis* and its allies. R. D. Trotter of Brin House, Inverness, discovered by chance in the early twentieth century that the little perennial *Meconopsis delavayi* could be propagated by root cuttings. The story goes that Trotter accidentally broke off a length of taproot and from this two lengths were successfully rooted and grew on into new individuals. This method has been little tried in recent years, often due to the lack of suitable material; however, root cuttings may be the way ahead for the rare perennial species, especially those that are reluctant to set seed in cultivation. The method involves cutting sections of taproot (say 2–3 cm long) and placing them in pans of a gritty-sandy mixture, ensuring that they are placed top upright. Late spring would probably be the best time to try or immediately after flowering, using a modest amount of heat in a propagating frame.

ESTABLISHING IN THE GARDEN AND GROWING-ON

Whether you are growing *Meconopsis* in a woodland glade or in more open positions in the garden, then group planting is by far the most effective. Groups of plants of a single species look far better than isolated plants: they enhance one another and, perhaps more importantly, group sowing is far more likely to result in a good seed set. If the species is to remain pure in the garden, then the different species should be kept apart to reduce the chance of hybridisation happening: this occurs all too readily with the large evergreen monocarpic species such as *M. paniculata*, *M. regia* and *M. staintonii* and, while some of the hybrids will undoubtedly be very attractive, the species identity can be quickly lost. This is especially important for those wishing to maintain a collection of species, botanic gardens and national collections in particular.

Siting of the large evergreen monocarpic species requires some thought. The handsome evergreen leaf-rosettes can take up quite a lot of space (some can reach a metre across at maturity) and they will be in place for several years before flowering. An open woodland glade or gaps in shrubberies seem to suit them admirably, but the site should be sheltered because when the plants come into flower they can reach 2 m tall or more and are easily damaged by the wind, in fact they can be blown over. *Meconopsis* are quite brittle plants in general and the leaves are easily bruised and the petioles snapped by buffeting winds.

COMPATIBILITY

The question of compatibility (see p. 24) in *Meconopsis* is a fascinating one and a subject of particular interest to the gardener. It has to be stressed that little is known of compatibility in the genus, either within or between species and what is known has been primarily gleaned from the comments of gardeners and others who have observed them in cultivation. While unintentional hybrids have arisen on numerous occasions in gardens, it shows clearly that some species at least are interfertile and there are some well-known examples (for instance *M. baileyi* × *M. grandis* = *M.* ×*sheldonii*; *M. grandis* × *M. integrifolia* = *M.* ×*beamishii* Prain; *M. baileyi* × *M. integrifolia* = *M.* ×*sarsonsii* Sarsons ex *Gard. Chron.*; *M. simplicifolia* × *M. integrifolia* = *M.* ×*harleyana* G. Taylor). Compatibility within species is less well established. There have been numerous comments in print over the years since different species of *Meconopsis* have come into cultivation referring to a lack of seed set, or plants (both species and clones) apparently losing their fertility. However, this is mostly conjecture and there may be other factors that play an important role. For instance there may be a lack of suitable pollinators around when the plants are in flower, or the weather conditions may upset fertilisation and subsequent fruit production as it can with many different cultivated plants. In the wild the majority of species of *Meconopsis* receive ample water during the summer months, the prime season when they flower and when the fruit capsules are developed. In Western Europe and North America for the most part, the summer months tend to be dry and extended dry periods without irrigation can cause the young fruits to abort or fail to develop after pollination, or they may develop and contain little or no viable seed. It is often said that solitary cultivated plants of *Meconopsis* either fail to set any seed or sets very little, even under ideal growing conditions: this would seem to indicate that within some species at least there is a self-incompatibility system at work. Self-incompatibility enhances outcrossing and is advantageous to the species, allowing a free mixing of genes between individuals, and this may help to explain why hybrids between species happen quite readily in the garden setting. Why this is less likely to occur in the wild is explained in the section on p. 26.

This has an important bearing on maintaining a collection of *Meconopsis* species in cultivation, whether this is at a botanic garden or in a private garden. Under good growing conditions species are far more likely to set ample fruits, and subsequently plenty of seed, if several individuals of that species are grown in close proximity — group planting is essential. At the same time, if several species are grown close together then hybridisation is very likely to occur, leading to hybrid offspring and loss of species identity. While this is not so critical for perennial species which can be propagated by division, it is critical for those, the majority of species, that are monocarpic (flowering and fruiting just once) and can only be propagated from seed. If this seed is of hybrid origin, or the species fails to set seed for one reason or another, then the species is quickly lost to cultivation. The only way to successfully maintain a fine collection of *Meconopsis* species in the garden is to grow plants in groups and to isolate species from one another. In botanic gardens and large gardens there is space to separate and isolate species for seed production while at the same time bringing species together for the purpose of flora display and education.

This also means that if seed is purchased or exchanged its authenticity has to be questioned, unless that is, it comes from a reputable source where species maintenance is a priority. Seed of wild source origin, when available, is far more reliable, although the identity of the species may sometimes be in doubt until plants have been grown on and flowered. It is vitally important that seed of wild origin be adequately labelled with the collector and number (if it was given one) and country and place of origin. Botanic gardens have a scrupulous procedure for wild-collected seed of all sorts and it is easy to trace back any particular collection to its origins. At the same time responsible botanic gardens often place some seed from each collection in a gene bank for future reference and use, a good insurance policy.

One might wonder why, with their propensity to hybridise, the members of the *Meconopsis napaulensis* aggregate (subgenus *Meconopsis*, sections *Polychaetia* and *Superbae*) have not acquired a plethora of hybrid names. In fact only two are recorded: *M.* ×*coxiana* G. Taylor (*M. baileyi* × *M. violacea*) and *M.* ×*musgravei* (*M. baileyi* × *M superba*) both described in the early 1930s, the first appearing spontaneously in a batch of *M. violacea* seedlings at Glendoick in Perthshire, the second as a deliberate cross by C. T. Musgrave, Godalming, Surrey. Neither is now in cultivation despite the fact that George Taylor (1934) pointed out that "*M. musgravei* produced fertile seed, and there appears to be no reason why it should not become established in gardens". The fact that numerous hybrids have arisen in cultivation amongst the members of subgenus *Meconopsis* but few have been scientifically described, results from the fact that the majority have arisen in the past in collections of a number of species (notably *M. paniculata*, *M. regia*, *M. staintonii* (formerly recorded under the *M. napaulensis* umbrella) and *M. wallichii*) and it is uncertain which were the parents of a particular hybrid. Added to this is the fact that hybrids in this group tend to be fertile and this means that in succeeding generations, backcrosses and other species or hybrids could be involved, so that hybrids could have a very mixed genetic base. From a horticultural point of view such hybrids are best referred to the *M. napaulensis* Group.

In an ideal world it should be possible to keep species apart and distinct in cultivation and it is certainly an important goal in collections, especially those of botanic gardens. However, from a gardeners' point of view the aims are less restricting. What most gardeners want is a range of first rate *Meconopsis* that are reliable and colourful and many of those of hybrid origin fit the bill, generally making excellent garden plants. While the purist may scoff at these plants, many are very fine hybrids as witnessed by the numerous exuberant clones and cultivars amongst the big perennial blue poppies (those based primarily on a hybrid complex involving *M. baileyi* and *M. grandis*). At the same time the *M. napaulensis* aggregate has enhanced gardens with their impressive and varied evergreen leaf-rosettes and their pyramid of flowers in various shades of yellow, pink, red, apricot, mauve and blue.

COLLECTING AND STORING SEED

Unless stored under optimum conditions, *Meconopsis* seed is not long-viable, in fact most viability can be lost in a twelve month period; however, for most, seed collected in the summer will be fine for sowing the following spring but here a word of caution. As has been noted under cultivation, the viability of seed of certain species is lost very quickly indeed and is therefore best sown the moment it is ripe or at the latest by the same autumn. This applies particularly to *M. punicea*, *M. simplicifolia* and *M. quintuplinervia*, three closely related species. It is also generally advised that wild-collected seed be sown as soon as possible; that generally means in the autumn of the year it was havested. If you have sufficient seed then it is worth sowing some in the autumn and some the following spring, thereby increasing the chance of success, particularly as winter can be a testing time for young seedlings.

Collecting seed in the garden is a fairly simple process. The moment the fruit capsules are ripe, pores open at the top and the seeds can be easily shaken out into an envelope or other container. Alternatively, whole seed capsules can be snipped off carefully and placed in large envelopes for later sorting and cleaning This is best done when the weather is dry as damp seed can readily attract moulds. Select the best and plumpest capsules and collect only from good forms if this is possible. Beware that where several species or cultivars are grown in close proximity then there is a good chance of hybridisation having occurred and the resultant seedlings may be unpredictable. Each packet is best carefully labelled as to its contents to avoid confusion. This is especially important when passing on seed to other growers or placing them in seed exchanges. See record keeping below.

Seed can be cleaned of chaff by using graded sieves or by a simple system of winowing over a large piece of newspaper in a wind-free environment. The seed tends to be heavy and the chaff light so that the two are readily separated. Care should be taken to avoid contamination from one batch of seeds to the next. In addition, the chaff will probably contain a lot of dust and broken hairs and bristles from stems and capsules and this can be an irritant to skin and eyes and it should certainly not be inhaled, so care is advisable.

The optimum conditions for storing seed are in a seed bank where temperature, humidity and light can be stringently controlled. Such conditions are really only available to institutes such as botanic gardens and agricultural institutes. Under ideal conditions the viability of the seeds of countless species, including *Meconopsis*, can be maintained for many years. However, for most people this is simply not practical, but certain measures can be taken to maximise seed viability. The first essential is that the seed is dry and free from fungal infection. The second is that the seed is kept cold. These conditions can be best met by putting the batches of dry seed in airtight containers with a little fungicide, and placing them in the top of a domestic refrigerator. The optimum conditions for storage and germination have been well described by D. Thompson in 'Germination responses of *Meconopsis*' published in the *Journal of the Royal Horticultural Society* in 1968 (93: 336).

RECORD KEEPING

If you are making a collection of *Meconopsis* then it is important, as with any group of plants in the garden (private or public), to keep accurate records. Such records should include the name of the plant, species or cultivar, its origin (from wild collected seed, another grower or nursery, for instance), the date acquired and perhaps most important of all whether it is second or third generation, i.e. grown from seed from the original plant. This information can then be handed onto anyone else who receives your plants, so that their origins can be readily traced back. If a particular plant or batch of plants has a collector's number then this should always be used in conjunction with the name (for instance *M. grandis* NAPE178 = Nagaland and Arunachal Pradesh Expedition 2003). Second, third etc generation seedlings from such collections should always be indicated by an ex, i.e. *M. grandis* ex NAPE 178 to avoid confusion.

HYBRIDS

Meconopsis seldom hybridise in the wild, or at least very few hybrids have been reported. Just two have been authenticated: *M.* ×*cookei* G. Taylor (*M. punicea* × *M. quintuplinervia*) which can be found occasionally in the wild in north-western China wherever the parent species grow in close proximity to one another. The second is *M.* ×*kongboensis* Grey-Wilson (*M. simplicifolia* × *M. sulphurea*), a very localised hybrid found in one or two places in south-eastern Tibet (see p. 203).

Other hybrids have been reported from cultivation, even as chance seedlings or as deliberate crosses; however, several of these cannot be substantiated with any accuracy. They are as follows:

M. baileyi × *M. grandis* subsp. *grandis* = *M.* ×*sheldonii*
    G. Taylor

M. baileyi (*M. betonicifolia* hort.) × *M. integrifolia* = *M.* ×*sarsonsii* Sarsons ex *Gard. Chron.* Not authenticated.

M. baileyi (*M. betonicifolia* hort.) × *M. paniculata* = *M.* ×*auriculata* Stapf. Not authenticated.

M. baileyi (*M. betonicifolia* hort.) × *M. superba* = *M.* ×*musgravei* G. Taylor

M. baileyi (*M. betonicifolia* hort.) × *M. violacea* = *M.* ×*coxiana* G. Taylor

M. dhwojii × *M. napaulensis* (hort.) = *M.* ×*ramsdeniorum* G. Taylor

M. grandis (probably subsp. *grandis*) × *M. integrifolia* = *M.* ×*beamishii* Prain

M. grandis (probably subsp. *grandis*) × *M. simplicifolia* = *M.* ×*hybrida* Puddle ex *Gard. Chron.*

M. integrifolia × *M. simplicifolia* = *M.* ×*harleyana* G. Taylor

M. integrifolia × *M. quintuplinervia* = *M.* ×*finlayorum* G. Taylor

M. latifolia × *M. napaulensis* (hort.) = *M.* ×*decora* Prain. Not authenticated.

Other unnamed hybrids have been reported from cultivation including *Meconopsis sherriffii* × *M. simplicifolia*, *M. delavayi* × *M. quintuplinervia* ('Kaye's Compact'), *M. sherriffii* × *M. superba* and *M. napaulensis* (hort.) × *M. grandis*. In addition, numerous unnamed hybrids have been reported within the *M. napaulensis* aggregate, producing a medley of garden hybrids probably involving *M. paniculata*, *M. regia*, *M. staintonii* and *M. wallichii* to some extent, but this complex includes inadvertant backcrosses and crosses between hybrids.

It is interesting that nearly all the authenticated hybrids recorded are between closely related species from within the same subgenus. There are no recorded hybrids between members of subgenus *Cumminsia* or subgenus *Discogyne* and any species from the other subgenera. On the other hand hybrids within the species of subgenus *Meconopsis* and subgenus *Grandes* are quite common. It is clear that, within cultivation at least, it is often very difficult to be precise about the parentage of a particular hybrid unless the cross was made deliberately. In the wild crosses within the species of a particular subgenus are uncommon primarily because either they do not grow in the same vicinity or they inhabit different habitats.

The potential of carefully planned hybridisation for gardeners is profound, especially if the range of colours can be increased, but perhaps more importantly if vigorous, floriferous hardy, soundly perennial hybrids can be produced. While I find this potential fascinating it falls outside the remit of this present work and for this reason only those hybrids that are found in the wild are dealt with in detail in the pages that follow.

Overleaf:

Meconopsis wallichii var. *wallichii* in cultivation at Logan Botanic Garden, Scotland. PHOTO: CHRISTOPHER GREY-WILSON

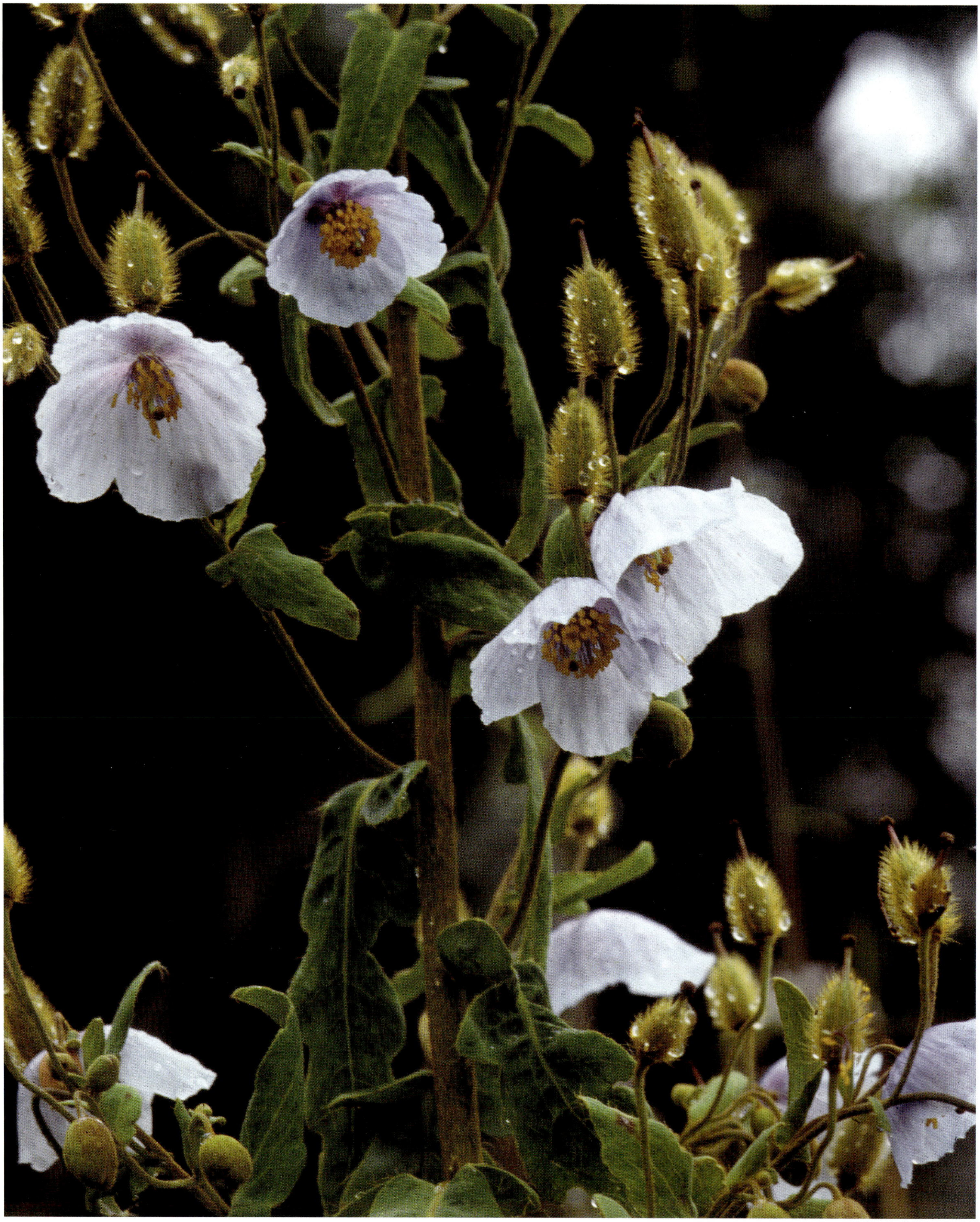

5. DISTRIBUTION AND HABITAT

The genus *Meconopsis* has a decidedly Sino-Himalayan distribution. Species are found the full length of the Himalaya from Chitral in NW Pakistan, through Kashmir and northern India to Nepal, Sikkim, Bhutan, NE India (Arunachal Pradesh) and northern Myanmar. In China they are found in central, southern and eastern Tibet (Xizang), Yunnan, W Hubei and Sichuan northwards through western China to Qinghai, Gansu and Shaanxi (Map 1).

The genus has evolved in the summer rainfall monsoon region with all the species being found at mid or high altitudes. Only two species, the widespread *Meconopsis horridula* and *M. torquata* (confined to the mountains close to Lhasa), have adapted to the drier regime of the Tibetan Plateau away from the influence of the Himalaya and the higher mountains of southeastern Tibet and western China: this is almost certainly a secondary adaptation, because the former, which shows a great range of variability, is also found in the wetter parts of the northern Himalaya.

While a few species, *Meconopsis horridula* sensu lato, *M. integrifolia* and *M. paniculata* in particular, are found over a considerable range, the majority of species have quite a limited distribution, some such as *M. balangensis*, *M. superba*, *M. venusta* and *M. pseudovenusta* being confined to just a few locations in a very limited area. Two main areas of diversification can be recognised: (1) central Nepal eastwards to Bhutan has 23 endemic species, while (2) western Sichuan and the neighbouring region of north-western Yunnan has 22 endemic species. Overall, China, including Tibet, has 44 species of *Meconopsis* of which 32 are endemic, while Nepal has 23 species of which 12 are endemic. Overall some 45 species are to be found in the Himalaya range of which a high percentage (37 species) are endemic.

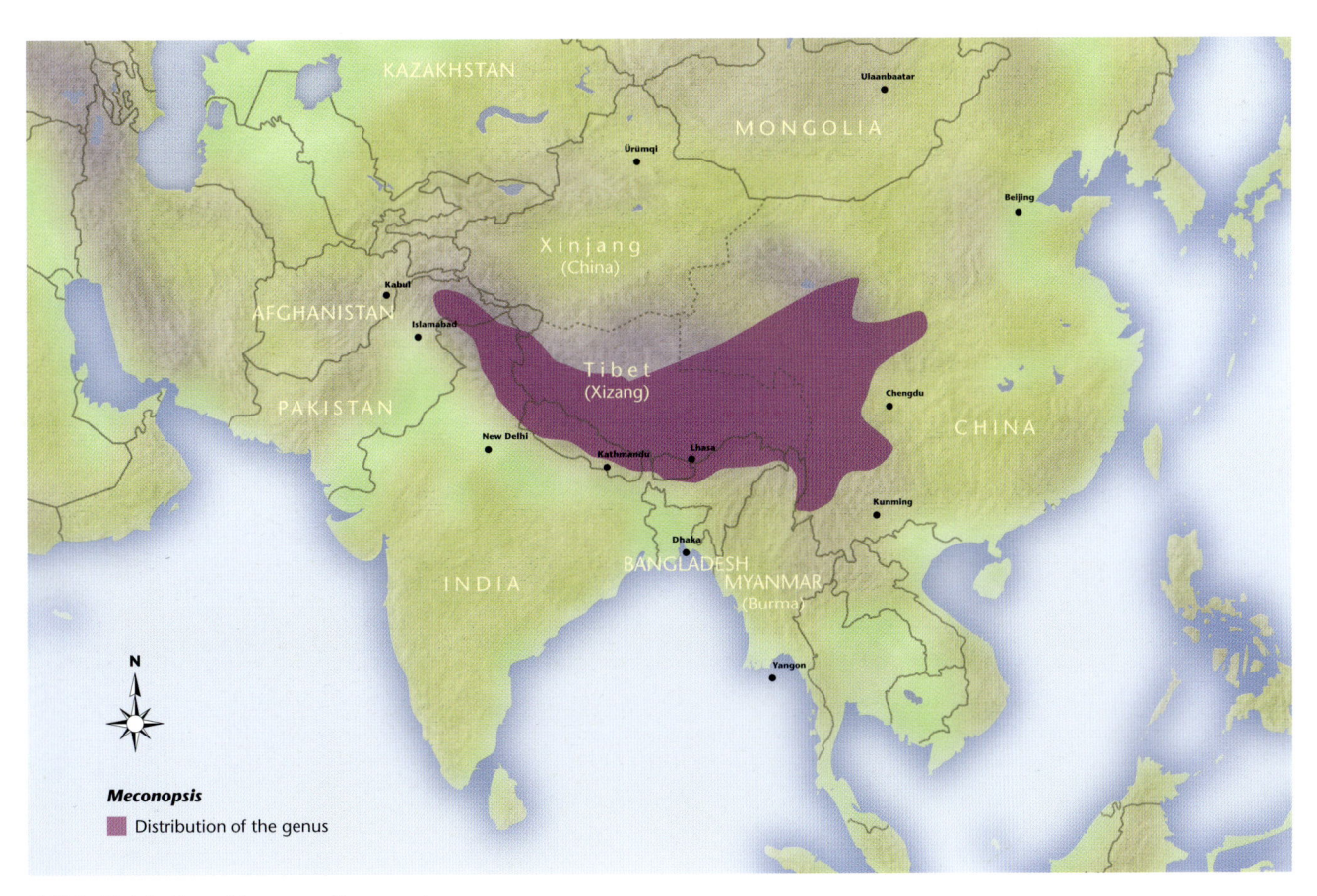

MAP 1. Distribution of the genus *Meconopsis*.

MAPS

The maps accompanying the species descriptions in this work show the overall distribution of each species as far as it is known. The areas outlined for each taxon are based on details from herbarium specimens, field data and the evidence of photographs, when the locality has been recorded. It is evident that in some regions data is very limited and this applies especially to eastern Tibet. For example, there are very few, if any, records for the huge area stretching north-east of Lhasa towards Qinghai and north-western Sichuan where one might reasonably expect *Meconopsis integrifolia* and *M. racemosa* to grow, yet these areas appear blank on the maps.

Initially dot maps based upon herbarium data were envisaged for each species, however, a great deal of additional information is available today, especially from field observations and photographs from the wild. In many places, for instance Bhutan and western and south-western China, the gathering of specimens is now prohibited unless special permission has been granted, so that quite often interesting locations and sightings are not accompanied by herbarium collections. I realise that dot maps, in which each recorded occurrence of a species (generally a herbarium specimen), are ideal, giving as they do a very good impression of the known distribution and occurrence of a particular taxon. However, many collections and observations are imprecise and identification is not always reliable. In addition, although I have been able to study some Chinese material (Kunming, Yunnan for instance) much Chinese material has not been seen except via the Chinese Virtual Herbarium, and this would have led to more accurate and complete distribution maps. As a result distribution on the maps is indicated by shading within a closed area in which the particular taxon is to be found. No attempt

Meconopsis integrifolia subsp. *integrifolia* with *M. punicea*; Xueshanliang, Huanglong, NW Sichuan, 4100 m. PHOTO: HARRY JANS

has been made to indicate frequency of occurence as would be apparent from dot maps and it can be assumed for widely distributed taxa that their frequency diminishes towards the limits of the distribution indicated.

Comments on the distribution within individual species are mentioned under the species concerned in the main text. While indicating the overall distribution of a species or subordinate category, shade maps are particularly useful in comparing species distributions and relating sections and series one to another.

The maps should be viewed with a degree of caution: the plants may not be found throughout the region defined except in the appropriate habitats and within specified altitudes. I am always interested to learn of new authenticated sightings of any of the species, particularly if they fall outside the present known distibution or at an altitude other than indicated. I am well aware that any monograph can only present facts as they are known at the time of publication. Ongoing field work and research in various disciplines is certain to add greatly to our knowledge in the coming years and that can only be to our advantage.

HABITATS

Meconopsis occupy quite a wide range of montane habitats from the middle and upper forest zones, both coniferous and evergreen, including mossy forests, shrubberies, river and stream margins and a range of habitats above the treeline. These latter include alpine meadows and pastures, alpine steppe, boulder fields, moraines, screes and rock crevices and ledges. While many grow on limestone mountains, particularly those found in Sichuan and Yunnan, others are adapted to more neutral or acid substrates over acid rocks or areas where the calcium has been leached out of the surface soil. While species like *M. venusta* and *M. pseudovenusta* are exclusively found growing on limestone, *M. baileyi* and *M. bella* are adapted to acid substrates. The widely distributed and very variable *M. horridula*, on the other hand, relishes a wide range of soil types from mildly acid to alkaline.

Meconopsis grow primarily in regions of high summer rainfall, with most of the rain falling during late June, July, August and early September. The spring and autumn months tend to be the driest, while during the winter most species, especially those growing at or above the treeline will be blanketed in a deep layer of snow. This means two things in general. First, plants can only start into growth once the snow has melted and this may not happen until late April or May or even early June after particularly heavy snow the previous winter. This

A medley of *Meconopsis*, *M. psilonomma*, *M. integrifolia* and *M. punicea*; Xueshanliang, Huanglong, NW Sichuan, 4100 m. PHOTO: HARRY JANS

THE GENUS MECONOPSIS
DISTRIBUTION AND HABITAT

Left:

Meconopsis integrifolia subsp. *integrifolia*; Xueshanliang, Huanglong, NW Sichuan, 4100 m. PHOTO: HARRY JANS

Below:

Meconopsis integrifolia subsp. *souliei* Wolong form, with *Primula longipetiolata*; Balangshang, Wolong, W Sichuan, 4110 m. PHOTO: HARRY JANS

in turn means that, like many montane plants, *Meconopsis* species tend to have a relatively short growing season, as winter comes back to the higher elevations in September and October. Second, the summer months should be the hottest months of the year but the summer monsoon can have a significantly depressing effect on the temperatures as the mountain slopes can be enveloped in heavy mist, sometimes for days or weeks on end. The atmosphere can be warm and very humid and it is to these conditions that *Meconopsis* are adapted and the period during which they put on most of their growth and flower. It is little surprising therefore that the majority of species have nodding or semi-nodding flowers which thus protect the important fertile parts from being soaked, the petals acting as a natural umbrella. There are exceptions: some forms of *M. horridula* and *M. integrifolia* generally bear ascending to upright flowers but they tend to inhabit drier regions. Most importantly, the humid atmosphere that envelops many of the species during the summer months greatly aids in the development of the fruit capsules and subsequent seed production. It is noticeable in the wild that, during unseasonably dry summers, fruit production can be poor. In a similar way dry summers can greatly affect seed production when species are grown in cultivation.

While many species of *Meconopsis* grow in isolation from one another, there are many instances of two or three, occasionally more species growing in a close association in the wild. I have seen this on a number of occasions, especially in the border lands between Gansu, Sichuan and Qinghai provinces in China where *M. integrifolia* subsp. *integrifolia* often grows together with *M. punicea* and *M. psilonomma*, and occasionally also with *M. sinomaculata* or *M. quintuplinervia*.

Right, from top:

The misty slopes of the Barun Khola valley, where, in the riverside meadows *Meconopsis grandis* is common.
PHOTO: CHRISTOPHER GREY-WILSON

The Big Snow Mountain, Daxueshan, in NW Yunnan on whose rocky slopes several species of *Meconopsis* can be found, including *M. lancifolia* subsp. *eximia* and *M. lijiangensis*.
PHOTO: CHRISTOPHER GREY-WILSON

Stone Mountain, Jigzhi, SE Qinghai, consists of yak-grazed moorland, open scrub and exposed rocky habitat where several species of *Meconopsis* are common including *M. integrifolia*, *M. punicea* and *M. quintuplinervia* and occasionally the hybrid between the latter two species, *M.* × *cookei*. PHOTO: CHRISTOPHER GREY-WILSON

ALTITUDES

Meconopsis species are found over a considerable altitudinal range in the wild, in fact from 2300 to 5300 m. The lowest record is for *M. quintuplinervia* which, however, has a wide range of altitudes in the wild, 2300–4600 m. Other species found at lower altitudes include *M. gracilipes*, *M. robusta* and *M. wallichii*, all found as low as 2440 m but extending upwards to around about the 4000 m level for the higher records. The widely distributed Himalayan *M. impedita* has a impressive range from 2825 to 4878 m, while *M. paniculata* is almost as profound, 2700 to 4570 m. Both *M. racemosa* and the closely related *M. zhongdianensis* are found as low as 3000 and 3230 m respectively, reaching 3910 and 4700 m. However, the majority of species are found in the wild between about 3200 and 4700 m. The most widespread of all *Meconopsis* species, *M. horridula*, has an impressive altitudinal range from 3700 to 5590 m and is almost certainly found as high as 5790 m, although this is not authenticated; however, it was noted at that high altitude by the 1921 Mount Everest Expedition. This wide altitudinal span is only rivalled by *M. integrifolia* which can be found in the wild from 3100 to 5300 m. While most of the higher altitude species are denizens of open, generally rocky alpine terrain or bleak moorlands, those at lower altitudes generally inhabit the upper forest zone. Some like *M. baileyi*, *M. napaulensis*, *M. paniculata* and *M. wallichii* often reaching out above the treeline onto yak pastures and other types of meadow, or along shrubby stream banks.

The high bleak grass moorland of the Kongbo Pa La (Mi La), well known to plant explorer Frank Kingdon Ward, with *Meconopsis pseudointegrifolia*. PHOTO: ANNE CHAMBERS

The upper Marsyandi Valley, C Nepal, above Braga. On the rocky slopes both *M. bella* and *M. horridula* are locally common.
PHOTO: CHRISTOPHER GREY-WILSON

6. TAXONOMIC TREATMENT

SPECIES DELINEATION

There is no doubt that *Meconopsis* presents the taxonomist with considerable difficulties of interpretation and that despite the fact, or perhaps in spite of the fact, that ample material exists today for a detailed analysis. While this present study is hugely biased towards a strictly morphological approach, I have taken into account the various phylogenetic and DNA-based studies published by various researchers in recent years, notably Jørk & Kadereit 1995; Kadereit *et al.* 1997 and Carolan *et al.* 2006. These studies, although by no means complete as far as *Meconopsis* is concerned, have thrown considerable light on the relationship of various entities within the Papaveraceae, not least the circumscription of the genus *Papaver* itself. While it would be pointless repeating in detail the results of these various studies it is worth stressing their prime findings.

• The sole European representative of *Meconopsis*, *M. cambrica*, is well separated from the Asian representatives of the genus in molecular analysis, suggesting that two distinct lineages exist within *Meconopsis*. It is suggested that incongruence exists with previous taxonomic classifications as regards the position of *M. cambrica*. If this taxon is included within *Papaver* (a suggestion that finds strong favour with many) then a new monotypic section would need to be found for it as there is a lack of apparent morphological similarities with extant species of *Papaver*. The alternative would be to create a monotypic genus for *M. cambrica*; however, although there is much to be said in favour of the latter, there is a real need for a realignment of all the genera in Papaveraceae. The taxonomic implications of the removal of *M. cambrica* from *Meconopsis* are discussed on p. 8.

• Within Asian representatives of *Meconopsis* a marked incongruence exists, demonstrating that there are two distinct clades: the smaller, represented by *M. chelidonifolia*, *M. oliveriana*, *M. smithiana* and *M. villosa* is a sister group (included in the genus *Cathcartia* in this work) to the remainder of *Meconopsis* (more than 70 species as currently defined). While the intricacies of these findings can be interpreted in a number of ways, with separation of *Papaver* and related groupings into a series of closely allied genera at one extreme or the wholesale lumping together of existing genera, including *Meconopsis*, into *Papaver* at the other.

• Within *Papaver*, section *Meconella* reveals a significant amount of morphological differencies with other sections within *Papaver* (excluding section *Argemonidium*). Section *Meconella* represents a cluster of species widely distributed from Canada and Greenland to Scandinavia, Siberia and Japan, as well as the higher mountains of central Europe, including species such as the *Papaver alpinum* aggregate, *P. anomalum*, *P. croceum*, *P. involucratum*, *P. miyabeanum* and *P. radicatum*. These species are characterised by small tufts of simple to pinnatisect leaves, scapose flowers, pale anthers and filaments, yellow, orange or white petals and bristly valvate fruit capsules in which the stigma radiates over the cap, often with deep incisions between the rays. Phylogenetic analysis (Kadereit, Carolan *et al.*) show that section *Meconella* is a sister group to those Asian *Meconopsis* used in analysis. In simple terms (and I urge readers to read and analyse the papers referred to) several courses of action can be determined. First, if section *Meconella* is retained in a redefined *Papaver*, then *Meconopsis* would also have to be included within *Papaver*. Second, in order to maintain *Meconopsis* then section *Meconella* would have to be removed from *Papaver*: as considerable morphological differences exists between *Papaver* s.s. and section *Meconella*, then this interpretation is likely to gain favour. Third, it could be concluded that section *Meconella*, and possibly *Meconopsis* (Asian representatives), are derived from an early lineage that separated prior to that which gave rise to the other sections of *Papaver* (excluding section *Argemonoidium*). Fourth, it could also be concluded that section *Meconella* could be raised to generic level or indeed included in *Meconopsis*.

However, there are clear morphological differences between *Papaver* section *Meconella* and *Meconopsis*, not least the presence of a well-developed style in the majority of species of the latter: in a few species the style is very short or obsolete but this can be considered a derived character. In the majority of species of *Meconopsis* there is a well-defined stem and only one or two species (for instance *M. bella*, *M. delavayi* and *M. venusta*) can be truly called acaulous, while the tendency in these scapose species for agglutinisation of some of the scapes into a pseudoraceme is unique to the Papaveraceae. In section *Meconella* the filaments and anthers are pale while in the majority of species of *Meconopsis* they are dark, the filaments usually the same colour as the petals, often darker. While white, yellow and orange petals predominate in the former, yellow, red, blue, lavender, mauve, purple and pink, rarely white, characterise *Meconopsis*. There is also a strong tendency in *Meconopsis* to a multiplication of petals beyond the basic four.

It is perhaps in their geography and habitat that the two show greatest divergence: section *Meconella* consists of northern temperate hemisphere species where the summer months, when they are in flower, are the driest and for this reason they bear ascending to erect flowers; in contrast *Meconopsis* is a genus that has evolved predominantly in the summer monsoon belt of the Sino-Himalayan region and the flowers are mostly lateral-facing, half- or fully nodding, presumably an evolutionary adaptation to ensure that the fertile centres of the flowers are kept dry. It is perhaps no coincidence that in the drier parts of the north-western Himalaya and northern China and Siberia, where section *Meconella* species are to be found, *Meconopsis* are absent.

It can be argued that an expanded *Papaver*, that includes as it does at present section *Meconella*, with the addition of *Meconopsis* and various other small genera (e.g. *Stylomecon*) would, in my view at least, serve little purpose other than to make a large and complicated genus even more cumbersome. On the other hand, to combine *Papaver* section *Meconella* and *Meconopsis* into a single genus does not seem to make sense, either on morphological or geographical criteria and would make the genus very hard to define, besides which the handful of section *Meconella* species look very different from *Meconopsis* as currently defined, and no one who knew them would find them confusing. With the removal of *Meconopsis cambrica* from the genus and the four species centred on *M. villosa*, separated off as *Cathcartia*, then the core of Asiatic *Meconopsis* forms a homogeneous and readily definable genus and this is how it is treated in this monograph.

MECONOPSIS Vig. emend. Grey-Wilson

Monocarpic or polycarpic perennials with evergreen or herbaceous basal rosettes of leaves, caulous or acaulous. *Stock* often taprooted, the taproot dauciform to napiform, simple or sometimes branched, or occasionally the rootstock more diverse and supporting several growing points, occasionally entirely fibrous-rooted. *Plants* adorned in most parts with multicellular-multiseriate, simple or barbellate hairs, the hairs thin and weak, sometimes brittle, to stout and bristly or spine-like, rarely glabrous. *Leaves* all basal or basal and cauline, petiolate, the lamina simple to pinnatisect or pinnatifid, bipinnatisect or bipinnatifid, pinnately veined, sometimes with 3 or 5 (–7) parallel veins reaching part way or the whole length of the lamina, the margin entire to finely toothed or variously lobed; cauline leaves, when present, spirally arranged, similar to the basal leaves but generally short-petiolate or sessile, the upper generally merging with the bracts, the uppermost leaves/bracts sometimes aggregated into a false whorl. *Flowers* solitary, scapose or borne in simple racemes or panicles, more rarely in a pseudoumbellate raceme, in paniculate types the upper part of the inflorescence is generally racemose, the lower part with cymules of 2–11(–13) flowers; flowers campanulate to cup- or saucer-shaped, sometimes more or less flat, generally nodding or half-nodding, but sometimes ascending to upright. *Buds* generally nodding or half-nodding, rarely upright. *Sepals* 2, rarely 3, imbricate with a lobe on the upper left margin normally, and a hyaline right margin, caducous. *Petals* biseriate, 4–11, rarely more, yellow, red, pink, mauve, lavender, blue, purple, occasionally white, the margin entire to variously erose. *Stamens* numerous usually, rarely as few as 12–24, the filaments linear, sometimes dilated for part or the whole of their length, glabrous, often the same colour as, or darker than, the petals, sometimes whitish; anthers basifixed, yellow, orange, greyish or purplish; pollen tricolpate, polyporate or inaperturate, generally yellow or orange, occasionally cream or buff. *Gynoecium* polycarpellate, unilocular, ovoid to obovoid to oblong or narrow-subcylindric; style usually present, sometimes more or less obsolete, the style linear or expanded towards the base, sometimes markedly expanded to form a disk covering the top of the ovary/fruit capsule; stigmas free or confluent to form a capitate to clavate structure, often decurrent onto the upper part or the style, sometimes decurrent onto the upper part of the ovary, 3–14(–17), lobed, sometimes obscurely so. *Fruit capsules* erect, ellipsoid, ovoid or obovoid to narrow fusiform or subcylindric, variously pubescent with weak or brittle hairs or more robustly pungent, often densely so, occasionally glabrous, valvate, slightly to markedly longitudinally grooved on the outside, splitting at maturity with 3–14 valves for a short distance (up to one third the length of the capsule) from close to the apex. *Seeds* reniform to more or less ovoid or subfalcate, smooth but generally with longitudinal rows of shallow pits.

Conserved type species: *M. regia* G. Taylor

77 species and 2 naturally occurring hybrids in the Sino-Himalaya

INFRAGENERIC CLASSIFICATION

Subgenus MECONOPSIS

[Syn. *Meconopsis* subgenus *Eumeconopsis* (Prain) Fedde in Engl., *Pflanzenr.* 4, 104: 251 (1909) emend. excl. sect. *Cambricae*, sect. *Eucathcartia* and subsect. *Cumminsia*; *M.* subgenus *Eumeconopsis* sect. *Polychaetia* subsection *Eupolychaetia* G. Taylor, *The Genus Meconopsis*: 30 (1934)].

Monocarpic plants often forming a substantial persistent (evergreen) leaf-rosette, sparsely to densely covered in most parts by barbellate hairs or bristles of one or two types, occasionally with simple bristles on the fruit capsules. *Inflorescence* erect, racemose or part or wholly paniculate, bracteate and bracteolate. *Petals* 5, rarely 6. *Fruit* capsule always bearing a persistent style.

section MECONOPSIS

[Syn. *Meconopsis* subgenus *Eumeconopsis* section *Polychaetia* subsection *Eupolychaetia* series *Superbae* Kingdon-Ward, *Gard. Chron.*, Ser. 3, 79: 252 (as '*Surperbae*' (April 1926), nom. nud., op. cit.: 460 (June 1926) emend.; *M.* ser. *Robustae* Prain, *J. Asiat. Soc. Bengal, Pt. 2, Nat. Hist.* 64 (2): 315 (1896) pro parte; *M.* sect. *Robustae* (Prain) Fedde in Engl., *Pflanzenr.* 4, 104: 267 (1909), pro parte].

Leaf-lamina elliptic to oblanceolate or oblong-ovate, unlobed, the margin finely toothed, densely sericeous on both surfaces with long barbellate hairs underlain by shorter hairs of a similar type, especially dense and felt-like when young.

(3 species [1–3]: *M. regia*, *M. superba*, *M. taylorii*). Type species: *M. regia* G. Taylor.

section POLYCHAETIA Prain, *Ann. Bot.* 20: 352 (1906) emend., excl. ser. 7 & 9.

[Syn. *Meconopsis* subgenus *Eumeconopsis* section *Polychaetia* subsection *Eupolychaetia* series *Robustae* Prain, *J. Asiat. Soc. Bengal, Pt. 2, Nat. Hist.* 64 (2); 315 (1896) emend. excl. sp. 6; *M.* section *Robustae* (Prain) Fedde in Engl., *Pflanzenr.* 4, 104: 267 (1909) pro parte].

Leaf-lamina deeply lobed often bipinnatifid to bipinnatisect, the margin toothed or not, sparsely to densely covered in spreading barbellate hairs or bristles, underlain or not by shorter hairs of a similar type, sometimes dense when young. Type species: *M. paniculata* (D. Don) Prain.

series Robustae Prain, *J. Asiat. Soc. Bengal, Pt. 2, Nat. Hist.* 64: (2): 315 (1896) emend. excl. *M. paniculata*, *M. napaulensis* and *M. violacea*.

[*Meconopsis* subgenus *Eumeconopsis* sect. *Polychaetia* subsect. *Eupolychaetia* series *Robustae* Prain, *J. Asiat. Soc. Bengal, Pt. 2, Nat. Hist.* 64 (2): 315 (1896), pro parte]

Plants to 1.5 m tall in flower, often less than 1 m. *Pubescence* of stems and leaves consisting of simple barbellate hairs all of the same type. *Petals* usually yellow, occasionally red.

(4 species [4–7]: *M. chankeliensis*, *M. dhwojii*, *M. gracilipes*, *M. robusta*). Type species: *M. robusta* Hook. f. & Thomson.

series Polychaetia

[*Meconopsis* subgenus *Eumeconopsis* sect. *Polychaetia* subsect. *Eupolychaetia* series *Robustae* Prain, *J. Asiat. Soc. Bengal, Pt. 2, Nat. Hist.* 64 (2): 315 (1896), emend., pro parte]

Plants generally large and robust in flower often 1.5 m tall and reaching as much as 2.5 (–3) m. *Pubescence* of stems and leaves consisting of long barbellate hairs of bristles underlain by a sparse to dense felt of similar though much shorter hairs. *Petals* pink, blue, purple, wine-purple, red, yellow (in *M. autumnalis* and *M. paniculata*), occasionally white.

(8 species [8–15]: *M. autumnalis*, *M. ganeshensis*, *M. napaulensis*, *M. paniculata*, *M. staintonii*, *M. violacea*, *M. wallichii*, *M. wilsonii*).

Subgenus DISCOGYNE G. Taylor, *The Genus Meconopsis*: 107 (1934).

[Syn. *Meconopsis* section *Polychaetia* Prain, *Ann. Bot.* 20: 352 (1906), pro parte; *M.* series *Torquatae* Prain op. cit.: 355 (1906); *M.* subgenus *Polychaetia* (Prain) Fedde in Engl., *Pflanzenr.* 4, 104: 262 (1909), pro parte; *M.* section *Torquatae* (Prain) Fedde op. cit.: 265 (1909); *M.* section *pinnatifoliae* C. Y. Wu & N. Zhang, *Bull. Bot. Lab. N.E. Forest. Inst., Harbin* 1980 (8): 98 (1980)].

Monocarpic plants forming a persistent (evergreen) leaf rosette, sparsely to densely covered in barbellate bristles. *Leaves* elliptic-oblanceolate to oblanceolate or occasionally wedge-shaped, entire or incisely lobed or 3–5 (–7) toothed at the apex. *Inflorescence* simple, part-bracteate, racemose, occasionally branching from the base. *Petals* 4, rarely 5. *Style* expanded at the base into a broad disk immediately surmounting the ovary (and later the fruit capsule) and overlapping at the perimeter. *Fruit capsules* bristly, surmounted by a broad, somewhat overlapping, shallowly lobed or somewhat angled disk and a persistent (occasionally subobsolete) style.

(7 species [16–22]: *M. bhutanica, M. discigera, M. manasluensis, M. pinnatifolia, M. simikotensis, M. tibetica, M. torquata*). Type species: *M. discigera* Prain.

Subgenus GRANDES (Prain) Grey-Wilson, **subgenus nov.**

[Syn. *Meconopsis* subgenus *Eumeconopsis* section *Polychaetia* subsection *Cumminsia* pro parte]

Monocarpic or polycarpic plants with an indumentum of simple barbellate hairs, non-evergreen, persisting overwinter as a bristly bud or buds, surrounded by the remains of the old petiole bases and bristles, at or just below the soil surface. *Flowers* solitary and scapose or subumbellate and then bracteate. *Petals* 4–9, occasionally more. *Fruit capsule* always bearing a persistent style. Type species *M. grandis* Prain.

> **section GRANDES** (Prain) Fedde in Engl., *Pflanzenr.* 4, 104: 262 (1909) pro parte, excl. sp. 17–19.
>
> [Syn. *Meconopsis* subgenus *Eumeconopsis* section *Polychaetia* subsection *Cumminsia* series *Grandes* Prain, *J. Asiat. Soc. Bengal, Pt. 2, Nat. Hist.* 64 (2): 320 (1896)].
>
> *Plants* bearing a subumbellate to subracemose inflorescence with up to 7 flowers (occasionally more), sometimes with a solitary terminal flower but then always accompanied by several bracts, the uppermost leaves (bracts) aggregated into a false whorl and from whose axils most of the flowers arise. *Flowers* semi-nodding, sideways facing or ascending.
>
>> **series Grandes**
>> [syn. *M.* sect. *Racemosae* C. Y. Chuang & H. Chuang, *Acta Bot. Yunnan.* 2 (4): 374 (1980) series *Grandes* Prain pro parte]
>> Polycarpic plants with blue, pink, purple or wine-purple flowers; petals usually 4, occasionally 5–6 in the terminal flower. Fruit capsule 4–7-valved.
>> (4 species [23–26]: *M. baileyi, M. betonicifolia, M. grandis, M. sherriffii*).
>>
>> **series Integrifoliae** Grey-Wilson, **series nov.**
>> [syn. *M.* sect. *Racemosae* C.Y. Chuang & H. Chuang, *Acta Bot. Yunnan.* 2 (4): 374 (1980) series *Grandes* Prain pro parte]
>> *Monocarpic plants* with yellow flowers. Petals 6–11. Fruit capsule (4–)7–11-valved.
>> (4 species [27–30]: *M. integrifolia, M. lijiangensis, M. pseudointegrifolia, M. sulphurea*). Type species: *M. integrifolia* Franch.
>
> **section SIMPLICIFOLIAE** (G. Taylor) C. Y. Wu & H. Chuang, *Acta Bot. Yunnan.* 2 (4): 375 (1980) emend., excl. series *Henricanae* and series *Delavayanae*]
>
> [Syn. *Meconopsis* subgenus *Eumeconopsis* section *Polychaetia* subsection *Cumminsia* series *Simplicifoliae* G. Taylor, *The Genus Meconopsis*: 49 (1934); *M.* series *Grandes* Prain, *J. Asiat. Soc. Bengal, Pt. 2, Nat. Hist.* 64 (2): 320 (1896), pro parte; *M.* section *Grandes* (Prain) Fedde in Engl., *Pflanzenr.* 4, 104: 262 (1909), pro parte].
>
> *Plants* with all leaves basal and bearing solitary scapose, often nodding, occasionally sideways-facing, flowers. Type species: *M. simplicifolia* G. Don.

series Simpliciﬂoliae
[incl. section *Nyingchienses* L. H. Zhou, *Bull. Bot. Lab. N.E. Forest Inst., Harbin* 1980 (8): 98 (1980)]

Leaves pinnately veined or weakly 3-veined towards the base, the margin with a tendency to irregular toothing or somewhat lobed. *Flowers* saucer-shaped to cupped or flattish. *Petals* normally 5 or more.

(2 species [31–32, plus 2 hybrids 33–34 (involving *M. integrifolia* and *M. sulphurea* respectively)]: *M. nyingchiensis*, *M. simplicifolia*)

series Puniceae Grey-Wilson series nov.
[syn. *Meconopsis* subgenus *Eumeconopsis* section *Polychaetia* subsection *Cumminsia* series *Simplicifoliae* G. Taylor, pro parte]

Leaves with 3 or 5 well marked longitudinal veins, the margin entire. *Flowers* campanulate to campanulate-infundibuliform. *Petals* normally 4.

(2 species [35–36, plus 1 hybrid 37]: *M. punicea*, *M. quintuplinervia*). Type species: *M. punicea* Maxim.

Subgenus CUMMINSIA (Prain) Grey-Wilson, stat. nov.
[Syn. *Cathcartia* sect. *Cumminsia* Prain, *Ann. Bot.* 20: 368 (1906); *Meconopsis* subgenus *Eumeconopsis* section *Polychaetia* subsection *Cumminsia* (Prain) G. Taylor, *The Genus Meconopsis*: 48 (1934), pro parte, emend. excl. series *Grandes* and series *Simplicifoliae*; *M.* series *Aculeatae* Prain, *J. Asiat. Soc. Bengal, Pt. 2, Nat. Hist.* 64 (2): 313 (1896); *M.* section *Aculeatae* (Prain) Fedde in Engl., *Pflanzenf.* 4, 104: 255 (1909)]

Monocarpic or occasionally polycarpic plants with or without a well-developed stem. *Leaves* in basal rosettes, dying away to a resting bud in the autumn at or close to the soil surface, present at flowering time. *Indumentum* variable from stout to weak non-barbellate bristles, to weak, soft or brittle hairs. *Flowers* solitary and scapose or borne in a poorly to well-developed racemose inflorescence in which the lower flowers are often bracteate. *Petals* 4–10, often blue, violet, pink or purple but also on occasions yellow, red or white. *Fruit capsule* bearing a persistent style, often bristly or hairy, sometimes glabrous. Type species: *M. lyrata* (H. A. Cummins & Prain) Fedde ex Prain.

section ACULEATAE
[Syn. *Meconopsis* subgenus *Eumeconopsis* section *Polychaetia* subsection *Cumminsia* series *Aculeatae* pro parte]

Monocarpic plants with relatively dense rosettes of leaves, armed on all parts except the petals with thin to moderately stout bristles, sometimes sparsely so. *Leaf-lamina* pinnatisect to pinnatifid to bipinnatifid, sometimes shallowly so. *Inflorescence* racemose, occasionally scapose. *Petals* often 4, sometimes 5–8.

(4 species [38–41]: *M. aculeata*, *M. latifolia*, *M. neglecta*, *M. speciosa*). Type species: *M. aculeata*.

section RACEMOSAE C. Y. Chuang & H. Chuang emend. excl. series *Grandes*.
[Syn. *Meconopsis* subgenus *Eumeconopsis* section *Polychaetia* subsection *Cumminsia* series *Aculeatae* pro parte; *M.* sect. *Simplicifoliae* series *Delavayanae* sensu C. Y. Wu & H. Chuang pro parte, *Acta Bot. Yunnanica* 2 (4): 376 (1980)]

Monocarpic plants armed on all parts except the petals with stiff, often stout, pungent, persistent bristles. *Leaf-lamina* entire at the margin to sinuous or with a few irregular and generally uneven lobes. *Inflorescence* a simple raceme, or flowers solitary and scapose, the scapes (pedicels) sometimes partly agglutinosed in the lower part. Type species: *M. racemosa* Maxim.

series Racemosae
[incl. *Meconopsis* section *Racemosae* C. Y. Wu & H. Chuang, *Acta Bot. Yunnan.* 2 (4): 374 (1980) emend., quoad *M. racemosa*; *M.* sect. *Simplicifoliae* series *Delavayanae* sensu C. Y. Wu & H. Chuang pro parte, *Acta Bot. Yunnan.* 2 (4): 376 (1980)]

Stamens uniform with linear, non-dilated filaments. *Petals* 4–9, blue, lavender, purple, pink or wine-purple, occasionally yellow (in *M. georgei* and *M. prainiana*) or white.

(9 species [42–50]: *M. bijiangensis*, *M. georgei*, *M. horridula*, *M. lhasaensis*, *M. prainiana*, *M. prattii*, *M. racemosa*, *M. rudis*, *M. zhongdianensis*).

series Heterandrae Grey-Wilson, series nov.
Stamens dimorphic, the inner with dilated (air-filled), incurving filaments, the outer linear. *Petals* 4–7, blue, mauve, pinkish or pale purple.

(2 species [51–52]: *M. balangensis*, *M. heterandra*). Type species: *M. heterandra* Tosh. Yoshida, H. Sun & Boufford.

section IMPEDITAE Grey-Wilson, section nov.
[Syn. *Meconopsis* subgenus *Eumeconopsis* section *Polychaetia* subsection *Cumminsia* series *Aculeatae* pro parte].

Monocarpic or occasionally polycarpic plants with lax basal rosettes of entire to sinuate or pinnatifid (occasionally bipinnatifid) leaves, with a sparse to moderately dense indumentum of simple weak hairs, or glabrous or subglabrous. *Flowers* one to many per leaf-rosette, scapose, occasionally several scapes (pedicels) partly agglutinosed; petals often 4 but up to 10, blue, dark violet, violet-blue, purple, wine-purple or blackish purple. *Stamens* with linear or partly dilated filaments. Type species: *M. impedita* Prain.

series Impeditae
[syn. *M.* sect. *Simplicifoliae* series *Delavayanae* sensu C. Y. Wu & H. Chuang pro parte, *Acta Bot. Yunnan.* 2 (4): 376 (1980)]

Monocarpic plants with a thin weak indumentum or more or less glabrous. *Leaves* all basal in a small, often lax rosette, withering away to fibrous leaf bases in the autumn, the plants overwintering as a small resting bud (or buds) at or just below the soil surface. *Flowers* solitary, scapose, sometimes numerous per plant; petals usually 4–5, but up to 10 in *M. pseudovenusta*.

(7 species [53–59]: *M. concinna*, *M. impedita*, *M. musicola*, *M. pseudovenusta*, *M. pulchella*, *M. venusta*, *M. xiangchengensis*).

series Henricanae C. Y. Wu & H. Chuang, *Acta Bot. Yunnan.* 2 (4): 376 (1980) emend. excl. *M. concinna*, *M. neglecta* and *M. wumungensis*.
[syn. *Meconopsis* subgenus *Eumeconopsis* section *Polychaetia* subsection *Cumminsia* series *Aculeatae* pro parte; *M.* sect. *Simplicifoliae* (Taylor) C. Y. Wu & H. Chuang. pro parte]

Monocarpic plants, with a sparse to moderately dense indumentum of rather brittle, weak hairs. *Leaves* all basal in a small, lax rosette. *Flowers* solitary, scapose, one or several from a single leaf-rosette; petals 4–9. *Stamens* with filaments dilated in the middle or towards the base, otherwise linear.

(3 species [60–62]: *M. henrici*, *M. psilonomma*, *M. sinomaculata*). Type species: *M. henrici* Bureau & Franch.

series Delavayanae G. Taylor, *Genus Meconopsis*: 76 (1934).
[Syn. *Meconopsis* series *Primulinae* Prain, *J. Asiat. Soc. Bengal, Pt. 2, Nat. Hist.* 64, 2: 319 (1896) pro parte quoad sp. *M. delavayi*; *M.* series *Primulinae* Prain, *J. Asiat. Soc. Bengal, Pt. 2, Nat. Hist.* 64, 2: 319 (1896) pro parte, quoad sp. *M. delavayi*; *M.* sect. *Primulinae* (Prain) Fedde in Engl., *Pflanzenr.* 4, 104: 259 (1909) pro parte, quoad sp. 12; *M.* sect. *Simplicifoliae* series *Delavayanae* sensu C. Y. Wu & H. Chuang pro parte, *Acta Bot. Yunnan.* 2 (4): 376 (1980)]

Polycarpic plants, glabrous or sparsely hairy, with a branching rootstock, each branch bearing a tuft of leaves and a solitary scapose, semi-pendent flower. *Leaves* entire, glabrous to sparsely pubescent. *Pedicels* erect to ascending in fruit. *Petals* 4 (–6) deep purple to violet-purple. *Fruit capsule* subcylindrical to narrow-oblong, 3–5-valved.

(1 species [63]: *M. delavayi*). Type species: *M. delavayi* Franch. ex Prain.

section FORRESTIANAE C. Y. Wu & H. Chuang, nomen emend.
[Syn. *Meconopsis* subgenus *Eumeconopsis* section *Polychaetia* subsection *Cumminsia* series *Aculeatae* pro parte; *M.* section *Forrestii* C. Y. Wu & H. Chuang, *Acta Bot. Yunnan.* 2 (4): 375 (1980)]

Monocarpic plants with small lax, generally few-leaved rosettes, the leaves and stems sparsely to densely covered with patent, thin bristles. *Leaves* entire, often rather undulate at the margin, withering in the autumn, unflowered plants residing overwinter as a bud close to the soil surface. *Inflorescence* an ebracteate raceme, sometimes just 2-flowered, rarely reduced to a solitary flower. *Petals* 4–8, pale blue to purplish blue or pinkish purple, to deep purple. *Fruit capsule* narrow-obovoidal to oblong-ellipsoidal or subcylindrical, 2–5 (–6)-valved, rigidly erect, glabrous to moderately bristly.

(3 species [64–66]: *M. forrestii*, *M. lancifolia*, *M. yaoshanensis*). Type species: *M. forrestii* Prain.

section CUMMINSIA (Prain) Grey-Wilson **comb. nov.**
[*Cathcartia* sect. *Cumminsia* Prain, *Ann. Bot.* 20: 368 (1906); *Meconopsis* series *Cumminsia* (Prain) Prain, *Bull. Misc. Inform., Kew* 1915: 142 (1915); *M.* subgenus *Eumeconopsis* sect. *Polychaetia* subsect. *Cumminsia* (Prain) G. Taylor, *The Genus Meconopsis*: 48 1934)]

Monocarpic tap-rooted plants with a short to well-developed stem, occasionally this sub-obsolete, without basal leaf-rosettes, the basal leaves few and generally withered by flowering time, the plants with a weak, often rather sparse, indumentum of sub-barbellate bristles or more or less glabrous. *Flowers* usually borne in the axils of the uppermost leaves, occasionally solitary terminating a spasely leafy stem; bracteoles often present. *Fruit capsule* oblong-ellipsoidal to subcylindrical or narrow-obovoidal, glabrous to sparsely bristly (more densely so in *M. argemonantha*). Type species: *M. lyrata* (Cummins & Prain) Fedde ex Prain.

series Primulinae Prain, *J. Asiat. Soc. Bengal, Pt. 2, Nat. Hist.* 64 (2): 319 (1896) emend. excl. *M. lyrata*.
[Syn. *Meconopsis* subgenus *Eumeconopsis* section *Polychaetia* subsection *Cumminsia* series *Primulinae* Prain, *J. Asiat. Soc. Bengal, Pt. 2, Nat. Hist.* 64 (2): 319 (1896); *M.* sect. *Primulinae* (Prain) Fedde in Engl., *Pflanzenr.* 4, 104: 259 (1909) pro parte., quoad sp. 9; *M.* sect. *Simplicifoliae* series *Delavayanae* sensu C. Y. Wu & H. Chuang pro parte, *Acta Bot. Yunnan.* 2 (4): 376 (1980)]

Taproot dauciform, slender and tapering. *Flowers* lateral-facing to half-nodding; petals (4–)5–9 yellow, white, blue or purple-blue. *Fruit-capsule* narrow-obovoid to oblong-ellipsoidal, often hairy or bristly.

(3 species [67–69]: *M. argemonantha*, *M. florindae*, *M. primulina*). Type: *M. primulina* Prain.

series Cumminsia
[syn. *Meconopsis* subgenus *Eumeconopsis* sect. *Primulinae* (Prain) Fedde in Engl., *Pflanzenr.* 4, 104: 259 (1909), pro parte].

Taproot napiform, short. *Flowers* nodding, 2–3 or solitary, rarely more; petals 4, occasionally 5–6 in the terminal flower if several flowers present, pale bluish lilac to pale rose. *Fruit capsule* narrow-oblong to subcylindrical, glabrous.

(9 species [70–78]: *M. bulbilifera*, *M. compta*, *M. exilis*, *M. lamjungensis*, *M. ludlowii*, *M. lyrata*, *M. polygonoides*, *M. sinuata*, *M. wumungensis*).

section BELLAE (Prain) Fedde in Engl., *Pflanzenr.* 4, 104: 261 (1909).
[Syn. *Meconopsis* subgenus *Eumeconopsis* section *Polychaetia* subsection *Cumminsia* series *Bellae* Prain, *J. Asiat. Soc. Bengal, Pt. 2, Nat. Hist.* 64 (2): 321 (1896); *M.* sect. *Simplicifoliae* series *Delavayanae* sensu C. Y. Wu & H. Chuang pro parte, *Acta Bot. Yunnan.* 2 (4): 376 (1980)]

Polycarpic plants, with a long dauciform rootstock and a short stem densely beset with the fibrous basal remains of old leaves; leaves simple to pinnately or bipinnately lobed. *Flowers* solitary, scapose, up to 18 per tuft of several congested rosettes. *Pedicels* stiffening and arching to strongly recurved as the fruits develop. Petals 4 (–6), pink, pale mauve, blue or purple-blue. *Fruit capsule* pear-shaped to obovoidal, generally very shortly stiped, 4–7-valved.

(1 species [79] *M. bella*). Type species: *M. bella* Prain.

KEY TO THE SUBDIVISIONS OF THE GENUS MECONOPSIS

1. Ovary and fruit capsule surmounted by a fringed, lobed or angled, somewhat projected disc . . . . subgenus **Discogyne**
+ Ovary and fruit capsule not surmounted by a disk 2

2. Leaf-rosettes large and evergreen, dense and symmetrical (subgenus **Meconopsis**) 3
+ Leaf-rosettes small or leaves tufted, not evergreen 5

3. Leaf-lamina unlobed, with a finely toothed margin, generally densely sericeous with barbellate hairs . . . . section **Meconopsis**
+ Leaf-lamina lobed, often bipinnatisect or bipinnatifid, sparsely to densely barbellate pubescent but not sericeous (section **Polychaetia**) 4

4. Pubescence of stems and leaves consisting of simple barbellate hairs all of the same type series **Robustae**
+ Pubescence of stems and leaves consisting of long barbellate hairs underlain by a felt of much shorter barbellate hairs . . . . series **Polychaetia**

5. Plants adorned with barbellate hairs (subgenus **Grandes**) 6
+ Plants adorned with simple hairs or bristles 9

6. Plants with scapose flowers; bracts absent; petals blue, lavender purple or red (section **Simplicifoliae**) 7
+ Plants with a well-developed stem and bracts (uppermost in a pseudowhorl), the inflorescence with 1–7 flowers, occasionally more (section **Grandes**) 8

7. Leaf-lamina pinnately veined or weakly 3-veined towards the base; flowers saucer-shaped to cupped or flattish; petals normally 5 or more . . . . series **Simplicifoliae**
+ Leaf-lamina with 3 or 5 well marked longitudinal veins; flowers campanulate to campanulate-infundibuliform; petals normally 4 . . . . series **Puniceae**

8. Polycarpic plants; petals 4 (5–6), blue, pink, purple or wine-purple; fruit capsule 4–7-valved . . . . series **Grandes**
+ Monocarpic plants; petals 6–11, yellow; fruit capsule (4–)7–11-valved . . . . series **Integrifoliae**

9. Flowers borne in the axils of stem leaves (stem sometimes very short); bracteoles often present; basal leaves few, generally withered by flowering time; plants with brittle or stiff hairs or subglabrous (section **Primulinae**) 10
+ Flowers borne in an ebracteate or bracteate inflorescence or on basal scapes, the scapes sometimes agglutinosed in the lower part; basal leaves in a lax to dense rosette, present at flowering time; plants with subpungent or pungent bristles, sometimes densely so, or with soft, rather weak hairs (subgenus **Cumminsia**) 11

10. Flowers saucer-shaped to flattish, the petals (4–)5–9, yellow, white, blue or purple-blue; fruit capsule narrow-obovoid to oblong-ellipsoid . . . . series **Primulinae**
+ Flowers campanulate, the petals usually 4 (occasionally 5–6 in the terminal flower), pale blue to lilac, lavender or pale rose; fruit capsule narrow-oblong to subcylindrical . . . . series **Lyratae**

11. Plants adorned with thin to stout pungent bristles; fruit capsule densely patent pungent-bristly (section **Aculeatae**) 12
+ Plants adorned with soft, rather weak, never pungent, hairs; fruit capsule with sparse to dense, often rather weak bristles, subglabrous or glabrous 14

12. Leaf-lamina pinnatisect to bipinnatifid, occasionally shallowly so . . . . section **Aculeatae**
+ Leaf-lamina undivided, the margin sometimes sinuous or shallowly and unevenly lobed (section **Racemosae**) 13

13. Filaments all linear . . . . series **Racemosae**
+ Filaments of inner stamens strongly incurved and dilated . . . . series **Heterandrae**

14. Flowers borne in an ebracteate, racemose inflorescence, sometimes only 2–3-flowered . . . .
. . . . section **Forrestianae**
+ Flowers solitary, scapose, some scapes occasionally agglutinosed in the lower part (sections **Impeditae** & **Bellae**) 15

15. Polycarpic plants 16
+ Monocarpic plants 17

16. Leaves all entire, not evergreen; flowers one per leaf-rosette; fruit capsule narrow-oblong to subcylindrical, borne on an erect scape . . . . series **Delavayanae**
+ Leaves often dimorphic, evergreen, entire to pinnately or bipinnately lobes, often on the same plant; flowers several per leaf-rosette; fruit capsule obovoidal to pear-shaped, borne on a curved or deflexed scape . . . . section **Bellae**

17. Filaments of all stamens dilated in the middle or bottom half . . . . series **Henricanae**
+ Filaments of all stamens linear . . . . series **Impeditae**

KEY TO SPECIES OF MECONOPSIS
(naturally occuring hybrids not included)

1. Ovary and fruit capsule surmounted by a fringed, lobed or angled, somewhat projected, often red or purplish, disk, the style broadly and abruptly expanded at the base, always present 2
+ Ovary and fruit capsule without a surmounting disk, the style not abruptly expanded at the base, occasionally absent or very short 8

2. Leaves more or less entire; flowers relatively small, the petals not more than 38 mm long; stigma capitate, 1–4 mm diameter 3
+ Leaves variously lobed; flowers relatively large, the petals 40–67 mm long; stigma linear, 4–7 mm long . . 4

3. Inflorescence with a single main axis; petals blue, purple or mauve; ovary with whitish bristles and a 2–4 mm long style . . . . 19. **M. simikotensis** (W Nepal)
+ Inflorescence with several main axes; petals scarlet; ovary with fawn or orange bristles and a 7–12 mm long style . . . . 22. **M. manasluensis** (C Nepal)

4. Style more or less obsolete: flowers in a spike-like inflorescence with very short pedicels or subsessile; petals partly soft-bristly on the reverse . . . . 21. **M. torquata** (CS Tibet)
+ Style distinct, 3–7 (–10) mm long; flowers in a racemose inflorescence, the pedicels always apparent; petals glabrous 5

5. Stylar-disk pentagonal in outline, shallowly-lobed, not projecting beyond the edge of the ovary; petals maroon; stem leafy below inflorescence . . . . .18. **M. tibetica** (S Tibet)
+ Stylar-disk sinuate or 8-angled, lobed or fringed at the margin, projecting beyond the edge of the ovary; petals pale yellow, bluish-purple to deep purple, more rarely crimson or crimson-purple; stigma capitate or linear; stem leafy or leafless below inflorescence 6

6. Stems leafy (excluding leaf-like bracts); leaves pinnately-lobed for full length with 3–5 pairs of segments .... 20. **M. pinnatifolia** (CN Nepal, S Tibet)
+ Stems leafless below inflorescence; leaves only lobed in the upper third, often trilobed, or 4–5-lobed near the apex 7

7. Flowers pale yellow, the petals with recurved margins; leaf apex usually 3-lobed, occasionally entire (uppermost rosette leaves); style 4–6 mm long................. 16. **M. discigera** (E Nepal, Sikkim)
+ Flowers blue to purple, the petals with flat margins; leaf apex mostly 5(–7)-lobed (except young plants); style 2–4 mm long... 17. **M. bhutanica** (W Bhutan)

8. Plants beset with barbellate hairs.. 9
+ Plants beset with simple, non-barbellate hairs 35

9. Plants monocarpic with large evergreen leaf-rosettes; flowers usually borne in paniculate inflorescences (racemose in *M. robusta* and *M. violacea*) 10
+ Plants monocarpic or polycarpic with non-evergreen leaf-rosettes; flowers solitary and scapose or borne in subumbellate inflorescences 24

10. Lamina of basal leaves with a serrate, rarely shallowly-lobed, margin, densely sericeous 11
+ Lamina of basal leaves pinnatifid to bipinnatifid, to pinnatisect or pinnately-lobed, covered with sparse to dense, spreading stiff hairs 13

11. Indumentum of leaves and stems with simple barbellate hairs; flowers yellow, rarely red ....
  .... 1. **M. regia** (C Nepal)
+ Indumentum of leaves and stems with long simple barbellate hairs underlaid with shorter barbellate hairs; flowers white, rose-pink to red... 12

12. Flowers white, borne singly on the inflorescence; stigma purple.. 2. **M. superba** (W Bhutan, S Tibet)
+ Flowers rose-pink to red, mostly borne in 3–5-flowered cymules in the inflorescence; stigma dark brown... 3. **M. taylori** (C Nepal)

13. Leaves with a sparse to dense indumentum of simple, more or less uniform, barbellate hairs, sometimes glabrescent at maturity.. 14
+ Leaves with an indumentum of long simple, barbellate hairs underlain with short simple or branched hairs, often densely so 17

14. Flowers red or purple; basal leaves often bipinnatifid .... 7. **M. chankheliensis** (W Nepal)
+ Flowers yellow, rarely white; basal leaves pinnatifid to pinnatisect 15

15. Plants 120–180 cm tall in flower, the flowers borne singly on lateral pedicels, the inflorescence confined to the upper half of the plant .... 4. **M. robusta** (N India; Kumaon; W Nepal)
+ Plants rarely exceeding 90 cm tall in flower, the flowers mostly borne in 2–3-flowered cymules in the inflorescence from the base of the plant upwards... 16

16. Leaves and stems sparsely bristly, the bristles uniformly pale, the leaves with 3–5 pairs of lateral segments; fruit capsule narrow-oblong or flask-shaped, the style 6–10 mm long ....
  .... 5. **M. gracilipes** (C Nepal; S Tibet)
+ Leaves and stems densely bristly, the bristles with a purple-black base, the leaves with 6–9 (–10) pairs of lateral segments; fruit capsule broad-oval or oval-oblong, the style 3–6 mm long.................
  .... 6. **M. dhwojii** (E Nepal; S Tibet)

17. Flowers yellow 18
+ Flowers red, pink, blue, lilac, violet, occasionally white.. 20

18. Leaves and stems with a sparse underlying indumentum of short barbellate hairs; stigma and ovary usually with 5 divisions, the stigma greenish; plant rarely exceeding 1 m tall in flower .... 8. **M. napaulensis** (C Nepal)
\+ Leaves and stems with a dense underlying indumentum of barbellate hairs; stigma and ovary with 6–12 divisions, the stigma yellow or reddish-purple; plant often 1.1–1.85 m tall in flower 19

19. Stigma reddish-purple, 1–4 mm diameter; bracts patent; fruit capsule with ascending to appressed bristles. .... 13. **M. paniculata** (Nepal, Sikkim, Bhutan, N Arunachal Pradesh (Assam), S Tibet)
\+ Stigma yellow, 3–8 mm diameter; bracts deflexed downwards; fruit capsule with patent bristles...... .... 14. **M. autumnalis** (C Nepal)

20. Mature fruit capsules narrowed at the base; flowers with red or pink (sometimes with a violet tinge) or violet petals.. 21
\+ Mature fruit capsule rounded at the base; flowers with blue, violet-blue, lilac, lavender, mauve, purple or wine-purple petals... 23

21. Inflorescence racemose, the flowers borne singly in the inflorescence, the petals violet; basal leaves narrow, linear-oblong in outline...................... 15. **M. violacea** (Myanmar, SE Tibet)
\+ Inflorescence paniculate, the flowers mostly (except the uppermost) borne in axillary cymules in the inflorescence (lateral branches with 2–5 flowers, more in *M. wallichii*), especially in the lower part of the inflorescence, the petals pink or red, occasionally purplish blue or white; basal leaves oblong to lanceolate in outline 22

22. Flower buds with patent barbellate hairs; fruit with patent, simple, long barbellate hairs; fruit-capsule style 4–8 mm long, to 1 mm diameter at the base .... 9. **M. ganeshensis** (C Nepal)
\+ Flowers buds with erect or incurved barbellate hairs; fruit capsule with ascending long barbellate hairs plus an underlying indumentum of short barbellate hairs; fruit-capsule style 8–14 mm long, 2–3 mm diameter at the base .... 10. **M. staintonii** (C Nepal)

23. Inflorescence with short, erect lateral branches, congested and spike-like; flowers violet wine-purple to purple, purplish red or crimson, the lateral cymules 1–5-flowered; fruit capsule (8–)11–15 mm diameter, with erect to ascending bristles...................... 12. **M. wilsonii** (SW China, E Myanmar)
\+ Inflorescence with long, spreading branches, not spike-like; flowers blue to lilac or pinkish-blue, occasionally wine-purple or deep red, the lateral cymules mostly 5–13-flowered; fruit capsule 6–10 mm diameter, with wide-spreading bristles. .... 11. **M. wallichii** (E Nepal, Sikkim, Bhutan, SE Tibet)

24. Flowers generally borne in umbels or pseudoumbellate racemes, accompanied by bracts, the uppermost bracts in a pseudowhorl; filaments nearly always white............................... 25
\+ Flowers solitary and scapose, without bracts: filaments nearly always the same colour as the petals, often darker 32

25. Flowers yellow; plants monocarpic.. 26
\+ Flowers blue, purple, pinkish-purple, pink or wine-red; plants usually polycarpic 29

26. Stigma broad and sessile, the stylar arms markedly decurrent on to the top of the ovary/fruit capsule; flowers and buds ascending to upright............... 27. **M. integrifolia** (W & NW China, E Tibet)
\+ Stigma narrow and with a short to long style, the stylar arms not decurrent on to the top of the ovary/fruit capsule; flowers and buds generally nodding or half-nodding 27

27. Plants subscapose, often with solitary flowers; style short-bristly....................... .... 29. **M. pseudointegrifolia** (NW Yunnan, SE Tibet)
\+ Plants with a well-defined stem, bearing 2 or more flowers; style glabrous 28

28. Flowers generally nodding to half-nodding with usually 6–9 petals; leaves pinnately-veined (only 3-veined at the base of the lamina); style distinct, 3–11 mm long the base of the stigma lobes not decurrent onto the style . . . . 30. **M. sulphurea** (SW China, SE Tibet)
+ Flowers lateral-facing to ascending with 5–6 petals; leaves 3-veined in the lower half to two-thirds; style short, not more than 4 mm long, the base of the stigma lobes decurrent onto the style . . . . 28. **M. lijiangensis** (SW China, E Tibet)

29. Lamina of the basal leaves, at least, cuneate, attenuate into the petiole. 30
+ Lamina of basal leaves at least shallowly cordate to truncated, abruptly contracted into the petiole . . 31

30. Fruit capsule 24–30 × 12–18 mm, always bristly, with a very short (not more than 2 mm long) to more or less obsolete style; flowers solitary, the petals normally 6–8, flesh to rose pink . . . . 26. **M. sherriffii** (N Bhutan and adjacent Tibet)
+ Fruit capsule 27–58(–68) × 6–15 mm, bristly or glabrous, with a pronounced style 4–8 mm long; flowers generally 2–5 in a subumbellate inflorescence, occasionally solitary, the petals normally 4–6, blue, purple-blue to wine or wine-purple . . 25. **M. grandis** (Nepal to Sikkim, Bhutan, NE India and adjacent Tibet)

31. Uppermost 3–5(–7) leaves aggregated into a false whorl; mature leaf-lamina with 8–19 pairs of teeth; mature fruit capsule 10–16 mm diameter, moderately to densely bristly, the style 2–5.5 mm long with a stigma 4–5 mm diameter . . . . 24. **M. baileyi** (SE Tibet)
+ Uppermost leaves alternate, not aggregated into a false whorl; mature leaf-lamina with 5–9 pairs of teeth; mature fruit capsule 8–9 mm diameter, glabrous or rarely with a few scattered bristles, the style 5–8 mm long with a stigma 2–3 mm diameter . . . . 23. **M. betonicifolia** (SW China)

32. Flowers blue, purple or lavender, occasionally reddish purple, lateral-facing to half-nodding, saucer-shaped to broad-campanulate. 33
+ Flowers red blue or lavender, pendent, narrow-campanulate 34

33. Flowers saucer-shaped or shallowly cupped, horizontal to ascending, normally with 5–9 petals; scapes and fruits with reflexed bristles; plants usually monocarpic; stigma clavate. . . . . 31. **M. simplicifolia** (C Nepal to Sikkim, Bhutan and SE Tibet)
+ Flowers campanulate, ascending to nodding, normally with 4–6 petals; scapes and fruits glabrous or with spreading to ascending bristles; plants polycarpic; stigma cupuliform . . 32. **M. nyingchiensis** (SE Tibet)

34. Petals crimson, rarely white, considerably longer than broad . . 35. **M. punicea** (W China and E Tibet)
+ Petals blue, purple-blue, lavender, lilac or violet, rarely much longer than broad . . . . 36. **M. quintuplinervia** (W & NW China, NE Tibet)

35. Flowers borne in the axils of stem leaves (stem sometimes very short); bracteoles often present; basal leaves few, generally withered by flowering time; plants with brittle or stiff hairs or subglabrous (section *Primulinae*). 36
+ Flowers borne in an ebracteate or bracteate inflorescence or on basal scapes, the scapes sometimes agglutinosed in the lower part; basal leaves in a lax to dense rosette, present at flowering time; plants with subpungent or pungent bristles, sometimes densely so, or with soft, rather weak hairs 47

36. Flowers saucer-shaped to flattish, the petals (4–)5–9, yellow, white, blue or purple-blue; fruit capsule narrow-obovoid to oblong-ellipsoid 37
+ Flowers campanulate, the petals usually 4 (occasionally 5–6 in the terminal flower), pale blue to lilac, lavender or pale rose; fruit capsule narrow-oblong to subcylindrical 39

37. Flowers pale blue, lavender-blue mauve-blue or pink . . . . 69. **M. primulina** (Bhutan and neighbouring parts of Tibet)
+ Flowers white, cream, or yellow. 38

38. Fruit capsule densely bristly, 16–25 mm long; petals white, cream or lemon-yellow . . . . 67. **M. argemonantha** (SE Tibet)
+ Fruit capsule glabrous to sparsely bristly, c. 27 mm long; petals lemon-yellow . . . . 68. **M. florindae** (Bhutan and neighbouring parts of Tibet)

39. Leaves all petiolate, shallowly to deeply trilobed with a much larger terminal lobe, or trifoliate; stem decumbent towards the base, with adventitious buds or fibrous-roots often borne at the lower nodes **40**
+ Leaves all petiolate or the upper sessile or subsessile, entire to pinnately divided, not trilobed; stem erect to ascending, rarely decumbent at the base, without adventitious buds or fibrous-roots at the lower nodes **41**

40. Delicate plants with thread-like, somewhat zigzag, stems, narrowed and extended at the base, often bearing tiny bulbils at the nodes; lower-surface of leaves densely covered with minute stellate patterning; flowers solitary, generally smaller, the petals 12–17 mm long . . . . . 71. **M. bulbilifera** (Nepal, Sikkim)
+ More robust plants with shorter and firmer stems more or less uniform throughout, without bulbils at the nodes; lower surface of leaves not stellately patterned; flowers 1–5, generally larger, the petals 13–22 mm long . . . . 70. **M. lyrata** (Darjeeling District of India)

41. Ovary and fruit capsule bristly; flowers 3–8 borne in an evenly spaced bracteate raceme (bracts often displaced so that the flowers may appear to be ebracteate) . . . . 78. **M. sinuata** (C & E Nepal, NE India (Sikkim), Bhutan, SE Tibet)
+ Ovary and fruit capsule glabrous or subglabrous; flowers solitary or several clustered at the top of the stem, bracteate **42**

42. Plants 6–25 cm tall in flower; upper leaves petiolate or subsessile; lamina pinnately lobed, coarsely toothed or crenate, rarely entire **43**
+ Plants 20–50 cm tall in flower; upper leaves sessile; lamina entire, sinuate or crenate, not lobed **45**

43. Flowers 2–8, the petals 13–22 mm wide; stamens c. 36; style 2–4 mm long . . . . 77. **M. ludlowii** (E Bhutan, NE India — Arunachal Pradesh)
+ Flowers 1–3, the petals 7–12 mm wide; stamens 10–24; style 1–3 mm long **44**

44. Inflorescence often scapose with elongate pedicels and a reduced stem; leaf-lobes often rounded with contracted bases (especially in scapose plants); petals with an obtuse to acute apex . . . . 75. **M. wumungensis** (SW China — N Yunnan)
+ Inflorescence not scapose, usually with a single terminal flower borne on a leafy stem; leaf-lobes broadly ovate to oblong; petals with a rounded to obtuse apex . . . . 76. **M. compta** (SE Tibet, N Myanmar, SW China — NW Yunnan)

45. Flowers solitary, terminal; stem leaves (including bracts) 3–4, not subtending shoots; lamina of upper leaves with semi-amplexicaul base; petals narrow-obovate to narrow-elliptic, the apex acute to acuminate . . . . 74. **M. polygonoides** (W Bhutan, S Tibet)
+ Flowers both terminal and axillary; stem leaves (including bracts) 3–8, often subtending a lateral shoot with a flower and 0–2 leaves (bracteoles); lamina of upper leaves with or without a semi-amplexicaul base; petals obovate or broad-obovate, the apex rounded to subacute **46**

46. Uppermost 2–3 leaves subopposite or in a pseudowhorl normally; lamina of lower leaves with a rounded to attenuate base, those of the upper leaves with a semi-amplexicaul base . . . . 72. **M. exilis** (SW China — NW Yunnan, NE Myanmar)
+ All leaves alternate; lamina of lower leaves with shallowly cordate to truncated base, those of the upper leaves without a semi-amplexicaul base . . . . 73. **M. lamjungensis** (C Nepal — Lamjung Himal)

47. Plants adorned with thin to stout pungent bristles; fruit capsule densely patent pungent-bristly..... 48
+ Plants adorned with soft, rather weak, never pungent, hairs; fruit capsule with sparse to dense, often rather weak bristles, subglabrous or glabrous.................................... 63

48. Leaf-lamina pinnatisect to bipinnatifid, occasionally shallowly so........................ 49
+ Leaf-lamina undivided, the margin sometimes sinuous or shallowly and unevenly lobed......... 53

49. Leaf-lamina undulate to shallowly lobed........................40. **M. latifolia** (N Kashmir)
+ Leaf-lamina shallowly to deeply pinnatifid.. 50

50. Flowers all borne on basal scapes................... 39. **M. neglecta** (NW Pakistan — Chitral)
+ Flowers borne in part or wholly bracteate inflorescences................................... 51

51. Leaves toothed to shallowly pinnatifid, 7.9.2 cm wide; only the uppermost flowers ebracteate; style of fruit capsule not exceeding 6 mm long.........................40. **M. latifolia** (N Kashmir)
+ Leaves deeply pinnatifid, not more than 5 cm wide; only the lowermost flowers in the inflorescence bracteate; style of fruit generally more than 6 mm long..................................... 52

52. Style stout; ovary bristles reddish-brown............... 41. **M. speciosa** (SE Tibet, NW Yunnan)
+ Style slender; ovary bristles straw-coloured to golden.. 38. **M. aculeata** (N Pakistan, Kashmir, NW India)

53. Filaments of inner stamens strongly incurved and dilated................................. 54
+ Filaments all linear.. 55

54. Flowers all scapose, the petals rounded to broadly oval, widely overlapping..................
.... 52. **M. heterandra** (W Sichuan — Mianning)
+ Flowers borne in a well-developed raceme, the petals elliptic-obovate, little or not overlapping......
.....................51. **M. balangensis** (W Sichuan — Balangshan, Siguniangshan, Xaiojinshan)

55. Dwarf plants 6–20(–27 cm tall in fruit); flowers scapose or pedicels partly agglutinosed.......... 56
+ Medium to tall plants 14–95 (–120) cm tall; flowers primarily borne in a racemose inflorescence, occasionally on basal scapes on the same plant... 57

56. Leaves usually with dark dots (base of bristles); flowers lateral-facing to nodding; style broadened at base in fruit, sometimes markedly so............. 42. **M. horridula** (C & E Himalaya & S & SE Tibet)
+ Leaves thin, papery, without dark dots; flowers lateral-facing to ascending; style linear in fruit.......
.... 43. **M. lhasaensis** (S Tibet centred on Lhasa)

57. Leaves and stems dotted with raised dark bases to the bristles............................. 58
+ Leaves and stems without dark dots, the bristles concolorous................................ 59

58. Leaves glaucous above, the basal 15–42 mm wide; fruit capsule not more than 15 mm long........
.... 48. **M. rudis** (SW China — NW Yunnan, SW Sichuan)
+ Leaves deep green or yellowish green, sometimes purple-flushed, the basal 7–14 mm wide; fruit capsule 25–38 mm long............. 49. **M. bijiangensis** (NW Yunnan — Biluoxueshan, Gaoligongshan)

59. Petals 4 (rarely the terminal flower in the inflorescence 5–6-petalled), blue, pale yellow or white.....
.... 47. **M. prainiana** (N Myanmar, SE Tibet)
+ Petals 5–9, blue, blue-purple, ruddy purple, wine-purple, crimson-purple or crimson, more rarely pale yellow or white.. 60

60. Fruit capsule (minus style) 15–34 mm long; stigma green, greenish yellow or purplish............ 61
+ Fruit capsule (minus style) 9–17 mm long; stigma white, occasionally greyish.................. 62

61. Inflorescence generally with more than 20 crowded flowers; anthers cream or white; petals blue, blue-purple, purple or claret. . . . . 46. **M. zhongdianensis** (NW Yunnan)
+ Inflorescence with 18 flowers or fewer; anthers orange-yellow; petals blue-purple, purple-crimson, dark red or yellow. . . . . 50. **M. georgei** (NW Yunnan)

62. Inflorescence not secund; stigma white, protruding beyond the boss of stamens; fruit capsule with erect to ascending spines; anthers white. . . . . 44. **M. racemosa** (W China, C, E & SE Tibet)
+ Inflorescence secund or semi-secund; stigma greenish or yellowish-green, included within the boss of stamens; fruit capsule with spreading spines; anthers yellow, grey-yellow. . . . . 45. **M. prattii** (SW China & SE Tibet)

63. Flowers solitary, scapose, some scapes occasionally agglutinosed in the lower part 64
+ Flowers borne in an ebracteate (occasionally the lowermost flowers bracteate), racemose inflorescence, sometimes only 2–3-flowered. 75

64. Polycarpic plants, with clustered leaf-rosettes. 65
+ Monocarpic plants, with a solitary leaf-rosette. 66

65. Leaves all entire, not evergreen; flowers one per leaf-rosette; fruit capsule narrow-oblong to subcylindrical, borne on an erect scape . . . . 63. **M. delavayi** (NW Yunnan)
+ Leaves often dimorphic, evergreen, entire to pinnately or bipinnately lobed, often on the same plant; flowers several to many per leaf-rosette; fruit capsule obovoidal to pear-shaped, borne on a curved or deflexed scape . . . . 79. **M. bella** (C & E Himalaya, SE Tibet)

66. Filaments of all stamens dilated in the middle or bottom half 67
+ Filaments of all stamens linear. 69

67. Flowers deeply cupped, the petals with a dark basal blotch; ovary and young fruit finely stellate-pubescent . . . . 62. **M. sinomaculata** (W China — NW Sichuan, SE Qinghai)
+ Flowers saucer-shaped, the petals unmarked at base; ovary and young fruit with scattered bristles. . . 68

68. Scapes slender, 2–4 mm diam. towards the base; petals 1.4–2.6 cm wide; anthers bright yellow; mature fruit capsules 7–11 mm diam. . . . . 60. **M. henrici** (W China — W, SW Sichuan)
+ Scapes relatively stout, 4–7 mm diam. towards the base; petals 2.3–3.5 cm wide; anthers buff; mature fruit capsules 11–18 mm diam. . . . . 61. **M. psilonomma** (S Gansu, NW Sichuan, SE Qinghai)

69. Leaves glabrous or subglabrous. 70
+ Leaves moderately to fairly densely pubescent 72

70. Mature fruit capsules 36–93 mm long, at least 5 × longer than wide; petals 4 . . . . 58. **M. venusta** (NW Yunnan)
+ Mature fruit capsule generally less than 28 mm long, less than 4 × longer than wide; petals 4–8 (–10) . . 71

71. Flowering and fruiting scapes not more than 7 per plant; fruit capsules not more than 5 mm diam., the persistent style 1.5–5 mm long . . . . 55. **M. concinna** (SW Sichuan, SE Tibet & NW Yunnan)
+ Flowering and fruiting scapes at least 8, and up to 17, per plant; fruit capsules 7–9 mm diam., the persistent style 4–6 mm long . . . . 59. **M. pseudovenusta** (NW Yunnan)

72. Flowers solitary, all borne on basal scapes 73
+ Flowers borne in a racemose, sometimes agglutinosed, inflorescence, sometimes accompanied by one or several basal scapes. 76

73. Petals 5–11, rarely 4. 74
+ Petals normally 4, rarely 5–6 on the first flower to open 75

74. Plant generally more than 15 cm tall in flower, bearing 7–20 flowers; pedicels (scapes) generally decumbent towards the base; ovary and style usually blackish purple, the style 4–7 mm long in flower . . . . 53. **M. impedita** (N Myanmar, SW Sichuan, SE Tibet & NW Yunnan)
+ Plant not exceeding 10 cm tall in flower, bearing up to 9 flowers; pedicels not decumbent at the base; ovary and style cream, the style 2–3 mm long . . 54. **M. xiangchengensis** (SW Sichuan, NW Yunnan)

75. Leaf-lamina variable from entire to pinnately lobed, sometimes on the same plant; fruit capsule 4–5-valved, the persistent style up to 2.6 mm long. . . . . 55. **M. concinna** (SW Sichuan, SE Tibet & NW Yunnan)
+ Leaf-lamina always entire; fruit capsule 3–4-valved, the persistent style not more than 3 mm long . . . . 57. **M. pulchella** (S Sichuan — Mianning Xian)

76. Leaf-lamina pinnately lobed, occasionally crenate-margined; flowers lateral-facing, borne in an agglutinosed raceme . . . . 56. **M. muscicola** (NW Yunnan (Laojunshan) and SW Sichuan (E of Yunning))
+ Leaf-lamina entire; flowers lateral facing or nodding 77

77. Flowers borne only towards the top of the inflorescence: fruit capsule narrow-clavate to subcylindric, not more than 6 mm diam., with a very short or obsolete style, not more than 1.5 mm long. . . . . 66. **M. forrestii** (SW Sichuan, NW Yunnan)
+ Flowers scattered along much of the length of the inflorescence, often accompanied by one or several basal 1-flowered scapes; fruit-capsule ellipsoid to obovoid or oblong-cylindric, 5–11 mm diam., with a 2–7 mm long persistent style 78

78. Inflorescence with 8–16 flowers, the lowermost bracteate; fruit-capsule 35 mm long or more . . . . 65. **M. yaoshanensis** (NE Yunnan)
+ Inflorescence with up to 8 flowers, often 3–5, all ebracteate; fruit-capsule not more than 29 mm long . . . . 64. **M. lancifolia** (SW Gansu, W & SW Sichuan, NW Yunnan)

Subgenus MECONOPSIS

This striking subgenus contains all the known monocarpic species of *Meconopsis* that overwinter in the years before flowering as an increasingly large evergreen rosette. All the species are beset with barbellate hairs, from short to long, which adorn most parts of the plant with the exception of the petals. Most produce large numbers of flowers in a terminal paniculate or subpaniculate inflorescence, very occasionally racemose (in *M. robusta* and *M. violacea*), in which the flowers in the lower half of the inflorescence are borne in distinctive lateral cymules.

KEY TO SECTIONS OF SUBGENUS MECONOPSIS

1. Lamina of basal leaves with a serrate, rarely shallowly-lobed, margin, densely sericeous . . . . section **Meconopsis**
+ Lamina of basal leaves pinnatifid to bipinnatifid, to pinnatisect or pinnately-lobed, covered with sparse to dense, spreading stiff hairs . . . . section **Polychaetia**

section MECONOPSIS

Section *Meconopsis* consists of three species, all local endemics, *M. regia* and *M. taylori* to central Nepal, *M. superba* to western Bhutan. They are characterised by possessing large and handsome evergreen rosettes, the leaves more or less spear-shaped and with a finely toothed, rarely slightly lobed, margin and a dense indumentum that gives the foliage, especially when young, a silky sheen. The flowers are borne in leafy panicles with the lowermost flowers often borne in lateral cymules, the upper flowers solitary from the axils of the upper bracts. All three species tend to have large multi-valved (up to 17-valved in *M. taylorii*) capsules with *M. taylorii* bearing the largest capsules of any known *Meconopsis* species. Horticulturally all three species are highly desirable, both for their handsome leaf-rosettes but also for the arresting sight that they make in flower. Unfortunately, only *M. superba* is currently in cultivation as far as I am aware, although various plants are found in gardens and nursery catalogues erroneously under the name *M. regia*.

KEY TO SPECIES OF SECTION MECONOPSIS

1. Indumentum of leaves and stems with simple barbellate hairs; flowers yellow, rarely red . . . . .1. **M. regia** (C Nepal)
+ Indumentum of leaves and stems with long simple barbellate hairs underlain with shorter barbellate hairs; flowers white, rose-pink to red 2

2. Flowers white, borne singly on the inflorescence; stigma purple. . . . .2. **M. superba** (W Bhutan, S Tibet)
+ Flowers rose-pink to red, mostly borne in 3–5-flowered cymules in the inflorescence; stigma dark brown. . . . . 3. **M. taylorii** (C Nepal)

1. MECONOPSIS REGIA

Meconopsis regia G. Taylor, *J. Bot.* 67: 259 (1929). Type: W Nepal, Barpak, 12–15,000 ft, *Lall Dhwoj* 18 (BM, holotype; E, K isotypes); also W Nepal, Michet, 14–15,000 ft, *Lall Dhwoj* 195 (paratypes BM, E, K), fruiting specimen.

DESCRIPTION. *A large monocarpic plant* to 1.75 m tall in flower, with most parts covered in dense, soft, silvery or, more often golden, barbellate hairs, giving the leaves in particular, a silky appearance. *Stem* erect, very leafy and densely sericeous. *Basal leaves* forming a substantial dense, evergreen rosette up to 1 m across in the year prior to flowering, the lamina elliptic, tapered at both ends, 31–47.5 × 6.5–11.4 (–14) cm at maturity, the margin evenly and finely serrate, decurrent below into a broad, often rather short, petiole to 6–18 (–22) cm long, densely felted on both surfaces by long appressed, very slender barbellate hairs; lower cauline leaves similar to the basal but shorter petioled, the upper cauline leaves progressively smaller, similar to the basal but sessile, half-clasping the stem, often rather more coarsely toothed, merging into the bracts. *Inflorescence* a many-flowered, open, rather narrow panicle occupying the upper third of the plant, with the lower flowers borne in lateral cymules of 2–4 generally, while the upper flowers are solitary; flowers deeply cupped, horizontal to half-nodding, 10–13 cm across. *Peduncles* 8–18 cm long, densely felted, elongating up to 29 cm long in fruit. *Pedicels* 2.5–16 cm long, pubescent like the peduncles, elongating in fruit up to 34 cm. *Buds* narrow-ovoidal, sericeous, nodding to half-nodding. *Petals* 4 (–6), soft butter yellow, suborbicular, 5–6.6 × 4.8–7 cm, partly overlapping one another, slightly recurved and undulate at the margin. *Stamens* numerous, the filaments filiform, pale yellow, the anthers orange. *Ovary* ovoid, covered with dense, appressed, slender barbellate bristles, the style relatively stout, 5–8 mm long, terminating in a broad capitate, 7–12 shallowly-lobed, reddish or blackish purple, stigma to 6 mm diameter. *Fruit capsule* oblong-ellipsoid, 18–45 × 12–16 mm, 7–12-valved, covered with dense, appressed, barbellate bristles, the persistent style 7–14 mm long, 3–5 mm wide at the base, ridged. *Seeds* ellipsoid-reniform with a fine reticulated testa, the reticulations set in rows. $2n = 56$. Figs 5A, 6B, Plate 1.

DISTRIBUTION. C Nepal: (Gorkha, Kaski & Lamjung districts; centred on the Gurkha, southern Manaslu and Lamjung Himals; 3660–4268 m (Maps 2 & 3).

HABITAT. Growing at the foot of cliffs or amongst boulders, rocky meadows, sometimes amongst scattered shrubs.

FLOWERING. Late June to early August.

SPECIMENS SEEN. *Dhwoj* 18 (BM, E, K), 195 (BM, E, K), 234 (E, K), s.n. 1927 (E); *K.N. Sharma* E64 (BM), E367 (BM), E368 (BM); *Stainton, Sykes & Williams* 4627 (BM, E), 6139 (BM, E), 8621 (BM, E), 8622 (BM, E), 8623 (BM, E), 8637 (BM, E).

Meconopsis regia is without question one of the handsomest of the large *Meconopsis*, not least in its remarkable large golden, occasionally silvery, silky leaf-rosettes and its

THE GENUS MECONOPSIS
1. MECONOPSIS REGIA

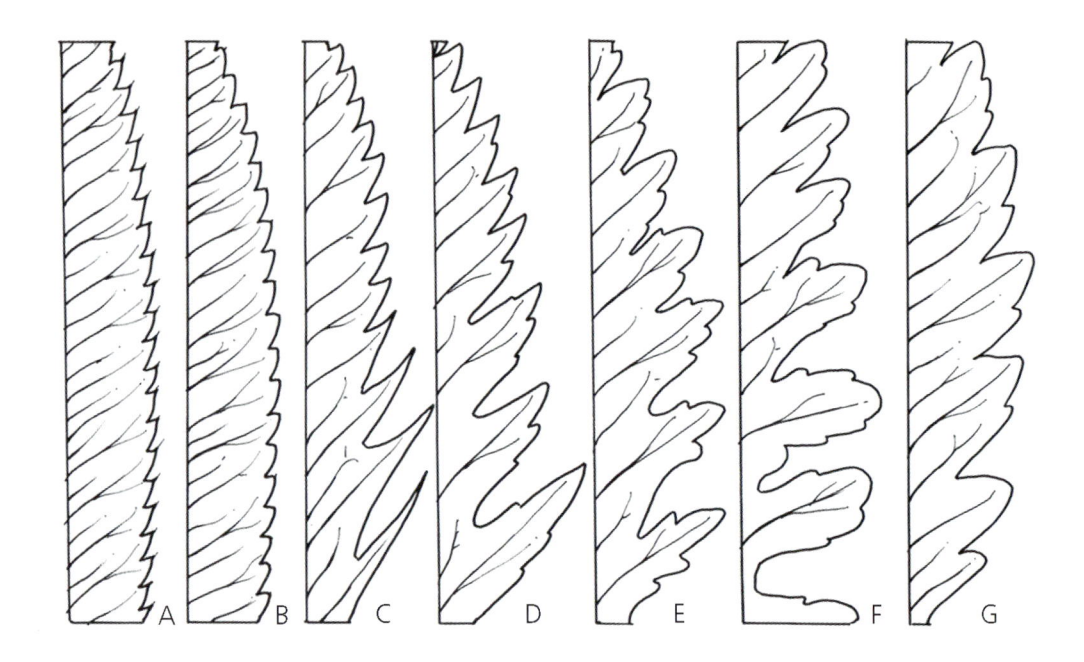

FIG. 5. Lower leaf margin detail, pubescence not shown: **A**, *Meconopsis regia*; **B**, *M. superba*; **C–D**, *M. paniculata* subsp. *pseudoregia*; **E–F**, *M. paniculata* subsp. *paniculata*; **G**, *M. taylorii*. All × ²/₃.
DRAWN BY CHRISTOPHER GREY WILSON

Meconopsis regia leaf-rosettes; Lamjung Himal, C Nepal.
PHOTO: DIETER SCHACHT

finely saw-edged leaves. George Taylor writing in the *Journal of Botany* in 1929 states that it was first sent to the United Kingdom in 1928, its introduction due to Col. Sir Clive Wigram and Mr T Hay, Superintendent of Central Parks, London. George Taylor goes on to say that "The affinity of the species is clearly with *Meconopsis superba* King ex Prain. In their ovary characters the two species are practically identical, and they further resemble each other in the serration of the leaves and sericeous indumentum. *M. regia* has, however, yellow flowers borne in an upwardly branched inflorescence". As noted under the following species, *M. superba*, other morphological differences are apparent.

Records for *Meconopsis regia* cited in the new *Flora of Nepal* for the western districts of Humla, Jumla and Rukum are almost certainly erroneous. These specimens (e.g. *Polunin, Sykes & Williams* 3589, 4359 & 4627) are referable to the widely variable *M. paniculata* and included in this work as *M. paniculata* subsp. *pseudoregia*. Photographs of plants from these areas by Alan Dunkley, Margaret North, Toshio Yoshida and others support this view. This leaves *M. regia* confined to a rather limited area of the central Nepal Himalaya as recorded above.

Meconopsis regia has perhaps the handsomest rosettes of any *Meconopsis* which can, in cultivated specimens at least, reach as much as 1–1.5 m in diameter with individual leaves up to 60 cm long being not unusual, while in flower, plants can reach as much as 2 m. As the young plants increase in size the outermost leaves brown and die away to be replaced by those above that expand to fill the space. Like *M. superba*, plants do not generally reach flowering maturity until their third or fourth year and almost certainly take longer in the wild due to a rather protracted growing season. Plants set seed in abundance but young fruits are prone to rotting in cold wet weather. Unfortunately, the species has apparently disappeared from cultivation, being overwhelmed by hybridisation in gardens (including botanic gardens), particularly with *M. paniculata* and *M. staintonii*. All the specimens purporting to be *M. regia* that I have examined in recent years turn out to be such hybrids, generally readily identified by their rougher, more hoary, looking leaves with variable amounts of lobing along the margin. There is always the hope that the species itself is still maintained in cultivation somewhere. It is yet one more example of the need for the legal re-introduction of seed from the wild, both for scientific study as well as to enhance live collections. *M. regia* hybridises in gardens very readily with other members of subgenus *Meconopsis* and the resultant hybrids, many of which are very vigorous and handsome plants, are fully fertile. Hybridisation in the wild is generally excluded as the species grow in isolation from one another, or there is a disparity in flowering time or altitudes if they do inhabit the same vicinity.

In the wild the hairs on the autumn rosettes of *M. regia*, as well as those of mature plants, take on a rich golden-brown hue, the indumentum with a silky sheen and this feature has also been noted in cultivated plants in the past. This feature clearly distinguishes it from *M. paniculata* with which it is sometimes confused, especially in herbarium specimens. However, George Taylor notes that "It appears that there are two strains in cultivation [of *M. regia*]; one in which the golden hairs are present throughout the life of the plant, and the other more robust form in which the hairs are silvery at first but ultimately golden-brown". It should be noted, however, that the velvety golden-brown characteristic appears to be lost in most instances in hybrids involving *M. regia* as one of the parents.

The leaf hairs are very interesting on close inspection: very long and thread-like they arise rather obliquely backwards from the leaf surface, but quickly curve forwards so that all the hairs face the leaf apex. The mass of parallel, forward-directed, hairs give the leaves their characteristic appearance and impart a noticeable sheen at the same time, this being more marked in *M. regia* than in any other species.

Below left: *Meconopsis regia*; overwintering leaf-rosettes, Lamjung Himal, C Nepal. PHOTO: DIETER SCHACHT

Below right: *Meconopsis regia*; spring leaf-rosettes, Lamjung Himal, C Nepal. PHOTO: DIETER SCHACHT

60 | THE GENUS MECONOPSIS
1. MECONOPSIS REGIA

Meconopsis regia first flowered in Britain in 1931 and was then 'the object of much admiration', particularly the large, dense, over-wintering leaf-rosettes which greatly enhance the garden merit of this fine species. As noted, the species is doubtfully in cultivation today in western gardens and plants offered under the name are in my experience at least, always hybrids. The species was known to hybridise readily in gardens within years of its introduction, particularly with members of section *Robustae*, especially *M.* 'napaulensis' (probably *M. staintonii*) and *M. paniculata*. While *M. paniculata* × *M. regia* produced, as expected, yellow-flowered offspring, the *M. regia* × *M. staintonii* hybrids produced yellow-, pink- and red-flowered offspring, bicolours and picotees. However, the picture soon became far more complicated as the hybrids showed a great deal of vigour and were fertile, crossing and backcrossing to produce the hybrid muddle seen in gardens today, although this situation has certainly been known since as far back as the 1950s, possibly even earlier. Added to the lack of authenticated records on the development of these garden hybrids is the fact that other species could also have been, and probably were, involved within sections *Meconopsis* and *Robustae*, notably *M. wallichii*.

Interestingly, the ready occurrence of hybrids in cultivation may throw some understanding of the way the species perform in the wild. It is known that so-

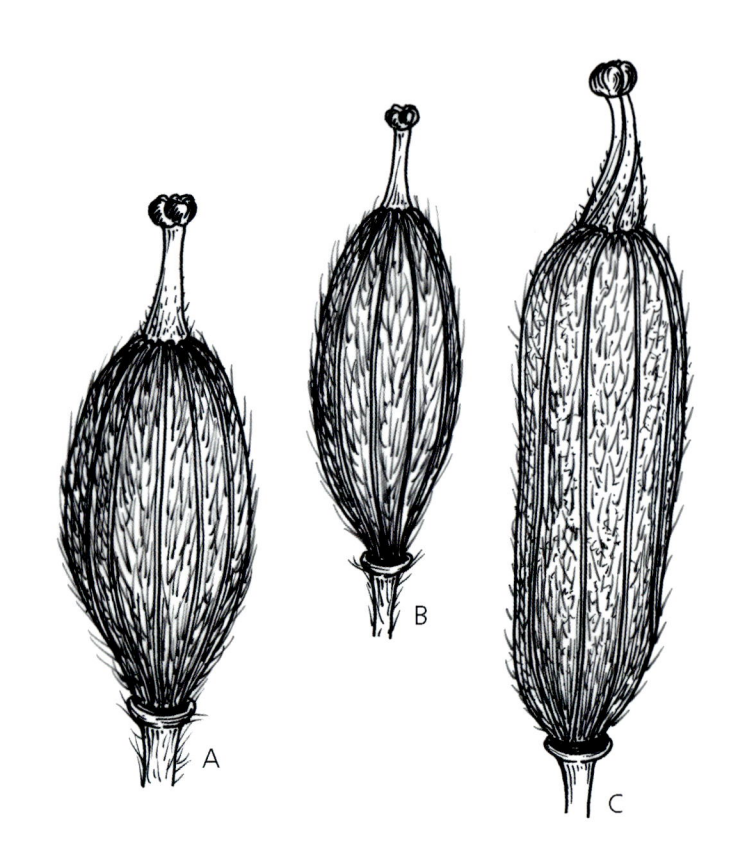

FIG. 6. Fruit capsules: **A**, *Meconopsis superba*; **B**, *M. regia*; **C**, *M. taylorii*. All × 1¹⁄₃.
DRAWN BY CHRISTOPHER GREY-WILSON

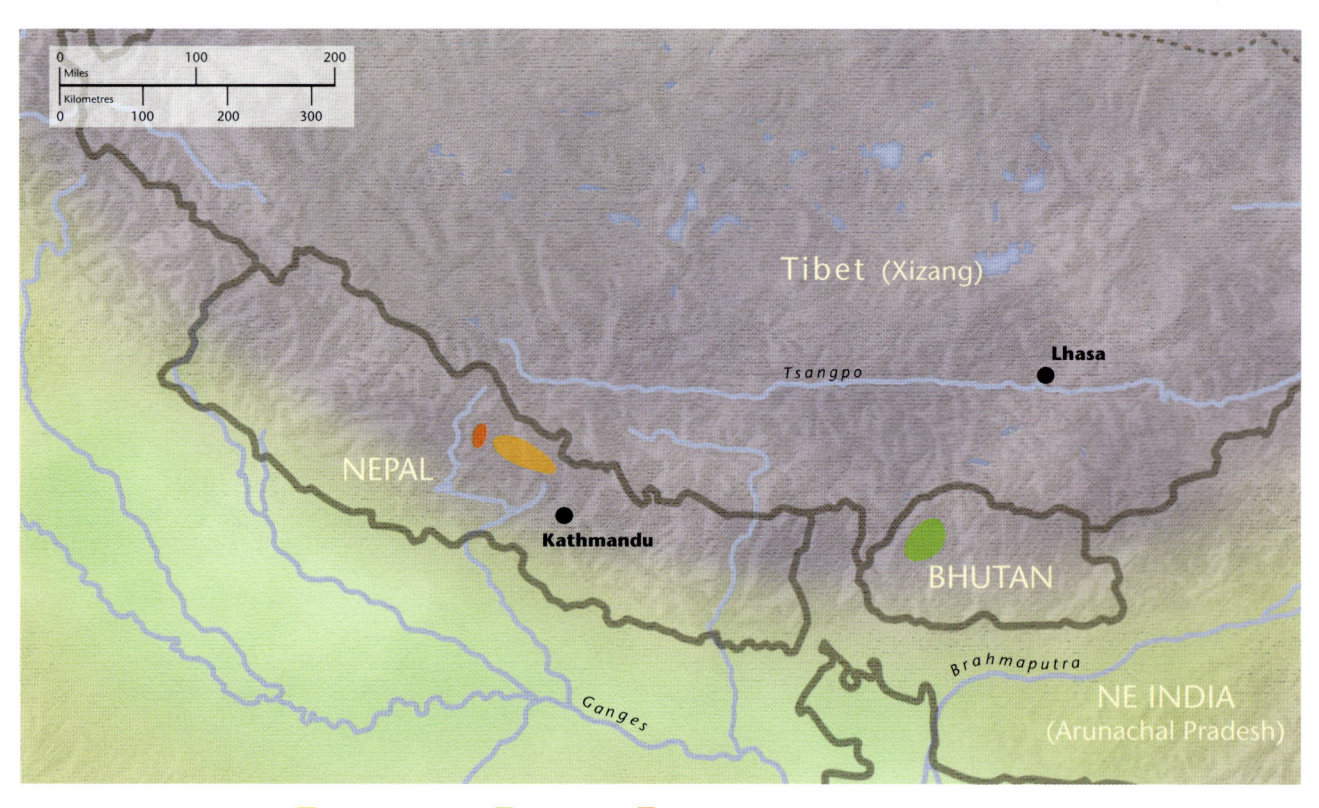

MAP 2. Species distribution: ■ *Meconopsis regia*, ■ *M. superba*, ■ *M. taylorii*.

called red-flowered forms of *M. regia* are to be found in the wild, primarily in the western Lamjung Himal, Nepal, and a number of collections were made by the team of Stainton, Sykes and Williams in the 1950s. S, S & W 5998 (BM, E), 6028 (BM, E) and 6282 (BM, E), all collected in the vicinity of Rambrong in the Lamjung Himal, are red-flowered forms with a brown stigma. Seed was collected under the number 8620 and introduced into cultivation in 1954. Plants flowered successfully but later generations were swamped in a hybrid swarm (presumably with *M. paniculata* and *M. staintonii*), as has already been noted had the yellow-flowered *M. regia*. As a result, today none of these plants are in cultivation, although many hybrids purporting to be *M. regia* are in circulation but they invariably have hybrid tell-tale signs, partly lobed leaves and green stigmas and petals varying from pure yellow to pink, red or even bicoloured.

Stainton, Sykes & Williams clearly had doubts about the red-flowered plants that they came across in the wild. S, S & W 5998 (two sheets BM, one sheet E) was cited by the collectors as a possible hybrid between M. 'napaulensis' [*M. staintonii*!] and *M. regia*. The locality in the Lamjung Himal is to the east of established *M. staintonii* localities and to the immediate west of the upper

Meconopsis regia; summer rosette; E Rupina La, Gorkha Himal, C Nepal, c. 3500 m.
PHOTO: ALAN DUNKLEY

Meconopsis regia; S Rupina La, Gorkha Himal, C Nepal, c.3150 m, flowering in mid July. PHOTO: ALAN DUNKLEY

1. MECONOPSIS REGIA

Marsyandi valley and adjacent to the Annapurna Himal. The flowers are cited as "petals red, anthers orange". The plant has large, scarcely lobed lower leaves more akin to the red-flowered *M. taylori* found in the upper Seti and Mardi Kholas (southern Annapurna Himal), both reasonably close to the S, S & W locality. However, the Lamjung Himal is also the *locus classicus* for the yellow-flowered *M. regia* Tayl. In addition, the yellow-flowered *M. paniculata* (D. Don) Prain is also recorded from the Lamjung Himal (i.e. S, S & W 5996 & 8648) and the possibility of hybrids between two or more of these species cannot be ruled out. This matter requires further careful analysis. However, the trail does not end here, for *M. staintonii* has been collected in the upper Mardi Khola (S, S & W numbers 8461, 8479, 8480, 8482 & 8483), close to the *locus classicus* of *M. taylorii*. There is the intriguing possibility that the so-called 'red regias' are in fact hybrids and that the collections represent a hybrid swarm between *M. regia* and *M. taylorii*, perhaps also with the inclusion of *M. staintonii*. There is no evidence as far as I can judge that the widespread *M. paniculata* is involved; any hybrids between it at *M. regia* would be yellow-flowered and none of the plants in the supposed hybrid swarm are yellow.

The collections involved reveal an interesting array of characters. They tend to be robust plants and, while the leaf-laminas of both *M. taylorii* and *M. regia* are entire with a serrated margin, or slightly lobed towards the base in the former, those of the 'hybrid plants' are much more variable from entire and scarely toothed to partly and shallowly lobed and toothed or untoothed, to those that are more regularly lobed halfway to the midrib in the lower half of the lamina, less markedly so in the

Meconopsis regia; S Rupina La, Gorkha Himal, C Nepal, c. 3500 m.   PHOTO: ALAN DUNKLEY

Table 3
COMPARISON OF MORPHOLOGY OF THE SPECIES OF SECTION MECONOPSIS AND THEIR PUTATIVE HYBRID

CHARACTERS	*M. regia*	*M. staintonii*	*M. taylorii*	Putative hybrid
Leaf lamina of basal leaves	entire, finely toothed	with pronounced lobes	entire, subentire to coarsely toothed,	entire, subentire to coarsely lobed, entire occasionally slightly lobed towards the base
Fruit capsule (mm)	18–45 × 12–16	22–35 × 8–13	35–60 × 11–22	35–40 × 14–15
PERSISTENT STYLE				
Length (mm)	7–14	7–14	12–16	11–13
Diameter at base (mm)	3–5	2–3	3.5–7	3–6
Capsule valves	7–12	7–11	7–17	9–14

THE GENUS MECONOPSIS | 63
1. MECONOPSIS REGIA

PLATE 1

Meconopsis regia: the original painting by Stella Ross-Craig, used as the basis of the lithograph in *Curtis's Botanical Magazine* t. 9438

THE GENUS MECONOPSIS
1. MECONOPSIS REGIA

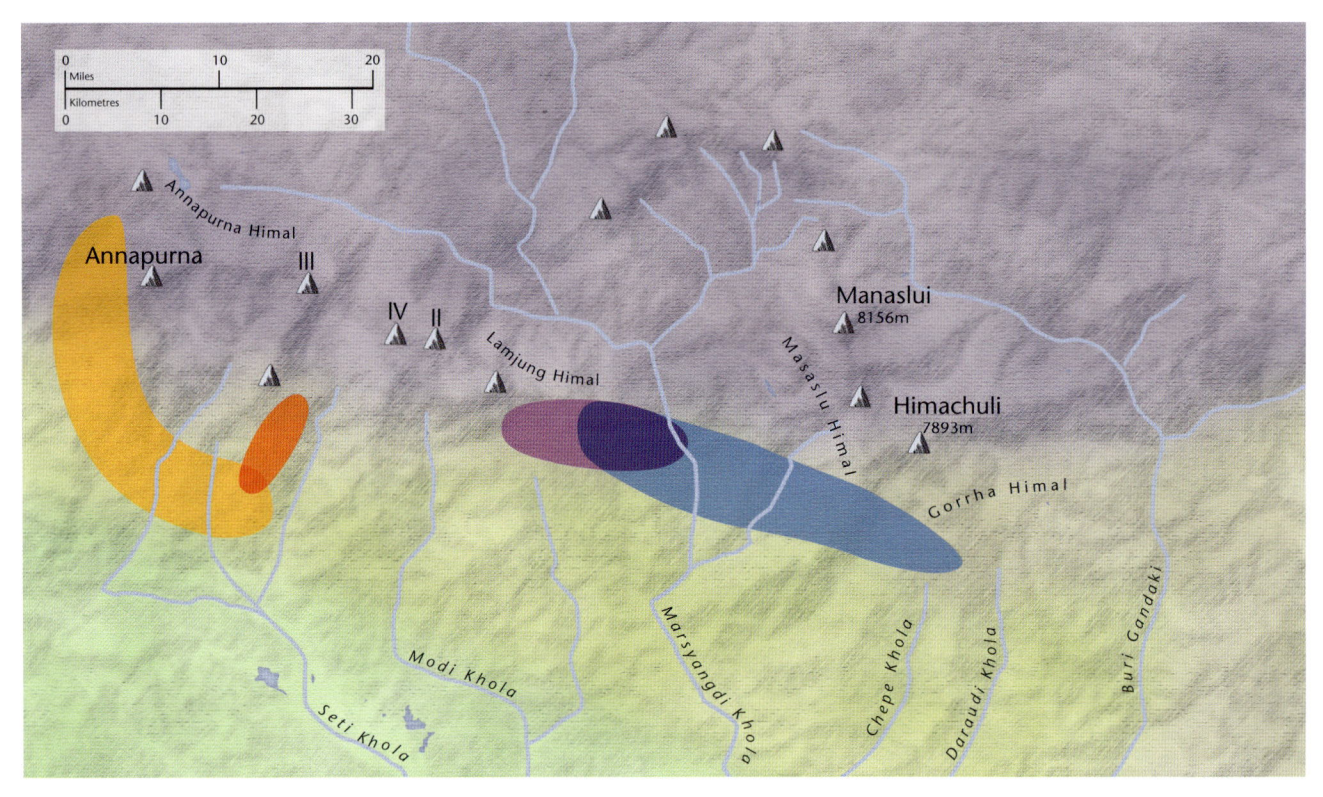

MAP 3. The speculative distribution of putative hybrids in the Lamjung Himal: ■ *Meconopsis regia*, ■ *M. taylorii*, ■ *M.* putative hybrids, ■ *M. staintonii*.

upper part. *M. staintonii* bears deeply lobed leaves and the influence of this species cannot be ruled out. The pubescence also varies greatly: while most of the plants in question bear a fairly dense layer of short barbellate hairs, the addition of much longer barbellate hairs varies from sparse to quite dense from one collection to another (listed below), but nowhere near as dense as in *M. regia*. Flower colour can also be indicative of hybridity in *Meconopsis*: S, S & W 6282 bears "petals red paler at base" and purported to have been found "growing with yellow and intermediate 6281"; 6281 "petals yellow, pink towards tips" bearing leaves like a coarsely toothed *M. regia*.

If these plants represent hybrids, which in my view is highly likely, then it is only the third record of *Meconopsis* species hybridising in the wild (see also *M.* ×*cookei* and *M.* ×*kongboensis*) and the only instance in the Himalaya. All the sightings (see below) are apparently only recorded in the vicinity of Rambrong (28°22'N, 84°12'E) in the Lamjung Himal. It has to be stressed that neither *M. staintonii* nor *M. taylorii* have been found in the Lamjung Himal, being found further west in the southern part of the Annapurna Himal and any suggested hybrids must be considered to be putative at present. However, the west-east distance between the two is only about 20 km and the very rugged high ridges that dissect the area inbetween have not been much explored botanically. While the Lamjung Himal is contiguous with the Annapurna Himal, they are separated from the Manaslu and Gorkha Himals to the east by the long deep valley of the Marsyandi river (Map 3).

It is interesting to compare some details of the species and the putative hybrid (see Table 3).

SPECIMENS SEEN (of the putative hybrid). *Stainton, Sykes & Williams* 5998 (BM, K), 6028 (BM), 6281 (BM), 6282 (BM, E, K), 8620 (BM, E), 8648 (BM) (Photo, p. 68).

Stainton, Sykes & Williams 6281 (BM, E, K) appears to represent an intermediate type that was collected in July 1954 in the Lamjung Himal along with *Stainton, Sykes & Williams* 6282 with deep red flowers. *Stainton, Sykes & Williams* 6281 (BM) is recorded as having pale yellow flowers with the petals pink towards the margins. Such plants may represent simple intermediates between the two colour forms of *Meconopsis regia*, or they may be more complex introgressive hybrids between *M. regia* and *M. taylorii* but this is pure conjecture at the present time and requires a proper detailed investigation in the field. However, *M. taylorii* has never been found in the Lamjung Himal to my knowledge, although it is found in the adjacent district of Kaski to the west. The only other species that could be involved is *M. staintonii* which is common in the Kali Gandaki valley area and which ventures east into Lamjung district: this species can have red, pink or white flowers. Again there is no evidence that it grows in proximity to *M. regia*, indeed this seems very unlikely due to the fact that the two species do not share the same habitat: *M. regia* is a plant of open, exposed mountain habitats while *M. staintonii* is a plant of wooded valleys and shrubberies, stream and river margins.

2. MECONOPSIS SUPERBA

Meconopsis superba King ex Prain, *J. Asiat. Soc. Bengal, Pt. 2, Nat. Hist.* 64 (2): 317 (1896). Type: Bhutan, Ha District, Ho-ko Chu, *Dungboo* 280 (E, K, P).

DESCRIPTION. *A bold monocarpic plant to 1.2 m tall in flower, forming a large dense, symmetrical, silvery-grey evergreen rosette to 75 cm across (to 90 cm in cultivated plants) in the year prior to flowering, although generally less. Stem erect, covered in soft silvery, appressed, barbellate hairs. Leaves silvery-sericeous with soft appressed, simple, barbellate hairs on both sides giving the surface a velvety sheen (pubescence consisting of long barbellate hairs intermixed with shorter, more markedly barbellate ones); basal leaves oblanceolate to elliptic-oblong or ovate-oblong, 20–36 × 7–9.6 cm, gradually tapering at the base and with an acute apex, the margin regularly and rather finely serrate to serrate-dentate, the petiole broad, 5–15 cm long; lower cauline leaves like the basal but shorter petioled and gradually decreasing in size up the stem, the middle and upper cauline leaves similar but sessile and with an auriculate base. Inflorescence more or less a simple raceme borne on the upper half of the stem, rarely with more than 30 flowers, the lower flowers born in lateral cymules of 3, occasionally 4–5, the middle and upper flowers borne singly from the axils of the bracts. Bracts similar to the upper leaves, decreasing in size up the inflorescence; flowers lateral-facing, deeply cupped, 5–6 cm (to 9 cm when flattened) across. Pedicels fairly stout, 5–14.5 cm long (to 25 cm long in fruit), appressed-sericeous. Buds ovoid-elliptic, appressed sericeous, ascending to half-nodding. Petals 4 (sometimes 6 in the terminal flowers), white, obovate to suborbicular, 48–62 × 44–62 mm, somewhat ruffled at the margin. Stamens numerous, with filiform, white filaments and deep yellow anthers. Ovary ovoid to ellipsoid-oblong, densely appressed-tomentose, terminating in a rather stout 7–9 mm long style with a prominent deep purple or purple-black capitate stigma, exceeding the stamens. Fruit capsule oblong to ellipsoid-oblong, 36–50 × 15–22 mm, 9–16-valved normally,*

Meconopsis superba; Tshongsikha, Haa La, W Bhutan, 4100 m. PHOTO: DAVID & MARGARET THORNE

Above left: *Meconopsis superba*; Tshongsikha, Haa La, W Bhutan, 4100 m. PHOTO: DAVID & MARGARET THORNE
Above right: *Meconopsis superba*; leaf-rosette (winter) in cultivation. PHOTO: CHRISTOPHER GREY-WILSON

prominently ribbed, appressed-tomentose, the style 9–11 mm long, 3–4 mm diameter at the base. *Seeds* more or less reniform with a finely reticulated surface. Figs 3D, 5B, 6A.

DISTRIBUTION. CW Bhutan (Ha La and Kyu La); 3900–4300 m (Map 2).
HABITAT. Rocky slopes, screes, generally amongst scattered shrubs.
FLOWERING. June–August.
SPECIMENS SEEN. *Dungboo* 280 (E, K, P); *Gould* 1381 (K); *Ludlow & Sherriff* 93 (BM, E), 3573 (E); *Ludlow, Sherriff & Hicks* 17562 & 17562A (BM, E), 19617 (BM, E).

Meconopsis superba is endemic to Bhutan where it is restricted to, as far as it is known, a relatively small area of the western part of the country centred upon Ha district. The original specimen was collected at 'Ho-Ko-Chu' which was sometimes given as Chumbi but it is generally now thought to be from the Ha district. The Chumbi valley lies in southern Tibet, squashed between north-eastern Sikkim and north-western Bhutan. There is no authenticated record of the species having been found in Tibet and the recently published Papaveraceae for the *Flora of China* does not include it.

Meconopsis superba undoubtedly finds its closest ally in the central Nepalese *M. regia*. Both species share in common the sericeous indumentum of stems and leaves and the finely serrated leaf margin; however they differ in important respects. While *M. superba* has white flowers, those of *M. regia* are yellow. The fruit details also differ, those of *M. regia* being on the whole rather smaller. Generally speaking, *M. superba* is a less substantial plant than its close cousin. In the wild the leaf-rosettes, which are silvery, silky and smooth, can reach 75 cm across, although often smaller (in cultivation they can reach as much as 90 cm). In contrast, the leaf-rosettes of *M. superba* are larger (certainly to 120 cm diameter in cultivated plants) silky and golden, the colour most pronounced in over-wintering rosettes.

It is difficult to tell how old plants are when they come into flower in the wild. In cultivation most flower in their third and fourth year from seed, but in the wild, under much harsher conditions five or six years may be the norm. Like *Meconopsis regia*, both species inhabit exposed regions above the tree line and plants will be buried in a protective blanket of snow during the winter months. In contrast, in cultivation plants can be subject to frost damage through a lack of winter snow, and this applies particularly to plants as they come into growth, especially after mild snowless winters, so that some form of winter protection is advisable, this kept in place until the spring frosts have relented. Plants are slow to germinate from seed, longer than that of most other species, even when sown fresh. Thin sowing is essential as the young seedlings are very prone to damping-off and other fungal infections. Even when well grown, plants are unlikely to come into flower until their third or fourth year. The inflorescences arise quite early in the year and at this stage the young leaves and flower buds are particularly prone to frost damage. Plants that have flowered well usually set copious seed, especially from fruits set on the lower part of the infructescence. *M. superba* received an Award of Merit (AM) in 1940 when exhibited at the Royal Horticultural Society in London.

The history of the species is an interesting one. It was described in 1896 from fragmentary specimens made by a native collector, Dungboo, in 1884 and, at that time said to be from Chumbi (see above). After that nothing was known of the species; however, the plant came into cultivation mysteriously in the 1920s from an unknown source, the first flowering taking place in 1927. Six years later, in 1933, that indefatigable team of Frank Ludlow and George Sherriff (accompanied by Hicks on occasions) made a series of collections in western Bhutan including some fine herbarium specimens of

Right: *Meconopsis superba*, details of flowers; Tshongsikha, Ha La, W Bhutan, 4100 m. PHOTO: DAVID & MARGARET THORNE

Below left: *Meconopsis superba*; Tshongsikha, Haa La, W Bhutan, 4100 m. PHOTO: DAVID & MARGARET THORNE

Below right: *Meconopsis superba*, mature dehisced fruit capsule; Tshongsikha, Ha La, W Bhutan, 4100 m. PHOTO: DAVID & MARGARET THORNE

M. superba which provided the basis of our knowledge. Further collections were made by Ludlow & Sherriff in 1937 and 1949 also in Ha district. In recent times the species has been photographed by a number of trekkers in the region including a Japanese tour group in June 2006 (one member Hideo Takahashi took some fine photographs of it in flower) and these show what a distinct and beautiful species it is. *M. superba* is reasonably well established in cultivation today, the plants stemming primarily from those early collections of Ludlow & Sherriff who managed to gather an ample supply of seed from the wild. Its status in the wild is questionable: human activity, especially grazing, in the area has put the population under great stress and it must now be considered to be very vulnerable.

Toshio Yoshida has reported plants seen in Ha district bearing pinkish or pale yellow flowers and has attributed these to hybrids between *Meconopsis superba* and *M. paniculata* which is very common throughout mountainous Bhutan; however, these plants may be attributable to the pink- and red-flowered forms of *M. paniculata* found in other parts of Bhutan and Arunachal Pradesh in NE India. This requires further close examination. It is very easy to assign variants seen in the field to hybrids but this is often not so and often difficult to prove otherwise without detailed field work and laboratory investigation.

3. MECONOPSIS TAYLORII

Meconopsis taylorii L. J. H. Williams, *Trans. Bot. Soc. Edinburgh* 41: 347 (1972). Type: C Nepal, Annapurna Himal, Seti Khola, 4115 m, 2 Aug. 1954, *Stainton, Sykes & Williams* 6593 (BM, holotype; BM, E, isotypes).

DESCRIPTION. *A large monocarpic plant* 1.2–1.8 m tall in flower, forming a substantial, regular, grey-green, evergreen rosette, to 1 m across, sometimes more, in the years before flowering, most parts of the plant covered in a medium to rather dense felt of long, straw-coloured, barbellate, 10–15 mm long, hairs underlain with a short indumentum of similar hairs that appear to be substellate, these 0.5–1 mm long. *Basal leaves* numerous, dense, the lamina elliptic-oblong to oblong-oblanceolate, 38–60 × 8–16.5 cm, narrowed at the base into a broad, winged petiole, 12–25 cm long, the apex acute to subacute, the margin subentire to finely and regularly toothed, sometimes more coarsely so, or somewhat lobed, the lamina covered on both surfaces with long barbellate hairs, especially dense along the midrib beneath (these hairs up to 15 mm long); cauline leaves smaller than the basal, gradually decreasing in size up the stem and merging with the bracts, elliptic-oblong, 24–41.5 × 6.2–11 cm, with a coarsely toothed margin which generally becomes shallowly lobed towards the semi-clasping leaf base, pubescent like the basal leaves, only the lowermost petiolate. *Inflorescence* a pyramidal panicle occupying the upper half of the plant or sometimes most of its height, with the lower flowers in cymules of 2–5 normally, while the upper flowers are solitary or sometimes paired; flowers deeply cupped, 5.5–8 cm across, horizontal to half-nodding. *Bracts and bracteoles* like the upper cauline leaves but increasingly smaller up the inflorescence, sessile, toothed to subentire. *Peduncles* 18.5–24.5 cm long, elongating in fruit, covered with long barbellate hairs with a short substellate underlay. *Pedicels* 8.5–12.6 cm long (elongating up to 21.5 cm in fruit), pubescent like the peduncles, especially densely so towards the top. *Buds* nodding, subglobose to ovoid, 22–30 mm long. *Petals* 4, pale to deep pink or dark red, obovate to almost rounded, 30–51 × 30–42 mm, slightly recurved at the margins. *Stamens* numerous, with white filaments and orange anthers. *Ovary* ovoid, covered with dense ascending to appressed, barbellate bristles, the style 5–10 mm long, terminating in a dark brown, capitate stigma, 5–8 mm diameter. *Fruit capsule* oblong-clavate, 35–60 ×

Meconopsis aff. *taylorii*; Marci Bugli above Taglung, Kali Gandaki Valley, C Nepal, c. 4000 m. PHOTO: JAN BURGEL

11–22 mm, strongly ribbed, 7–17-valved, covered with ascending barbellate bristles at first but these mostly breaking off near the base at maturity, but leaving a substellate underlay, the persistent style ribbed, 12–16 mm long, 3.5–7 mm wide at the base, with short barbellate hairs, especially towards the base. *Seeds* black, oblong, 1–1.2 mm long, 0.6–0.7 mm wide, finely tuberculate. (Fig. 5G; 6C)

DISTRIBUTION. C Nepal (Kaski District; Annapurna Himal, upper Mardi (Madi) and Seti Kholas; 3658–4575 m (Map 2).

HABITAT. Open grassy and rocky slopes, often amongst boulders.

FLOWERING. July–August.

SPECIMENS SEEN. *Stainton, Sykes & Williams* 6525 (BM, E), 6554 (BM), 6593 (BM, E), 8459 (BM), 8505 (BM), 8506 (BM, E), 8507 (BM), 8611 (BM), 8612 (BM).

The field notes accompanying the type collection read: "On open slopes. Height 4–5 ft. Leaves pale green. Whole plant with straw coloured hairs. Petals pink. Filaments white, anthers orange. Stigma dark brown."

Meconopsis taylorii was named in honour of Sir George Taylor in recognition of his considerable contribution to the study and classification of the genus, culminating in his monograph *The Genus Meconopsis* published in 1934. This work had been initiated while George Taylor was at the Royal Botanic Garden Edinburgh and continued after his appointment as Keeper of Botany at the British Museum (Natural History). While at the Museum he had greatly promoted the botanical exploration of the Nepalese Himalaya which resulted in a number of important expeditions in the 1950s and 1960s, those of Stainton, Sykes and Williams and Polunin, Sykes and Williams perhaps being the best known, but there were many others. As a result of these expeditions a great deal more information on *Meconopsis* was gleaned.

All the known gatherings of *Meconopsis taylorii* were made on a knife-edged southern spur of the famous Fishtail Mountain, Machhapuchare, to the north of Pokhara, central Nepal, in the upper reaches of two rivers the Mardi (Madi) Khola and the Seti Khola (Mardi Khola-Seti Khola confluence), by the team of Stainton, Sykes and Williams during their 1954 Nepal expedition. Seed collections were made under the numbers 8459, 8506, 8507, 8611, while the following collections had leaves added from adjacent seedling rosettes, 8459 and 8506. Since then no further herbarium material or seed collections have been made. However, intriguingly, the Czech Jan Burgel, led an expedition to the Dhaulagiri-Annapurna region in 2000 and photographed several plants of a *Meconopsis* placed at the time under the umbrella name of *M. napaulensis*. These were located on the flanks of Mt Nilgiri above the village of Taglung at about 4000 m and, although herbarium specimens were not gathered, the photos are probably of *M. taylorii* in flower and young fruit in early July. Interestingly, this location lies to the north-west of Machhapuchare, the *locus classicus*, with the peak of Annapurna itself and Annapurna South separating the two localities.

Meconopsis taylorii is a little known but distinctive species closely allied to *M. regia* but with coarser, sometimes shallowly lobed leaves, pink or red flowers. It is also distinguished in bearing the largest fruit capsules known in the genus, these latter being strongly ribbed and very distinctive at maturity. It is, as far as it is known at present, restricted to a relatively small area of the Annapurna Himal in central Nepal; however, it may be present further to the north-east in the Lamjung Himal: see the note under *M. regia*.

Interestingly, the pollen of *Meconopsis taylorii* shows a surprising divergence from that found in either *M. regia* or *M. superba*. In his 1965 paper (*Grana Palynologica* 6: 191–209) Douglas Henderson reveals this unexpected diversity. The pollen grains of *M. taylorii* are spheroidal, about 25 μ diameter, without apertures and covered with conical spinules of varying size; the pollen type fits into Henderson's Betonicifolia-type. In contrast, those of *M. regia* and *M. superba* are of the Primulina-type, being 3-6-colpate and covered in minute spinules. This disparity is difficult to explain but it may be, as James Cobb (1989) remarks, that "if closely related species that are interfertile have radically different pollen then the characteristics of the pollen are a later and secondary development and tell nothing about relatedness".

As has been stated seed was introduced into cultivation by Stainton, Sykes and Williams, the discoverers, in the 1950s under various numbers (see above) but, although plants were brought to maturity, the species did not persist in cultivation for very long. This is a great pity as this is quite clearly a very imposing plant and there have been no subsequent introductions, indeed as far as I am aware no botanists have ventured to the type locality since its first discovery.

The area of the Annapurna Himal where *Meconopsis taylorii* was discovered requires further exploration and a careful analysis in the field: the area is in fact not very far from Pokhara in central Nepal so that access should be possible. The link of this area to the adjacent Lamjung Himal to the east would repay further exploration, especially as far as *Meconopsis* is concerned. The relationship between *M. regia*, *M. staintonii* and *M. taylorii* certain requires detailed field analysis.

section POLYCHAETIA

This section, although clearly very closely related to section *Meconopsis*, is readily distinguished by the character of the leaves which, instead of being finely toothed along the margin, are deeply lobed often bipinnatifid to bipinnatisect, the margin toothed or not. In addition, the leaves are covered in a sparse to dense indumentum of spreading barbellate hairs or bristles, underlain or not by shorter hairs of a similar type, this sometimes forming a thick underfelt, especially so when young.

KEY TO SERIES OF SECTION POLYCHAETIA

1. Leaves with a sparse to dense indumentum of simple, more or less uniform, barbellate hairs, sometimes glabrescent at maturity . . . . series **Robustae**
+ Leaves with an indumentum of long simple, barbellate hairs underlain with short simple or branched hairs, often densely so . . . . series **Polychaetia**

series Robustae

The four species in this series are characterised by having hairs only of one type, which are especially noticeable on the leaves; these are long and finely barbellate. While three of the species bear yellow flowers, the fairly recently described western Nepalese *Meconopsis chankeliensis* has red flowers. Of the species *M. dhwojii* is the smallest, attaining no more than 60 cm in height in flower, while the others can easily reach a metre, occasionally more.

KEY TO SPECIES OF SERIES ROBUSTAE

1. Flowers red or purple; basal leaves often bipinnatifid . . . . .7. **M. chankheliensis** (W Nepal)
+ Flowers yellow, rarely white; basal leaves pinnatifid to pinnatisect 2

2. Plants 120–180 cm tall in flower, the flowers borne singly on lateral pedicels, the inflorescence confined to the upper half of the plant . . . . 4. **M. robusta** (N India — Kumaon, W Nepal)
+ Plants rarely exceeding 90 cm tall in flower, the flowers mostly borne in 2–3-flowered cymules in the inflorescence from the base of the plant upwards 3

3. Leaves and stems sparsely bristly, the bristles uniformly pale, the leaves with 3–5 pairs of lateral segments; fruit capsule narrow-oblong or flask-shaped, the style 6–10 mm long . . . . .5. **M. gracilipes** (C Nepal, S Tibet)
+ Leaves and stems densely bristly, the bristles with a purple-black base, the leaves with 6–9 (–10) pairs of lateral segments; fruit capsule broad-oval or oval-oblong, the style 3–6 mm long . . . . 6. **M. dhwojii** (E Nepal, S Tibet)

4. MECONOPSIS ROBUSTA

Meconopsis robusta Hook. f. & Thomson, *Fl. Ind.* 1: 253 (1855) and in Hook. f., *Fl. Brit. Ind.* 1: 118 (1872) *pro parte*. Type: N India, Garwhal, Kumaon, *Wallich* 8124 (K, holotype; BM[4], E, K, P, cotypes).

DESCRIPTION. *A monocarpic plant* to 1.5 m in flower, although generally less and as little as 60 cm, forming a medium-sized grey-green leaf-rosette to about 50 cm across in its first two or three years, most of the plant covered in simple, often rather sparse, barbellate hairs, becoming semi-glabrescent as the hairs generally break off eventually to leave a hispid base. *Basal leaves* rather lax and fern-like, elliptic to elliptic-oblong in outline, the lamina 14.5–30 × 4.8–7.5 cm, pinnatisect in the lower part but pinnatifid towards the apex, with 3–6 (–7) pairs of elliptic-oblong to elliptic-oblanceolate, acute to

---

[4] There is an unnumbered *Wallich* specimen in the British Museum (Natural History) inscribed *Meconopsis robusta* by Joseph Dalton Hooker, which closely matches the specimen under the number 8124 in the Wallich Collection at Kew. I see no reason why it should not be considered to be a cotype and have accorded it such status above. In addition, there is a sheet marked *Wallich* '1836' and a second sheet marked 8124 (the same as the number assigned in the Wallich Herbarium) in the main Kew Herbarium.

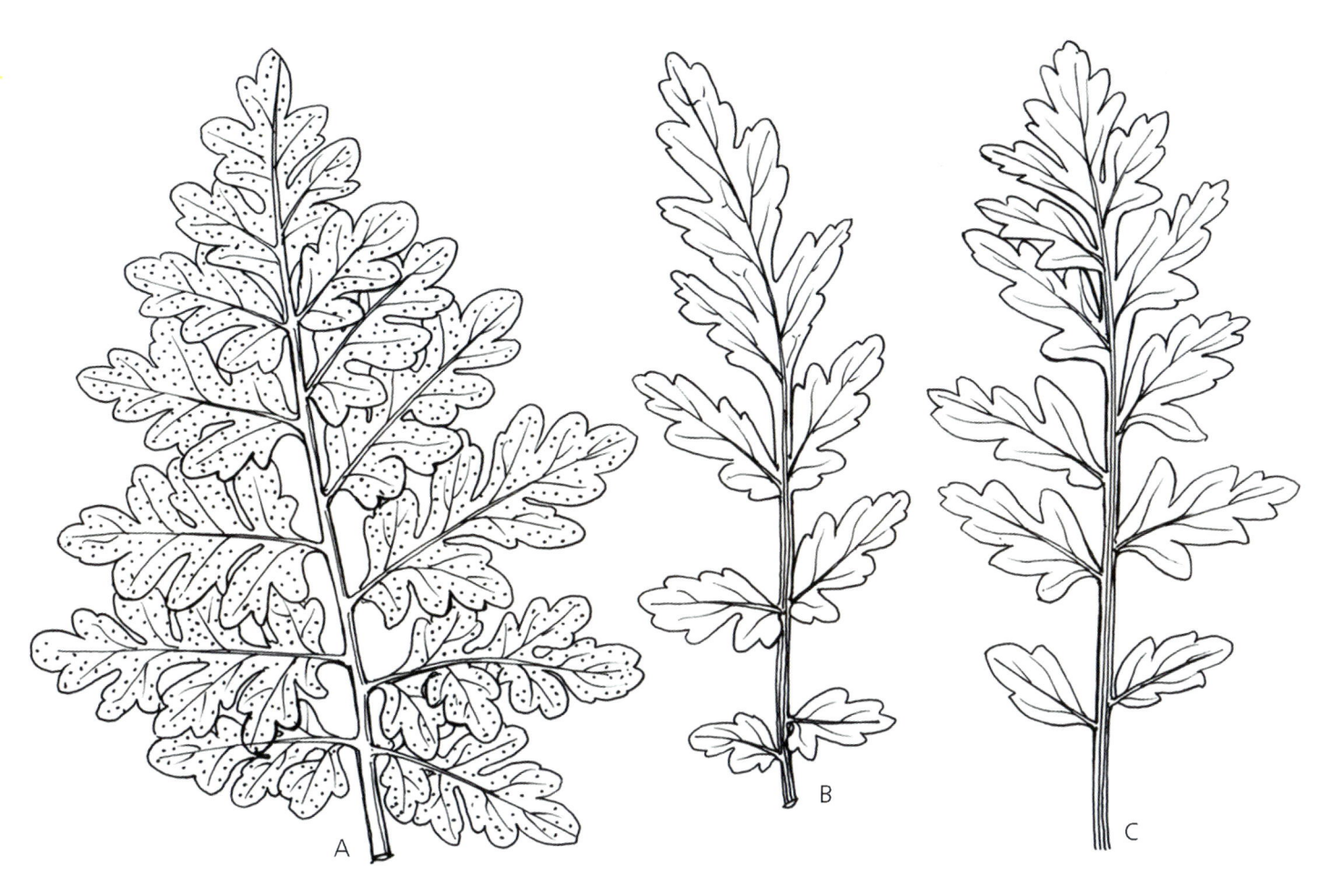

FIG. 7. Basal leaves, lower petioles and pubescence not shown: **A**, *Meconopsis dhwojii*; **B**, *M. gracilipes*; **C**, *M. robusta*. All × ²/₃. DRAWN BY CHRISTOPHER GREY-WILSON

subacute, segments, the lower pairs rather distant from one another, these generally with a few short lobes or coarse teeth (often 3–5 on each side), those towards the leaf apex often entire, the lamina narrowed below into a slender petiole 8.5–15 cm long; cauline leaves similar to the basal but increasing shorter-petioled up the stem, the uppermost sessile and merging with the bracts, semi-amplexicaul due to the lowermost leaf-segments. *Inflorescence* racemose, occupying the upper half to one third of the stem, the flowers bowl-shaped, half-nodding to nodding, 5–6 cm across. *Pedicels* 3.6–8.2 cm long, patent hispid with barbellate hairs. *Buds* nodding, ovoid to ovoid-oblong, covered in long ascending bristles. *Petals* 4, cream to sulphur-yellow, obovate, 28–40 × 17–30 mm, ruffled at the margins, sometimes slightly erose. *Stamens* numerous, about one third the length of the petals, with filiform filaments the same colour as the petals, 9–11 mm long, and yellow or orange-yellow anthers, 1.5–2 mm long, browning with age. *Ovary* ellipsoid to obovoid, densely covered in appressed to ascending barbellate bristles, the style 5.5–7 mm long, terminating in a capitate cream or yellow stigma, c. 2.5

Meconopsis robusta; Api Himal, upper Chamilaya Nadi, Darchula District, W Nepal, 3375 m.
PHOTO: COLIN PENDRY

mm diameter. *Fruit capsule* ellipsoid to ellipsoid-obovoid, 18–28 × 9–13 mm, covered in rather sparse ascending to semi-appressed bristles or their persistent bases, splitting by 6–9 short valves for a short distance from the top at maturity, terminating in a 7–10 mm long persistent style which broadens towards the base to 2–5 mm diameter. Figs 7C, 8E & F, 9J1,2.

DISTRIBUTION. N India (Kumaon, centred on the Pindari Valley), W Nepal (Bajhang, Doti & Himal districts; almost certainly in Baitadi and Darchula districts as well); 2440–4100 m (Map 4).

HABITAT. *Abies* woodland, shrubby and rocky meadows.

FLOWERING. June–August.

SPECIMENS SEEN. *F. M. Bailey* s.n. 1936 (E); *Mrs G. E. Benham* s.n. 1924 (BM); *Col. Biskam* s.n. 23/06/1933 (E); *Ikeda et al.* 20913119 (E, KATH), 20917021 (E), 20917053 (E, KATH), 20919047 (E, KATH); *Rajbhandari & Roy* 4112 (KATH); *Range Officer* s.n. 'Pindari' (2 sheets) (BM); *Shakya, Sharma & Amatya* 6300 (KATH); *J. L. Stewart* 636 (K); *Strachey & Winterbottom* 1 (K); *Wallich* 8124 (BM, E, K), s.n. (BM).

Meconopsis robusta is the most westerly species in subgenus *Meconopsis*, those monocarpic species with handsome evergreen leaf-rosettes. It is a little known species formerly believed to be endemic to Kumaon (part of the state of Uttar Pradesh) in N India but it has been recorded from a number of sites in the extreme west of Nepal in recent years. This is scarcely surprising as the two areas lie adjacent to one another. *M. robusta* is often confused with *M. paniculata*, differing most clearly in its racemose rather than paniculate inflorescences, but also significantly in the simple indumentum (i.e with hairs all of one type) of the leaves and stems. *M. robusta*, however, is an altogether slighter plant, the flowers smaller with petals that do not, or only slightly, overlap. Furthermore, the flowers are generally pendent rather than lateral-facing or semi-pendent.

As far as affinities go *Meconopsis robusta* finds its closest ally in the central Nepalese *M. gracilipes*, differing primarily in the simple, non-paniculate inflorescence and in the larger, relatively broader fruit capsules beset with ascending to semi-appressed, not reflexed, bristles. Both species and the related *M. dhwojii* bear yellow flowers and an indumentum of rather uniform barbellate hairs.

Meconopsis robusta is occasionally seen in collections in cultivation, although some of those that I have examined appear to be *M. paniculata* or perhaps hybrids between the two species. Being a slighter and less floriferous plant it is unlikely to become as widely grown as some of the other evergreen rosetted species, although it will always find a place in specialist collections and botanic gardens.

Plants form a rather modest leaf-rosette compared with some of its allies, being rarely more than 50 cm across in cultivation, this despite the fact that a robust flowering plant can reach beyond 1.5 m in height. In cultivation it is generally dismissed as a poor cousin of *M. paniculata*, however despite is size, it has a certain grace and charm

Meconopsis robusta; Khaptad National Park, Sukhedaha to Ghodadaune, Doti District, W Nepal, 3075 m. PHOTO: COLIN PENDRY

Meconopsis robusta; Api Himal, upper Chamilaya Nadi, Darchula District, W Nepal, 3375 m. PHOTOS: COLIN PENDRY

as recent photographs taken in the west of Nepal have revealed. It is said to tolerate rather drier conditions in cultivation than some of its brethren, coming as it does from the relatively drier western end of the Himalaya; however, it certainly requires ample moisture during the summer months and, like all the evergreen monocarpic species, a compost rich in organic matter. Seed is sometimes offered by some of the Indian seed houses that collect seed from the wild. In gardens it is prone to hybridise with its allies if grown in close proximity. Records suggest that *M. robusta* comes into flower often in its third year from seed.

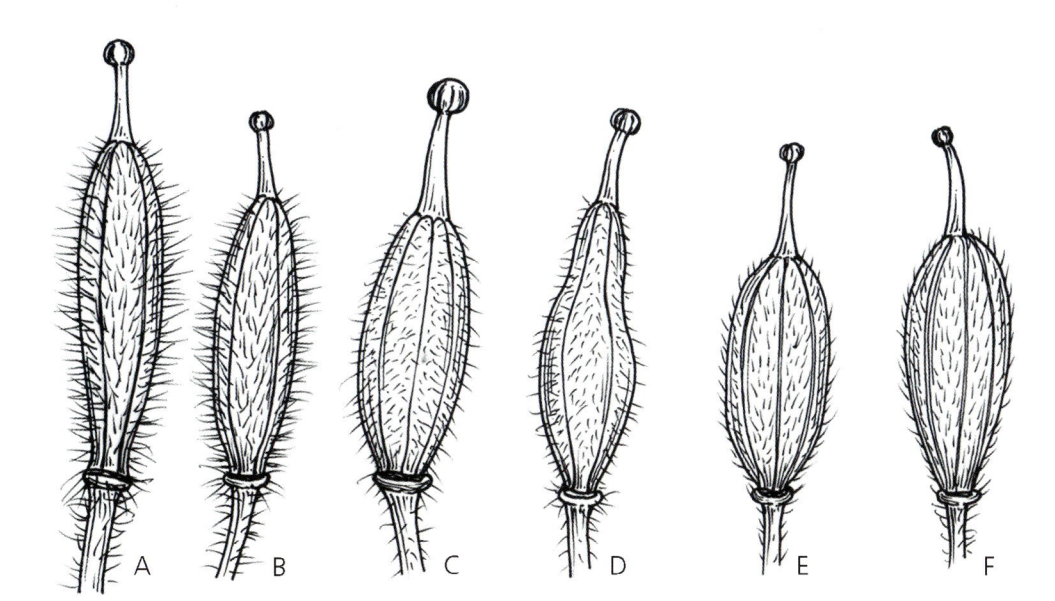

FIG. 8. Fruit capsules: **A–B,** *Meconopsis dhwojii*; **C–D,** *M. gracilipes*; **E–F,** *M. robusta*. All × 1⅓.
DRAWN BY CHRISTOPHER GREY-WILSON

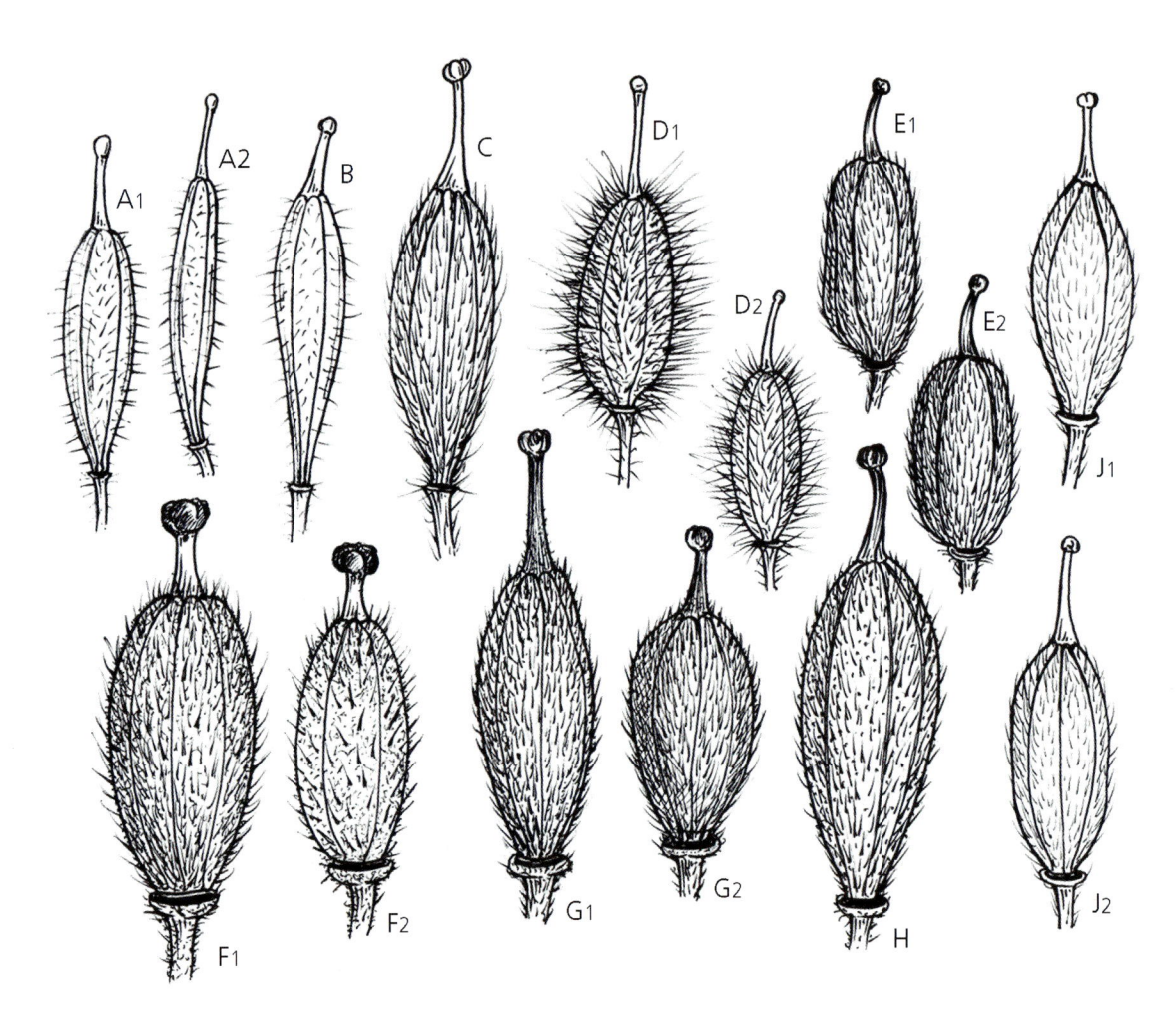

FIG. 9. Fruit capsules: **A1–2**, *Meconopsis napaulensis*; **B**, *M. ganeshensis*; **C**, *M. staintonii*; **D1–2**, *M. wallichii*; **E1–2**, *M. wilsonii* subsp. *australis*; **F1–2**, *M. autumnalis*; **G1–2**, *M. paniculata* subsp. *paniculata*; **H**, *M. violacea*; **J1–2**, *M. robusta*. All × $1^1/_3$. DRAWN BY CHRISTOPHER GREY-WILSON.

5. MECONOPSIS GRACILIPES

Meconopsis gracilipes G. Taylor, *The Gen. Meconopsis*: 38 (1934). Type: C Nepal, Buri Gandaki, Khorlak (Gorlak), 1928, *Lall Dhwoj* 17 (BM, holotype).

[syn. *Meconopsis rebeccae* Debnath & M. P. Nayar, *J. Jap. Bot.* 64 (5): 157 (1989). Type: C Nepal, S of Annapurna, above Siklis, 28°07'N, 84°06'E, 28 Aug. 1976, *Troth* 947 (BM, holotype); *Troth* 980 (BM, paratype), same locality].

DESCRIPTION. *A monocarpic plant* to 1 m tall in flower forming a lax rosette in the years before flowering, with a narrow dauciform root; stem erect, sparsely to moderately covered with stiff barbellate hairs, becoming glabrescent. *Basal leaves* generally absent at flowering time, petiolate, the lamina to 22 × 5 cm, pinnatisect, with 3–5 pairs of alternate to subopposite lateral segments, these ovate to oblong, 15–33 × 14–22 mm, with 3–5 shallow lobes, the terminal lobe pinnatifid, 2.6–5 × 2–3.8 cm, the lamina sparsely covered on both sides with long barbellate hairs, these generally breaking off in mature plants to leave a stubble of hispid bases; cauline leaves gradually decreasing in size up the stem, the lower similar to the basal leaves but with increasingly shorter petioles, the upper cauline leaves sessile, with an auriculate base, merging with the bracts. *Inflorescence* subpaniculate with the upper flowers solitary but the lower in 2–5-flowered cymes; flowers semi-nodding to lateral-facing, broad bowl- or saucer-shaped, 4.5–6 cm across. *Bracts* oblong to lanceolate, 4–7.2 × 2.1–4 cm, similarly divided to the upper leaves, often rather less deeply lobed, with an auriculate base and similar pubescence. *Peduncles* 4–12.5 cm long, moderately to sparsely covered in stiff slender barbellate hairs, becoming gradually glabrescent. *Pedicels* slender, 4–5.8 cm long, pubescent as the peduncles and elongating somewhat in fruit. *Buds* semi-nodding, obovoid to subglobose, covered in short incurved barbellate bristles. *Petals* 4, pale yellow, obovate to suborbicular, 18–31 × 16–25 mm, not or only slightly

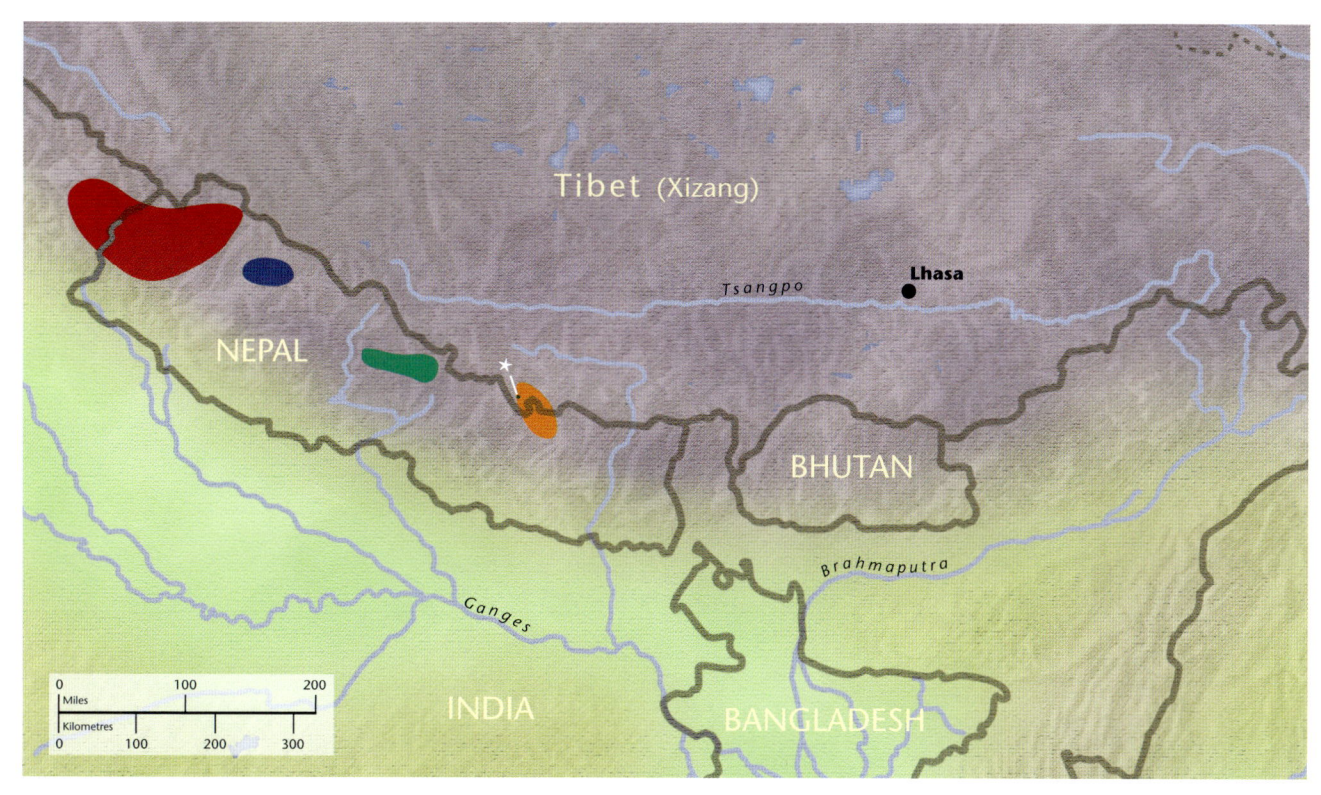

MAP 4. Species distribution: ■ *Meconopsis robusta*, ■ *M. dhwojii*, ■ *M. gracilipes* (* A single record of *M. gracilipes* from Nylam, Tibet), ■ *M. chankheliensis*.

overlapping. *Stamens* numerous, with pale yellow filaments and orange-yellow anthers, these latter browning with age. *Ovary* covered in dense appressed to ascending slightly barbellate bristles; style 4–6 mm long with a small capitate stigma. *Fruit capsule* flask-shaped to narrow ovoid-ellipsoid, broadest towards and above the base or just below the middle, 18–25 × 7–9 mm, splitting for a short distance from the apex at maturity into 4–6 (–7) valves, covered with rather sparse spreading to slightly deflexed bristles or their persistent hispid bases, and with a 6–10 mm long persistent style. *Seeds* reniform, 0.5–0.8 mm long. 2n = 56. Figs 7B, 8C–D.

DISTRIBUTION. Central Nepal (Gorkha, Kaski & Lamjung districts; Annapurna Himal, Lamjung Himal, Gorkha (Gurkha) Himal), S Tibet[5]; 2440–4085 (–5210[6]) m. Endemic to C Nepal (Map 4).

HABITAT. Rocky slopes, moist forests and forest margins, shrubberies, stream margins, generally in shade or partial shade.

FLOWERING. June–early August.

SPECIMENS SEEN. *F. M. Bailey* s.n. 1937 (E); *Dhwoj* 17 (BM, E), 20 (BM); *EKSIN* 163A (E); *Polunin* 792 (BM, E); *Sharma* E446 (BM), E450 (BM), E456 (BM) and E457 (BM); *J. A. Ratter* C5252 (L); *K. N. Sharma* 12/194 (BM); *Stainton, Sykes & Williams* 4879 (BM, E), 6417 (BM, E), 6689 (BM), 8473 (BM, E); *Troth* 947 (BM), 980 (BM).

Meconopsis gracilipes has been consistently confused with *M. dhwojii*, both in herbaria and in literature; however, the two species, although clearly closely related, are readily distinguished, not only by geography but also in various details of the leaves and fruit capsules. The basal and lower leaves of *M. gracilipes* are rather shallowly lobed, with relatively few (3–5) broad, obtuse segments, while *M. dhwojii* has more deeply cut, narrower pairs (6–9 normally). The leaves of the former tend to be a rather pale green and unmarked, those of the later deep grey-green, often purple flushed, with the base of the bristles purplish black.

Meconopsis gracilipes was formerly quite common in cultivation, but is now rare and not currently listed in the *RHS Plant Finder*. It is one of the smaller members of the *Polychaetia* and an elegant plant in flower. Its flask-shaped fruit capsules with their long styles and

---

[5] There is a single photographic record of this species taken in the region of Nyalam on the Tibet-Nepal highway just north of the Tibetan border.

[6] Many high altitude records for *Meconopsis* in Nepal are based on collections by K. N. Sharma (e.g. 12/94 (BM), 17,100 ft (5210 m)); none of these can be substantiated and they need to be viewed with some scepticism. The collector may have had a faulty altimeter or one inappropriately set.

the pale, often rather sparse, pale leaf bristles (not dark-based) readily serve to distinguish it from *M. dhwojii* which is found further east in Nepal. The two also differ significantly in the dissection of the leaves and number of divisions as stressed above. The fruit capsules of *M. dhwojii* are relatively longer and thinner, often with a shorter persistent style. One other important difference can be observed in habitat preferences: while *M. gracilipes* prefers shaded or semi-shaded places, often in forest or shrubberies, or along bushy streamsides, *M. dhwojii* prefers more open rocky habitats.

In cultivation *Meconopsis gracilipes* forms quite a large evergreen rosette up to 75 cm across, generally maturing in the third or fourth year from seed. In cultivation plants can be larger than in the wild, perhaps reaching as much as 1.2 m. Flowers are produced over quite an extended season, although only a few open at a time. It is a fine plant for part-shaded places such as woodland glades or pockets in a shrub border. The stubbornness of the species to set viable seed in cultivation has contributed to its decline and fresh seed legally collected from the wild is badly needed to replenish lost stocks. Seed of the true species is rarely on offer, if at all, today, but should it be, then the utmost care should be made to keep it away from its brethren: like all members of subgenus *Meconopsis*, *M. gracilipes* tends to hybridise readily in the garden environment and the genuine species can soon be swamped by hybrid progeny.

The *Flora of China* account includes *Meconosis gracilipes* from S Tibet (Nyalam, see footnote 4) but there is almost certainly some confusion here as both *M. gracilipes* and *M. dhwojii* are recorded from close to Nyalam. Both species are very closely related and easily confused at first glance. See under the following species, *M. dhwojii*.

6. MECONOPSIS DHWOJII

Meconopsis dhwojii G. Taylor ex Hay, *New Fl. & Silva* 4: 225, fig. 82 (1932) & *Gard. Chron.*, Ser. 3, 92: 409, figs. 198–99 (1932). Type: E Nepal, Sangmo, 12,000 ft, *Dhwoj* 0297[7] (BM, holotype, isotypes).

DESCRIPTION. *A small monocarpic plant* reaching 60 cm tall in flower, forming a rather dense rosette to 30 cm across, with a dauciform taproot. *Leaves* grey-green, often flushed with purple, especially in the autumn and winter, 16–33 × 4.2–10.3 cm, including the petiole, covered with rather sparse straw-coloured bristles that arise from a purplish-black somewhat raised base, the bristles 6–11 mm long; basal leaves bipinnatisect with generally 6–9 (–10) primary pairs of lateral segments, the lowermost segments rather distant, the uppermost more

Meconopsis gracilipes; S Tibet, Kathmandu-Lhasa highway close to Nepalese border. PHOTO: DAVID HASELGROVE

Meconopsis dhwojii; two year old leaf-rosette in cultivation. PHOTO: CHRISTOPHER GREY-WILSON

[7] There are three sheets under this number at the British Museum (Natural History), one of a specimen in flower, the other two in fruit. George Taylor noted, upon receiving Dhwoj's specimens, that the flowers were pale yellow, despite the fact that the field notes accompanying them indicated 'white'; however, several subsequent collections from the same region indicate that the flowers are 'cream'.

crowded with the terminal segment pinnatisect, all the lobes obtuse to subacute, often terminating in a bristle as well as being sparsely bristly on both sides; cauline leaves with up to 9 pairs of lateral segments, similar to the basal but becoming gradually smaller and with fewer segments, merging with the sessile bracts, the middle and upper sessile with the lowermost pair of segments forming an auricle-like base; petiole of basal leaves up to 16 cm long, slender, those of the lower stem leaves considerably shorter to subsessile. *Inflorescence* a rather narrow cone-like panicle occupying at least two thirds the height of the plant, the lower flowers in lateral cymules of 2–5, the upper flowers solitary, the rachis bristly; flowers numerous, cup- to saucer-shaped, horizontal to half-ascending, 40–65 mm across. *Peduncles* ascending, to 9.5 cm long, sparsely bristly. *Pedicels* ascending to erect, 5.5–13 cm long (to 19 cm long in fruit), sparsely bristly in the lower part but densely so immediately beneath the flowers. *Buds* ovoid to subglobose, covered with spreading bristles. *Petals* 4, creamy-yellow, obovate to suborbicular, 25–37 × 22–37 mm. *Stamens* numerous, with whitish filiform filaments and orange-yellow anthers. *Ovary* ellipsoid to ovoid, densely bristly, the bristles at first appressed but later spreading as the fruit develops; style slender, 2.5–6 mm long, terminating in a capitate green or yellowish-green stigma. *Fruit capsule* ellipsoid, 20–41 × 7–8 mm, 5–6-valved, covered in spreading, stiff bristles up to 8 mm long, the persistent linear style 6–8 mm long. *Seeds* ovoid-ellipsoid, 0.6–0.8 mm long. 2n = 56. Figs 7A, 8A–B.

DISTRIBUTION. CE Nepal, Dolakha District (centred on the Rolwaling Himal) and S Tibet (Nyalam); 2950–4755 m[8] (Map 4).

HABITAT. Rocky places, open shrubberies and stream banks.

FLOWERING. June–August.

SPECIMENS SEEN. *F. M. Bailey's Collectors* 15 (BM) & 71 (BM); *Dhwoj* 0297 (BM); *K. N. Sharma* 9 (BM), 31 (BM), 41 (BM), K; *Stainton* 4680 (BM) and 4700 (BM).

---

[8] According to the original description and information on the type sheets (*Dhwoj* 297) the species is found from 12,000 to 18,000 ft (i.e. 5487 m); this latter altitude is almost certainly incorrect for the plant in question.

Below, from left:
Meconopsis dhwojii; S of Panch Pokhari, E Nepal, 4000 m, flowering in mid-August.   PHOTO: TOSHIO YOSHIDA
Meconopsis dhwojii; Matsang, Tsangpo Valley, S Tibet, 3100 m.   PHOTO: DAVID & MARGARET THORNE
Meconopsis dhwojii basal leaves; Matsang, Tsangpo Valley, S Tibet, 3100 m.   PHOTO: DAVID & MARGARET THORNE

6. MECONOPSIS DHWOJII

Meconopsis dhwojii; Matsang, Tsangpo Valley, S Tibet, 3100 m.
PHOTO: DAVID & MARGARET THORNE

Lall Dhwoj, who discovered this charming species and in whose honour it is named, was an officer in the Royal Nepalese Army. Seed from Dhwoj's collections proved successful and the species first flowered in Britain in 1932. *M. dhwojii* has very attractive, rather small, evergreen rosettes of crisply dissected leaves which often turn an attractive purplish bronze during the winter months. The mature leaves are very distinctive with their dark purple pimples which surround the base of each bristle.

Meconopsis dhwojii has been systematically confused with other species in the wild, particularly with similar cream or yellow-flowered species from further west in Nepal, especially in the Langtang and Ganesh Himal regions. Thus specimens of the true *M. napaulensis* and *M. gracilipes* have been labelled *M. dhwojii*. Although the former two certainly grow in close proximity, *M. dhwojii* is found much farther to the east and was thought to be restricted to the Rolwaling Himal to the west of Mt Everest; however, recent photographic evidence from David & Margaret Thorne, David Haselgrove and others show that it is also present close to Nyalam in southern Tibet, indeed close to the Kathmandu-Lhasa highway. Nyalam in fact lies adjacent to, and immediately west of, the Rolwaling Himal. The known localities of the closely related *M. gracilipes* lie significantly further to the west. The record of this latter species from Nyalam in S Tibet should therefore be transferred to *M. dhwojii*.

Meconopsis dhwojii is closely related to *M. gracilipes* (see above) and like it is one of the smaller and daintier members of the *Polychaetia*. The finely dissected, very neat leaf-rosettes often have a purplish or reddish flush, while the purple or blackish base to the leaf and stem bristles are very distinctive and reminiscent of the distantly related *M. horridula* Hook. f. & Thomson in some of its forms. In *M. gracilipes* the basal leaf laminas are less finely divided, while the indumentum is one of uniformly pale, soft bristles. In addition, differences are readily seen in the mature fruits: those of *M. dhwojii* are broad oval to oblong in outline with a 3–6 mm long persistent style, while those of *M. gracilipes* are relatively shorter compared to length, more flask-shaped and with a longer persistent style, 6–10 mm long. Both species are readily distinguished from *M. napaulensis* on account of their thinner, uniform indumentum of simple, long, barbellate hairs (a denser indumentum of long barbellate hairs underlain with a felt of much shorter barbellate hairs in *M. napaulensis*).

George Taylor (1934) draws comparisons particularly with *Meconopsis robusta* found in N India (Kumaon) and the extreme west of Nepal. Although both species belong to the same series they really are very different. Most notably, the inflorescence of *M. robusta* is a simple raceme, that of *M. dhwojii* paniculate, at least in the lower half. However, there are further differences: the basal and lower leaves of the former have just 3–6 pairs of lateral segments, those of the latter 6–9; the fruit capsules of *M. robusta* are 9–13 mm in diameter and 6–9-valved, those of *M. dhwojii* just 7–8 mm diameter and 5–6-valved; in addition the fruit capsule style of *M. robusta* broadens distinctly towards the base, while that of *M. dhwojii* is linear throughout.

Meconopsis dhwojii is reasonably well known in cultivation, forming an attractive, finely dissected leaf-rosette like some exotic cut-leaved lettuce, flushed and blotched with purple. Although plants can flower in the second year from seed, they normally do so in the third and fourth years. Well grown, the species can make a handsome cone of pale yellow or cream flowers from near ground level and up to a metre in height. Despite having been in cultivation continuously since 1932, it is not particularly common in collections. The species can hybridise in cultivation with other members of section Polychaetia. *M.* ×*ramsdeniorum* G. Taylor (see *J. Roy. Hort. Soc.* 83: 18 (1962)) is said to be a hybrid between *M. dhwojii* and *M. napaulensis*. This is unlikely as, as far as it is known, the true *M. napaulensis* was not in cultivation at the time and nor has it been in cultivation until very recently. The other parent is likely to be *M. paniculata* or *M. staintonii* or another hybrid. In any event *M.* ×*ramsdeniorum* seems to be unremarkable, James Cobb (1989) describing it as "a sterile shadow of either parent".

Meconopsis dhwojii, immature fruits; Matsang, Tsangpo Valley, S Tibet, 3100 m. PHOTO: DAVID & MARGARET THORNE

7. MECONOPSIS CHANKHELIENSIS

Meconopsis chankheliensis Grey-Wilson, *Curtis's Bot. Mag.* 23 (2): 203 (2006). Type: W Nepal, Jumla District, Maharigaon, 15,000 ft, 20 July 1952, 'damp rock outcrops, partial shade', *Polunin, Sykes & Williams* 224 (BM, holotype, isotype).

DESCRIPTION. *A monocarpic plant* to 1.5 m tall, but sometimes as little as 40 cm in flower, with a dauciform taproot, the whole plant covered in golden-brown or yellowish, simple, rather rigid, finely barbellate hairs, densely so when young; stem 11–16 mm diameter at the base. *Leaves* in a basal, evergreen rosette in the first year or two; basal leaves present at flowering time, pinnatisect or bipinnatisect, oval in outline, to 24 × 9.5 cm, with 4–7 primary segments, these oval to elliptic, 25–50 × 17–26 mm, with one or several shallow marginal lobes, stiffly hairy on both surfaces, the hairs even, 3–5 mm long; petiole unwinged and rather slender, 7–16 cm long, stiffly hairy. *Cauline leaves* similar to the basal but sessile, the lowermost sometimes shortly petiolate, decreasing in size up the stem and merging gradually into the bracts, the uppermost with 3–4 primary divisions only. *Inflorescence* a simple raceme, or semi-panicle with only the lowermost flowers in lateral cymules of 2–3, the others solitary in the inflorescence which occupies more than half the height of the plant. *Bracts* sessile, similar to the uppermost leaves, with 2–5 pairs of primary divisions, semi-amplexicaul at the base. *Peduncles*, when present, 8.5–23.5 cm long, adorned with spreading stiff hairs. *Pedicels* slender, 4.5–18 cm long, pubescent like the peduncles, densely so towards the top. *Buds* nodding, ovoid to subglobose, the sepals covered in sparse patent, stiff hairs, these 4–6 mm long. *Petals* 4, purple to dull dark red, subrounded to obovate, 21–40 × 21–42 mm. *Stamens* numerous, with purple filaments and yellow or orange anthers. *Ovary* covered with ascending orange bristles; style slender, 4–8 mm long, glabrous, terminating in a c. 3 mm diameter subcapitate, yellowish, obscurely 6-lobed, stigma. *Fruit* unknown, but certainly bristly and probably 6-valved.

DISTRIBUTION. W Nepal (Jumla & Mugu districts: Maharigaon, Chankheli Lagna and Chankheli Lekh); 3100–4575 m (Map 4).

HABITAT. In *Quercus-Abies* forest, in partially shaded rocky places and amongst streamside shrubs.

FLOWERING. Late May–July.

SPECIMENS SEEN. *F. M. Bailey* 181 (E), s.n. 6 June 1936 (E); *Polunin, Sykes & Williams* 224 (BM), 4324 (BM, E), 5942 (BM); *Stainton* 6333 (BM).

This interesting, recently described species remained undiscovered for many years since it was first collected by Capt. Bailey in the 1930s, followed by subsequent collections, principally by Polunin, Sykes & Williams in the early 1950s: most of the sheets were filed under *Meconopsis napaulensis* at both the British Museum and Edinburgh. However, *M. chankeliensis* is very distinct and in the details of leaves and indumentum it comes closest to *M. dhwojii* and *M. gracilipes*, especially the former, differing from both most obviously in its purple or dark red, rather than yellow flowers. The normally bipinnately-lobed basal leaves are also very distinctive. From *M. dhwojii* it also differs in the pale rather than purplish-black base to the leaf hairs and from *M. gracilipes* in having more acute leaf divisions which are more densely hairy and less glabrescent at maturity.

Meconopsis chankeliensis is not in cultivation and comes from the remote under-botanised west of Nepal in an area that is politically sensitive at the present time. Despite this, there have been several Edinburgh Botanic Garden inspired collecting trips to western Nepal in recent times but, to my knowledge, none have come across this interesting, restricted species.

Interestingly, the Jumla-Dolpa region of western Nepal appears to be particularly rich botanically, or at least it acts as an isolated outlier for various plants: *Meconopsis grandis* Prain s.l. and *M. paniculata* (D. Don) Prain s.l. are both found in the vicinity but nowhere else in western Nepal, in fact the nearest known records are to the east; east of the Kali Gandaki valley for *M. paniculata* and east of the Everest region for *M. grandis*. Other plants such as *Incarvillea himalayensis* Grey-Wilson also have their only known western location near Jumla, the species being generally found much further east in southern Tibet and south-western China. This intersting phytogeographical phenomenon clearly requires further investigation.

series Polychaetia

This important series contains some eight species that have been the subject of a great deal of confusion and misinterpretation over the years, this primarily centred around the identity of De Candolle's *Meconopsis napaulensis*. The series is important horticulturally, containing as it does some of the finest evergreen monocarpic species renowned both for their large and handsome leaf-rosettes as well as their large panicles of flowers, making them some the tallest members of the genus. The series is thus characterised by generally large and robust plants often 1.5 m tall in flower, sometimes reaching as much as 2.5 (–3) m. The pubescence of stems and leaves consists of long barbellate hairs or bristles underlain by a sparse to dense felt of similar though much shorter hairs. The flowers have a wide range of colours, with blue, purple, wine-purple, red and yellow (in *M. autumnalis* and *M. paniculata*) predominating, although occasionally white-flowered plants can be seen in wild populations, particularly in *M. staintonii*. In cultivation there is a marked tendency for the species to hybridise; however, in the wild this is generally precluded as the species are on the whole isolated from one another geographically, while in the few places where two species overlap (e.g. *M. napaulensis* and *M. paniculata* in the Langtang region of central Nepal) no authenticated hybrids have been recorded.

KEY TO SPECIES OF SERIES POLYCHAETIA

1. Flowers yellow 2
+ Flowers red, pink, blue, lilac, violet, occasionally white 4

2. Leaves and stems with a sparse underlying indumentum of short barbellate hairs; stigma and ovary usually with 5 divisions, the stigma greenish; plant rarely exceeding 1 m tall in flower . . . . 8. **M. napaulensis** (C Nepal)
+ Leaves and stems with a dense underlying indumentum of barbellate hairs; stigma and ovary with 6–12 divisions, the stigma yellow or reddish-purple; plant often 1.1–1.85 m tall in flower 3

3. Stigma reddish-purple, 1–4 mm diameter; bracts patent; fruit capsule with ascending to appressed bristles . . . . 13. **M. paniculata** (Nepal, Sikkim, Bhutan, N Arunachal Pradesh (Assam), S Tibet)
+ Stigma yellow, 3–8 mm diameter; bracts deflexed downwards; fruit capsule with patent bristles . . . . 14. **M. autumnalis** (C Nepal)

4. Mature fruit capsules narrowed at the base; flowers with red or pink (sometimes with a violet tinge) or violet petals 5
+ Mature fruit capsule rounded at the base; flowers with blue, violet-blue, lilac, lavender, mauve, purple or wine-purple petals. 7

5. Inflorescence racemose, the flowers borne singly in the inflorescence, the petals violet; basal leaves narrow, linear-oblong in outline. . . . . 15. **M. violacea** (Myanmar, SE Tibet)
+ Inflorescence paniculate, the flowers mostly (except the uppermost) borne in axillary cymules in the inflorescence (lateral branches with 2–5 flowers, more in *M. wallichii*), especially in the lower part of the inflorescence, the petals pink or red, occasionally purplish blue or white; basal leaves oblong to lanceolate in outline 6

6 Flower buds with patent barbellate hairs; fruit with patent, simple, long barbellate hairs; fruit-capsule style 4–8 mm long, to 1 mm diameter at the base . . . . .9. **M. ganeshensis** (C Nepal)
+ Flowers buds with erect or incurved barbellate hairs; fruit capsule with ascending long barbellate hairs plus an underlying indumentum of short barbellate hairs; fruit-capsule style 8–14 mm long, 2–3 mm diameter at the base. . . . . 10. **M. staintonii** (C Nepal)

7. Inflorescence with short, erect lateral branches, congested and spike-like; flowers violet wine-purple to purple, purplish red or crimson, the lateral cymules 1–5-flowered; fruit capsule (8–)11–15 mm diameter, with erect to ascending bristles . . . . 12. **M. wilsonii** (SW China, E Myanmar)
+ Inflorescence with long, spreading branches, not spike-like; flowers blue to lilac or pinkish-blue, occasionally wine-purple or deep red, the lateral cymules mostly 5–13-flowered; fruit capsule 6–10 mm diameter, with wide-spreading bristles. . . . . 11. **M. wallichii** (E Nepal, Sikkim, Bhutan, SE Tibet)

8. MECONOPSIS NAPAULENSIS

Meconopsis napaulensis DC., *Prodr.* 1: 121 (1824). Type: *Wallich*, Nepal 1821[9] (G-DC, holotype; K isotypes); Grey-Wilson, *Curtis's Bot. Mag.* 23 (2): 186 (2006), fig. 4, 16.
[syn. *Papaver paniculatum* D. Don, *Prodr. Fl. Nepal*: 197 (1825), pro parte; *Meconopsis nipalensis* Hook. f. & Thomson, *Fl. Ind.* 1: 253 (1855); *M. napaulensis* sensu Taylor pro parte in *Meconopsis*: 44 (1934); *M. longipetiolata* G. Taylor ex Hay, *New Fl. & Silva* 4: 226, fig. 83 (1932) & *Gard. Chron.* 1932, Ser. 3, 92: 41 (1932). Type: Described from a cultivated plant raised from seed originating in Nepal (BM, specimen lost). C Nepal, Langtang, 11,500–12,500 ft, 21 June 1949, *Polunin 485* (lectotype BM, isolectoypes BM), neotypified by Debnath & Nayar (1987) placing the type locality in the Langtang valley, Rasuwa district of central Nepal].

The true identification and typification of De Candolle's *Meconopsis napaulensis*, the first Asiatic species of *Meconopsis* to be described, is pivotal in the correct understanding of the members of series *Polychaetia*. I researched this problem in depth several years ago and the results were published in *Curtis's Botanical Magazine* in 2006. Since then little has changed by way of interpretation; however, the information contained in that article was important and I make no excuse for repeating much of it in this work.

While assessing the true identity of *Meconopsis napaulensis* DC. it was necessary to re-evaluate the whole of the *M. napaulensis* aggregate, a single species according to George Taylor (1934) that stretched in a narrow band from west Nepal through Sikkim, Bhutan, NE India (Assam) and northern Myanmar (Burma) to western

---

[9] Note by David Prain (dated 20 March 1915) attached to *Wallich* 8121 in the main Kew Herbarium reads: "Dr Wallich obtained this plant in Nepal in 1821. One specimen was sent to Geneva and was described as *Meconopsis napaulensis* DC. in 1824. That specimen is now in the Prodromus Herbarium. In 1830 Wallich issued the remaining specimens as his no. 8121. Therefore *Wallich* n.8121 is *M. napaulensis* DC. In 1855 *Wallich* no.8121 was not taken up by Hooker and Thomson in *Flora Indica*. In 1872 H & T included *Wallich* n.8121 in *M. robusta* Hk.f. & Th. which was based in 1855 on *Wallich* no.8124. In 1896 and in 1906 Prain treated *Wallich* n.8121 as distinct from *Wallich* n.8124 and as identical with *M. wallichii* Hook. f. var. *fusco-purpurea* Hook.f., Bot. Mag. T.6790 (1884). If 1909 Fedde has reverted to the view of Hooker and Thomson in the *Flora of British India* (1872). If the view of Hooker and Thomson of 1872 be correct then their name *M. robusta* (1855) must be replaced by the original name *M. napaulensis* DC. (1824)."

8. MECONOPSIS NAPAULENSIS

Meconopsis napaulensis; flowering plant and young leaf-rosettes, Langtang Valley, C Nepal. PHOTO: DIETER SCHACHT

Meconopsis napaulensis; leaf-rosette, Langtang Valley, C Nepal. PHOTO: DIETER SCHACHT

China. This necessitated revising the whole of Taylor's subsection *Eupolychaetia* (i.e. subgenus *Meconopsis*) and, as a result, four new species were described: *M. chankheliensis*, *M. ganeshensis*, *M. staintonii* and *M. wilsonii*, while *M. wallichii* Hook. f. was reinstated. Since publication of the 2006 paper an additional species, *M. autumnalis*, has been described by Paul Egan, after careful field analysis.

De Candolle described *Meconopsis napaulensis* in 1824 based on material sent to him by Nathanial Wallich which he had collected in Nepal under the number 8121: this specimen resides in the *Prodromus* Herbarium in Geneva. The type specimen, which is rather poorly preserved and somewhat fragmentary, is a portion of a fruiting plant; the fruit capsule characteristics are clearly discernible. Later, in 1830, Wallich issued the remaining specimens of the same collection as his number 8121. Specimens under this number reside in the main herbarium at the Royal Botanic Gardens, Kew, as well as in the Wallich Collection (also at Kew). Both sheets are annotated in pencil by Joseph Dalton Hooker. Subsequently no mention was made of *Wallich* 8121 in Hooker & Thomson's *Flora Indica* of 1855. However, confusion was caused when Hooker & Thomson equated *Wallich* 8121 with another species, *M. robusta* Hook. f. & Thomson, described in 1855 and based on another Wallich collection, number 8124. To back this up both *Wallich* 8121 and 8124 (one sheet of each) in the Wallich Collection at Kew are inscribed in pencil by Hooker as *M. robusta*; however, the sheet in the main Kew Herbarium is inscribed *M. napaulensis*. This is rather odd in view of the fact that had Hooker & Thomson considered the two collections to be the same then the early name *M. napaulensis* would have been valid and not *M. robusta*. The confusion with the name *M. napaulensis* can however be attributed to David Don who, in 1825 (*Prodr. Fl. Nepal.*) equated *M. napaulensis* with *Papaver paniculatum* D. Don as a synonym. *P. paniculatum* was based on yet another Wallich specimen, number 8123b. This was clearly an error on Don's part, yet the true identity of his species was not realised until 1896 when David Prain transferred it to the genus *Meconopsis* (i.e. *M. paniculata* (D. Don) Prain).

David Prain (1896 and 1906) took the view that *Wallich* 8121 and 8124 represented two distinct species, a view with which I concur, but he equated the former with *Meconopsis wallichii* Hook. f. var. *fusco-purpurea* Hook. f. described in *Curtis's Botanical Magazine* in 1884 (t. 6790). Three years later, in 1909, Fedde reverted to the view of Hooker & Thomson's *Flora of British India* (1872) by including *M. napaulensis* within *M. robusta*.

There is no doubt that the species centred around

Meconopsis napaulensis and *M. paniculata* have presented taxonomists with a complex of forms and variations that have proved difficult to fathom. Yet over the years a number of assumptions have been made that have clouded, rather than helped clarification. In his much-lauded monograph of 1934 (*An Account of the Genus Meconopsis*), George Taylor states quite emphatically that (under *M. napaulensis*) 'In this widespread and variable species I have been unable to find any satisfactory characters by which to differentiate the several strains. Here, as in other members of the genus, there is a flux of forms which so intergrade as to make precise definition impossible. It appeared that there might be some correlation between the geographical distribution and the form assumed, but investigation did not substantiate this.' In addition he states that 'Specimens have been sent from that country [sic Nepal] which are identical with the type-collection of *M. napaulensis*, and the flower colour is now definitely known to be red, purple, blue, or occasionally white.'

Earlier authors, prior to the work of George Taylor, placed a lot of emphasis on flower colour as an important and defining character; however, Taylor places no such importance to it, this despite the fact that the colour of *M. napaulensis* has never been established (the type specimens are all in fruit and there are no helpful field notes). Interestingly, other characters that might have proved useful, particularly pubescence, leaf characteristics and, perhaps most significantly, fruit characters have been greatly under-used. No doubt there are those who will come along in the near future to carry out DNA studies on this group, but the sampling would need to be both extensive and thorough if it is to be of any practical use. In the meantime this present study is based unashamedly on morphological characters.

In preparing this new monograph of *Meconopsis* I have tried to look at species delineation afresh. This has been greatly helped by the large amount of data that has been collected in the wild since George Taylor wrote his monograph. There has been a great deal of new herbarium material added to the existing collections; foremost amongst these are the magnificent collections of Polunin, Sykes & Williams (in Nepal) and those of Ludlow & Sherriff (primarily in Bhutan and southern Tibet), exemplary collections for the most part with full and informative field notes, with details of colour of all the flower parts meticulously recorded, no doubt at the behest of George Taylor. In addition, in recent times, renewed exploration of western and south-western China and south-eastern Tibet (Xizang) has added a great deal more material, as well as allowing plants to be studied in situ, often in well-known localities visited in

Meconopsis napaulensis; flowering plant, with young fruits, Langtang Valley, C Nepal.
PHOTO: DIETER SCHACHT

the early part of the twentieth century by some of the great plant collectors, particularly George Forrest, Frank Kingdon Ward, Ernest Henry Wilson and Abbé Delavay, but there were many others. One huge asset has been the chance to examine numerous colour photographs taken in the wild by various photographers and this information has been of great help in subsequent analysis of herbarium material. Over the years I have been able to travel extensively in Nepal and western China and have had the benefit of seeing many species of *Meconopsis* growing in the wild. This has given me the chance to re-examine all the different entities in detail and the results have been rather surprising, particularly as concerns *M. napaulensis* and its allies.

8. MECONOPSIS NAPAULENSIS

One factor not mentioned so far is *Meconopsis* in cultivation. There is no doubt that gardeners' views of the different species have been almost wholly governed by what they have grown and seen in cultivation. Large monocarpic plants with red or pink flowers are always assigned to *M. napaulensis*, while similar plants with yellow flowers are placed in *M. paniculata*. In recent times plants with blue, lilac or purple flowers formerly included in *M. napaulensis*, have been distinguished as *M. wallichii*, thus reconfirming Hooker's view of as long ago as 1852. However, as most *Meconopsis* growers will tell you *M. napaulensis* (of hort.) and *M. paniculata* hybridise freely in the garden when grown in close proximity to one another so that the species boundaries are quickly blurred. Fortunately, *M. wallichii* tends to flower later and the chances of crossing are much reduced, especially if seed can be collected from the later maturing fruits. For the former two species, and indeed others like the fabled *M. regia*, which also hybridises freely with *M. 'napaulensis'* and *M. paniculata*, the only way to be certain of what is being grown is to raise fresh stock from wild collected seed. For this reason it is important to ensure that at least some seed from wild sources is preserved in bone fide seed banks. At the same time botanic gardens and others growing the monocarpic evergreen species should ensure that they are isolated from one another if the species are not to be corrupted by hybrisiation.

Despite George Taylor's views freely expressed in his 1934 monograph, I have found that there are marked regional differences to be seen in stature, leaf and inflorescence characteristics, as well as details of the fruits, this backed up to a greater or lesser extent by flower colour. As a result I proposed a series of different, although obviously closely connected, species, based on clear morphological characters, as well as geography: this is all presented in the key to species concerned (see p. 80) and the descriptions that follow.

This brings me back to my starting point and that is the assumption that the plant in cultivation with red or pink flowers called *Meconopsis napaulensis* in gardens is in fact correctly identified and, if not, what is it? Authenticated cultivated specimens under the name *M. napaulensis* closely match herbarium material and photographs taken in western Nepal centred on the Kali Gandaki valley which divides the great Dhaulagiri massif from that of the Annapurna Himal. These specimens represent, along with *M. paniculata* and *M. regia*, the largest and most robust of the evergreen monocarpic species. The former differs markedly in stature, leaf and fruit characters from the type of *M. napaulensis*; so much so that I was forced to the conclusion that they

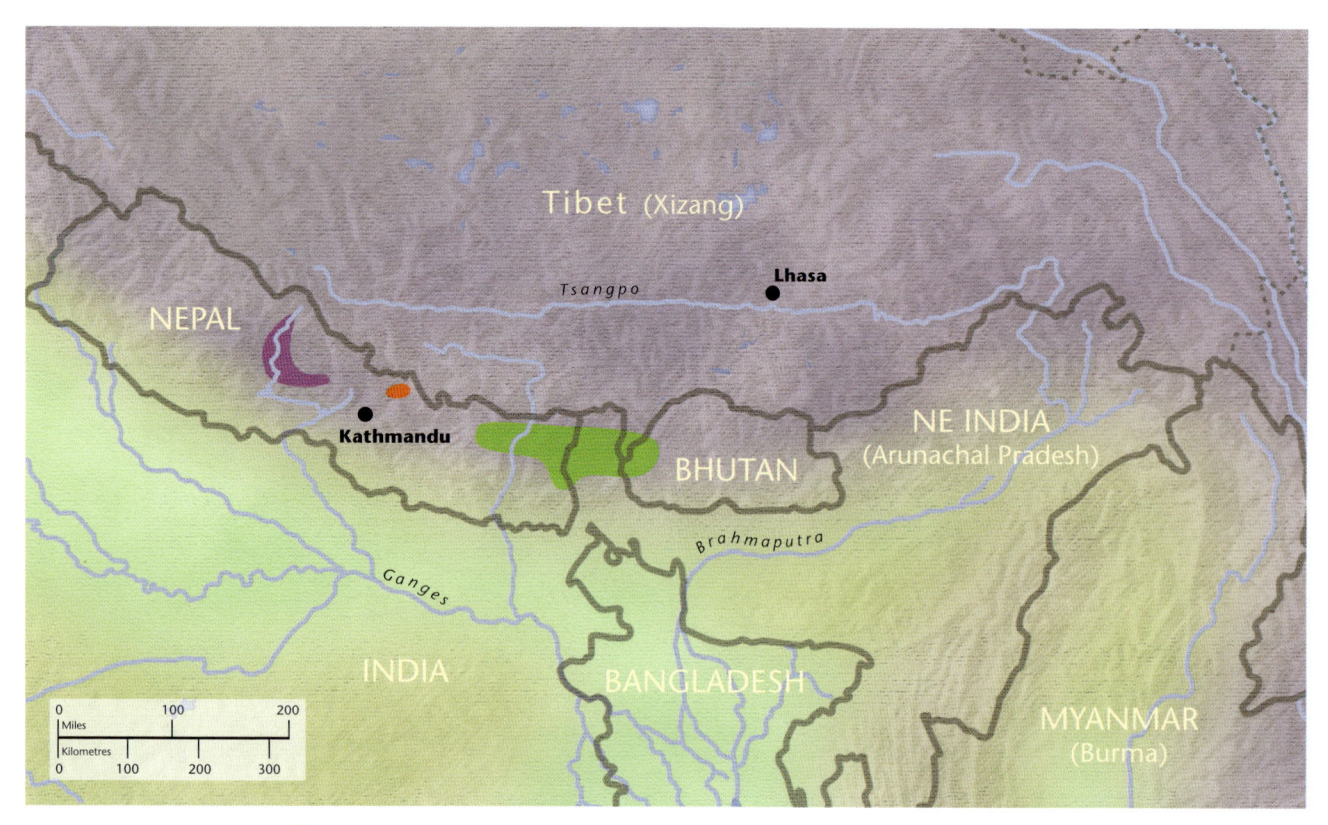

MAP 5. Species distribution: ■ *Meconopsis napaulensis*, ■ *M. staintonii*, ■ *M. wallichii*.

represented a distinct and hitherto undescribed species which I named *M. staintonii*. This will no doubt prove aggravating to gardeners who have long cherished plants in cultivation under the name *M. napaulensis*.

So what is Wallich's *Meconopsis napaulensis*? The original description in the De Candolle *Prodromus* 1: 21 is brief as one might expect from the rather inadequate and fragmentary nature of the type collection. "*M. napaulensis*, capsulis valde echinatis, stylo ovarii fere longitudine, stigmato crassissimo, foliis plurimis sinuato pinnatifidis, summins sesselibus, caule pedunculus sepalisque setosis — in Napaulia."

Interestingly, the locality marked on the sheet in the Wallich collection in the Kew Herbarium is Ghossain Than which equates with Gossain Than, Goshai Kand or Gossaingstan, a locality known in Nepal today as Gossainkund, a place with an attractive lake not far to the north of Kathmandu and on one of the prime trekking routes to the well-known and botanically rich area of the Langtang Valley (Rasuwa District), today one of the most botanised regions of Nepal. Nearly all the subsequent collections of *Meconopsis napaulensis* are from the same general vicinity which is easily accessible from Kathmandu and presumably one of the easiest parts of the Nepalese Himalaya to reach, even in Wallich's day; for instance one of Bailey's 1935 collections filed under '*M. napaulensis*' at the British Museum (Natural History) closely matches the type material and was also collected at Gossainkund. It is notable that no Wallich collections are recorded in the mountains (the Annapurna Massif in particular) north of Pokhara, 120 km west of Kathmandu, where he would surely have seen and collected other *Meconopsis*. In addition, modern photographs from the region (particularly those of Eiko Chiba, Dieter Schacht and Toshio Yoshida) taken in the Gosainkund-Langtang area, back up herbarium observations.

One final point to consider is whether or not all the material of *Wallich* 8121 is in fact *Meconopsis napaulensis* DC. At least three sheets exist: the holotype at Geneva consists of two fragments of stem with leaf remnants, one piece with three attached fruit capsules (two immature, the other almost mature), while the other fragment has a piece of stem and bract, with a pedicel and part of a petal; no flower colour was recorded. The second in the Wallich Collection at Kew consists of two individuals, one of the upper part of the plant and inflorescence with several immature fruits, the other consists of a stem base and the lower part of the inflorescence, but without flowers or fruits. The third specimen is located in the main herbarium at Kew and to this is attached a lengthy hand-written note (dated 20 March 1915) by David Prain (see footnote 9) outlining the origin and whereabouts of *Wallich* 8121. This latter sheet again bears two specimens, one of the upper part of an infructescence with three semi-mature fruit capsules, the other, also the upper part of the plant but with only petiole, bracts and pedicel remnants. None of the specimens is well preserved and all are incomplete and, as is the norm for Wallich collections, there is no field data of significance. Despite this, enough can be gleaned from this material to be able to get a clear picture of the plant in question and, more importantly,

Meconopsis napaulensis; E of Gosainkund, C Nepal, 3500 m. PHOTO: TOSHIO YOSHIDA

to be able to link it up with more recent twentieth century collections. Thus the leaf and fruit characters and details of the indumentum can all be qualified and, furthermore, match a handful of specimens which have all been collected in a relatively limited area of central Nepal between the Buri Gandaki and the Langtang Valley. In addition, these specimens reveal that the flower colour is yellow and not red, purple or blue as had been previously assigned to *M. napaulensis* DC.

DESCRIPTION. *Monocarpic plant* to 1.1 m tall, sometimes as little as 50 cm in flower, with a slender dauciform taproot, the whole plant covered or more or less covered with long barbellate bristles with a generally sparse underlying layer of much shorter, but similar, hairs, this latter eventually caducous. *Basal and lower cauline* leaves absent or partly present at flowering time, long-petioled, to 34 × 5 cm long; lamina oblong to lanceolate in outline, pinnatisect towards the base, pinnatifid towards the apex, with 6–9 pairs of opposite to subopposite segments, the segments forward directed, oval to elliptic or obovate, 14–23 × 8–13 mm, shallowly and obtusely, sometimes acutely, lobed, to subentire, sparsely to moderately bristly; petiole slender, unwinged, 6.2–26 cm long, bristly, the lowermost broadening to a wide sheathing base; middle and upper leaves similar to the basal leaves but becoming gradually smaller upwards to merge with the bracts, with a short or rather indistinct petiole that is clearly auriculate at the base and clasping the stem; auricle formed from the lowermost pair of leaf-segments, half-clasping the stem. *Inflorescence* semi-paniculate, occupying the top two thirds of the plant, sometimes extending to the base, with the uppermost flowers solitary but the lower in lateral cymules of up to 11; flowers deeply cupped, 7–9 cm across. *Bracts* sessile, similar to the upper leaves, decreasing in size up the inflorescence, pinnatifid, the lowermost with 3–4 pairs of lobes, semi-amplexicaul at the base. *Peduncles* 5–9 cm long, pubescent like the stems, with a mixture of long barbellate hairs underlain with similar but short ones. *Pedicels* slender, 2–8.5 (–13) cm long, pubescent like the peduncles. *Buds* ovoid to obovoid, 14–16 mm long at maturity, covered in stiff, spreading barbellate hairs. *Petals* 4, primrose-yellow, obovate, 24–32 × 20–23 mm. *Stamens* numerous, with pale yellowish filaments and yellow to orange-yellow anthers. *Ovary* covered in ascending bristles; style slender, 4–8 mm long, the stigma greenish or greenish yellow, capitate, 5–6 mm diameter. *Fruit capsule* erect, narrow-obovoid, 26–32 × 6–11 mm, clearly broadest above the middle and with a short stipe at the base, 5 (–6)-valved normally, covered with rather sparse spreading to somewhat reflexed bristles, 5–7 mm long, these often snapping off at maturity to leave peg-like projections, the persistent style 6–9 mm long in fruit, somewhat broadened and 0.8–2 mm diameter at the base. $2n = 56$. (Fig. 9A1,2).

DISTRIBUTION. C Nepal; Bagmati Zone (Pulchok and Rasuwa Districts), Ganesh Himal; 3200–4500(–4880) m (Map 5).

HABITAT. Rocky and grassy slopes, meadows, open shrubberies, stream margins.

FLOWERING. Late May–Aug.

Meconopsis napaulensis; Gosainkund, C Nepal, 3800 m. PHOTO: TOSHIO YOSHIDA

SPECIMENS SEEN. *F.W. Bailey's collectors* s.n. 15 June 1935, Gosainkund (BM) & 7 June 1937, Dije Gompa (BM),14 (BM); *Halliwell* 157 (K); *Miyamoto et al.* (University of Tokyo) 94-40030 (BM, E), 94-40053 (E), 94-00059 (E), 94-00076 (BM, E); *E. H. Nicholson* 2637 (BM); *Polunin* 485 (BM, E), 1420 (BM); *Schilling* 433 (K), 933 (K); *K. N. Sharma* 33/94 (BM), 42/94 Kalora (sheet marked Tibet!) (BM), E60 (BM), E62 (BM), E372 & 373 (BM), E378 (BM), E446 (BM), E448 (BM), E457 (BM), 495 (BM) E530 (BM); *Stainton* 4026 (BM), 6360 (BM); *Wallich* 1821 (G-DC, K).

Meconopsis napaulensis is not in cultivation to my knowledge at the present time. However, it certainly has been in the past on several occasions under the name *M. longipetiolata*, but even though it has been listed in several botanic garden seed catalogues in recent years under the latter name, plants do not appear to be correctly identified. However, recently, while I was examining photos in the collection at the Royal Botanic Garden Edinburgh I came upon a number of glass plate photos including several of the rare *M. violacea* and two labelled *M. gracilipes* that are clearly *M. napaulensis*: these photos, presumably of plants growing at the garden, are undated but were almost certainly taken in the 1930s. This species grows fairly close to Kathmandu so it should be possible to acquire fresh seed, even in the politically turbulent Nepal of today. The fact that it somewhat resembles a slighter, more dwarf *M. paniculata* (the most widespread species in section *Polychaetia*) may have meant that it has be ignored by collectors in recent years. That is as it may be, but to have true *M. napaulensis* in cultivation would clearly be an asset, besides which it would make a fine plant for a woodland glade with its pretty over-wintering rosettes.

Meconopsis longipetiolata G. Taylor was based upon a living specimen grown from wild collected seed of uncertain provenance, but derived from Nepal. Unfortunately, the flowering specimen (which was subsequently dried) upon which the description was based, was deposited at the British Museum (Natural History) but subsequently lost: George Taylor (1934) relates that "….. by an unfortunate misunderstanding, the living plant on which the description was based was removed and has since been lost." I have subsequently made a thorough search of all the *Meconopsis* material at the British Museum and have also not been able to locate the type specimen. However, there are two sheets amongst the cultivated herbarium sheets at the British Museum (BM) clearly marked *M. longipetiolata*. These were supplied by Andrew Harley esq, Devonhall, Crook of Devon in Perthshire on the 6th June 1936. There is no reason to doubt the authenticity of these specimens and it is particularly interesting to note that they were supplied just four years after the plant was first described and it is highly likely that they go back to the same seed source gathered from an unknown location in Nepal. The alternative view is that they are second generation progeny produced from the original introduction, although the time span may not allow for this.

The original description is not very full and no fruit details are given. However, details of the indumentum are included, although the main distinguishing character given was the presence of "the unusually long petioles", but no actual length is given for these. However, the description and details given in the *Gardener's Chronicle* (loc. cit.) in 1932, when it first flowered, are accompanied by half-tone photographs of the plant in question: one of a leaf-rosette in a pot, the other, also in a pot, of a young flowering specimen. These photographs, along with the botanical details given show that *Meconopsis longipetiolata* fits comfortably within the circumscription of *M. napaulensis* which can have petioles up to 26 cm long in some individuals, although half that length is more normal. As George Taylor was unaware of the complications surrounding the true identity of *M. napaulensis* then it is not perhaps surprising that he failed to make the connection and as a result described a new species.

9. MECONOPSIS GANESHENSIS

Meconopsis ganeshensis Grey-Wilson, *Curtis's Bot. Mag.* 23 (2): 188 (2006). Type: C Nepal, Ganesh Himal, Ankhu Khola, 28°12'N, 85°05'E, 13,500 ft, *Stainton* 3990 (BM holotype; E isotype)

DESCRIPTION. *A rather slender monocarpic plant* to 1.2 m tall in flower, generally less, the stem reddish or green flushed with red, 8–16 mm diameter at the base, the stems and leaves sometimes with a mauvish-pink blush, the whole plant covered in 4–9 mm long, pale orange or fawn-coloured, stiffish, barbellate hairs, generally accompanied by some underlying simple barbellate hairs, but never densely so, especially on the upper stem, pedicels and sepals. *Basal leaves* in a lax rosette, with a slender petiole 6–13 cm long, scarcely expanded at the base, the lamina narrow-oblong in outline, to 14 × 4.2 cm, pinnately-lobed, with 5–8 pairs of rather distant, alternate to subopposite leaflets, these oval to elliptic or elliptic-ovate, 12–22 × 5–13 mm, subentire or with up to 3 coarse teeth or short lobes on either side, thinly to moderately pubescent overall, the end leaflet pinnatifid. *Cauline leaves* similar to the basal but gradually narrower and smaller towards the top of the stem and merging with the leaf-like bracts, pinnately-lobed in the lower

Meconopsis ganeshensis; Ganesh Himal, C Nepal. PHOTO: BILL BAKER

half but increasingly pinnatifid towards the apex, the lobes entire to subentire, sometimes slightly toothed, the lowermost leaves petiolate, but the middle and upper sessile, with the lowermost pair of leaflets extended to the base as a wedge-shaped, flange-like wing. *Inflorescence* subpaniculate, occupying most of the stem, sometimes produced sub-basally as well, with the lowermost flowers generally in small cymules of 2–3, sometimes solitary, but the uppermost flowers always solitary, the rosette leaves generally partly withered by flowering time; flowers cup-shaped, nodding to half-nodding. *Peduncles* 4–11 cm long, extending somewhat as the fruits develop, moderately to densely pubescent. *Pedicels* rather slender, 2.5–9 cm long (to 12.2 cm long in fruit), pubescent like the peduncles, densely so towards the top. *Flower buds* ovoid to subglobose, covered with dense long, ascending, somewhat incurved, stiff barbellate hairs, these mostly 6–9 mm long. *Petals* 4, dark red to pinkish-red, occasionally pink, sometimes with a purplish flush, oval to oval-obovate, 22–40 × 20–35 mm, with the margin slightly recurving. *Stamens* numerous, with filiform filaments the same colour as the petals and 2 mm long, orange-yellow anthers. *Ovary* with dense suberect barbellate bristles, the style glabrous slender, 6–8 mm long, with a capitate, rather obscurely 5–6-lobed stigma. *Fruit* capsule narrow obovoid to ellipsoid-oblong, 15–20 × 6–10 mm, 5–6-valved, narrowed to a distinct and slender stipe at the base, with a sparse covering of long spreading to somewhat deflexed bristles (rather dense in immature fruits), but becoming glabrescent at maturity, terminating in a 5–8 mm long persistent filiform style, that is slightly broadened at the base. Fig. 9B.

DISTRIBUTION. Central Nepal; Ganesh Himal (N Dhading, Nuwakot & Rusawa districts: Ankhu Khola, Mailung Khola, Paldol area, Pongsing); 3658–4600 m.

HABITAT. Stony alpine slopes, amongst rocks, rocky meadows, often growing in association with lower perennial herbs.

FLOWERING. July–August.

SPECIMENS SEEN. *Dhwoj* 21 (BM) (2 sheets, one in flower, the other with the previous year's fruits), 129★ (BM, E); *EKSIN* 163B (E); *Miyamota et al.* (University of Tokyo) 94-00059 (BM, E, TI) and 94-00076 (BM, E, TI), 94-40083 (TI); *K. N. Sharma* E365 (BM); *Stainton* 3990 (BM, E).

NOTE. ★*Dhwoj* 129 is marked 'Purple' on the field notes without qualification, or any other field data apart from the locality 'Pongsung'. Pongsung lies in the Ganesh Himal south of Paldol peak along the Mailung Khola (a tributary of the Trisuli River, a locality just to the east of the type locality of *Meconopsis ganeshensis* in the upper Ankhu Khola). The rather inaccessible Ganesh Himal are normally approached either up the Ankhu Khola or from the Trisuli; however, there is a longer approach entering the Ganesh from the north-west from the upper reaches of the Buri Gandaki, a little known region botanically and certainly one that would repay further exploration.

Meconopsis ganeshensis is a graceful member of series *Polychaetia* characterised by rather finely dissected foliage and an indumentum of slender bristles that gives the leaves, stems and drooping buds a rather shaggy appearance. Unlike the more robust species, *M. staintonii* and *M. paniculata* in particular, the indumentum is not accompanied by a dense underfelt of shorter hairs, although there is some rather sparse underlying indumentum present. It is almost certainly closely related to *M. napaulensis*, but is readily distinguished on account of its flower colour and in the details of the dissection and indumentum of the leaves. It is, in addition, a slighter plant, the leaf segments more finely dissected, the stem leaves with a wedge-shaped rather than auriculate base.

Unfortunately, this recently described species is not in cultivation and is unlikely to be so in the near future. Although the Ganesh Himal is fairly close to Kathmandu, the area is remote and not readily accessible, except by an intrepid expedition. It is also an exceptionally wet area during the monsoon when the plants are in flower; seed is normally ripe towards the end of the monsoon period, in September and October. I am very grateful to Dr Bill Baker, RBG Kew, who mounted an extremely successful expedition to the Ganesh Himal in August 1992, for allowing me access to his photographs, one of which is included in this work. This expedition, which gathered a host of interesting plants, including red forms of *Roscoea purpurea*, revealed that the Ganesh Himal and surrounding mountains are an important region of biodiversity and one that requires more detailed exploration, particularly those mountains that border the Tibetan frontier.

Meconopsis ganeshensis is an extremely dainty and attractive species and would make a valuable addition to the smaller evergreen monocarpic species grown in gardens. Further material from the wild would be valuable in making a more comprehensive overview of the species.

10. MECONOPSIS STAINTONII

Meconopsis staintonii Grey-Wilson, *Curtis's Bot. Mag.* 23 (2): 190 (2006). Type: C Nepal, Kali Gandaki Valley, above Lete, 12,000 ft, 4 June 1954, *Stainton, Sykes & Williams* 5583 (BM, holotype; BM, E, isotypes).
[syn. *M. napaulensis* sensu Taylor pro parte in *Meconopsis*: 44 (1934)].

DESCRIPTION. *A robust monocarpic plant* to 2 m tall, generally less, with a stout dauciform root up to 30 mm diameter, forming a substantial leaf rosette, up to 1 m across, in the years before flowering. *Stem* stiffly erect, to 30 mm diameter near the base, moderately to densely covered (especially in younger plants) in long yellow or golden-brown spreading, barbellate hairs with an underlying felt of short barbellate hairs which appear to be substellate. *Basal and lowermost leaves* long-petiolate, the petiole 7–16 cm long, unwinged, pubescent, the lamina oblong-lanceolate to lanceolate, 24–52 × 6.5–18 cm, pinnatisect in the lower half, increasingly pinnatifid towards the apex, with 4–8 pairs of lateral segments, these oval to oblong or ovate, subacute to obtuse, serrately toothed to subentire, moderately to densely covered on both surfaces with a mixture of long yellow, golden or occasionally tawny-coloured barbellate hairs underlain with a felt of short barbellate hairs. *Middle and upper cauline leaves* similar to the basal but short-petiolate, the upper stem leaves increasingly smaller, sessile, with a wedge-shaped to rounded or semi-auriculate base, otherwise similar to the basal leaves. *Inflorescence* a large multi-flowered panicle forming a broad cone shape at maturity, pubescent like the stem, most flowers

Meconopsis staintonii, white-flowered form; S of Annapurna Sanctuary, C Nepal, 3200 m.
PHOTO: TOSHIO YOSHIDA

in lateral cymules of 3–11, occasionally more on the lowermost branches of the inflorescence; flowers cup-shaped to campanulate, nodding. *Bracts* similar to the upper leaves, sessile, decreasing in size towards the top of the inflorescence. *Peduncles* spreading widely apart to ascending, 7.5–17 cm long, pubescent like the stems. *Pedicels* slender, 7.8–18 cm long. *Buds* nodding, ovoid, densely covered in ascending to incurving bristles giving the buds a rather golden appearance. *Petals* 4, red, pink or pinkish-red, sometimes with a hint of purple, occasionally white, obovate to suborbicular, 36–60 × 40–75 mm, somewhat crumpled in appearance and generally with an undulate margin. *Stamens* numerous, the filaments generally the same colour as the petals, the anthers yellow or orange-yellow. *Ovary* densely covered with appressed or slightly spreading orange-yellow bristles, with a stout 5–8 mm long style; stigma capitate, dark green. *Fruit capsule* ellipsoid to obovoid, 22–35 × 8–13 mm, densely covered with semi-appressed to ascending barbellate, 6–7 mm long bristles, 7–11-valved, splitting for a short distance from the top, slightly waisted towards the base, the apex narrowed into a stout 8–14 mm long style which is 2–3 mm diameter at the base and puberulous or with short bristles in the basal half. 2n = 56 (Ratter 1968, as *M. napaulensis*). Fig. 9C.

DISTRIBUTION. C Nepal (Baglung, Kaski, Myagdi & S Mustang districts; primarily Annapurna Himal, Kali Gandaki Valley and the Lamjung Himal; 2590–4300 m (Maps 3 & 5).

HABITAT. Open oak-fir forest, rocky slopes, ravines, shrubberies.

FLOWERING. Late May–July.

SPECIMENS SEEN. *Ohba et al.* 8331987 (E); *Schilling* 1028 (K); *Seventh Botanical Exped. Himalaya* (1983) 8331987 (E); *Stainton* 6357 (BM), 6359 (BM); *Stainton, Sykes & Williams* 747 (BM, E), 970 (BM, E), 1192 (BM), 1651 (BM, E), 1719 (E), 3099 (BM, E), 3488 (BM), 3996 (E), 5583 (BM, E), 6359, (BM), 7405 (BM), 7943 (BM, E), 8461 (BM), 8479 (BM), 8480 (BM), 8481 (BM), 8482 (BM), 8483 (BM).

Above left: *Meconopsis staintonii*, pale pink-flowered form; Marci Bugli above Taglung, Kali Gandaki Valley, c. 4000 m, C Nepal. PHOTO: JAN BURGEL
Above right: *Meconopsis staintonii*, flower detail; Marci Bugli above Taglung, Kali Gandaki Valley, c. 4000 m, C Nepal. PHOTO: JAN BURGEL

Meconopsis staintonii, red-flowered form; Yak Kharka, western terrace of Kali Gandaki, NE of Mt Dhaulagiri, C Nepal, 3700 m.
PHOTO: TOSHIO YOSHIDA

This robust and extremely handsome species has a limited distribution centred on the Kali Gandaki valley of west central Nepal, where it is a feature of woodland margins and shrubberies at mid-altitudes flowering in June and July. It is a familiar species to those who grow *Meconopsis* with its large striking golden leaf-rosettes and huge panicles of nodding pink or red flowers. When it came into cultivation is problematic, mainly due to the confused use of the name *M. napaulensis*. George Taylor states in his 1934 monograph of the genus that it " …… was known in cultivation before 1831, but the evidence for this is unconvincing, and it appears that the first record of the species flowering in this [sic UK] country is in 1852." However, all these early records seem to be for collections made in eastern Nepal by Joseph Dalton Hooker and are referrable to *M. wallichii* Hook. It is likely that *M. staintonii* came into cultivation in the 1950s through the collections of Stainton, Sykes & Williams. Other so-called red-flowered '*M. napaulensis*' are almost certainly referable to the red form of *M. wallichii* or to possible red-flowered forms of *M. regia*, these latter requiring further detailed investigation. In any case, when grown in close proximity all these species, along with *M. paniculata*, tend to hybridise so that in cultivation species boundaries are soon blurred unless the species are kept well isolated from one another.

Meconopsis staintonii is most closely related to *M. paniculata*, differing most obviously in the pink or red, rather than yellow flowers, but also in the dissection of the leaves, in the presence of a green rather than purple stigma, and in the narrower, more ellipsoid fruit capsules. From *M. regia* it differs both in the lobed rather than tooth-margined leaves, in the rougher less silky indumentum, and in the generally narrower, less markedly ribbed fruit capsules. White-flowered forms are seen fairly frequently in the wild, especially in the middle reaches of the Kali Gandaki valley.

11. MECONOPSIS WALLICHII

Meconopsis wallichii Hook., *Curtis's Bot. Mag.* 78, t. 4668 (1852). Type: *Bot. Mag.* t. 4668 (1852); *J. D. Hooker* s.n. Sikkim (BM, K) neotype chosen here[10].
[syn. *M. wallichii* var. *typica* Prain, *Bull. Misc. Inform.*, Kew 1915: 176 (1915); *M. napaulensis* sensu Taylor pro parte, *The Genus Meconopsis*: 44 (1934)].

DESCRIPTION. *A robust monocarpic plant* to 1.4 m tall, forming a large evergreen rosette up to 80 cm across in the first two or three years before flowering; stem erect, covered with stiff barbellate, patent to ascending hairs, with an underlay of shorter but similar hairs. *Basal leaves* to 29 cm long overall, including the petiole which can be up to 10 cm, occasionally longer, the lamina oval to oval-oblong in outline, pinnately-lobed below, pinnatifid towards the apex, with 5–8 pairs of primary, opposite or subopposite segments, these well-spaced generally, oval to elliptic, with a subentire to toothed or slightly lobed margin, sparsely to moderately hairy on both surfaces with long, fawn-coloured or brownish, barbellate hairs underlain with far shorter, but similar, hairs, sometimes becoming glabrescent, especially above. *Cauline leaves* similar to the basal but becoming progressively smaller

---

[10]*Meconopsis wallichii* was based on cultivated plants grown from seed collected by Joseph Dalton Hooker in Sikkim in 1848.

and gradually merging with the bracts, the lower with a short petiole, the upper sessile or subsessile. *Inflorescence* a substantial candelabra-like panicle taking up the upper half to one third of the stem, the lowermost branches with flowers in cymules of 3–11 (–15), the uppermost usually solitary; flowers cup-shaped, nodding, 4–7 cm across. *Peduncles* 4.5–9 (–12.5) cm long. *Pedicels* 6–14.5 cm long, clothed with erect to patent hairs. *Buds* subglobose to ovoid or ellipsoid, nodding, covered with short to long (2–8 mm long) patent to incurved, barbellate, stiff hairs, underlain by much shorter, but similar, hairs. *Petals* 4, subrounded to obovate, 20–35 × 23–38 mm, somewhat overlapping, blue, mauve-blue, purplish-pink, lavender-blue, maroon-purple, to deep red, occasionally whitish with a blue or lilac tint, generally somewhat ruffled towards the margins. *Stamens* with filaments the same colour as the petals, although often darker, the anthers yellow or orange. *Ovary* narrow ellipsoidal, covered with appressed to ascending barbellate bristles, and with a slender style, 5–9 mm long, terminating in a small capitate greenish stigma, 1–2 mm diameter. *Fruit capsule* flask-shaped, with a rounded base, slightly narrowed at the top, 11–25 × 7–12 mm, 5–7-valved, covered in moderately dense patent long barbellate bristles with an underlay of short but similar bristles, terminating in a very slender persistent, glabrous style 7–12 mm long, not more than 1 mm diameter at the base. (Fig. 9D1,2).

DISTRIBUTION. E Nepal (Ilam, Ramechhap, Sankhuwasabha, Solukhumbu & Taplejung districts; Arun Valley and Thudam eastwards), Sikkim and CW Bhutan (Chelai La, Paro, rare!); 2440–4300 m (Map 5).

HABITAT. Forest clearing, margins and pathways, shrubberies, rocky alpine meadows, ravines, stream margins.

FLOWERING July–September.

SPECIMENS SEEN. *T. Anderson* 363 (K); *Bannerjee & Shakya* 5574 (BM); *Beer* 8246★[11] (BM), 25417[11] (BM); *Cave* s.n. 29 July 1914 (E), 6582 (E); *C. B. Clarke* 13460 (K), 27522 (K); *Dhwoj* 129[11] (BM), 250[11] (BM), 504[11] (BM); *Grey-Wilson et al.* 4527 (K); *D. J. Long et al.* 728 (E); *East Nepal Exped. Koshi Zone* (1991) 9120391 (E); *Flora of East Nepal* (1995) 9584061 (E), 9584100 (E), 9596095 (E); *J. S. Gamble* 120A (K), 9481 (K); *J. D. Hooker* s.n. Sikkim (BM, K); *Dr King's Collector* May 1896 (K), Aug. 1887 (K); *Ludlow, Sherriff & Hicks* 19855 (BM); *Mijamoto et al.* 9584061 (E), 9596095 (E); *Ohashi et al.* 775177 (BM); *Ohba et al.* 8571040 (E), 8572313 (E), 8580659 (E); *D. Probert* s.n. Sandakphu[12] (K); *Rabongla* s.n. (K); *Rohmoo Lepcha* 1054[11] (E); *Shilling* 1028[11] (K); *Stainton* 844 (BM), 975[11] (BM); *Wallich* 8121 (K); *Williams & Stainton* 8438 (BM, K).

Above, from top:

Meconopsis wallichii var. *wallichii* leaf-rosette; in cultivation. PHOTO: CHRISTOPHER GREY-WILSON

Meconopsis wallichii var. *wallichii*; W of Topke Gola, Jaljale Himal, E Nepal, 3700 m. PHOTO: TOSHIO YOSHIDA

[11] Specimens indicated bear red or wine-purple flowers and are attributable to var. *fuscopurpurea* (below); however, some of the other specimens cited above have no colour indication on the field notes and may also belong to this variety. Apart from flower colour, no quantifiable differences have been noted between the two colour forms, although apparently they do not grow together in the wild.

[12] This specimen is recorded as 'India, W Bengal, Sandakphu, 600 m' which is at odds, both geographically and altitudinally, with all other records of *M. wallichii* and is therefore regarded as an erroneous record.

Joseph Dalton Hooker brought this graceful and very beautiful plant to the attention of gardeners, along with *Meconopsis simplicifolia* (D. Don) Walp., although Nathaniel Wallich had first discovered the latter some thirty years previously. Hooker had collected seed of *M. wallichii* in Sikkim in 1848 and it first flowered in Britain in June 1852. A plant grown at Kew was illustrated in *Curtis's Botanical Magazine* that same year and formed the basis of the description of the species. Interestingly, no herbarium specimen of the plant figured seems to exist (it is certainly not in the herbarium at Kew) so that the original watercolour drawing (the basis of the lithographic plate 4668) becomes the type of the name *M. wallichii*. Oddly, there exist in the Kew Herbarium four sheets of the species collected in Sikkim Himal by J. D. Hooker, all without number but all annotated as such in Hooker's own handwriting, while one sheet carries a pencil sketch of the fruit details. These specimens are not mentioned in the rather short account in the *Botanical Magazine*. Similar specimens, presumably from the same gatherings and also unnumbered, reside in various other herbaria apart from Kew (BM, E, FI, M, MANCH).

William Jackson Hooker, who described *Meconopsis wallichii*, states that "The plant, with us, grown in pots in a frame, attains a height of two and a half to three feet. The whole herb is pale subglaucous green, everywhere hispid with long spreading ferrugineous setae. Radical leaves large, petiolate, lyrato-pinnate, or pinnate below and pinnatifid above, the pinnae and lobes ovate-oblong ……. corolla of four subrotundo-obcordate, spreading, pale blue petals, having sometimes a slight tinge of green".

Henry Elwes, who visited Sikkim many years later remarked that *Meconopsis wallichii* (as '*M. napaulensis*') "was the most beautiful herbaceous plant in the world." In his *Memoirs of Travel, Sport and History* (1930), he goes on to say, "Imagine a large rosette of leaves clothed with long golden hairs, which, when covered with raindrops, glisten in the sunshine, running up into a branching spike of golden green buds covered with similar hairs, and opening from the top downwards into large poppy-like flowers, normally of a bright pale purple, whose centre is filled with a mass of golden anthers."

Although George Taylor (1934) was firmly of the belief that this species was little more than a colour variant of '*Meconopsis napaulensis*' opinion has shifted in the intervening years and gardeners who had long thought it to be distinct were, for once, pleased when the old name *M. wallichii* was resurrected. This change has not been universally accepted (even the *Flora of Bhutan* still doggedly adheres to *M. napaulensis*!). However, *M. wallichii* as here defined is only found in the eastern Himalaya, from E Nepal to central Bhutan. In cultivation, as in the wild, it tends to come into flower later than most other section *Polychaetia* species so that in a group the seed is likely to be pure rather than of hybrid origin; however, there is often an overlap in flowering so that the first flowers of *M. wallichii* may coincide with those of other species or hybrids and thus some hybridisation may occur. This can be circumvented by gathering seed only from those fruits derived from the later flowers in an inflorescence when the chance of hybridisation is much reduced. Having said this, the only true way of ensuring purity is to isolate species in gardens; copious seed is normally set with these large monocarpic species. This not only extends the *Meconopsis* flowering season but helps to ensure that the species (or at least the later flowers) do not hybridise with other species growing in close proximity. In the wild the only other species with which it could possibly grow is *M. paniculata*, but I cannot confirm this.

Meconopsis wallichii var. *wallichii*; W of Topke Gola, Jaljale Himal, E Nepal, 3500 m.
PHOTO: TOSHIO YOSHIDA

94 THE GENUS MECONOPSIS
11. MECONOPSIS WALLICHII

PLATE 2

Meconopsis wallichii var. *wallichii*; the painting by W. H. Fitch for *Curtis's Botanical Magazine*, t. 4668 (1852), which represents the type of the name *M. wallichii* in the absence of any validated herbarium specimen.

Meconopsis wallichii in various colour forms is well-established in collections, especially in the north of Britain, mainly from gatherings made in east Nepal in the past twenty or so years; the pale blue forms are especially pleasing.

Joseph Dalton Hooker described var. *fusco-purpurea* in 1884, assigning to it forms of *Meconopsis wallichii* found in the wild that had purple or wine-purple flowers. George Taylor (1934) extended this to include red-flowered forms as well. These latter are quite often encountered in eastern Nepal, for instance in the region encompassing Ghunsa, Panch Pokhari and the Yalung Glacier, but they also to be found in Bhutan. As so many herbarium specimens do not have the colour of the flowers recorded it is impossible to assign them accurately to one or other variety.

var. fusco-purpurea

var. **fusco-purpurea** Hook. f., *Curtis's Bot. Mag.* 110: tab. 6760 (1884). Type: Cultivated specimen grown in Weybridge, by G. F. Wilson = *Curtis's Bot. Mag.* tab. 6760 (K, holotype).
[syn. *M. wallichii* var. *rubrofusca* Prain, *J. Asiat. Soc. Bengal pt. 2, Nat. Hist.* 64 (2): 315 (1896), *laps. pro fusco-purpurea*; *M. wallichii* forma *purpurea* Bulley, *Fl. & Silva* 3: 84 in obs.(1905); *M. wallichii* forma *fusco-purpurea* Bulley, loc. cit. (1905)].

DESCRIPTION. Flowers wine-purple, purple-red or red.
DISTRIBUTION. East Nepal (Sankhuwasabha District), W Bhutan (Haa District), possibly elsewhere in E Nepal (see footnote 11 above).

All specimens with purple or red flowers are attributed to this variety by George Taylor (*Meconopsis*: 47 (1934)). Interestingly, in the wild var. *fusco-purpurea* appears to grow in isolation from the typical plant (var. *wallichii*) and I have no authenticated record of the two colour forms growing in mixed colonies or indeed in very close proximity to one another. Photos of both taken by Toshio Yoshida in eastern Nepal show what a beautiful and graceful plant this can be in the wild. His exquisite photos of var. *wallichii* growing in a misty woodland setting are especially evocative and reveal splendidly the sort of conditions that the plant grows in when it is in full flower, this coinciding with the summer monsoon which is particularly heavy in the eastern Himalaya. In recent years var. *fusco-purpurea* has been observed and photographed in western Bhutan (Sele La, Haa).

Above, from top:
Meconopsis wallichii var. *wallichii*; Milke Danda, S of Topke Gola, E Nepal, 3200 m. PHOTO: CHRISTOPHER GREY-WLSON
Meconopsis wallichii var. *wallichii* in cultivation at Logan Botanic Garden, Scotland. PHOTO: CHRISTOPHER GREY-WLSON

96 | THE GENUS MECONOPSIS
11. MECONOPSIS WALLICHII

Clockwise from above:

Meconopsis wallichii var. *fusco-pupurea* coming into flower; Sela La, W Bhutan, 3550 m. PHOTO: DAVID & MARGARET THORNE

Meconopsis wallichii var. *fusco-purpuea* leaf-rosette; Sela La, W Bhutan, 3550 m. PHOTO: DAVID & MARGARET THORNE

Meconopsis wallichii var. *fusco-purpurea*; S of Ghuna, Kanchenjunga Himal, E Nepal, 4100 m. PHOTO: TOSHIO YOSHIDA

Meconopsis wallichii var. *fusco-purpurea*; S of Panch Pokhari, Dolakha District, E Nepal, 3800 m. PHOTO: TOSHIO YOSHIDA

THE GENUS MECONOPSIS | 97
11. MECONOPSIS WALLICHII

PLATE 3

Meconopsis wallichii var. *fusco-purpurea*; a painting by Matilda Smith for *Curtis's Botanical Magazine*, t. 6760.

12. MECONOPSIS WILSONII

Meconopsis wilsonii Grey-Wilson, *Curtis's Bot. Mag.* 23 (2): 195 (2006). Type: China, W Sichuan, SE of Moupin (Baoxing), 11–13,000 ft, *E. H. Wilson* 1152 (K, holotype; E, K isotypes).

[*M. napaulensis* sensu G. Taylor pro parte, *The Genus Meconopsis*: 44 (1934)].

DESCRIPTION. *A rather robust monocarpic plant* to 1.5 m tall, with a dauciform root, the whole plant covered in stiff greyish to straw-coloured or ginger barbellate hairs with an underlay of similar short barbellate or substellate hairs. *Stem* erect green, sometimes flushed with red or purple, patent to appressed pubescent, striate, up to 1.7 cm diameter at the base. *Leaves* in a substantial evergreen basal rosette, 35–70 cm across, in the years prior to flowering, bluish- or grey-green, sometimes with a purplish or reddish flush, especially along the midrib and main veins: basal leaves wide-spreading, often absent at flowering time, long-petiolate, the lamina lanceolate to lanceolate-elliptic or elliptic-oblong, to 30 × 10 cm, pinnatifid in the lower part, pinnatisect towards the apex, with 6–12 (–15) pairs of opposite or subopposite lateral segments, the lowermost segments rather distant, the upper crowded, these ovate to oval-oblong, 25–70 × 10–45 mm, pinnatisect (with 1–4 obtuse or subobtuse lobes of teeth on each side) to subentire (the uppermost) with obtuse to subobtuse lobes, covered on both sides with rather sparse greyish stiff barbellate hairs with a denser underlay of much smaller but similar hairs; lower and middle cauline leaves similar to the basal but with a decreased petiole, lax to crowded, but the upper stem leaves decreasing gradually in size upwards, sessile, pinnatifid, with fewer segments, half-clasping the stem, merging with the bracts; petiole of basal and lower cauline leaves 9–20 cm long. *Inflorescence* fastigiate-paniculate or subpaniculate with short erect to sharply ascending branches, the lowermost flowers generally in cymules of 1–5, the uppermost solitary; flowers numerous, half-nodding or lateral-facing, deeply cup-shaped, occupying the upper half of the plant. *Bracts* similar to the upper leaves, decreasing in size up the inflorescence, sessile, semi-amplexicaul; bracteoles oval to elliptic, 9–25 mm long, entire or sometimes slightly lobed. *Peduncles* stout, 0.6–8 cm long (to 10.4 cm in fruiting specimens) erect to ascending, often held more or less parallel to the main rachis, pubescent like the stem. *Pedicels* 0.4–10 cm long (to 12 cm in fruiting specimens), pubescent like the stem. *Buds* nodding to half-nodding, narrow-ovoid, with patent to ascending or appressed, stiff barbellate hairs underlain with a felt of similar but shorter hairs. *Petals* 4, purple, to wine-purple, deep wine-crimson, 'light maroon', violet or violet-blue, purple-blue, occasionally white flushed with wine-purple, often with deeper veining, obovate to more or less orbicular, 22–42 × 22–42 mm, generally rather crumpled. *Stamens* numerous, the filaments the same colour as the petals, somewhat uneven, half-spreading, the anthers orange to golden-yellow. *Ovary* ovoid, covered in ascending to appressed bristles, the style greenish or purple, 4–7 mm long, with a capitate purple, occasionally white, stigma. *Fruit capsule* ovoid to ellipsoid, 14–34 × 8–15 mm, the uppermost fruits in the infructescence or secondary ones in cymules, often smaller than the others, the capsules generally 5–7-valved, covered in ascending to subappressed barbellate bristles, and with a 5–8 mm long persistent style. *Seeds* ovoid to ovoid-elliptic, 0.8–1.2 × 0.5–0.8 mm, somewhat asymmetric, mid-brown to deep reddish brown, finely rugose.

DISTRIBUTION. W & SW China (W Sichuan (Baoxing area in particular), W & NE Yunnan (Cangshan, Yulongxueshan, Mekong-Salween, Maikha-Salween and Shweli-Salween divides); NE Myanmar (Kachin); 2745–3965 m (Map 6).

Meconopsis wilsonii subsp. *wilsonii*; Mianning, S Sichuan, 3000 m. PHOTOS: TOSHIO YOSHIDA

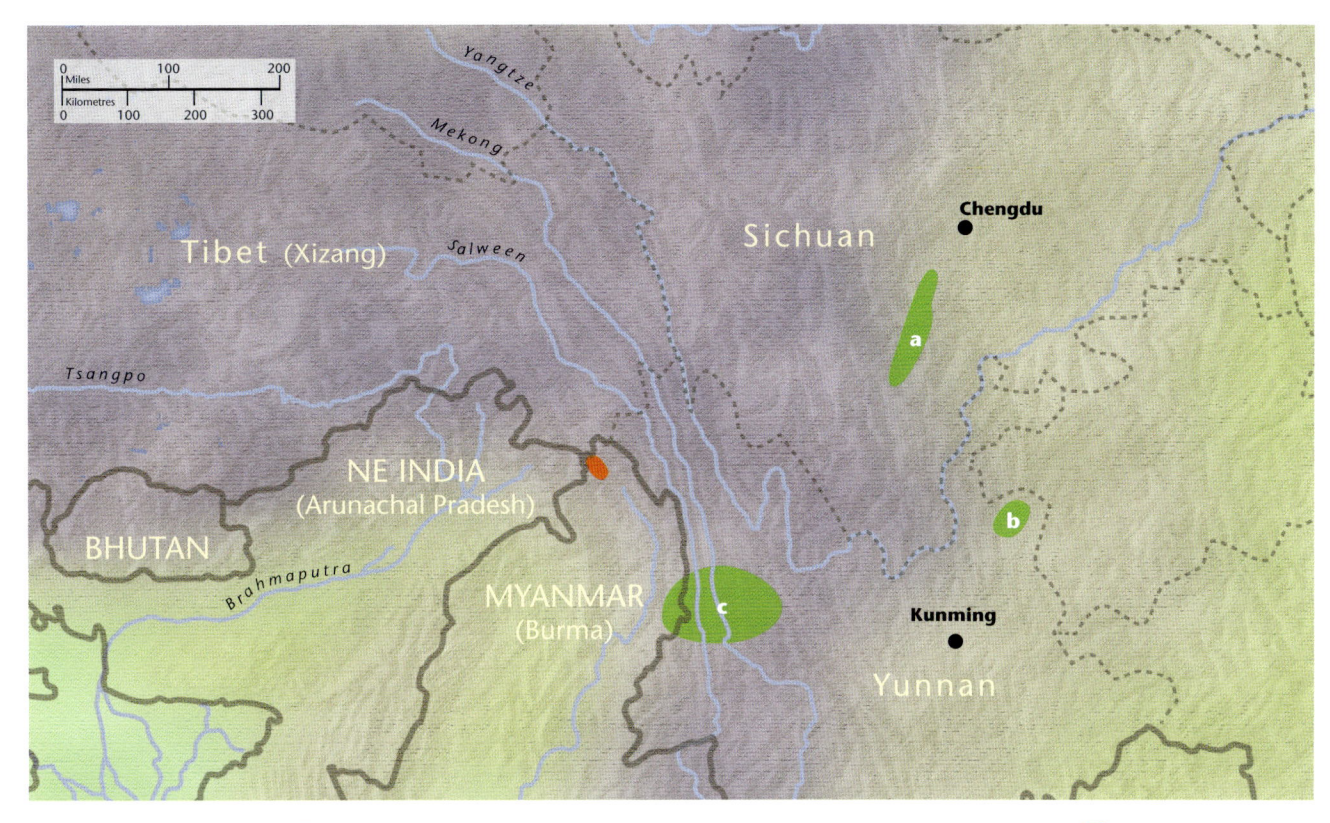

MAP 6. Species distribution: ■ Meconopsis wilsonii a – subsp. wilsonii, b – subsp. orientalis, c – subsp. australis, ■ M. violacea.

HABITAT. Forest margins, moist shrubberies, open stony or grassy slopes, cliff ledges and other rocky places.
FLOWERING. June–early August.

It is perhaps surprising that this taxon had remained unrecognised for so many years, harbouring as it did under the *Meconopsis napaulensis* umbrella, yet it is distinctive in both character and in its distribution. It is generally a large and rather coarse plant with handsome greyish or bluish-green leaf rosettes. Furthermore, it is very distinctive in flower with its spire-like inflorescences with their rather short-branches which give the whole inflorescence a narrow 'fastigiate' appearance. The species is the only member of the subgenus *Meconopsis* to be found in China (excluding Tibet) and there it has a rather curious distribution: E. H. Wilson collected it in western Sichuan, south of Moupin (now Baoxing) in the early years of the twentieth century, while Forrest, Kingdon Ward and others collected it in north-western Yunnan and the neighbouring part of Burma (Myanmar) in succeeding years. In more recent years it has also been found in NE Yunnan in the Wumengshan. Curiously, it has been seen very little in recent years and few expeditions have come across it, although some fine photographs of it in the wild have been taken. However, plants in cultivation at the present time stem from a collection made by the late James Taggart in the vicinity of Mt Wumung (Wumengshan), in north-eastern Yunnan, and incorrectly distributed as *M. wumungensis* K. M. Feng. Since then the plant has been photographed in flower in the wild by Duncan Coombs and Joseph Atkin and others in Yunnan in 2005 and, more recently, in several localities by Toshio Yoshida.

Previous authors (Taylor 1934 in particularly) are rather disparaging about this plant, indicating that it is not particularly attractive in flower, the petals being described variously as dull purple or wine-purple or even 'wine-crimson'. However, photographs of it taken in the wild show it to be a very striking plant which, in its best forms, has flowers of violet, violet-purple or violet-blue. In addition, the handsome leaf-rosettes are a soft blue-grey, the leaves sometimes enhanced still further by reddish rachises and petioles. Thus it is quite the match of its handsome Himalayan cousins. It has been confused with *M. violacea* on a number of occasions but the two are very different in both the details of the leaves and inflorescence, while the fruit capsules of *M. violacea* are proportionately longer and narrower than those of *M. wilsonii* sensu lato. The former is a narrow endemic found in the border regions of northern Myanmar and the adjacent part of Tibet.

The species is named in honour of the intrepid traveller, plant collector and author Ernest Henry Wilson who collected many interesting plants in western China and elsewhere, particularly trees and shrubs, but notable herbaceous plants including the first introduction of *Lilium regale* which he discovered in the Ming valley of northern Sichuan. Tragically he was killed along with his wife in a motorcar accident in the USA in 1930.

Meconopsis wilsonii finds its closest ally in *M. wallichii* Hook. f., differing markedly in having a narrow 'fastigiate' inflorescence and infructescence with short, erect to suberect branches, in the few-flowered lateral cymules, and in the broader fruit capsule with erect to ascending, not patent, bristles. *M. wallichii* has a more westerly distribution being found in eastern Nepal (from the Arun Valley eastwards), Sikkim and western Bhutan where it is rare.

The inflorescence (and subsequent infructescence) of *Meconopsis wilsonii* is of particular interest, being paniculate in the lower part but racemose in the upper. The lower branches of the inflorescence are short and stiffly ascending, giving the whole a distinctly columnar appearance. The lateral cymules are often 2–5-flowered but on occasion, especially in subsp. *orientalis*, they may be reduced to a single flower, sometimes because the secondary buds abort at an early stage. In these instances the inflorescence appears to be entirely racemose; however, close examination will show that the lower flowers are bracteolate, thus inferring a reduced paniculate state. Interestingly, in section *Polychaetia* only two species possess simple racemose inflorescences, these being the western Himalayan, yellow-flowered, *M. robusta*, and the eastern Himalayan, violet-flowered, *M. violacea*. In *M. wilsonii* the reduced state of the inflorescence is also revealed at the fruiting stage, when quite frequently all the infrustescences may bear single-fruited 'pedicels'.

The Yunnan representatives of this new species differ enough to be considered as distinct subspecies (subsp. *australis* and subsp. *orientalis*) as is outlined in the key below. The Sichuan subspecies (subsp. *wilsonii*) is altogether more elegant in the dissection of its leaves and poise of the flowers but, alas, it is not in cultivation; indeed it has only rarely been seen or recollected in recent years to my knowledge.

KEY TO SUBSPECIES OF MECONOPSIS WILSONII

1. Basal and lower cauline leaves closely spaced, acutely angled (less than 45°) to the main stem, with 9–12 (–15) pairs of primary segments . . . . subsp. **orientalis** (NE Yunnan)
+ Basal and lower cauline leaves widely separated, widely angled (up to 90°) to the main stem, with 4–8 (–10) pairs of primary segments 2

2. Basal and lower cauline leaves with 6–8 (–10) or pairs of primary segments, the middle and upper cauline leaves and lower bracts deeply divided with up to 8 pairs of segments; peduncles and pedicels with ascending to subappressed hairs. subsp. **wilsonii** (W Sichuan and NW Yunnan)
+ Basal and lower cauline leaves with 4–6 (–8) pairs of primary segments, the middle and upper cauline leaves and lower bracts shallowly divided, with up to 5 pairs of segments; peduncles and pedicels with patent hairs . . . . subsp. **australis** (W Yunnan, Myanmar)

subsp. wilsonii

DESCRIPTION. *Basal and lower cauline leaves* with 6–8 (–10) pairs of lobes, the cauline leaves rather lax, about 10 per 50 cm of stem, held at an angle of about 90° to the stem; upper leaves and bracts deeply divided, with up to 8 pairs of lobes, although the uppermost few-lobed. *Peduncles and pedicels* with ascending to subappressed yellowish grey barbellate hairs. *Buds* with subappressed-pubescence. *Flowers* with purple, purple-blue or violet-purple coloured petals. *Stamens* tightly clustered around the ovary, with purple filaments. *Ovary* with a purple stigma. *Fruit capsule* ovoid-ellipsoid. (Photos, p. 98)

DISTRIBUTION. W China: W Sichuan (SE Baoxing (Moupin) area, Mianning County (Yele Nature Reserve)); 3355–3965 m (Map 6).
HABITAT. Grassy places and open scrub.
FLOWERING. June–July.
SPECIMENS SEEN. E. H. Wilson 1152[13] (K), 3165, (K).

In flower subsp. *wilsonii*, like subsp. *australis*, bears rather lax cauline leaves, calculated at about 10 per 50 cm of

---

[13] There are five fine, well-prepared sheets of this number in the Kew Herbarium.

stem. The leaves bear an intermediate number of lobes between the other two subspecies, while the bracts are more deeply divided than either subsp. *australis* or subsp. *orientalis*. It is perhaps the least well known of the subspecies stemming primarily from the fine specimens collected by Ernest Henry Wilson.

subsp. australis Grey-Wilson, *Curtis's Bot. Mag.* 23 (2): 1907 (2006). Type: China, Yunnan, Shweli-Salween Divide, 25°30'N, August, *Forrest* 15883[14] (E, holotype; BM, K, isotypes).

DESCRIPTION. Similar to subsp. *wilsonii* in the density and disposition of the cauline leaves, but differing from that subspecies in having more coarsely divided laminas with normally no more than 6 pairs of lateral segments, the upper leaves and bracts shallowly divided. *Peduncles and pedicels* with patent barbellate hairs. *Buds* patent-pubescent. *Flowers* with rose-pink to purple to wine-purple, deep, often dull, wine-crimson, occasionally maroon petals. *Stamens* held closely around the ovary, with purple filaments. *Ovary* with a purple or white stigma. *Fruit capsule* ovoid-ellipsoid. (Fig. 9E1,2).

DISTRIBUTION. W & NW Yunnan (Biluoxueshan, Cangshan, Maikha-Salween divide, Mekong-Salween divide, Shweli-Salween divide), NE Myanmar (Upper Burma, Kachin; Chimi-li, Myitkyina, above Hpawte); possibly also in SE Tibet (Xizang); 2745–3685 m (Map 6).

HABITAT. Rocky and stony places, cliffs, forest and scrub margins, shrubberies.

FLOWERING. late June–August.

SPECIMENS SEEN. China (W & NW Yunnan): *Forrest* 6790 (BM, E), 8957 (BM, E), 11971 (BM, E), 15883 (E, BM, K), 17614, (E, K), 18212 (BM, E, K), 18363 (BM, E, K), 24602 (E, K), 24767 (E, K), 26817 (E, K), 26878 (E, K), 27301 — as 26817 but in fruit (K), 27344 (E, K), 29883

---

[14] A typographical error in the original publication cited the type specimen as *Forrest* 15833 when it should have been, as was clearly cited in the list of 'specimens seen', 15883.

Left: *Meconopsis wilsonii* subsp. *australis*; Pianma, NW Yunnan, 2800 m. PHOTO: JOE ATKIN

Above centre: *Meconopsis wilsonii* subsp. *australis*; Cangshan, Yangbi County, NW Yunnan, 3200 m. PHOTO: TOSHIO YOSHIDA

Above right: *Meconopsis wilsonii* subsp. *australis*, upper part of an infructescence; Bijiang, NW Yunnan. PHOTO: DAVID RANKIN

(E), 29884 (E), 30116 (BM, E); *Kingdon Ward* 3281 (E); *McLaren* 27 (BM, E, K), 64 (BM, K); *Naw Mu Pa* 17429 (K); *T. T. Yü* 16906 (BM, E)

Subsp. *australis* is the most widely distributed of the three subspecies of *Meconopsis wilsonii*, being found in a narrow band of NW Yunnan encompassing the rugged mountains of the Yangtse-Mekong-Salween divides as well as the adjacent region of northern Myanmar. In recent years it has been photographed in the wild by various people including Duncan Coombs, Joe Atkin and Toshio Yoshida. It is certainly a very attractive plant in flower with its columnar inflorescence: in the most striking forms the flowers are a rich purple or purple-blue.

subsp. orientalis Grey-Wilson, D. W. H. Rankin & Z. K. Wu, *Curtis's Bot. Mag.* 28 (1): 32–46, t. 700 (2011). Type: cultivated specimen, Lasswade, Scotland, from a James Taggart introduction from Wumengshan, NE Yunnan (E, holotype; Kunming, isotype).

DESCRIPTION. *Stem, petioles and pedicels* strongly flushed with red. *Basal and middle leaves* with 9–12 (–15) pairs of segments, the lower part of the leaf pinnately divided, the upper part pinnatisect; cauline leaves densely arranged (c. 40 in a 50 cm length of stem), the upper mostly held at an angle of <45° to the stem. *Bracts* shallowly divided. *Peduncles and pedicels* with patent barbellate hairs, the hairs pale yellowish ginger. *Buds* patent-pubescent. *Flowers* with violet to bright purple-violet petals. *Stamens* with rather uneven, half-spreading filaments. *Ovary* with a purple stigma. *Fruit capsule* ovoid-ellipsoid to ellipsoid.
DISTRIBUTION. NE Yunnan, Jiao Zi Xueshan (Wumengshan); 3450–3750 m (Map 6).
HABITAT. Open damp scrub on rocky slopes, often in partial shade.
FLOWERING. June–July.
SPECIMENS SEEN. *Peng Hua et al.* 9587 (K); *J. Taggart* s.n., cult. (E, KUN); *T. Yoshida* K28 (KUN), 47 (KUN).

Subsp. *orientalis* is a distinctive plant especially in the density of its leaves when it comes into flower. In addition, the characteristic red stems, petioles and leaf rachises give the plant an added attraction. Whether this coloration is common to the entire population of subsp. *orientalis* it is difficult to say; however, this subspecies is well isolated from other populations of *Meconopsis wilsonii*. The inflorescence can appear to be racemose rather than paniculate at first glance; however, close examination will show that even if all the lower lateral branches in the inflorescence bear solitary flowers, rather than cymules of 2–3, then bracteoles are always present, revealing that the inflorescence is in effect a reduced panicle.

Meconopsis wilsonii subsp. *orientalis* leaf-rosette; Wumengshan, NE Yunnan, 3200 m. PHOTO: TOSHIO YOSHIDA

Meconopsis wilsonii subsp. *orientalis* leaf-rosettes; Wumengshan, NE Yunnan, 3100 m. PHOTO: DAVID RANKIN

Above: *Meconopsis wilsonii subsp. orientalis*, lower part of a flowering plant; Laowoshan, Wumeng Shan, NW Yunnan, 3100 m. PHOTO: DAVID RANKIN

Left, from top:
Meconopsis wilsonii subsp. *orientalis*, leaf details; Laowoshan, Wumeng Shan, NW Yunnan, 3100 m. PHOTO: DAVID RANKIN
Meconopsis wilsonii subsp. *orientalis*; Laowoshan, Wumeng Shan, NW Yunnan, 3100 m. PHOTO: DAVID RANKIN

Subsp. *orientalis* was brought into cultivation by the late Jamie Taggart who collected seed on Mt Wumung or Wumengshan (Wumung Shan) in 1995 and has been maintained in cultivation since that date. This subspecies has proved to be readily straightforward in cultivation, plants having been grown for more than ten years in the west of Scotland. Those growing it have found that seedlings overwintered under cover, then planted out in the spring thrive without further protection. Fortunately, those grown by James Taggart and David Rankin have been kept well separated from related taxa to reduce the risk of hybridisation.

NOTE. In all three subspecies the fruit capsule size varies considerably: most apparent is the fact that those in the upper part of the infructescence are generally considerably smaller than those towards the base, although the terminal fruit capsule (i.e. the first flower in the inflorescence to open), may also have a large capsule. In addition, sometimes only the terminal flower in the lateral cymules will produce a fruit, the other flowers failing to do so; this applies particularly to the secondary flowers in any cymule. This has been noted both in the wild and in cultivated specimens, but it is by no means a constant feature and specimens can be found in which all the flowers go on to develop fruits. Fruit set may be determined by conditions at the time of flowering and flowers may fail to set fruit if it is too dry.

13. MECONOPSIS PANICULATA

Meconopsis paniculata (D. Don) Prain, *J. Asiat. Soc. Bengal, Pt. 2 Nat. Hist.* 64 (2): 316 (1896). Type: Nepal, 'Gop Tan', *Wallich* 8123b (K, holotype).

[syn. *Papaver paniculatum* D. Don, *Prodr. Fl. Nepal.*: 197 (1825), nomen illegit, pro parte; *Stylophorum nepalense* (DC.) Spreng. in L., *Syst. Veg.* ed. 16, 4 (2): 203 (1827) pro parte; *Meconopsis napaulensis* Walp., non DC., *Gen. Syst. Gard. and Bot.* 1: 135 (1831), nom. illegit., pro parte; *M. nipalensis* sensu Hook. f. & Thomson, *Fl. Ind.* 1: 253 (1855); *M. wollastoni* Regel, *Gartenfl.* 25: 291 (1876) in obs., nom. nud.; *M. paniculata* var. *typica* & var. *elata* Prain, *J. Asiat. Soc. Bengal, Pt. 2 Nat. Hist.* 64 (2): 316 (1896); *M. robusta* Prain, non Hook. f. & Thomson, *J. Asiat. Soc. Bengal, Pt. 2 Nat. Hist.* 64: 315 (1896) pro parte; *Polychaetia paniculata* Wall. ex Prain, *J. Asiat. Soc. Bengal, Pt. 2 Nat. Hist.* 64 (2): 316 (1896); *Meconopsis nepalensis* var. *elata* B., *Irish Gard.* 13: 103 (1918). Type: India, 1882–1887, Dr King's collector (P); India, 1892, Gammie (P); *M. himalayensis* (*Himalayense*) Hort. ex Hay, *New Fl. and Silva* II: 170 (1930), in obs., fig. 67].

DESCRIPTION. *Robust monocarpic plant* to 2.5 m, but generally 1–1.5 m tall, with a thick dauciform taproot to 20 mm diameter, the whole plant covered in a pubescence of golden or fawn, stiff barbellate hairs, these underlain by a sparse to rather dense felt of much shorter barbellate hairs; stem erect, 26 mm diameter near the base. *Leaves* rather bright yellow-green or lime-green, occasionally more bluish-green, forming a substantial rosette in the years before flowering, to 1 m across, sometimes more; basal leaves petiolate, the lamina lanceolate to narrow-elliptic to oblanceolate, 25–39.5 × 6.5–11.8 cm, pinnatifid overall or, more usually pinnatisect in the lower half and gradually pinnatifid in the upper, with 6–9 (–10) pairs of primary segments, these ovate to elliptic to asymmetrically triangular, entire or somewhat lobed at the margin, with a dense indumentum especially when young; lower cauline leaves similar to the basal but with progressively shorter petioles, the middle and upper cauline leaves decreasing in size up the stem, similar to the basal leaves but sessile, with a rounded to semi-amplexicaul base; petiole of basal leaves 12–34 cm long, occasionally longer, broadly expanded at the base. *Inflorescence* a raceme or panicle, the latter with the lower flower in lateral cymules of 3–9, the upper flowers generally solitary, the panicle often rather narrow and congested at first but often broadening as the lower flowers open; flowers nodding to semi-nodding, or lateral facing, cupped to saucer-shaped, 6–11 cm across. *Bracts* transitional from the upper leaves and decreasing in size up the inflorescence, pinnatifid

Meconopsis paniculata; NW Bhutan. PHOTO: MARTIN WALSH

Meconopsis paniculata; NW Bhutan. PHOTO: DAVID & MARGARET THORNE

to deeply toothed, pubescent like the leaves. *Peduncles* 11–19 mm long, pubescent like the stem. *Pedicels* slender, 4.7–15.5 cm long (to 17 cm long in fruit). *Buds* ovoid to ellipsoid or more or less pear-shaped, covered in ascending to patent barbellate bristles underlain by a felt of much shorter, yet similar, hairs. *Petals* 4, rarely 5, cream to pale primrose yellow or butter-yellow,

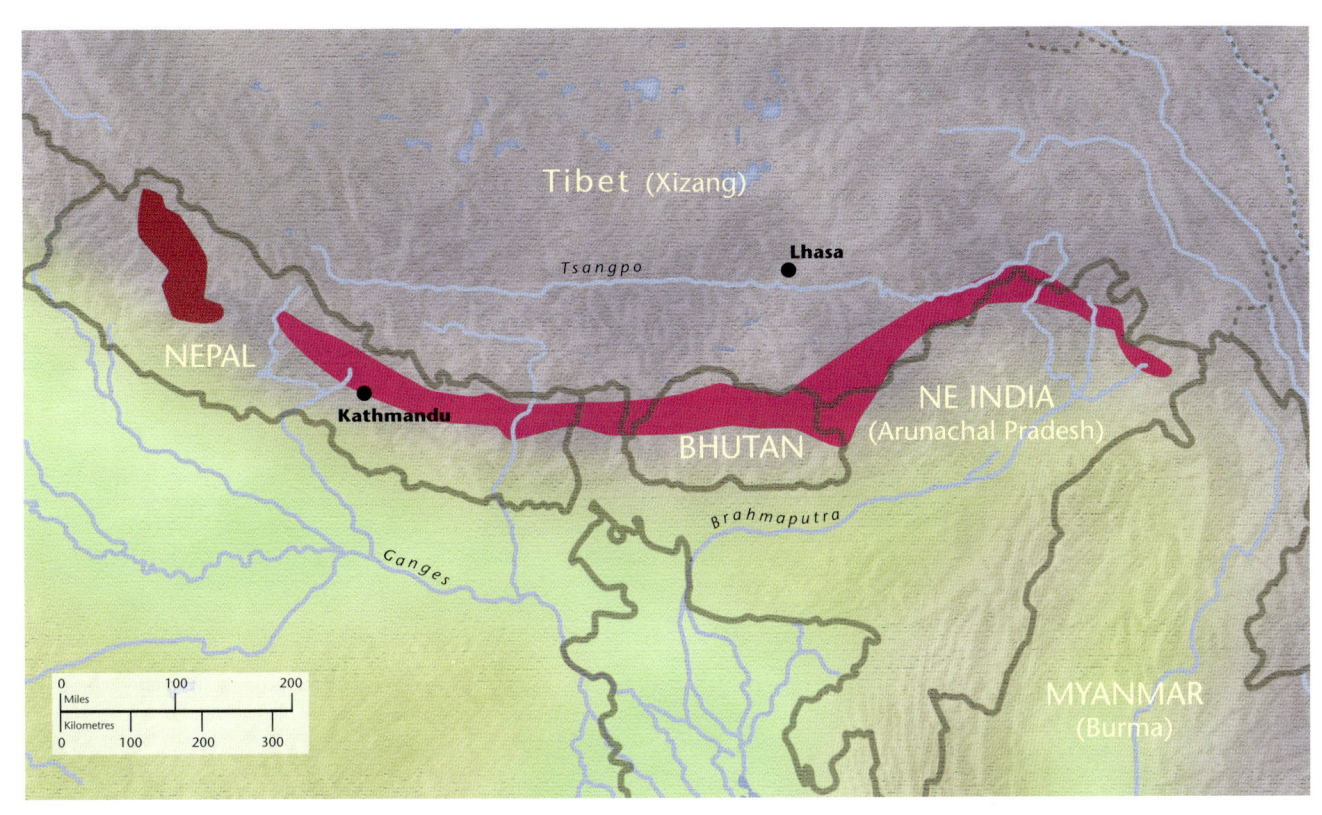

MAP 7. Distribution of *Meconopsis paniculata*: ■ *M. paniculata* subsp. *paniculata*, ■ *M. p.* subsp. *pseudoregia*.

rarely pink or deep red, obovate to suborbicular, 32–57 × 30–48 mm, generally with a slightly undulate margin. *Stamens* numerous the filaments cream to pale yellow, the anthers yellow to orange-yellow. *Ovary* subglobose to globose, densely covered with appressed golden-yellow barbellate bristles, the style 6–9 mm long with a capitate purple or reddish-purple stigma. *Fruit capsule* obovoid to ellipsoid or ellipsoid-oblong, 18–40 × 9–16 mm, (6–)8–12-valved and splitting for a short distance from the apex, densely covered with ascending barbellate bristles with a dense underlay of short barbellate hairs, the persistent style 8–14 mm long, generally 2–4 mm diameter, puberulous in the lower part or glabrous. *Seeds* subreniform, densely papillate at first with rows of tiny pits when dried. 2n = 56 (28, Sugiura 1940!).

DISTRIBUTION. Nepal (Himalayan districts of Humla, Jumla and Jajarkot, eastwards to Sankhuwasabha & Taplejung districts, but absent from Bajura, Mugu, Dolpa and Gorkha, perhaps also absent from Nuwakot and Sindhupalchok), C & N Bhutan, S Tibet (Xizang; Kyimdong, Mago, Tsari) and NE India (Arunachal Pradesh[15], Sikkim); (2135–)2700–4570(–4878) m (Map 7).
HABITAT. Open woodland and woodland margins, open shrubberies, meadows and grassy slopes, rocky alpine slopes, valley slopes, streamsides and ravines.
FLOWERING. Late May–August.

Meconopsis paniculata is extremely variable in the wild, both in the lobing of the leaves and in the character of the inflorescence. In some forms the basal leaves are deeply lobed, while in others they are only slightly lobed, or just toothed. The inflorescence can be broad candelabra-like panicles or narrow and pillar-like; however, both specimens and photographs can be misleading as inflorescences can spread out as successive flowers open, especially the later flowers which are borne on the lower branches of the panicle. In addition, individual flowers can vary from deeply cupped to saucer-shaped, and from nodding to ascending. This variation is not perhaps surprising as *M. paniculata* is by far the most widespread species in subgenus *Meconopsis*, with a distribution from west Nepal eastwards to Arunachal Pradesh in the

---

[15]Frank Kingdon Ward records this species as far East as the Mishmi Hills in Assam (Arunachal Pradesh), close to the Myanmar border, although it appears to be rare and localised this far east. "From behind a rock rose a pillar of sulphur-yellow flowers, and there was our first, and as it turned out, only *Meconopsis* from the Mishmi Hills — *M. paniculata* …… But as a botanist I was deeply impressed by my find, for there are other things about plants, besides novelty, which appeal to the botanist. *M. paniculata* for example was hitherto known to occur only in Sikkim and Nepal, and its appearance 400 miles further east had a special interest in relation to the spread of Himalayan plants towards China." (Kingdon Ward, *Plant Hunting on the edge of the World*: 262 (1985 edition). Apart from *M. horridula* sensu lato, *M. paniculata* has the greatest west-east range of any *Meconopsis* in the Himalaya.

Indian Himalaya and SE Tibet (Xizang). The large grey-green rosettes adorned with fawn or golden hairs are extremely handsome and in the wild large colonies can often be seen, the mountain slopes on occasion thick with the stately candelabras of soft yellow flowers. At the same time it is a very variable species; in the most attractive forms the flowers are large, pendent and borne in a spreading panicle, while in others the flowers can seem small and bunched. A particularly useful character is the presence of a purple or reddish-purple stigma to the otherwise all-yellow flower, but this is absent from the most westerly populations. Interestingly, this feature is always present in cultivated plants of wild origin but is soon lost when plant hybridise with other evergreen monocarpic species, especially *M. napaulensis*, *M. staintonii* and *M. wallichii*.

The species was introduced into cultivation in 1849 from Sikkim by Joseph Dalton Hooker, although the first authenticated record of it flowering in cultivation was 1863; from this a plate was figured in *Curtis's Botanical Magazine* published in 1866 under the erroneous name '*M. nipalensis*'. This stemmed from the confusion in the true identity of Wallich's type, first described as *Papaver paniculatum* by David Don but subsequently amalgamated by subsequent authors with *M. napaulensis*, a situation only satisfactorily resolved by George Taylor in his 1934 monograph of the genus.

Meconopsis paniculata is one of the commonest species in cultivation and various new accessions have been added in recent years, particularly from Nepal and Sikkim, including those with 'ginger foliage' and the so-called Ghunsa form (from E Nepal) with very neat rosettes of scarcely lobed leaves with a gingery-orange flush, which is especially prominent in the autumn and winter. James Taggart (1989) comments that "To most gardeners this species looks like a yellow *M. napaulensis* and indeed in many cases forms part of the latter's hybrid swarm". While this is undoubtedly true it has to be remembered that when this was written *M. napaulensis* was considered in a very broad sense following Taylor's (1934) monograph in which the flower colour was described as ranging from "red to purple or blue, occasionally white". Of course in the light of current research *M. napaulensis* is now considered to be a series of

Above: *Meconopsis paniculata*; leaf-rosette, spring, Phephe La, C Bhutan. PHOTO: CHRISTOPHER GREY-WILSON

Right, from top:

Meconopsis paniculata; leaf-rosette, summer, Ghunsa Khola, E Nepal. PHOTO: CHRISTOPHER GREY-WILSON

Meconopsis paniculata; leaf-rosette, late spring, Rudong La, C Bhutan. PHOTO: CHRISTOPHER GREY-WILSON

Meconopsis paniculata; Jule Tsho, Thimphu, C Bhutan, 4500 m. PHOTO: TIM LEVER

closely allied species with the true *M. napaulensis* bearing yellow flowers. However, this in no way detracts from the fact that in gardens these species have hybridised to such an extent in the most part that recognition of individual species in the swarm has become impossible. Needless to say, many of these hybrids are extremely fine garden plants.

When raised from wild-collected seed *Meconopsis paniculata* presents the grower with few problems and seedlings can be planted out in the autumn, although perhaps leaving them until the following spring is wiser. Given rich feeding, with plenty of well-rotted manure or compost, plants will grow away quickly in their second season, quickly forming their characteristic handsome rosettes that are especially attractive in the autumn and winter when they take on russet hues. Plants generally flower in their second or, more likely, third full season; the inflorescences emerge in the late spring but are generally not damaged by late frost as are some of the other species. In addition, unlike some of the other species in section *Polychaetia*, plants seldom suffer from crown rot during the winter months. Generally speaking it is a better plant to try in drier gardens away from the favoured *Meconopsis* areas of northern Britain and Ireland and elsewhere. Well grown plants can reach more than 2 m tall and carry in excess of 150 flowers, borne over many weeks. It is a plant for sunny or part-shaded situations but, like all the larger evergreen monocarpic species, shelter from strong wind is advisable otherwise flowering plants are easily knocked over.

KEY TO SUBSPECIES OF MECONOPSIS PANICULATA

1. Leaf-lamina pinnatifid, often pinnatisect in the lower half; stigma purple or reddish-purple . . . . subsp. **paniculata** (Nepal from Kali Gandaki eastwards, NE India (Arunachal Pradesh, Sikkim), Bhutan, S Tibet)
+ Leaf-lamina coarsely serrate, sometimes pinnatfid towards the base; stigma pale greenish . . . . .subsp. **pseudoregia** (W Nepal; W of the Kali Gandaki)

subsp. paniculata

DESCRIPTION. Details as in the key above. Figs 5E,F, 9G1,2.

DISTRIBUTION. Nepal (Himalayan districts of Mustang, Myagdi, Manang, Kashki and Lamjung, eastwards to Sankhuwasabha & Taplejung districts, but apparently absent from Gorkha, perhaps also absent from Nuwakot and Sindhupalchok), Bhutan, S Tibet (Xizang; Kyimdong, Mago, Tsari) and NE India (Arunachal Pradesh, Sikkim); (2135–)2700–4570(–4878) m (Map 7).

SPECIMENS SEEN. *AGSES* 253 (K), 269 (K); *F. M. Bailey* s.n. Gokar La (E); *R. Bedi* 202 (K), 391 (K), 606 (K); *L. W. Beer* 8244 (BM); *Bor & Ram* 20559 (K); *Botanical Exped. Himalaya* (1985) 8530246 (E), 8530434 (E), 8571968 (E); *Botanical Exped. West Nepal* (1991) 91041-1 (E); *Bowes-Lyon* 5086 (BM); *Cave* 667c (E), s.n. Chanago 20 Sept. 1916 (BM, E), s.n. 24 Oct. 1916 (BM), s.n. Aug. 1918 (K); *R. E. Cooper* 787 (E), 1605 (E, K), 2998 (E); *Cox, Hutchinson, Maxwell-MacDonald* 3087 (E); *DNEP* 3BY196 (E); *Curzon* 52 (K), 53 (K); *Cutting & Vernay* 4 (K); *Dhwoj* 19 (E), 235 (BM, E), 305 (BM, E); *Edinburgh Makalu Exped.* (1991) 465 (E); *Fifth Botanical Exped. Himalaya* (1972) 720527 (E); *Flora of East Nepal* (1994) 94-70378 (E), (1995) 95-84041 (E), 95-96530 (E); *Gammie* 150 (K); *Gould* 423 (K), 510 (K), 514 (K), 1137 (K); *Hara et al.* s.n. 18 May 1967 (BM); *Hobson* s.n. Yatung, SE Tibet (K); *Hooker* s.n. Lachen, 11,000 ft (K), s.n. Sikkim (K); *Kanai et al.* 72-0527 (BM, E), 92-05217 (E); *Kingdon Ward* 6960 (E), 8425 (K), 11868 (BM); *Kurasaki* 85-71968 (E); *Long et al.* 465 (E),

Below left: *Meconopsis paniculata*; Xierilong, Kama Valley, S Tibet, 4240 m. PHOTO: DAVID & MARGARET THORNE

Below right: *Meconopsis paniculata*; W of Topke Gola, E Nepal, 3500 m. PHOTO: TOSHIO YOSHIDA

517 (E); *Ludlow & Sherriff* 286 (BM), 747 (BM), 1039 (BM), 1171 (BM), 1869 (BM, E), 2111 (BM, E), 2631 (BM), 2646 (BM), 2678 (BM), 2809 (BM), 3295 (BM); *Ludlow, Sherriff & Hicks* 16450 (BM, E), 19106 (BM), 19156 (BM), 19799 (BM), 20733 (BM, E), 21163 (BM), 21168 (BM, E); *Ludlow, Sherriff & Taylor* 6351 (BM, E); *D. McCosh* 4125 (BM); *Minaki et al.* (Botanical Exped. W Nepal) 9104141-1 (E); *Miyamoto et al.* 94-40030 (E), 95-84041 (E); *Ohba et al.* 83-41182 (BM, E), 85-30246 (E), 85-30434 (E); *Polunin* 485 (BM), 1420 (BM), 1921 (BM); *Ribu & Rohmoo* 5090 (K), 5853 (E, K); *Rohmoo Lepcha* 1050 (E); *Schilling* 1001 (K); *Seventh Botanical Exped. Himalaya* (1983) 8341182 (E); *K. N. Sharma* 7/94 BM), 8/94 (BM), 33/94 (BM), 62 (BM, E), 530 (BM), E80 (BM), E86 (BM), E363 (BM), E373 (BM), E378 (BM), E380 (BM), E398 (BM), E495 (BM); *T. B. Shrestna & D. P. Joshi* 285 (BM); *Sinclair & Long* 5198 (E, K), 5261 (E); *Stainton* 649 (BM, E), 888 (BM), 4026 (E), 4700 (BM); *Stainton, Sykes & Williams* 3070 (BM), 5996 (BM, E), 6068 (E), 6518 (BM, E), 6658 (E), 7468 (E), 8591 (BM), 8648 (E); *Suzuki et al.* 94-70378 (E); *Third Darwin Nepal Fieldwork Training Exped.* BY196 (E); *Tshering & Nawang* s.n. 28 July 73 (E); *Wallich* 8123b (K); *L. H. Williams* 661 (BM), 764 (BM), 889 (BM); *Wollaston* s.n. (K); *T. Wraber* 395 (BM).

Subspecies *paniculata* is widespread in the central and eastern Himalaya and the mountains close by in adjacent parts of Tibet. It is variable plant in stature, but more especially in the dissection of the leaves and in the density of the indumentum and such differences can often be seen within a colony or by comparing adjacent colonies in the wild. In some, the leaves are deeply lobed for the most part, in others more shallowly lobed. In some plants the indumentum can consist of rather sparse long barbellate hairs, in others it can be dense. At the same time, the undercoat of much shorter, but also barbellate, hairs is generally dense although it can on occasions be quite sparse, especially in mature leaves. As a result of the indumentum the leaf colour can vary from pale greenish yellow to quite deep greyish green, while in the autumn the leaves can take on rich russet, golden or tawny hues as the hair colour changes, a feature also noticeable in cultivated plants. While the flower colour is consistently a rather pale, more or less primrose, yellow over most of its range the occasional occurrence of plants with deep crimson or pink (var. *rubra*) flowers has been noted in populations in both Bhutan and north-western Arunachal Pradesh. These sightings have been backed up by some fine photographs taken in the wild, although regretfully not by herbarium specimens. The surprise discovery of the pink-flowered variant in Bhutan in recent years by Walsh, Lever *et al.* and, in 2012, of both deep red- and reddish pink- flowered forms in Arunachal Pradesh by the same team requires explanation. In both instances these variants were growing in larger populations of the normal, predominately yellow-flowered form of *Meconopsis paniculata* subsp. *paniculata*. Commenting by email on the latter, David and Margaret Thorne remark that "On 2 July, we found yellow-flowered *Meconopsis paniculata* and then a tall red-flowered *Meconopsis* which did not have the golden foliage of the Bhutanese *M. wallichii* var. *fusco-purpurea*. We saw red-flowered plants

Meconopsis paniculata, form with large flowers and a racemose inflorescence; Tampe La, Bhutan, 4500 m. PHOTO: TOSHIO YOSHIDA

13. MECONOPSIS PANICULATA

Above, from top:

Meconopsis paniculata Ghunsa form; leaf-rosette in cultivation.
PHOTO: CHRISTOPHER GREY-WILSON

Meconopsis paniculata, semi-mature fruit capsules; C Nepal (SE Ganesh Himal). PHOTO: PAUL EGAN

twice more in well separated locations, mixed with yellow-flowered forms and we were left in no doubt that it is a red-flowered form on *M. paniculata*. The foliage was identical and there were no intermediate flower colours."

George Taylor (1934) records that "It should be noted that white forms occur in nature. The appearance of a white-flowered form in cultivation was recorded by Burbidge in 1898, and recently forms in which the flowers are spotted with bronze have been described".

While a number of collectors have described the colour as cream rather than yellow, none have noted white forms in the wild, or at least there is no herbarium collection in which white is included on the field notes. It is known, for instance, that *Meconopsis staintonii* exists in the wild on occasion as a white-flowered form (and these may well be what Taylor was referring to!) and in cultivation white-flowered plants are seen from time to time but they are nearly always hybrids between *M. paniculata* and other members of the *M. napaulensis* group. It may be suspected by many that the red- and pink-flowered variants observed in the wild are of hybrid origin, however, there is no evidence to back such an assumption, especially as where they were found no other possible species were involved. Deep reds are very uncommon in members of subgenus *Meconopsis* but, interestingly, having said this, four of the species in the subgenus can have red (deep crimson often) flowers. They are *M. staintonii* as already mentioned, *M. wallichii* (in var. *fusco-purpurea*) and *M. regia*, as well as *M. paniculata*. Of these only *M. wallichii*, whose normal colour is blue or lavender, appears to have colonies of solely red-flowered plants. In the other three species, red-flowered plants are found as isolated plants, or just a few, amongst colonies of the normal colour (yellow in the case of *M. paniculata* and *M. regia*, pink or pinkish red in the case of *M. staintonii*). The fact that intermediates are not observed (except in the case of *M. regia*, see p. 61) suggests that red is a recessive gene and, like an albino, can only reveal itself very occasionally. Many *Meconopsis* have been found to produce occasional white-flowered forms in the wild but they are very rare and scattered, while the only white-flowered species as such is *M. superba*. It is not known how many genes govern flower colour in *Meconopsis* but it can be surmised that several do and that the red mutation can only be revealed very occasionally. The lack of intermediates along with the very few pink or red-flowered plants in any population would suggest that offspring between red- and yellow-flowered all come out yellow. It also suggests that self-pollination of red-flowered plants does not occur, otherwise very many more red-flowered plants would appear in the colony: it is known that many *Meconopsis* are self-incompatible or, at least, only weakly self-compatible.

var. rubra Grey-Wilson, **var. nov.** Like the typical plant var. *paniculata* in all details except that the flowers are deep scarlet, not yellow. Type: based on the photo (p. 111), taken in Arunachal Pradesh, NW India (Pange La to Poto, 4120 m) by David & Margaret Thorne.

THE GENUS MECONOPSIS
13. MECONOPSIS PANICULATA

Above and top left:

Meconopsis paniculata subsp. *paniculata* var. *rubra*; Jule Tsho, Thimphu district, C Bhutan, 4500 m. PHOTOS: TIM LEVER

Var. *rubra* is restricted, as far as it is known, to the central Himalaya, specimens having been found and photographed in central Bhutan and north-western Arunachal Pradesh in the altitudinal range 4120 to 4500 m. The Bhutanese plant has flowers of soft rose-pink, while those of the Indian plant are a deep crimson. In both instance, plants were found growing in populations of the ordinary, common and widespread yellow-flowered plant, subsp. *paniculata* var. *paniculata*, while in Arunachal Pradesh one small colony consisted only of the red-flowered variant. The occurrence of the occasional red variant is an intriguing one and is discussed more fully on p. 110.

Below and bottom left:

Meconopsis paniculata subsp. *paniculata* var. *rubra*; Pange La to Poto, NE India, Arunachal Pradesh, 4120 m. PHOTOS: DAVID & MARGARET THORNE

subsp. pseudoregia Grey-Wilson, **subsp. nov.**

Differs from the typical plant, subsp. *paniculata*, in the leaves that are coarsely toothed rather than lobed, sometimes pinnatifid towards the base, and in the pale greenish rather than purple stigma. Type: W Nepal, Dolpa District, near Pudamigaon, Suli Gad, 13,500 ft, *Polunin, Sykes & Williams* 2238 (BM, holotype, isotypes). (Fig. 5C, D).

DESCRIPTION. Details as in the key above. *Petals* pale yellow.
DISTRIBUTION. W Nepal (W Dolpa, Jajarkot, Jumla, Mugu and Rukum districts); 2745–4300 m (Map 7).
HABITAT. Open woodland, scrub, riverbanks.
FLOWERING. June–July.
SPECIMENS SEEN. *J. J. Burnet* 33a (BM); *S. B. Malla* 10709 (BM), 10969 (BM); *Minaki et al.* 91-04141 (E); *Polunin, Sykes & Williams* 2238 (BM), 3589 (BM, E), 4359 (BM, E), 4687 (BM); *T. B. Shrestha* 5142 (BM); *Stainton, Sykes & Williams* 3589 (BM), 4359 (BM); 4687 (BM).

The western populations of *Meconopsis paniculata* are quite variable in leaf character. In the most extreme forms the leaves are wholly unlobed, the margin adorned instead by large, regular jagged teeth. Plants often have a golden silky sheen to the leaves and this together with the leaf-margin details has led many to believe that they are forms of *M. regia* and not *M. paniculata* and as such they are recorded in the *Flora of Nepal*. However, detailed analysis has shown that they are much closer to *M. paniculata*. In these western populations the leaf margin is coarsely serrated with up to 13 teeth on each side of the basal leaves. In contrast the leaves of *M. regia* are very finely serrated with many more teeth per side. In addition, and perhaps most significantly, the indumentum is very different: in *M. paniculata* subsp. *pseudoregia* the long hairs on the leaves are coarsely barbellate and can be sparse to rather dense, but they are always underlain by a dense underlay of short coarsely barbellate hairs (sometimes referred to as a 'substellate underlay'), while in *M. regia* the indumentum consists only of dense, very long, slender, finely barbellate hairs and it is this sericeous pubescence that gives the leaves, especially in their rosette form, their characteristic appearance. Differences can also be noted in the petioles of the rosette leaves, those of subsp. *pseudoregia* being considerably longer than those of *M. regia*. *M. regia* also has larger flowers than *M. paniculata* sensu lato.

Interestingly, the stigma of subsp. *pseudoregia* remains pale green while the plants are in flower but as the flowers fade following pollination they change to a purple-brown colour.

The field notes for *Polunin, Sykes & Williams* 4359, collected at Ghurchi Lagna, W Nepal at 11,000 feet, read: "Grassy slopes on south side of pass. Leaves pale green and covered with pale yellow hairs. Sepals a pale delicate yellow. Filaments pale yellow, anthers orange. Ovary densely covered with white hairs." The flowers are variously described as pale yellow or lemon yellow, and several collectors note that the whole plant is covered in fine 'golden' or 'cream-coloured' hairs.

Meconopsis paniculata subsp. *pseudoregia*; NE of Maharigaon, SSW Sisne Himal, Jumla District, W Nepal, 3700 m, flowering in mid July. PHOTO: ALAN DUNKLEY

Meconopsis paniculata subsp. *pseudoregia*; W of Kagmara Pass, Dolpo District, W Nepal. PHOTO: TOSHIO YOSHIDA

Meconopsis paniculata subsp. *pseudoregia*; near Jagdula, Dolpo District, W Nepal, 3900 m. PHOTO: TOSHIO YOSHIDA

The western regions of Nepal, west of the significant Kali Gandaki Valley, a river that feeds the upper reaches of the Ganges, is interesting floristically. With a generally lower rainfall and drier inner valleys, this part of Nepal is not nearly so rich as those to the east in the Himalaya and there is a significant decline in the number of species of prime Himalayan genera such as *Primula* and *Rhododendron*. At the same time, the area is also an outlier for a number of species found much further to the east such as *Incarvillea himalayensis* and the westernmost populations of *Meconopsis grandis* (subsp. *jumlaensis*), neither found in the central parts of the country. Apart from the widespread and variable *M. horridula* none of the other western Nepalese taxa of *Meconopsis* are found east of the Kali Gandaki: *M. chankeliensis*, *M. simikotensis* and *M. grandis* subsp. *jumlaensis* are all endemic to this western part of Nepal, while *M. robusta* is also found further west in India in Uttar Pradesh, primarily in Kumaon district.

NOTE. Plants recorded as *Meconopsis regia* in western Nepal in the *Flora of Nepal* are wrongly attributed and are all referable to *M. paniculata* subsp. *pseudoregia*. *M. regia* is only found in a restricted area to the east of the Kali Gandaki valley.

14. MECONOPSIS AUTUMNALIS

Meconopsis autumnalis P. A. Egan, *Phytotaxa* 20: 48–50 (2011). Type: Central Nepal, Ganesh Himal (Rasuwa District), Tulo Bhera Kharka – Jaisuli Kund, 28°12'N, 85°13'E. 4160 m, *Miyamoto et al.* 9440053 (E, holotype, 2 sheets; KATH, TI, isotypes); Ganesh Himal, Jagessor Kund, 28°14'N, 85°11'E, 13,000 ft, *Stainton* 4028 (BM, paratype, 4 sheets, one specimen).

DESCRIPTION. *A monocarpic plant* 0.9–1.6 m tall in flower, though as little as 0.7 m on occasions; rootstock an elongated taproot to 3 cm diameter at the greatest, yellow-orange lactiferous. *Stem* erect, weakly ridged, 2.2–3.1 cm diameter at the base, densely pubescent with fawn to yellowish or orange-coloured, 5–9 mm long, barbellate bristles underlain by much smaller, though similar bristles. *Basal leaves* forming a substantial rosette in the years before flowering, pubescent overall like the stem, becoming gradually glabrescent except for marginal bristles, these to 10 mm long; lamina oblong to oblong-ovate, 33–55 × 8–16 cm, pinnatisect in the lower part, pinnatifid at the apex, with 3–6 opposite or subopposite pairs of segments, these ovate to oblong, to 4 × 2.2 cm, weakly pinnately-lobed at the margin, the lobes obtuse to acute; petiole 11–25 (–31) cm long, expanded and part sheathing at the base. *Cauline leaves* numerous, reducing in size gradually up the stem, gradually merging with the bracts, the lowermost petiolate and very similar to the rosette leaves, the upper sessile. *Bracts* sessile, strongly deflexed, lanceolate to lanceolate-oblong, the base with broad, rounded, semi-amplexicaul lobes, apex subobtuse to subacute, the margin weakly pinnatifid, pubescent as for the leaves; bracteoles and bracteolules present on lower branches of the inflorescence, similar to the bracts but increasingly smaller, the smallest more or less entire. *Inflorescence* a large, dense, columnar panicle with up to 250 flowers covering at least two-thirds of the height of the plant, the lower borne in lateral cymules of 3–8 (–12) flowers, the uppermost flowers solitary; flowers bowl- to saucer-shaped, 8–10.5 cm across, lateral-facing to semi-nodding. *Peduncles* 11–21 cm long (elongating in fruit up to 29 cm), wide-spreading to ascending, pubescent as the stem; pedicels 6–17 cm long. *Buds* ovoidal to broad-ellipsoidal, moderately barbellate-bristly, nodding. *Petals* 4(–6), pale primrose yellow, suborbicular to obovate,

Meconopsis autumnalis habitat; SE Ganesh Himal, Rasuwa District, C Nepal. PHOTO: PAUL EGAN

4.3–5.6 × 4.1–5.5 cm, scarcely overlapping or only so towards the base. *Stamens* numerous, about one-third the length of the petals, with pale yellow filaments and orange-yellow anthers. *Ovary* ellipsoidal, densely covered with ascending orange bristles; style 6–8 mm long, stout, with a capitate yellow, 6–8-lobed, stigma, 5–8 mm diameter. *Fruit capsule* ovoidal to ellipsoidal, 15–27 × 9–14 mm, dehiscing by 6–8 valves a short distance from the top, moderately covered with ascending to spreading barbellate bristles. Fig. 9F1,2.

DISTRIBUTION. Central Nepal; restricted to the SE Ganesh Himal (centred on Rasuwa & N Dhading districts; Paldol to Jaisuli Kund and Tula Bhera); 3300–4200 m.

HABITAT. Subalpine pastures, and grassy slopes, stream margins, margins of *Abies* forest, forest glades, in association with coarse herbs and dwarf shrubs.

FLOWERING. July–early September.

SPECIMENS SEEN. *Ganesh Himal, Central Nepal Expedition* (*Miyamoto et al.*) 9440053 (E); *Stainton* 4028 (BM).

This, one of the most recently described species, has the general appearance of the widely distributed *Meconopsis paniculata*. Both occupy similar habitats, grow to more or less the same height and bear yellow flowers in multi-flowered paniculate inflorescences. It is not perhaps surprising that this particular taxon has been overlooked; however, credit for noting differences between the two must go to Adam Stainton who observed it in the wild during his collecting trip in Nepal in 1964.

A lengthy field note attached to *Stainton* 4028 (2 sheets at BM; Ganesh Himal, Jagesor Kund) reads as follows:

> "Growing on grass slopes over a distance of about 5 miles between 12,500–13,500 ft. Only other *Meconopsis* species in the vicinity *M. paniculata* (*Stainton* 4026), with which it intermingled at 12,500 ft. The two species were easily distinguished when growing together, and no intermediates forms were seen.
>
> Height of plants 3 ft; hairs of plant pale yellow; petals pale yellow. Distinguished from 4026 because:
> 1. Its flowers were larger and of a paler yellow.
> 2. Its hairs were pale yellow, not orange.
> 3. Its habit of growth was different. Flower buds were borne close to the stem and depressed so that when about to flower the stem has a marked columnar appearance. When in full flower the whole plant was much squatter and more compact than 4026, almost pyramidal in outline.
> 4. The pedicels of the flowers were markedly s-shaped, and the stem leaves of plants in flower were depressed and very auriculate.
>
> The overlap of 4026 and 4028 was at 12,500 ft. 4026 did not ascend above nor 4028 descend below this height."

Many of these remarks were later picked up in the description of *Meconopsis autumnalis* composed by Paul Egan and presented in a thorough and detailed analysis. He (2011) ran a discriminate analysis on this species, comparing it with *M. paniculata* with which it has been closely confused in the past, both in the field and in the herbarium. The result of this analysis, using various characters, showed that the two are quite distinct species. Both share the same ecology in the wild but, whereas *M. paniculata* has a very wide distribution in the Himalaya, the new species, *M. autumnalis*, appears to be restricted to a relatively small area of the Ganesh Himal in central Nepal. While they do overlap in distribution in the Ganesh and are known to grow in close proximity to one another, hybrids have not been recorded. This may be due to some inbuilt incompatibility system between the two species, although in cultivation members of this association hybridise fairly freely. However, although there may be a small amount of overlap in flowering time, hybridisation is generally precluded by the fact that the new species flowers rather later. In fact Paul Egan photographs taken in the wild clearly show *M. paniculata* in young fruit while *M. autumnalis* is just coming into full flower. In addition, it would appear from Adam Stainton's field notes that *M. autumnalis* tends to grow at a greater altitude, the two only overlapping at about 3810 m (12,500 ft). This again would help to preclude the possibility of hybridisation.

Paul Egan carried out a "careful examination of the morphological relationship with the closely related *Meconopsis paniculata*; a partially sympatric and highly variable species widely distributed over a large geographic area". Despite the highly variable nature of the latter the prime differences between it and *M. autumnalis* can be summarised as follows:

1. A generally smaller plant which in flower forms a characteristic columnar appearance with a denser inflorescence occupying much of the height of the plant: in *M. paniculata* the inflorescence is borne in the upper part of the plant.
2. Deflexed, auriculate, middle and upper cauline leaves/bracts; spreading (horizontal) to ascending in *M. paniculata*.
3. Stigma 5–8 mm diameter, yellow-green; 3–5 mm diameter and reddish purple in *M. paniculata* (except in the western Nepal's subsp. *pseudoregia*).
4. Fruit capsules relatively narrower, with long patent hairs; long hairs subappressed in *M. paniculata*.

In the adjacent populations studied, the flowers of *Meconopsis autumnalis* were generally somewhat larger and the mean length of the pedicels at anthesis decidedly longer. The long pedicels, particularly in the upper part of the inflorescence, which as Stainton observed are narrowly S-shaped, give the inflorescence its distinctive columnar appearance.

Above left: *Meconopsis autumnalis* leaf-rosette; SE Ganesh Himal, Rasuwa District, C Nepal. PHOTO: PAUL EGAN
Above right: *Meconopsis autumnalis*, leaf details; SE Ganesh Himal, Rasuwa District, C Nepal. PHOTO: PAUL EGAN

14. MECONOPSIS AUTUMNALIS

Meconopsis autumnalis, semi-mature fruit caspules; C Nepal (SE Ganesh Himal). PHOTO: PAUL EGAN

Meconopsis autumnalis is a very handsome species and in its more columnar flowering habit even more attractive than the closely related *M. paniculata*. Both species have great impact in flower because of the large number of blooms that a single mature plant can carry. It is possible that it could have been introduced in seed collections in the past, for the region in which it grows is fairly well explored especially by Adam Stainton, alone or in the company of Sykes and Williams on several important British Museum (Natural History) collecting expeditions in the 1950s. If it had been introduced into cultivation then it may well have had a part to play in the complex hybrid forms that represent '*M. napaulensis*' in cultivation today (see under that species). Without question *M. autumnalis* would be a grand addition to cultivation and it is to be hoped that authenticated and legitimately gathered seed can be introduced from the wild in the near future.

From *Meconopsis napaulensis*, with which it could also be confused, it differs primarily in the larger stigmas, relatively longer pedicels and deflexed middle and upper cauline leaves/bracts, as well as its larger stature, relatively longer and wider leaves. In addition, the fruit capsules of the two are markedly different: those of *M. napaulensis* are longer and narrower, narrow-obovoid and sparsely pubescent, 5 (–6)-valved, those of *M. autumnalis* ovoid to ellipsoid and rather densely pubescent, 6–8-valved.

The epithet *autumnalis* is not altogether appropriate: it is true that the species flowers somewhat later than *M. paniculata* but even so, its prime flowering season is July and August, mid-summer in the Nepalese Himalaya. Many members of section *Polychaetia* produce a few late flowers in the late summer and early autumn including *M. autumnalis* and *M. wallichii*, often on plants that have otherwise gone into the fruiting phase. The same applies to *M. napaulensis*.

Below left: *Meconopsis autumnalis*, lower part of inflorescence; SE Ganesh Himal, Rasuwa District, C Nepal. PHOTO: PAUL EGAN
Below right: *Meconopsis autumnalis* flower detail; SE Ganesh Himal, Rasuwa District, C Nepal. PHOTO: PAUL EGAN

15. MECONOPSIS VIOLACEA

Meconopsis violacea Kingdon Ward, *Garden* 91: 450, in obs. (1927) and *Ann. Bot.* 42: 857, tab. 16 fig. 2 (1928). Type: Upper Myanmar (Burma); Kachin State, Seinghku Wang, 11,000 ft, 12 June 1926, *Kingdon Ward* 6905[16] (K, holotype; BM, K, isotypes).

[*Meconopsis* sp. Kingdon Ward, *Gard. Chron.*, Ser. 3, 81: 433, fig. (1927)].

DESCRIPTION. *A monocarpic plant* to 1.8 m tall in flower, forming a bluish-green evergreen rosette, up to 60 cm across, in the years before flowering; stem erect covered with an indumentum of long, rather weak and short barbellate hairs. *Basal leaves* narrow linear-oblong in outline, narrowed below into a slender petiole to 8 cm long, the lamina linear-oblong, to 17 × 3.2 cm, pinnately-lobed, generally pinnatifid towards the apex, with 7–9 pairs of opposite or subopposite, oblong to lanceolate-oblong, lateral segments, these with an entire to irregularly toothed margin, covered on both surfaces with a mixture of fine long and short barbellate hairs, more especially along the margin and the midrib beneath; cauline leaves decreasing in size gradually up the stem, like the basal leaves, the lower with a short petiole but the upper sessile and with an auriculate base, merging with the bracts. *Inflorescence* a simple raceme carrying 12 or more flowers, occasionally the lowermost flowers in cymules of 2–3 (at least in cultivated plants); flowers bowl-shaped, nodding, 6–7 cm across. *Pedicels* slender, 4–8.5 cm long, pubescent, densely so towards the top. *Buds* elliptic-ovoid, pendent, finely hairy. *Petals* 4, occasionally 5–6, bluish-violet to purplish, suborbicular to obovate, 34–38 × 24–27 mm. *Stamens* numerous with filiform filaments the same colour as the petals, whitish towards the base, the stamens bright orange. *Ovary* ellipsoid, densely covered with pale golden, more or less appressed slender barbellate bristles, with a slender, slightly twisted style 6–7 mm long, terminating in a pale green clavate stigma. *Fruit capsule* oblong-ellipsoidal, 33–37 × 14–15 mm, 7–9-valved, splitting for a short distance from the apex, covered in appressed barbellate bristles or their bases, especially along the sutures, as well as short narrow cone-shaped, tubercule-like bristles, the style 7–9 mm long. Fig. 9H.

DISTRIBUTION. Upper Myanmar (Burma) and the adjacent region of Tibet (Xizang); the Diphukha La and Seinghku Valley; 3050–3965 m (Map 6).

HABITAT. Alpine meadows, gravelly slopes and forest margins.

FLOWERING. Late June–July.

SPECIMENS SEEN. *Kingdon Ward* 6905 (BM, K), 7207 (BM, K).

Meconopsis violacea is one of the most distinctive members of series *Polychaetia*, with its simple racemes of shot-silk-like violet or purplish flowers and narrow, evenly pinnately lobed, rather fern-like leaves, it cannot easily be mistaken for any other species. It is also relatively isolated from other members of the series in the wild, with the recently described *M. wilsonii* found to the east in eastern Myanmar and south-western China and *M. wallichii* found further west in eastern Nepal, Sikkim and central Bhutan. All three species bear flowers in the blue, purple, violet or red range. The widely distributed yellow-flowered Himalayan *M. paniculata* ventures further east than *M. wallichii*, being found east as far as NE India (Upper Assam) and the neighbouring region of southern Tibet (Xizang). Of these species, *M. violacea* probably finds its closest ally in *M. wilsonii* which also has narrow inflorescences, although in this case they are clearly part paniculate. In details of leaf and fruit the two species are very different and while *M. violacea* has nodding flowers, those of *M. wilsonii* generally face sideways or slightly nod. So little material of *M. violacea* is known from the wild that it is difficult to judge the degree of variability, nor do we know whether it is found in other parts of northern Myanmar, although according to the *Flora of China* it is found just north of the Myanmar border as well, although I have not seen any authenticated specimens or records from south-eastern Tibet; however, its presence there would not be unexpected.

Frank Kingdon Ward to whom we owe its discovery comments that "Early in June 1926, in the Seinghku valley, I noticed the buoyant rosettes of a large Poppy bursting vigorously through the snow. The pinnate leaves were sage green lined with long silken honey-coloured hairs, which glistened in the sunshine, as it might have been *Meconopsis paniculata*. Yet I felt certain it was not, and I watched it jealously as the tall leafy spire rose inch by inch from the deep crown until finally stopped by a flower-bud. Yet July came before it opened. By that time, from the axil of each bract lolled a fat bud, the topmost had thrown back their green hoods, and several silken violet flowers, piled with dark gold in the centre, fluttered out in the breeze. Such was *M. violacea*, a glorious column of bright colour, which flowered throughout July in the ice-worn valleys, being then about three feet high; throughout the rains it continued to grow however, till in September it was over four feet high and carried two dozen capsules, the topmost being

---

[16] There are two sheets of *Kingdon Ward* 6905 in the British Museum (Natural History), one of a fruiting specimen, the other of a leaf-rosette.

then ripe. *M. violacea*, though not so common as *M. betonicifolia* or *M. rubra* [*M. impedita* sic], was common enough, occurring in colonies of six or eight plants perhaps, on the steep meadow slopes, and not another anywhere near; nor does it climb much above 12,000 feet, or descend much below 11,000 feet. Although a true biennial, it should be perfectly hardy in Britain, and in beauty it is second only to the peerless *M. betonicifolia* [*M. baileyi* sic]." Kingdon Ward *Plant Hunting on the Edge of the World* (1930).

Kingdon Ward records that "in its first year it produces an enormous crown of sea-green leaves encased in golden hairs of silken texture .... leafy crown survives the winter under snow." Unfortunately, this very interesting and distinctive species has not been recollected from the wild in recent years: the inaccessibility of the region and political strife has made this impossible at the present time. With its narrow saw-like leaves, simple racemose inflorescences and bluish-violet flowers it is unlikely to be confused with any other members of the genus.

George Taylor (1934) records that "It is now well established in gardens in this country [UK sic] and the first record of flowering appears to be 1929"; however, although plants persisted in cultivation until the middle of the last century they eventually died out and the species is no longer in cultivation. *Meconopsis violacea* is an extremely handsome and beautiful species with elegant racemes of nodding flowers held above rather sparse, narrow, evenly lobed leaves. There is a tempting monochrome photograph of it in cultivation reproduced in the first edition of George Taylor's monograph *An Account of the Genus Meconopsis* (1934) and two fine half-tone glass plates in the collections at the Royal Botanic Garden Edinburgh. James Cobb (1989) commented that "It is certainly not straightforward in cultivation and was not recommended for gardens in the south of Britain even when it was common because of the particular requirements of summer humidity [sic northern Myanmar is subjected to a particularly heavy summer monsoon] .... This species appears to require a humid atmosphere in summer or it will not survive ....It is likely to be the one winter rosette-forming species that is fussy as to a strict winter dry, summer wet regime.....".

Meconopsis violacea; a half-tone photo taken when the species was in cultivation in the 1950s. PHOTO: COURTESY RBG EDINBURGH

Unfortunately, *Meconopsis violacea* comes from a remote part of the eastern Himalaya in the sensitive zone along the Myanmar-Tibet border, a region not explored in recent years, at least by western botanists. It is a region that holds great promise botanically and will certainly be a goal of scientists once the political situation in that region improves. In the meantime we can only read the enticing description of the region related to us by Frank Kingdon Ward in *Plant Hunting on the Edge of the World* which was published in 1930, just prior to Taylor's monograph. "Including *M. paniculata* from the Mishmi Hills, and the tiny *M. lyrata* [*M. compta* sic] from the Seinghku valley, no less than seven 'blue poppies' were collected on these ranges — not all of the 'blue' be it observed!" K.W.

Subgenus DISCOGYNE

[syn. *Meconopsis* sect. *Polychaetia* Prain, *Ann. Bot.* 20: 352 (1906) pro parte; *M.* ser. *Torquatae* Prain *op. cit.*: 355 (1906); *M.* subgen. *Polychaetia* (Prain) Fedde in Engl., *Pflanzenr.* 4, 104: 262 (1909) pro parte; *M.* sect. *Torquatae* (Prain) Fedde *op. cit.*: 265 (1909)]

Subgenus *Discogyne* was established by George Taylor in his 1934 monograph and included just two species, *Meconopsis discigera* Prain and *M. torquata* Prain, both described in 1906. The subgenus is readily distinguished by the presence of a stylar disk immediately surmounting the ovary. In the allied genus *Papaver* (the true poppies) there is no style and the stigmas extend across a disk on top of the ovary, almost like a starfish, and can be 5–9 (sometimes more) lobed. In contrast, in subgenus *Discogyne* there is a pronounced to very short style with the stigma at the tip, while the base of the style is expanded into a disk on top of the ovary, this extending to the margin and sometimes overlapping it. As such, this disk appears to be a derived character, yet within *Meconopsis*, and indeed the rest of the Papaveraceae, it is a unique one. In all other species of *Meconopsis* (subgenus *Meconopsis*, subgenus *Grandes* and subgenus *Cumminsia*) there is no such expansion of the base of the style, which is often of uniform thickness or somewhat expanded towards the base but not expanded across the top of the ovary. In the live state the stylar disk is brightly coloured, generally dark red or crimson, but occasionally pinkish or whitish, at least at first. Its function is not properly understood, but it may help to attract pollinating insects: all the species grow at or near extreme attitudes where pollinating insects can be rather scarce.

Since Taylor's monograph was written nearly 80 years ago much more material of these discoid species has been collected in the wild, primarily from Nepal, Bhutan and southern Tibet. Just as important has been the photographic evidence of recent years which has revealed subtle but clear differences between plants from different areas. As a result the subgenus has now extended to seven species: all have a very limited range in the wild as far as it is known, each is generally well isolated from the next and, furthermore, they can all be regarded as narrow endemics: in 1979 a third species, *Meconopsis pinnatifolia* C. Y. Wu & X. Zhuang, was added to the subgenus, having been discovered near Jilong in southern Tibet. In addition, and more recently (in 2006) I described a fourth species, *M. simikotensis*, based on material collected near Simikot in far western Nepal by that prodigious partnership of Stainton, Sykes & Williams during their 1952 expedition on behalf of the British Museum (Natural History). The addition of a fifth species, *M. tibetica* (as a result of the Alpine Garden Society Tibet 2005 tour), was both exciting and perhaps not altogether unexpected, for Tibet is a huge and still much underexplored area. A sixth species was discovered in central Nepal during the 2008 Flora of Nepal Manaslu Expedition of that year staged by the Royal Botanic Garden Edinburgh and this was described appropriately as *M. manasluensis* in 2011. Finally, in the following year *M. bhutanica* was added, recognising fundamental differences between the yellow-flowered eastern Nepalese and Sikkim plant (*M. discigera*) and the blue or purple-blue Bhutanese one, these formerly being included within a single species.

Subgenus *Discogyne* is confined to the Himalaya from Nepal to Bhutan and the neighbouring regions of Tibet north as far as the Lhasa area. They are high altitude plants found between about 3500 and 4900 m (11,480–14,900 ft), being primarily plants of rocky places and moraines, occasionally growing along stream and river margins, although one of the new species, *Meconopsis tibetica*, inhabits rocky low *Rhododendron* shrubberies. All seven species are monocarpic, spending their first two years or more as neat, hairy or rather bristly leaf-rosettes that gradually increase in size. In the autumn the leaves die and the plant overwinters as a large resting bud nestled deep at the base of the withered leaf-rosette. In some species such as *M. discigera*, the withered leaf rosettes can build up into a substantial wad over several years. At flowering stage, the centre of the rosette expands into a long leafy or leafless stem bearing a spike or raceme of flowers, only the lower of which are bracteate, varying in colour from pale yellow, to crimson, red, maroon, purple or blue, depending on the species. All the species of the subgenus bear barbellate hairs (soft, slightly barbed, bristles) and these they share in common with the members of subgenus *Meconopsis* and subgenus *Grandes*. As already stated, the single most distinctive feature of the subgenus is the presence of a flattened, toothed or lobed, often reddish or purplish, disk that sits immediately on top of the capsule. This in combination with the monocarpic herbaceous leafy-rosettes and the presence of barbellate hairs really define the subgenus.

There is no doubt that the members of subgenus *Discogyne* form a very close and distinctive alliance and they are unlikely to be confused with any other members of the genus. They are very beautiful plants in flower and distinctive elements in the high alpine flora of the central and eastern Himalayan region. Although seed of several has been introduced over the years (certainly as long ago as the early years of the twentieth century) and from various sources from Nepal in more recent times, few have been raised to flowering size and none have persisted for long in cultivation. They are without question very tricky plants to maintain in cultivation, requiring the specialist conditions of an alpine house, carefully controlled watering and an adequate rest period when the young rosettes wither away in the autumn. In the wild plants grow in severe and exposed habitats where the growing season is much contracted and plants spend a long winter buried beneath a blanket of snow. Seedlings build up slowly and it is not clear how long they take in the wild to reach maturity. Suffice it to say that three or four years or more is likely: in the wild young leaf rosettes of varying ages can often be observed growing in close proximity to those in flower. In fact they are quite difficult to find unless flowering or fruiting plants are seen.

KEY TO SPECIES OF SUBGENUS DISCOGYNE

1. Leaves more or less entire, occasionally with a few coarse uneven teeth; flowers relatively small, the petals not more than 38 mm long; stigma capitate, 1–4 mm diameter 2
+ Leaves variously lobed; flowers relatively large, the petals 40–67 mm long; stigma linear, 4–7 mm long . . 3

2. Inflorescence with a single main axis; petals blue, purple or mauve; ovary with whitish bristles and a 2–4 mm long style . . . . 19. **M. simikotensis** (W Nepal)
+ Inflorescence with several main axes; petals scarlet; ovary with fawn or orange bristles and a 7–12 mm long style . . . . 22. **M. manasluensis** (C Nepal)

3. Style more or less obsolete: flowers in a spike-like inflorescence with very short pedicels or subsessile; petals partly soft-bristly on the reverse. . . . . 21. **M. torquata** (C Tibet)
+ Style distinct, 3–7 (–10) mm long; flowers in a racemose inflorescence, the pedicels always apparent; petals glabrous 4

4. Stylar-disk pentagonal in outline, shallowly-lobed, not projecting beyond the edge of the ovary; petals maroon; stem leafy below inflorescence . . . . 18. **M. tibetica** (S Tibet)
+ Stylar-disk sinuate or 8-angled, lobed or fringed at the margin, projecting beyond the edge of the ovary; petals pale yellow, bluish-purple to deep purple, more rarely crimson or crimson-purple; stigma capitate or linear; stem leafy or leafless below inflorescence 5

5. Stems leafy (excluding leaf-like bracts); leaves pinnately-lobed for full length with 3–5 pairs of segments . . . . 20. **M. pinnatifolia** (CN Nepal, S Tibet)
+ Stems leafless below inflorescence; leaves only lobed in the upper third, often trilobed, or 4–5-lobed near the apex 6

6. Flowers pale yellow, the petals with recurved margins; leaf apex usually 3-lobed, occasionally entire (uppermost rosette leaves); style 4–6 mm long. . . . . 16. **M. discigera** (E Nepal, Sikkim)
+ Flowers blue to purple, the petals with flat margins; leaf apex mostly 5 (–7)-lobed (except young plants); style 2–4 mm long. . . . . 17. **M. bhutanica** (W Bhutan)

16. MECONOPSIS DISCIGERA

16. Meconopsis discigera Prain, *Ann. Bot.* 20: 356 (1906). Type: W Sikkim, Gucha La (Gocha La), 11,000 to 12,000 ft, *Cave* s.n., 1036 & 6628 (BM, E syntypes). Lectotype *Cave* 6628 (BM) chosen here.

[*M. discigera* sensu Taylor pro parte, *The Genus Meconopsis*: 108 (1934)].

DESCRIPTION. *Monocarpic plant*, 20–45 cm tall in flower, to 60 cm tall in fruit, covered for the most part with 4–12 mm long pale golden to straw-coloured barbellate hairs, the plant base with a thick felted wad, to 10 cm deep or sometimes more after a number of years, of closely imbricated, persistent dead leaves and leaf-bases, beset with long fawn-coloured hairs. *Taproot* elongate to 25 cm long or more, 7–12 mm diameter at the top, gradually narrowed to the tip. *Stem* erect, weakly to strongly ridged, 7–15 mm diameter near the base, pale green but sometimes suffused with red-purple. *Leaves* mostly in a basal rosette, the cauline leaves few and decreasingly smaller up the stem, the lowermost rosette leaves spreading, the middle and upper ascending; lamina flat, narrow oblanceolate, 2.5–10 × 1.2–3.0 cm, with a narrow wedge-shaped base decurrent onto the petiole, the apex usually 3-toothed, rarely 5-toothed, or occasionally entire in the upper rosette leaves; the upper-surface rather deep green, the lateral veins obscure, the lower surface paler; petioles narrow-winged, 3–10 cm long, expanded and sheathing at the base. *Inflorescence* racemose, with 9–18 broad bowl-shaped, nodding or half-nodding flowers, ebracteate except for the lowermost. *Bracts* leaf-like but increasingly smaller, sessile or subsessile, the apex entire, the base decurrent onto the stem/rachis. *Buds* oval; sepals 15–23 mm long, covered in dense patent barbellate bristles, greenish, sometimes partly suffused with red-purple. *Pedicels* 2–5 cm long, elongating to 12 cm long at the most in fruit. *Petals* 4, pale primrose-

Above left: *Meconopsis discigera*; Yalung Glacier margins, SW of Mt Kangchenjunga, E Nepal, 4500 m. PHOTO: TOSHIO YOSHIDA
Above right: *Meconopsis discigera*; S of Umbok La, Lumbasamba Himal, E Nepal, 4500 m. PHOTO: TOSHIO YOSHIDA

yellow or sulphur-yellow, obovate to orbicular, 40–55 × 35–55 mm, the apex rounded to truncate, the margin recurved outwards in the upper third, markedly erose. *Stamens* numerous, about one fifth the length of the petals, the filaments whitish or yellowish, the anthers 2–2.5 mm long, dull-orange in colour, browning with age. *Ovary* greenish, broadly ovoid, densely covered in rufous-coloured, erect or ascending bristles; style 3–5 mm long, broadened at the base into a pale green or yellowish disk-like appendage, capping the top of ovary with a recurved margin; stigma yellow-green to purple, clavate, 3.5–7 mm long (to 12 mm long in fruit including the style), protruding well beyond the mass of anthers. *Fruits capsule* narrow to broadly barrel-shaped, 15–20 mm long, 12–16 mm diameter, densely ascending bristly at first but becoming glabrescent or subglabrescent in the mature fruit, surmounted by a dark red, glabrous, 15–18 mm diameter, disk-like appendage, with a spreading, toothed margin. *Seeds* reniform, 0.8–1 mm long, with longitudinal rows of pits and slightly reticulate.

DISTRIBUTION. E Nepal (Ilam, Panchthar, Sankhuwasabha & Taplejung districts; Arun Khola eastwards, especially along the Yalung Glacier in the Kangchenjunga Himal, around Topke Gola in northern Jaljale Himal, and around Thudam in Lumbasamba Himal), W Sikkim (only known from Gucha La (Gocha or Guicha La)); (3100–) 3400–4800 m. Endemic to Nepal (Map 8).

HABITAT. Glacial valleys or on exposed stony hills that are often shrouded with drifting mists during the summer monsoon months, particularly cliff-ledges, rock-crevices, stable rocky high-alpine slopes, boulder fields and moraines, always in very well-drained situations, generally in humus-rich pockets.

FLOWERING. late June–August.

SPECIMENS SEEN. *AGSES* 608 (K); *L. W. Beer* 9448 (BM), 25639 (BM); *Beer, Lancaster & Morris* 10645 (E); *Briggs* 196 (K); *Cave* s.n., Guicha La (BM), 1 Sept. 1919 (E), s.n. 10 Oct. 1919 (E), 1036 (BM), 6628 (BM, E); *Curzon* 201 (K); *ESIK (Long et al.)* 394 (E); *Ribu* 37 (K); *Rohmoo Lepcha* 1036 (BM, E, K); *L. W. Beer* 9448

(BM), 25639 (BM); *Beer, Lancaster & Morris* 10645 (E); *Long, McBeath, Noltie & Watson* 394 (E); *Prain's Collector* 1115 (K); *Ribu* 37 (K); ; *T. B. Shrestha & D. P. Joshi* 373 (BM); *Stainton* 921 (E), 1707 (E), *H. Kanai et al.* 720715 (TI), *H. Ohashi et al.* 773792 (TI), 775131 (E, TI), *H. Ohba et al.* 9110423 (E, TI), 9120372 (TI), 9130287(E, TI), 9130304 (TI).

The first description of *Meconopsis discigera* by David Prain was based on the fruiting specimens collected in 1905 by G. H. Cave on the Gucha La in western Sikkim. Similar plants with flowers from the type locality were collected by a native collector on behalf of Cave including those in 1913 (*Cave* 6628), although the petals had been removed from these specimens and there was no record of the colour; David Prain originally recorded them as blue or purple. Prain revised the original description in 1915 (*Bull. Misc. Inform.*, Kew 1915: 170 (1915)) and added that the flowers were yellow; however, he corrected the colour in a subsequent publication in 1923 (*Country Life* liv: 111). George Taylor followed Prain's last description in his revision of the genus and excluded yellow from the flower colour of the species, recording them as 'dark-crimson, red, purple, or pale-blue'. This clearly resulted from a mix up with the Bhutanese material that had come to light in the intervening years. In Taylor's 1934 monograph of the genus he remarks that "Recent collections from Nepal and by the Mount Everest Expedition of 1922 have shown that the flower-colour is dark crimson, red, purple, or, as Cooper has recorded from Bhutan, blue". Red-flowered plants from the Everest region are referable to *M. tibetica* (see p. 127), blue ones from Bhutan, *M. bhutanica* (see p. 123). Interestingly, he also records that "A plant stated to be *M. discigera* and which proved to be monocarpic flowered at Edinburgh in 1930, but as the flowers were yellow its true identity is uncertain". With hindsight it is perhaps easy to see how this muddle occurred: today quite a few people have seen and recorded *M. discigera* in the wild, both in eastern Nepal and Sikkim where it is without exception yellow-flowered.

The name of the type locality Gucha La was corrected to 'Guicha La' on *Cave* 6628, but it is recorded as 'Gocha La' on recently published maps of the region. The Gocha La is located at the head of Onglaktang Glacier, south-east of the main peak of Mt Kangchenjunga. The altitude of the type locality was recorded 11,000–12,000 ft on the original type specimens, but *Cave* 6628 collected at the same locality in 1913 recorded an altitude of 15,000 ft. Toshio Yoshida, who was in the region in 1995, measured an altitude of 4500 m (14,764 ft) at the foot of Gocha La beside the glacier, but failed to find any plants of *Meconopsis discigera*.

Meconopsis discigera is found in a rather narrow band in eastern Nepal (east of the deep Arun valley), in areas subjected to a heavy summer monsoon. Although seed is sometimes available, and occasionally plants are listed in the RHS Plant Finder, this is by no means an easy species in cultivation and few have been lucky enough to flower it. However, there have been successes: a good yellow-flowered form flowered at Edinburgh Botanic Garden in 1930 and, far more recently, a similarly coloured plant flowered from seed collected by the AGS Sikkim expedition of 1983, this adding confirmation of the colour of the Sikkim plant as yellow. It is undoubtedly a tricky plant requiring a dry winter (but not a desiccating one!), ample moisture in the summer and a moisture-retentive, yet well-drained compost. Black fungal rot

Meconopsis discigera; Yalung Glacier moraines, SW of Mt Kangchenjunga, E Nepal, 4650 m. PHOTO: TOSHIO YOSHIDA

infection can be a problem in the autumn as plants go into their winter rest period, resulting in the quick death of young plants. Ample seed has been introduced over the years but apparently to little avail; however, the challenge is there for anyone wishing to try and there will always be those skillful enough to succeed in the long term.

In the wild *Meconopsis discigera* characteristically builds up a great wad of dead leaves under the living rosette and this no doubt affords some protection to the plant during the severe winter when, additionally, plants will be covered in a layer of snow.

With its pale primrose-yellow flowers borne in a compact raceme, *Meconopsis discigera* is a very beautiful species, its flower colour setting it apart from all other members of the subgenus. For a full comparison with the closely allied *M. bhutanica* see under that species.

17. MECONOPSIS BHUTANICA

Meconopsis bhutanica Tosh. Yoshida & Grey-Wilson, *The Plantsman* n.s. 11 (2): 96–101 (2012). Type: W Bhutan, Paro Chu, Kumathang, foot of Pangte La (Bhonte La), 12,500–13,000 ft, June 1949, *Ludlow, Sherriff & Hicks* 17471 (holotype, BM; isotypes BM).
[*M. discigera* sensu Taylor pro parte, *The Genus Meconopsis*: 108 (1934)].

DESCRIPTION. *Monocarpic plant*, 22–54 cm tall in flower, to 68 cm tall in fruit, covered for the most part by 3–7 mm long, whitish or fawn-coloured barbellate hairs, usually with a felted mat of old leaves forming a thick wad at the base of the living leaf-rosette, especially in older plants which may be a number of years old before flowering, these wads generally beset with stiff, fawn-coloured hairs. *Taproot* elongate, 12–30 cm long or more, 7–12 mm diameter at the top, gradually narrowed to the tip. *Stem* erect, usually weakly ridged, 7–20 mm diameter at the base, yellowish green, sometimes suffused with red-purple. *Leaves* pale or yellowish green, paler beneath, mostly clustered at or near the base, the cauline leaves few, gradually reduced in size upwards; the lowermost leaves spreading to somewhat recurved, the middle and upper leaves ascending to erect; lamina oblanceolate, 3–10 × (0.8–)1.2–3.1 cm, with a cuneate base formed by the attenuate lamina running down the petiole, the apex subacute to obtuse, the margin with (3–)5–7 coarse, somewhat oblique, subobtuse teeth towards the apex, somewhat rugose with depressed lateral-veins above; petioles of lowermost leaves to 8 cm long, narrowly winged, expanded towards the base and half-sheathing. *Bracts* only present at the lowermost flowers, sessile, leaf-like but smaller, elliptic-oblong, with a broad, somewhat decurrent base and an entire to trilobed apex. *Inflorescence*

Above left: *Meconopsis bhutanica*, habitat; Golung Chu, W Bhutan, 4550 m. PHOTO: DAVID & MARGARET THORNE
Above right: *Meconopsis bhutanica*, leaf-rosettes; MT Jhomolhari trail, W Bhutan, 4200 m. PHOTO: TIM LEVER

17. MECONOPSIS BHUTANICA

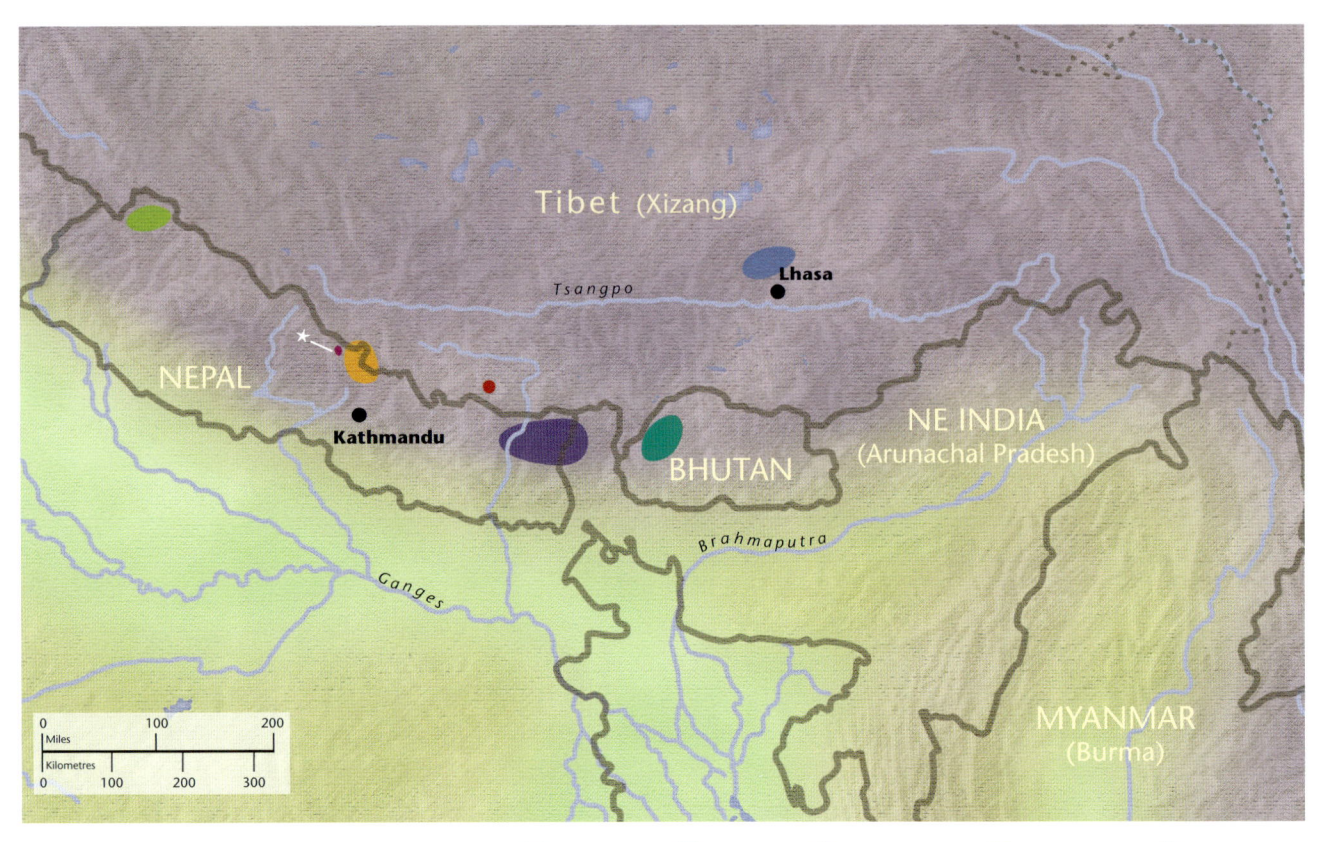

MAP 8. Species distribution: ■ *Meconopsis discigera*, ■ *M. bhutanica*, ■ *M. tibetica*, ■ *M. simikotensis*, ■ *M. pinnatifolia*, ■ *M. torquata*, ■ *M. manasluensis*.

racemose, ebracteate except for the lowermost flowers, bearing 6–19, saucer- to cup-shaped, half-nodding, but occasionally the uppermost erect or suberect, flowers. *Buds* oval, nodding to half-nodding; sepals 13–22 mm long, patent-bristly, greenish, often suffused with red-purple. *Pedicels* 12–32 mm long in flower, elongating to 30–92 mm long in fruit. *Petals* 4, rarely 5–6 in the terminal flower, blue-purple, occasionally reddish purple, broad-ovate to orbicular, 40–60 × 30–55 mm, the apex subobtuse to rounded, the margin flat, not recurved, entire or somewhat erose. *Stamens* numerous, about one quarter the length of the petals, with filaments similar in colour to, but darker than the petals; anthers yellow to orange-yellow, browning with age. *Ovary* pale green, obconic-cylindric, densely covered with erecto-appressed bristles, 5–6-valved; style 2–4 mm long, broadening at the base into an ovary-wide dark purple, somewhat lobed disk; stigma short-clavate, 2–3.5 mm long, pale green to purple, equal to or slightly exceeding the boss of anthers. *Fruit capsule* obconic-cylindric, 18–28 mm long, 12–18 mm diameter, densely patent-ascending hirsute, becoming gradually glabrescent at maturity, surmounted by a glabrous disc-like appendage with spreading, toothed margin, 15–22 mm diameter. (Fig. 10A).

DISTRIBUTION. W Bhutan (catchments of Golung Chu, Paro Chu and Thimpu Chu, Tso Phu; only known from the region around the northern watershed of these rivers); 3400–4480 m. Endemic to Bhutan (Map 8).

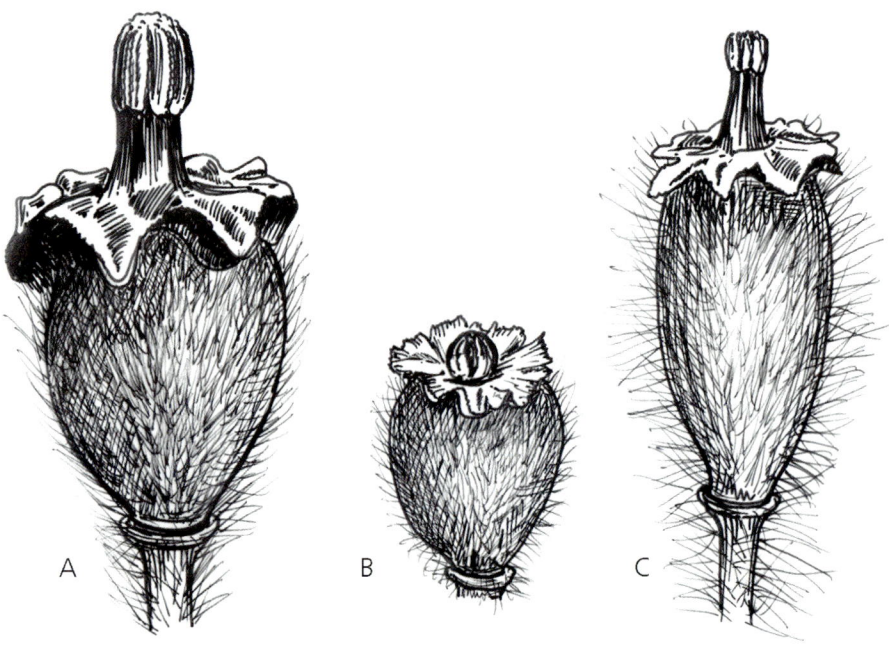

FIG. 10. Fruit capsules: **A**, *M. bhutanica*; **B**, *M. torquata*; **C**, *M. tibetica*. All × 2.
DRAWN BY CHRISTOPHER GREY WILSON

HABITAT. Unstable scree slopes, talus slopes, moraines, streamside gravels at valley-heads, generally growing in humus-rich pockets.

FLOWERING. Late June and July, occasionally later.

SPECIMENS SEEN. *Bowes Lyon* 15045 (E); *R. E. Cooper* 1990 (BM, E, K), 2944 (BM, E, K); *P. C. Gardner* 672 (BM); *Ludlow, Sherriff & Hicks* 16279 (BM, E), 16399 (BM); 17455 (BM, E), 17456 (BM), 17471 (BM, E).

NOTE. *Meconopsis bhutanica* was first collected by R. E. Cooper in 1914, on 27th July in flower (*Cooper* 1990) and on 28th September in fruit (*Cooper* 2944), near the head of the Thimpu Chu above Parshong (Barshong) at an altitude of 13,000 ft (3965 m). These specimens were, however, ill-preserved and it proved very difficult to examine the details of flowers and leaves. The famous four-member British team that explored Bhutan in 1949 (Frank Ludlow, George & Betty Sherriff and J. H. Hicks) collected the species four times at different stages: first in mid-May specimens were collected prior to flowering (*Ludlow, Sherriff & Hicks* 16279) on the southern slope of Pangte La ('Bhonte La' on the maps recently published and today on the well-trodden Chomolhari Round Trek) at 13,000 ft; additional pre-flowering specimens were also collected at the end of May (*L, S & H* 16399) on the southern slope of Yale La (probably the same area as the Cooper collections referred to above); thirdly specimens collected in flowers at the end of June (*L, S & H* 17471) on the northern side of Pangte La near the Tso Phu; finally specimens in mature fruits collected in mid-October (*L, S & H* 17455), also from the northern side of Pangte La.

Meconopsis bhutanica is perhaps the most beautiful member of subgenus *Discogyne*, a rather squat plant with a narrow cluster of substantial blooms which are a clear blue-purple in the most eye-riveting forms. The species, like most of its cousins, grows in rather severe and exposed habitats, taking a number of years to reach flowering size, in places where the growing season is limited by long winters and where the springs are dry and the summers wet.

Meconopsis bhutanica undoubtedly finds its closest ally in the eastern Nepal-Sikkim *M. discigera*, differing most obviously in flower colour, but other differences are observable on close inspection and these are outlined in Table 4 below.

Frank Ludlow came across the species in May 1949 on the Pangte La at about 13,000 feet: "*Meconopsis discigera* [*M. bhutanica* sic] (16279) today in bud .... It reminds me of *Meconopsis torquata* from Lhasa in one or two particulars. For example, the flower buds are closely adpressed together on the flower stalk, then there are numerous persistent dead leaves of previous years attached to the base and it is obvious that it takes several (2 or 3 or 4) years to reach flowering stage. In habitat it differs somewhat, being found along the banks of streams in the open, whereas *torquata* was always found on boulder scree, often overhung by boulders. Then again *torquata* occurs at 15,000 feet or more whereas *discigera* where I saw it was growing at not more than 13,000 feet at about the upper limit of the *Abies spectabilis* contour".

Meconopsis bhutanica; Jagothang, Thimphu district, W Bhutan, 4500 m. PHOTO: TIM LEVER

17. MECONOPSIS BHUTANICA

Table 4
COMPARISON OF THE PRIME CHARACTERISTICS SEPARATING *MECONOPSIS BHUTANICA* AND *M. DISCIGERA*

	Meconopsis bhutanica	*Meconopsis discigera*
Stem	usually weakly ridged	weakly to strongly ridged
Basal leaf mat	densely matted at the base, but never very thick or scarcely present	densely matted at the base, to 10 cm or more thick
Leaf posture	ascending to erect except basal ones	spreading to ascending
Leaf apex and petiole	usually 5–7-toothed toward apex; with broadly winged petioles	usually 3-toothed at apex; with narrowly winged petioles
Upper surface of leaves	pale or yellowish green, somewhat rugose with depressed secondary veins	deep green, flat, the secondary veins obscure
Hairs (all barbellate)	usually whitish, shorter, less dense	usually yellowish, longer, denser
Flower posture	half-nodding, occasionally the terminal one more or less erect	nodding to half-nodding
Flower shape	saucer- to cup-shaped	broadly bowl-shaped
Petal colour	blue-purple	pale-yellow
Petal shape	broad-ovate to orbicular, apex subobtuse to rounded, margin flat, entire to slightly erose	obovate to orbicular, apex rounded to truncate, margin recurved in the upper third, strongly erose
Stamen length	about a quarter of the petals	about a fifth of the petals
Disk-like appendage of style	dark-purple from the opening of the flower; similar to or slightly larger than fruit capsule in diameter	pale green or yellowish at first, dark-red after flowering; broader than fruit capsule in diameter
Anther colour	yellow to orange-yellow	dull-orange to brownish yellow
Style and stigma	4–7.5 mm long overall, the stigma short-clavate, 2–3.5 mm long, almost unchanged in fruit	6.5–12 mm long overall, the stigma clavate, 3.5–7 mm long, elongating to 12 mm long in fruit
Fruit capsule	obconic-cylindric	broadly to narrowly barrel-shaped

Below, from left:

Meconopsis bhutanica; Twin Lakes, Tshophu, W Bhutan, 4400 m. PHOTO: JOHN & HILARY BIRKS

Meconopsis bhutanica, flower detail; Jagothang, Thimphu district, W Bhutan, 4500 m. PHOTO: TIM LEVER

Meconopsis bhutanica, ovary shortly after fertilisation; Tso Phu, W Bhutan, 4400 m. PHOTO: DAVID & MARGARET THORNE

Harold Fletcher reports (*A Quest of Flowers*, 1975) that Ludlow never saw *Meconopsis bhutanica* (then included in *M. discigera*) east of the Paro watershed and this has been confirmed by subsequent observation.

Meconopsis bhutanica has not been successfully cultivated to my knowledge and it would certainly prove to be as tricky in that respect as its close cousin *M. discigera*. However, it has been seen and photographed in Bhutan in recent years and there are some very fine photographs of it in existence taken by, amongst others, Martin Walsh, Tim Lever, David & Margaret Thorne and Rosie Steele. With its squat habit yet ample flowers of semi-translucent, stained-glass blue-purple, it is a plant of haunting beauty, surely one of the loveliest high alpines in the rich flora of Bhutan. Interestingly, the Bhutanese have chosen a blue poppy, *M. grandis* subsp. *orientalis*, as their national emblem, but perhaps this equally beautiful, yet endemic species, would have been even more appropriate.

18. MECONOPSIS TIBETICA

Meconopsis tibetica Grey-Wilson, *Bull. Alpine Gard. Soc. Gr. Brit.* 74 (2): 222 (2006), Pl. p. 213, 223. Type: S Tibet (Xizang), between Dumba and Tsho Shau, near the latter (GPS: 28°03'47"N, 87°16'542"E to 28°01'504"N, 87°16'303"E), 19 July 2005, 4500 m, *Birks & Birks* s.n. (E, holotype).

DESCRIPTION. *A monocarpic plant* to 40 cm tall in flower, forming a lax rosette of leaves in the years before flowering but not building up a thick mat of dead or old leaves below the rosette. *Stem* erect, 6–7 mm diameter near the base, greenish or more often flushed with red, particularly towards the top and the rachis of the inflorescence, covered in dense, slender, silky barbellate hairs, these mainly 5–7 mm long, spreading to slightly deflexed. *Leaves* rather pale green, with a whitish midrib, the basal and lower stem leaves narrow-oblanceolate to elliptic-oblanceolate, the lamina 4–7.8 × 0.7–1 cm, attenuate below into the petiole, the apex subacute, moderately covered in hairs on both surfaces similar to those on the stem, the margin more or less entire or with several coarse teeth on each side; petiole slender, 2.5–5.4 cm long; upper stem leaves similar to the basal but gradually decreasing in size up the stem, short-petioled to sessile, generally spirally arranged but the upper three sometimes in a false whorl below the inflorescence. *Bracts* present in the lower part of the inflorescence, leaf-like, lanceolate-ovate, sessile, mostly 20–36 × 4–5 mm. *Inflorescence* a simple raceme, 7–14-flowered, the flowers half-nodding to horizontal, deeply cupped to wide saucer-shaped,

Meconopsis tibetica; Samchung La, S Tibet, 4400 m. PHOTO: DAVID & MARGARET THORNE

the buds subglobose, half-nodding, densely covered in spreading to ascending barbellate hairs. *Pedicels* 12–25 mm long, flushed with red, thickening somewhat at the top, covered in spreading hairs, particularly dense at the top and there 6–8 mm long. *Petals* brilliant maroon or claret, 4–5, broadly obovate to elliptic-oblanceolate, 31–65 × 19–58 mm, the margin finely and unevenly toothed. *Stamens* numerous, forming a distinct ring around the ovary but not enveloping it, the filaments linear, the same colour as the petals, the anthers bright yellow. *Ovary* greenish, beset with stiff ascending to patent barbellate bristles, 4–7-valved, with a glabrous

18. MECONOPSIS TIBETICA

Above from left:

Meconopsis tibetica; Samchung La, S Tibet, 4400 m. PHOTO: DAVID & MARGARET THORNE

Meconopsis tibetica leaf-rosette; Samchung La, S Tibet, 4400 m. PHOTO: DAVID & MARGARET THORNE

Meconopsis tibetica; Samchung La, S Tibet, 4400 m. PHOTO: DAVID & MARGARET THORNE

cap-like stylar disk at the top, the disk, often pentagonal in outline, bright red or pinkish before maturity, 5–6 mm across, slightly narrower than the ovary and with a shallowly lobed margin; style 5–7 mm long, glabrous, terminating in a white (brown at maturity), 4–5-lobed, capitate stigma. *Fruit capsule* narrow-obconic, 19–27 × 13–17 mm, densely patent hirsute, the stylar disk slightly wider than the capsule, with commonly 5 (but 4-7) triangular, undulate lobes, the persistent style 7–9 mm long. (Fig.10C).

DISTRIBUTION. S Tibet (Xizang); Kangshung trek, near Tsho Shau and the north side of the Sha-u La, Samchung La, Everest region (Tibetan side); 4208–4650 m. Endemic to S Tibet (Map 8).

HABITAT. *Rhododendron* thickets and *Rhododendron* heathland.

FLOWERING. Late June–July.

SPECIMENS SEEN. *Birks & Birks* s.n. (E); *Norton*[17] 177 (K); *Shipton*[18] 111 (K), 115 (K); *Wollaston* 177 (K), s.n. (K).

Meconopsis tibetica was described from a plant collected by John and Hilary Birks during the 2005 Alpine Garden Society trek to the region immediately north of Mt Qomolangma (known to everyone in the west as Mt Everest). A note sent with a letter from John and Hilary Birks reads: "This plant was found on the 19 July 2005 growing in *Rhododendron lepidotum*, *R. setosum* and *R. anthropogon* heath at 4500 m near Tsho Shau. Accompanying plants were *Cassiope fastigiata*, *Codonopsis thalictrifolia*, *Potentilla microphylla*, *P. fruticosa* and *Rhodiola* sp. (subsequently identified by Lawrie Springate as *R. himalensis*). *M. simplicifolia* was growing nearby. The new maroon *Meconopsis* was subsequently seen in two other populations in the area, including the lower slopes of

---

[17] 1922 Mt Everest Expedition.

[18] 1938 Mt Everest Expedition.

the Sha-u La (up to about 4600 m) on July 20. It was always growing in *Rhododendron* heath. Sterile plants could always be found near the flowering specimens. No specimens were seen on the other side of the Sha-u La and the plant was not seen subsequently on the trek. *M. horridula* occurred locally in the area, whereas *M. simplicifolia* was much rarer".

Meconopsis tibetica is a very striking species with its startlingly maroon flowers. It is the only species in subgenus *Discogyne* that does not grow amongst rocks or in bleak high moraines; rather it appears to prefer the sheltering confinement of low scrub, especially of *Rhododendron*. It is closely related to *M. discigera*, differing in its slightly toothed to entire leaves, in the leafy stems and bright maroon petals but also in the shallowly-lobed, pentagonal stylar disk and the presence of a capitate stigma.

Meconopsis tibetica appears to be a very local species as the keen-eyed and energetic AGS party walked through vast areas of seemingly suitable *Rhododendron* heath for seven days after crossing the Sha-u La without seeing any further sign of it. Interestingly, it transpires that the general area where *M. tibetica* was found is the most species-rich part on the whole Kangschung trek with nearly 300 species being recorded by the AGS party on the 19 and 20 July 2005.

However, since writing the original article in 2006 further interesting information has come to light. John Birks came across a book entitled *Mount Everest, the Reconnaissance 1921* by Lieut-Col C. K. Howard-Bury DSO, a fascinating account with excellent half-tone photographs and maps. This narrative tells of the attempt to conquer Everest from the north (i.e. from the Tibetan side). In the chapter entitled 'The Move to Kharta' (Kharta being the region to the east of Everest) is the following passage:

".....[We] then had a very steep and stony climb of nearly 3,000 feet to the Samchung Pass, 15,000 ft. As we approached the pass, and entered a moister climate, the vegetation increased rapidly. On these slopes there were rhododendrons 5 feet high, mountain ash, birch, willows, spiraeas and juniper. At the top of the pass there was not much of a view, but prowling round I came across some very fine saussureas with their great white woolly heads and a wonderful meconopsis of a deep claret colour that I had never seen before. There were fifteen to twenty flowers on each stem, and it grew from 2 to 3 feet high....."

Below from left:
Meconopsis tibetica flower detail; Samchung La, S Tibet, 4400 m. PHOTO: HARRY JANS
Meconopsis tibetica inflorescence; Samchung La, S Tibet, 4400 m. PHOTO: DAVID & MARGARET THORNE
Meconopsis tibetica semi-mature fruit capsules; Samchung La, S Tibet, 4400 m. PHOTO: DAVID & MARGARET THORNE

This account is certainly the first recorded sighting of the new species, *Meconopsis tibetica*. In fact the Samchung La is adjacent to the Sha-u La (the type locality of the new species) but at the head of the parallel valley to the east. Strangely, George Taylor makes no mention of this *Meconopsis* in his 1934 monograph despite the fact that A. F. R. Wollaston, the collector on the Everest expedition, records having collected it. In fact there are two specimens in the herbarium at the Royal Botanic Gardens, Kew relating to the north-eastern side of Mt Everest: the first, *Wollaston 177* is cited as the Karma Valley, Tibet at 12,000 ft, while the second, collected during the 1938 Everest Expedition, *Shipton 111*, is located as W of the Gyandar Range, 16,000 ft. Both specimens are clearly referable to *M. tibetica* and are recorded as having 'crimson flowers' and 'dark crimson with a gold centre'. There is an additional unnumbered collection at Kew inscribed "Mt Everest Expedition 1922 .... Karma Valley, Tibet, 12,000ft" which is probably a duplicate of *Wollaston 177*.

Since writing their note that accompanied the pressed specimen, John and Hilary Birks, with help from friends in the AGS and at the Herbarium, Royal Botanic Garden, Edinburgh, have identified the remaining species growing with or very near *M. tibetica*. These include the attractive *Androsace robusta* subsp. *purpurea*, *A. lehmannii*, *Codonopsis bhutanica*, *Primula primulina*, *P. tenella*, *P. walshii*, and *Lloydia serotina*, the pink-red-maroon louseworts *Pedicularis bella* and *P. nana*, the low-growing yellow *Cremanthodium oblongatum*, the roseroot-like *Rhodiola himalensis*, the elegant dwarf-willow *Salix lindleyana*, the common grass *Festuca tibetica* and the small sedge *Carex incurva*. The most surprising discovery was the parasitic broomrape *Orobanche coerulescens*, found by Rosie Steele growing near *Meconopsis tibetica* at 4500 m, about 1000 m above the recorded altitudinal limit of broomrapes in *Flora Xizangica* or the *Flora of China*.

19. MECONOPSIS SIMIKOTENSIS

Meconopsis simikotensis Grey-Wilson, *Bull. Alpine Gard. Soc. Gr. Brit.* 72 (2): 220 (2006). Type: W Nepal, Dozam Khola, near Simikot, 11,500 ft, June 1952, *Stainton, Sykes & Williams 4270* (BM, holotype).

DESCRIPTION. *A monocarpic plant* to 40 cm tall in flower, with some remains of old leaves at the base of the living leaf-rosette, most parts of the plant covered moderately to densely with pale yellowish, barbellate hairs, these 5–8 mm long. *Stem* erect, leafy, to 15 mm diameter at the base, covered with dense spreading hairs. *Leaves* pale green with a yellowish indumentum, spirally arranged, the basal and lower leaves elliptic-oblong, 15–19 × 1.4–2 cm, attenuate below into the petiole, the apex subacute, densely barbellate-pubescent on both surfaces, or more sparsely so beneath, the margin entire or with one or two teeth on each margin, occasionally with a few coarser teeth; petiole slender, 7.5–13.5 cm long, expanding somewhat towards the base; middle and upper stem leaves similar to the lower, gradually smaller up the stem and merging with the bracts, sessile or short-petiolate. *Bracts* leaf-like but smaller, only subtending the lower flowers in the inflorescence which is otherwise ebracteate. *Inflorescence* a rather congested raceme bearing up to 25 deeply cupped, half-nodding flowers, the rachis densely pubescent, with long spreading hairs. *Pedicels* short, 8–10 mm long, densely pubescent, particularly at the slightly expanded top. *Buds* nodding or half-nodding, more or less spherical, covered in dense spreading to ascending bristles. *Petals* 4, bright blue with a pinkish tinge, purple-

Meconopsis simikotensis; Simikot Lagna, Humla District, W Nepal, 3900 m.
PHOTO: COLIN PENDRY

Above left: *Meconopsis simikotensis*, part of inflorescence; Simikot Lagna, Humla District, W Nepal, 3900 m. PHOTO: COLIN PENDRY
Above right: *Meconopsis simikotensis*, flower detail; Simikot Lagna, Humla District, W Nepal, 3900 m. PHOTO: COLIN PENDRY

blue, purple or mauve, paler at the base, obovate to oblong-obovate, 23–39 × 17–37 mm, with a minutely toothed margin. *Stamens* numerous, the filaments the same colour as the petals, whitish towards the base, the anthers yellow. *Ovary* white, subspherical, covered in ascending to semi-appressed whitish barbellate bristles, terminating in a glabrous, fringed stylar cap; style 2–4 mm long, with a capitate purple or yellowish stigma, 1.5–2 mm diameter. *Fruit capsule* unknown.

DISTRIBUTION. W Nepal (Bajhang, Darchula & Humla districts; especially in the vicinity of Simikot; 3300–3965 m. Endemic to NW Nepal (Map 8).

HABITAT. Rocky and bouldery alpine slopes, grassy banks, riverside rocks.

FLOWERING. June–July.

SPECIMENS SEEN. *S. B. Malla* 14224 (BM, E); *Pendry et al.* A195[19] (E); *Polunin, Sykes & Williams* 4270 (BM, E), 4289 (BM, E), 4317 (BM, E).

Meconopsis simikotensis is found well to the west of the other members of subgenus *Discogyne*. Although specimens were collected some years ago the identity of the plant remained hidden and plants resided in herbaria at Edinburgh and the British Museum (Natural History) under the umbrella of *M. discigera* Prain. The Stainton, Sykes & Williams collections listed above are excellent, with ample field notes, and these show the species to be extremely attractive; however, for political and logistic reasons, this extreme western part of Nepal has been inaccessible for a number of years although in recent times several expeditions have ventured there. The chance of acquiring legitimate seed is unlikely at the present time and in the near future. *M. simikotensis* grows at rather lower altitudes than the other members of this association, so it might perhaps be relatively easier in cultivation. Although apparently endemic to the extreme north-western part of Nepal one would not be at all surprised if it were it to be found on the Tibetan side of the frontier, perhaps in the Changla Himal.

Meconopsis simikotensis is closely related to *M. tibetica* and *M. discigera*, differing from the former in its relatively small, blue, purple or mauve flowers and in the angled and fringed stylar disk and from the latter in its entire or slightly toothed leaves and leafy stems, as well as the presence of a capitate rather than linear stigma. *M. simikotensis*, like *M. tibetica* and *M. torquata*, grows in rather drier habitats than either *M. discigera* or *M. bhutanica* which being from the eastern Himal and found south of the main Himalayan divide receive very heavy monsoon rains during the summer months.

Until recently I have not seen any photos of this plant taken in the wild, nor in fact do they appear to exist in the Adam Stainton collection at the Royal Botanic Garden, Edinburgh, where one might have expected to find them; however, some recent photograhs taken by Colin Pendry (RBG Edinburgh) reveal the beauty of this little known species.

---

[19] Collected above Simikot, Karnali Zone, Humla District (29°59'14"N, 81°48'34"E) at 3900 m appears to belong here but differs in its taller stature, rather more toothed leaves and yellow stigma.

20. MECONOPSIS PINNATIFOLIA

Meconopsis pinnatifolia C. Y. Wu & X. Zhuang, *Acta Phytotax. Sin.*, 17 (4): 114 (1979). Type: Tibet, Jilong (Gyirong), Zadang-Xiapujing, 3500–3600 m, *Exped. Med. ad Tibet* 545 (PE holotype).

DESCRIPTION. *A monocarpic plant* to 58 cm tall in flower, forming a lax rosette of leaves in the years before flowering, the whole plant sometimes flushed with purplish-red; stem to 25 mm diameter at the base, moderately to densely bristly, the bristles barbellate, tawny-yellow, mostly 5–7 mm long, the stem base not enveloped in a thick wad of dead and old leaves. *Basal leaves* elliptic to elliptic-oblanceolate or elliptic-oblong in outline, pinnatifid, with 3–6 pairs of asymmetrically-elliptic, entire lobes, these up to 50 × 16 mm, moderately to sparsely bristly, not glaucous beneath, the lamina narrowed below into a 2.5–13 cm long sparsely bristly petiole; stem leaves similar to the basal ones, but decreasing in size upwards and merging with the bracts, less markedly lobed. *Inflorescence* racemose, with 9 or more flowers (rarely more than 20), the rachis with dense fawn or golden bristles, these 4–5 mm long generally; flowers deeply cupped, horizontal to semi-pendent, all but the uppermost bracteate, the bracts often located a little below the point of attachment of the flowers. *Pedicels* 8–25 mm long (terminal pedicel to 72 mm long in fruit), bristly like the stem, especially dense towards the expanded top. *Petals* 4, purple, mauve-blue, deep purplish-blue, violet or dark red, obovate, 42–72 × 38–68 mm, the margin shallow fringed. *Stamens* numerous, the linear filaments the same colour as the petals, often darker, the anthers yellow or golden. *Ovary* subglobose, with dense ascending bristles, capped by a fringed stylar disk 6–9 mm across, the style 5–11 mm long terminating in a clavate stigma 4–5 mm long. *Fruit capsule* 13–16 × 6.5–9 mm, with dense spreading bristles but becoming more or less glabrescent at dehiscence, probably 5–7-valved, the stylar disk 10–12 mm across, the persistent style up to 13 mm long.

DISTRIBUTION. NC & W Nepal (Dhading, Gorkha & Rasuwa districts; centred on the Ganesh Himal and Langtang), also in the adjacent region of S Tibet (Xizang; Jilong); 3600–4880 m (Map 8).

HABITAT. Alpine slopes, stony and grassy places, streamsides, rock ledges.

FLOWERING. June–Aug.

SPECIMENS SEEN. *Ikeda, Kawahara, Yano, Watson, Li, Subedi & Acharya* 20815156 (E); *Miyamoto et al.* (Univ. Tokyo) 94-40033 (BM, TI), 94-00045 (TI), 94-20073 (TI), 94-20093 (TI), 94-40033 (BM, E, TI); *Polunin* 647 (BM, E), 1334 (BM, E), 1465 (BM); *K. N. Sharma* E53 (BM), E97 (BM, E), E399 (BM, E), E401 (BM, E); *Society of Himalayan Botany Exped. Nepal* (1994) 9470343 (E); *Stainton* 4027 (BM, E); *Suzuki et al.* (Exp. Soc. Himal. Bot. Tokyo) 94-70343 (BM, E).

Meconopsis pinnatifolia has very distinctively lobed leaves which, unlike *M. discigera*, extend up the stem to the base of the inflorescence where they merge with the bracts. The species is not in cultivation but there is a good photograph of it taken in the wild in *The Alpine Plants of China* (Science Press, Beijing, 1982) edited by Zhang Jingwei, on p. 109.

The species was described from south-central Tibet to the west of Nyalam (Matsang Tsangpo valley) which lies on the main highway between Kathmandu and Lhasa. Since then further collections have been added, primarily from the Nepalese side of the border, particularly from the Ganesh Himal and Langtang valley regions which lie immediately west of Nyalam. These collections mostly predate the type collection, their significance not realised until after publication of the name *Meconopsis pinnatifolia* by Wu & Zhuang in 1979. Prior to that they had been assigned to *M. discigera* Prain from which they clearly differ not only in flower colour but also in details of the leaves and fruit capsules.

21. MECONOPSIS TORQUATA

Meconopsis torquata Prain, *Ann. Bot.* 20: 355, tab. 24, fig. 11 (1906). Type: S Tibet, Kyi-chu, near Lhasa, 1904, 11,500 ft, *Walton s.n.* (K holotype).

DESCRIPTION. *Monocarpic plant* forming a rosette of leaves in the years before flowering, the rosette building up gradually on the remnants of leaves (primarily lamina bases and petioles) from the previous year or years, these forming a dense bristly mass in time. *Stem* erect, 25–81 cm tall, to 22 mm diameter and somewhat expanded at the base, to 12 mm diameter above, densely beset with fawn-coloured barbellate bristles, as is the inflorescence rachis, the bristles up to 8 mm long. *Basal leaves* oblanceolate to linear-oblanceolate, 13.5–28.5 × 1.4–3.2 cm, tapered at the base into a narrow petiole, the apex obtuse to subacute, covered in pale stiffish, barbellate hairs which have a darkened, pimple-like base, giving a dotted appearance, the margin entire to rather irregularly and distantly sinuate; petiole 13–17 cm long, with a sheathing base; stem leaves similar to the basal but short-petioled to sessile, merging into the bracts which are generally irregularly lobed at the margin. *Inflorescence* a slender, rather congested, rarely lax, raceme of 13–22 flowers, occasionally more, the uppermost being ebracteate; flowers borne horizontally,

Above from left:

Meconopsis torquata, fruiting specimen; above Nangtse Jakhang, near Lhasa, S Tibet, 5000+ m. PHOTO: DAVID & MARGARET THORNE

Meconopsis torquata, basal leaf-rosette; above Nangtse Jakhang, near Lhasa, S Tibet, 5000+ m; above Nangtse Jakhang, near Lhasa, S Tibet, 5000+ m. PHOTO: DAVID & MARGARET THORNE

Meconopsis torquata, part of semi-mature infructescence; above Nangtse Jakhang, near Lhasa, S Tibet, 5000+ m. S Tibet. PHOTO: DAVID & MARGARET THORNE

saucer-shaped, 8–10.4 cm across. *Pedicels* 6–8 mm long (to 12 mm in fruit), occasionally subsessile, expanded at the top into a broad torus, pubescent. *Petals* 4 (–5), sky blue to pale red, oval to obovate, 38–60 × 28–45 mm, sparsely to moderately bristly on the exterior with pale fawn hairs, especially in the middle and towards the base. *Stamens* numerous, with linear filaments the same colour as, or slightly darker than, the petals, and yellow anthers. *Ovary* more or less obovoid, covered with dense oblique bristles, surmounted by a virtually sessile clavate stigma that is expanded at the base to 4–5 mm wide, 8-angled, somewhat sinuate reddish purple disk. *Fruit capsule*, obovoid to obovoid-oblong, 10–16 × 11–14 mm, 7–8-valved, with dense spreading bristles, splitting for a short distance beneath the 8–12 mm diameter stylar disk. (Fig. 10B).

DISTRIBUTION. SC Tibet (Xizang), especially in the hills to the west and north of Lhasa; 3505–4725 m. Endemic to SC Tibet (Xizang) (Map 8).

HABITAT. Screes and bouldery places, often in shade or part shade.

FLOWERING. June–early August.

SPECIMENS SEEN. *Ludlow & Sherriff* 8777 (BM, E), 9904 (E), 9059 (E), 9797 (BM, E), 9904 (E), 9957 (E); *Walton* s.n., Kyi Chu Valley, 15 miles E of Lhasa (K), s.n.1904 (K).

Meconopsis torquata is perhaps the oddest and least attractive member of the *M. discigera* association, found in the wild primarily in the hilly hinterland near Lhasa, especially just west of the Sera Monastery. The rather sinuate, unlobed, leaves and the narrow spike-like racemes are very distinctive. In addition, the petals bear some bristles[20] on the outside, a unique characteristic

---

[20] Petal bristles have apparently been observed on a number of occasions on different *Meconopsis* species, significantly from herbarium specimens; however, having observed some of these very closely with the aid of a microscope I can report that in all cases it is the ovary bristles or 'spines' that are being observed: in the pressing and drying process the bristles are forced through the thin petals and then flattened, thus giving the impression that they have arisen from the back of the petal. Only in *M. torquata*, and perhaps some forms of *M. horridula*, are petal bristles truly present.

in subgenus *Discogyne*. It is a rare and localised species, made so primarily because it has been collected over many years as a medicinal plant. Along with *M. horridula* sensu lato it probably inhabits drier and bleaker habitats than any other *Meconopsis*.

The climate of the Tibetan Plateau close to Lhasa is one of extremely cold dry winters and springs, sometimes with snow, while most rainfall (which is modest, about 400 mm p.a.) falls during the summer months, although this part of Tibet is shielded from the heavy excesses of the summer monsoon by the huge barrier of the Himalaya to the south, keeping much of southern Tibet in an effective rain shadow, except where gaps in the mountains allow the monsoon clouds to penetrate (e.g. Namcha Barwa in SE Tibet which records an impressive amount of rainfall, much falling as snow on the higher slopes).

Although not in cultivation at the present time, *Meconopsis torquata* has been in the past, most notably from seed brought to Britain by that intrepid collecting partnership of Ludlow & Sherriff. Reports of it flowering in cultivation (that of General Murray-Lyon in particular) are very few and unpromising, for the flowers "failed to open properly" and, like *M. discigera*, it failed to set seed, which does not bode well for a monocarpic plant.

A note attached to a Ludlow & Sherriff collection, number 8777, records that the "base of stem swollen and containing much watery fluid" which may help the plant remain hydrated during the flowering and early fruiting phase.

Frank Ludlow was based in Lhasa during the final years of the Second World War. Harold Fletcher reports (*A Quest or Flowers*: 235, 1975) that "Throughout his year in Lhasa Ludlow continued to collect plants. Sometimes he himself would make a short day expedition to the surrounding hills. At other times he would send his faithful Kashmiri servant, Ramzana, and a companion, to areas further afield. Their most dramatic find was *Meconopsis torquata*, a magnificent blue-flowered meconopsis growing at a height of 15,500 feet in boulder scree. This meconopsis had originally been found by Captain H. Walton on the 1904 expedition and the flower had been described as being red. Geordie and I found it again in July 1943 north of Lhasa growing in similar boulder scree. It was much more common here and grew between 14,000 and 16,000 feet — sometimes in open situations and sometimes in complete shade. Unfortunately it has not survived in cultivation". A half-tone plate accompanies the text.

22. MECONOPSIS MANASLUENSIS

Meconopsis manasluensis P. A. Egan, *Phytotaxa* 20: 50–53 (2011). Type: Central Nepal, Manaslu Himal (Gorkha District), east of Samdo, S side Sanam Khola, 28°38'N, 84°38'E, 4000 m, *Ikeda et al.* 20815156 (E holotype & isotype).

DESCRIPTION. *Monocarpic plant*, 30–60 cm tall in flower; taproot fleshy, twisted, 2.5–3 cm diameter at the greatest. *Stems* several, erect, agglutinosed (coalescing) close to the base, just above the crown of the plant, moderately covered, as are the leaves and pedicels, with fawn or orange barbellate hairs. *Basal leaves* borne in a lax rosette to 30 cm across in the years prior to flowering, spreading to ascending, the lamina elliptic-oblanceolate to oblanceolate, sometimes very narrowly so, 6.5–15.5 × 1.5–3 cm, cuneate into the petiole at the base, subacute to obtuse at the apex, margin entire, slightly undulate; petiole 3–5 cm long, narrowly winged; cauline leaves gradually decreasing in size upwards but otherwise

Meconopsis manasluensis, herbarium specimen prior to pressing; E of Samdo, S side Sanam Khola, Manaslu Himal, C Nepal, 4000 m.
PHOTO: MARK WATSON

similar to the basal leaves, the lower short-petioled, the upper sessile and merging into the bracts. *Inflorescence* several-branched from the base, with up to 30 flowers, the central axis with a single flower, the lateral branches subracemose with up to 5 part-bracteate flowers; flowers saucer-shaped, ascending to lateral-facing. *Bracts* alternate to subopposite, similar to the leaves but smaller and sessile. *Pedicels* 1.5–7.5 cm long, erect, curved during anthesis, later straight and elongating as the fruits develop. *Buds* nodding, ellipsoidal, moderately adorned with long flexuous bristles. *Petals* 4–5 (–8), scarlet with dark purple patches towards the base, obovate to broad-elliptic, 2–3 × 1–2 cm, the margin unevenly and rather coarsely erose. *Stamens* about one quarter the length of the petals, with reddish purple filaments and orange-yellow anthers. *Ovary* narrow-ovoidal, covered in dense fawn of orange erect bristles; style slender, 7–12 mm long, the basal stylar disk reddish purple, to 8 mm diameter, with an undulate, slightly lobed margin just overlapping the edge of the ovary; stigma capitate, yellow, 3–4 mm diameter, c. 6–8-lobed. *Fruit capsule* narrow cylindric-ovoidal, 8–16 × 4–7 mm, c. 8-valved, adorned with moderately sparse patent bristles.

DISTRIBUTION. Central Nepal (Gorkha District; Manaslu Himal, known only from the type locality at present); c. 4000 m. Endemic to NC Nepal (Map 8).

HABITAT. Boulder fields in moist, humus-rich soils, in otherwise herb-rich grasslands with scattered shrubs, 'locally common'.

FLOWERING. July–August.

SPECIMENS SEEN. *Ikeda, Kawahara, Yano, Watson, Li, Subedi & Acharya* 20815156 (E).

Meconopsis manasluensis is a distinctive member of subgenus *Discogyne*, unique in its multi-stemmed inflorescences but, at the same time, conforming to the prime characteristics of the subgenus, not least being the presence of a distinct stylar disk. As it has flowered successfully at least once in cultivation (see note below) there seems no reason why it should not do so again if seed were to become available. Apart from its decorative appeal, live plants would be interesting from a scientific point of view. The species was discovered during the 2008 Flora of Nepal Manaslu Expedition, and named after the corresponding mountain range within the Himalaya, the Manaslu Himal.

There is a fine pen and ink drawing by S. Bradley accompanying the original description in *Phytotaxa* in 2011, stressing both the habit of the plant as well as the flower and fruit details.

NOTE. Interestingly, there reside in the herbarium at the Royal Botanic Gardens Kew five sheets annotated

Meconopsis manasluensis, flower detail; E of Samdo, S side Sanam Khola, Manaslu Himal, C Nepal, 4000 m. PHOTO: MARK WATSON

'Cult. Roy. Bot. Gard. Edinburgh' and dated 1946 which are almost certainly *Meconopsis manasluensis* but determined as *M. discigera* Prain. All the specimens displayed bear more or less entire leaves, while three of the sheets demonstrate multi-racemose specimens, one with 4–5 branches, the most obvious character of *M. manasluensis*. The origin of these specimens is uncertain but clearly they were grown from seed collected in Nepal. There are no existing records at Edinburgh of this collection but it would be of great interest to know who collected seed and the location. At the time at the end of the First World War very few people were able to travel to Nepal which was virtually closed to the outside world. If plants flowered in 1946 then the seed would have been sown probably in the 1942–44 period. One intriguing possibility is that seed was collected by Capt. F. M. Bailey, the man who discovered *M. baileyi* in Tibet in 1913. However, Bailey, or his collectors, also collected in Nepal in the 1930s; one collection No. 13 present at the British Museum, is rather fragmentary (although clearly a subgenus *Discogyne* species) and could represent *M. manasluensis* but was collected farther east. The field notes read "Dudkhunda, 13,000 ft, 1935" with the addition of "seed taken". If this is so then the distribution of this recently described species might extend from the Manaslu area, roughly 84°34'E, to 86°36'E.

Subgenus GRANDES

Subgenus *Grandes* harbours some of the boldest and most colourful species of *Meconopsis*, a number of which are well-established in cultivation and well-known to gardeners. They include *M. grandis* itself as well as *M. baileyi*, *M. integrifolia*, *M. punicea* and the so-called harebell poppy, *M. quintuplinervia*. The subgenus is characterised by monocarpic or polycarpic plants with an indumentum of barbellate hairs. The leaf-rosettes are non-persistent, the leaves dying away in the autumn but often leaving parts of the lamina and petioles withered around the base of the plant which overwinters as a bud or series of buds at or just beneath the soil surface. Once spring arrives a fresh set of rosette leaves appear and the plant may or may not produce flowers at this stage or grow on for a year or two more before reaching the flowering phase. The flowers bear 4–9 petals which come in a wide span of colours including blue, purple, red and yellow. Apart from *M. integrifolia* sensu stricto and some forms of *M. punicea*, the fruit capsules always possess a distinct style. In subgenus *Grandes* the flowers are borne in two distinct manners which separate the two sections recognised here. In the majority of species plants produce a distinct stem with or without some cauline leaves but always with leaf-like bracts that are mostly borne in a lax cluster (pseudowhorl) at the top of the stem and from whose axils the majority of flowers arise. Alternatively, in the other condition, the flowers are always solitary and scapose, arising directly from the basal leaf-rosette. Despite this difference the members of the subgenus form a relatively homogeneous assemblage.

KEY TO SECTIONS OF SUBGENUS GRANDES

1. Flowers generally borne in umbels or psedoumbellate racemes, accompanied by bracts, the uppermost bracts in a pseudowhorl; filaments nearly always white. . . . . section **Grandes**
+ Flowers solitary and scapose, without bracts: filaments nearly always the same colour as the petals, often darker. . . . . section **Simplicifoliae**

section GRANDES

Within this section are included species with basically blue, pink or yellow flowers borne in a distinct subumbellate or subracemose inflorescence. On occasions the inflorescence is reduced to a solitary flower but with accompanying bracts. Sometimes plants produce a very short stem and, as a result, the bracts are difficult to observe, being hidden or partly hidden within the basal leaf-rosette; yet close inspection will nearly always reveal their presence.

The single-flowered inflorescence is an interesting state found in several species. In *Meconopsis sherriffii* and *M. pseudointegrifolia* the inflorescences are invariably solitary-flowered; whereas in *M. grandis*, *M. integrifolia* and *M. sulphurea*, while multi-flowered inflorescences are the norm, plants with solitary-flowered inflorescences are not uncommon. Interestingly, growing conditions can affect cultivated plants and it is possible that, even in the wild, seasonal fluctuations may affect plants in a similar manner. In George Taylor's (1934) monograph of the genus there is a fine half-tone plate (Pl. XVI) of *M. grandis* showing this condition: while the majority of the plants in the photograph bear solitary flowers several have two or three flowers and, while some clearly have a distinct stem, others have a reduced stem wholly or partly concealed by the basal leaf-rosette. Over the years it was not only George Taylor who pointed out the close similarity between *M. grandis* and *M. simplicifolia*, especially in the reduction of the inflorescence sometimes seen in the former: *M. simplicifolia* always bears solitary scapose flowers! While it can be argued from an evolutionary view-point that there has been a trend within *Meconopsis* from a well-defined inflorescence to the solitary scapose state and, while on occasions scapes can be agglutinated into a false inflorescence, the two states, scapose or not, are clearly defined and for that reason *M. grandis* and *M. simplicifolia* fall into distinct sections within the subgenus.

KEY TO SERIES OF SECTION GRANDES

1. Flowers blue, purple, pinkish-purple, pink or wine-red; plants usually polycarpic . . . . . . series **Grandes**
+ Flowers yellow; plants monocarpic. . . . . series **Integrifoliae**

series Grandes

The species of series *Grandes* bear blue, purple or pink, occasionally wine-coloured, flowers that generally have 4–6 petals, although up to 8 in the pink-flowered *Meconopsis sherriffii*. In all, the flowers are open bowl- or saucer-shaped, occasionally almost flattish, and are lateral-facing or half-nodding. The fruit capsules are invariably bristly and possess a distinct style. The series contains two of the most familiar and horticulturally important species in *M. grandis* and *M. baileyi*, indeed two of the finest and showiest species seen in gardens. Hybrids between the two in cultivation have led to a plethora of fine clump-forming perennials known to gardeners as the 'big perennial blue poppies' (see p. 158).

KEY TO SPECIES OF SERIES GRANDES

1. Lamina of the basal leaves, at least, cuneate, attenuate into the petiole. 2
+ Lamina of basal leaves at least shallowly cordate to truncated, abruptly contracted into the petiole 3

2. Fruit capsule 24–30 × 12–18 mm, always bristly, with a very short (not more than 2 mm long) to more or less obsolete style; flowers solitary, the petals normally 6–8, flesh to rose pink . . . . 26. **M. sherriffii** (N Bhutan and adjacent Tibet)
+ Fruit capsule 27–58 (–68) × 6–15 mm, bristly or glabrous, with a pronounced style 4–8 mm long; flowers generally 2–5 in a subumbellate inflorescence, occasionally solitary, the petals normally 4–6, blue, purple-blue to wine or wine-purple . . 25. **M. grandis** (Nepal to Sikkim, Bhutan, NE India and adjacent Tibet)

3. Uppermost 3–5 (–7) leaves aggregated into a false whorl; mature leaf-lamina with 8–19 pairs of teeth; mature fruit capsule 10–16 mm diameter, moderately to densely bristly, the style 2–5.5 mm long with a stigma 4–5 mm diameter . . . . 24. **M. baileyi** (SE Tibet, N Myanmar and NE India)
+ Uppermost leaves alternate, not aggregated into a false whorl; mature leaf-lamina with 5–9 pairs of teeth; mature fruit capsule 8–9 mm diameter, glabrous or rarely with a few scattered bristles, the style 5–8 mm long with a stigma 2–3 mm diameter . . . . 23. **M. betonicifolia** (SW China)

23. MECONOPSIS BETONICIFOLIA

Meconopsis betonicifolia Franch., *Pl. Delav.* 1: 42, t. 12 (1889). Type: NW Yunnan, *Delavay* 2152, China 1886–7, 'ad collum monts Kualapo, supra Hokin', 3200 m (P, holotype; K, isotype).

[syn. *Cathcartia betonicifolia* (Franch.) Prain, *Ann. Bot.* 20: 369 (1906); *Meconopsis betonicifolia* forma *franchetii* Stapf, *Curtis's Bot. Mag.* 153: sub tab. 9185, in obs. (1930); *M. b.* var. *franchetii* (Stapf) L. H. & E. Z. Bailey, *Hortus*: 390 (1930)]

DESCRIPTION. *Polycarpic plant* forming clumps or patches by means of slender, pencil-sized, underground stolons. *Stem* erect, to 1.5 m, often purple-flushed, slightly ridged, to 14 mm diameter near the base, usually glabrous, occasionally with a few scattered rufous, barbellate bristles towards the top. *Leaves* deep green, occasionally pale yellowish green, to somewhat glaucous, paler beneath, sometimes with a purplish midrib, the basal leaves petiolate, borne in a lax cluster, relatively small, the lamina ovate to paddle-shaped, 6.5–13.5 × 2.8–6.7 cm, the base cordate or subcordate, sometimes more or less truncated, somewhat decurrent onto the petiole, the apex subacute to subobtuse, generally with 5–9 coarse serrate-dentate teeth along the margin on each side, both surfaces adorned with spreading rufous, barbellate, rather weak bristles, especially along the margin; cauline leaves 5–7, all alternate (spirally arranged), spreading, sometimes somewhat recurved, the lowest short-petiolate, the others sessile, the lamina similar to those of the basal leaves, although often more oblong-ovate, the base of sessile leaves markedly clasping (auriculate), all alternate; petiole of basal leaves to 16.5 cm long, widening and sheath-like at the base. *Inflorescence* a raceme bearing up to 7 flowers, borne in the axils of the uppermost cauline leaves (bracts), the flowers semi-nodding to lateral-facing, saucer-shaped, 6.5–11 cm across. *Pedicels* 10–17 cm long in flower, glabrous or with a few patent bristles below the flower, generally glabrous in the lower part. *Buds* narrow-ovoid, nodding, the sepals patent-bristly. *Petals* 4, soft rose lavender to blue-violet, sometimes with a hint of purple, occasionally ruddy purple, ovate to ovate-elliptic, 3.4–5.3 × 2.7–4 cm, spreading, occasionally slightly reflexed and with slightly revolute margins, the margin somewhat undulate, often unevenly erose, occasionally somewhat unevenly, shallowly laciniate. *Stamens* numerous, about a quarter to one third the length of the petals, with linear white filaments 10–13 mm long and orange anthers 2–3 mm long. *Ovary* ellipsoid, along with the style about twice the length of the stamens, glabrous (rarely with a few bristles along the sutures), with a well-developed style 5–7 mm long, extending to 9 mm in fruiting specimens, the stigma 3.5–5.5 mm long, clavate, generally 5–6-lobed. *Fruit capsule* narrow spindle-shaped (fusiform), narrowed evenly at both ends, 2.5–3.3 × 8–9 mm, glabrous or rarely with a few short bristles along the sutures, 5–6-valved, splitting for about one quarter of its length from the top at maturity. (Figs 11D, 12A1–3).

DISTRIBUTION. SW China: NW Yunnan (primarily on the Mekong-Yangtse and Mekong-Chienchuan divides, also around Jianchuang west of Heqing (Hoking)); 3000–3963 m (Map 9).

HABITAT. Marshy woodland, rocky places, margins of alpine scrub, coniferous forests.

FLOWERING. June–July.

SPECIMENS SEEN (all from Yunnan). *Delavay* 2112 (BM), 2152 (K), s.n., 3 sheets, May 1889 (K), s.n. June 1888, s.n. (K), s.n. March 1889 (BM), s.n. 24 May 1889 (K), s.n. 25 May 1889 (BM), s.n. 2 May 1890 (K); *Forrest* 1923 BM), 1924 (BM), 1931 (BM), 21500 (E, K), 21573 (E, K), 21962 (E, K), 22286 (E, K), 23420 (BM, E), 25720 (BM, E, K), 25892 (E), 30082 (BM, E), 30096 (BM, E); *Gregory* 69 (BM); *Rock* 8579 (E, K), 8584 (E, K), 9487 (E), 9530 (E), 25167 (BM, E, K), 25720 (K).

Meconopsis betonicifolia was described in 1889 by Adrien René Franchet, in a publication devoted to the plant discoveries of one of the pioneering plant collectors in western China, Abbé Jean-Marie (Père) Delavay. Born in the Haute-Savoie in France in 1834, Delavay travelled to the province of Guangdong in 1867 as a missionary (for the Missions Etrangères) and botanist. During a brief return to France in 1881 he was persuaded by Armand David, another indefatigable collector and missionary, to collect specimens for the Paris Museum of Natural History. For the next fifteen years, with scarcely a break, Delavay amassed a considerable collection of dried plants

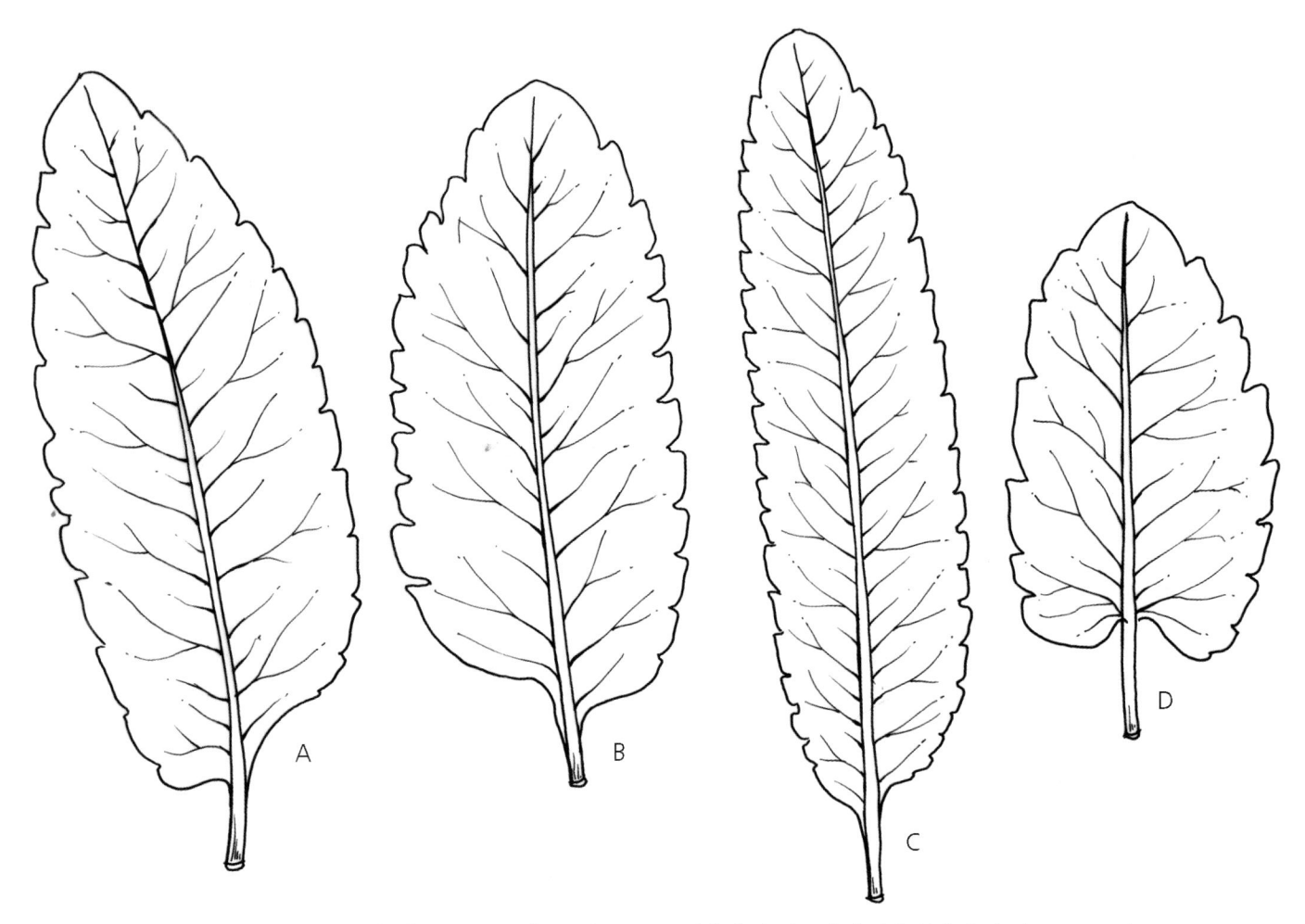

FIG. 11. Basal leaves, lower petioles and pubescence not shown. **A,** *Meconopsis baileyi* subsp. *baileyi*; **B,** *M. baileyi* subsp. *pratensis*; **C,** *M. baileyi* subsp. *multidentata*; **D,** *M. betonicifolia*. All × $^2/_3$. DRAWN BY CHRISTOPHER GREY-WILSON

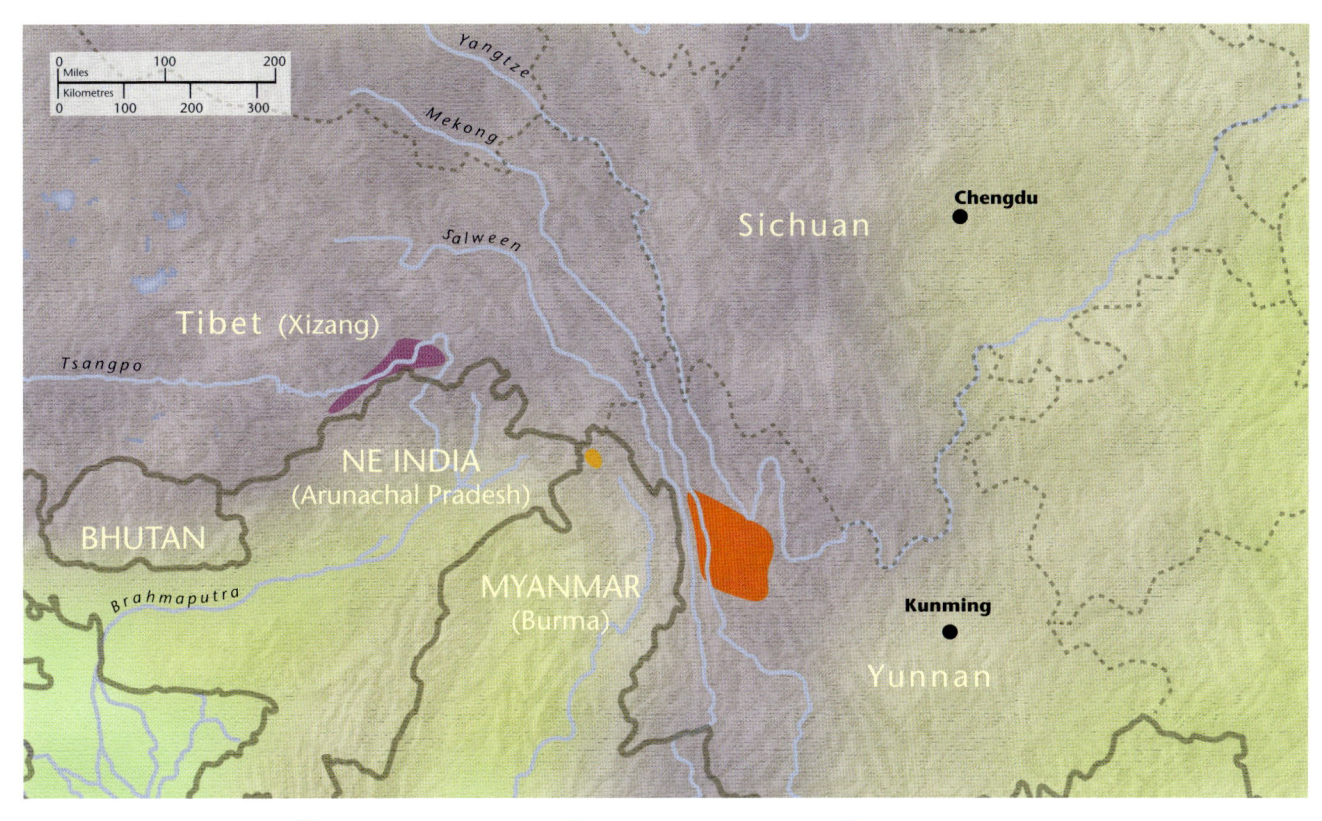

MAP 9. Species distribution: ■ *Meconopsis betonicifolia*, ■ *M. baileyi* subsp. *pratensis*, ■ *M. baileyi* subsp. *baileyi*.

(in excess of 50,000), a remarkable two-thirds of which were new to science. Delavay was not, however, an avid collector of living material and it was left to the famous Scotsman George Forrest and others to introduce many of his discoveries into cultivation, particularly from north-western Yunnan.

Among Delavay's discoveries there were a number of collections of a beautiful 'blue' poppy gathered near Hokin (or Hoching, now Heqing) and at San-tcha-ho in north-western Yunnan. The description drawn up three years later by Franchet was based upon these.

However, the glowing account of its introduction given by George Taylor was not directed to plants of Chinese origin but to collections made in south-eastern Tibet (Xizang). A brief résumé of its discovery is needed here in order to understand the subsequent evolution of the plant in cultivation and its subsequent treatment by botanists. In 1915 Sir David Prain described a new species of *Meconopsis*, *M. baileyi*, based on material collected during 1913 from Lunang, in the Rong Chu valley, Kongbo at 10,500 ft in south-eastern Tibet by Colonel F. M. Bailey in the company of Capt. Morshead. Unfortunately, the material was very fragmentary (consisting of little more than a flower stuffed into a notebook) but Prain considered he had enough to justify describing it. Yet, according to George Taylor nearly twenty years later, "…..the material was so incomplete and fragmentary that the author did not recognise the proper affinity of his species". There the situation remained until June 1924 when Frank Kingdon Ward ventured into the same region, indeed to the same locality, and was able to secure ample good material.

Kingdon Ward was able to relate his material directly to that of Bailey and, on subsequent expeditions, he made further gatherings, introducing at the same time ample seed under the name *Meconopsis baileyi*, although he failed to make any comparison with the Chinese *M. betonicifolia*. Kingdon Ward not only gathered material from Bailey's site but in various other localities along the divides and tributaries of the Tsangpo, and later in the adjacent region of northern Burma (now Myanmar).

"Beautiful as were the meadows of the rong ….. nevertheless, the finest flowers hid themselves modestly under the bushes, along the banks of the stream. Here among spiteful thickets of hippophae, barberry, and rose, grew that lovely poppy, *Meconopsis baileyi*, the woodland blue poppy …. Never have I seen a blue poppy which held out such hopes of being hardy, and of easy cultivation in Britain ….. It may be remarked in passing that the only known species of *Meconopsis* which bears any close resemblance to *M. baileyi* are the Chinese *M. betonicifolia* and the Bhutanese *M. superba*.'

(KW, *The Riddle of the Tsangpo Gorges* (1926)).

23. MECONOPSIS BETONICIFOLIA

Meconopsis betonicifolia; Jiangchuang, Lijiang County, NW Yunnan. PHOTOS: JOE ATKIN

However, once the material was brought back to Britain it was reappraised, as related by the collector: 'We [Kingdon Ward in the company of Earl Cawdor] found *Meconopsis baileyi* in many places; in fact it was by no means rare in this part of Tibet, and I was able to take back to England complete herbarium specimens and a large packet of ripe seed. The specimens were immediately identified at Kew as Prain's *M. baileyi*; they supplied all the missing parts and made a complete description possible ……Now that complete specimens and, before long, living plants of *M. baileyi* were available, they were soon recognised as being identical or near identical to a little-known species called *M. betonicifolia*. According to these initial findings the name *M. baileyi* therefore became superfluous; *M. baileyi* equalled the longer-known *M. betonicifolia*, and it is highly inconvenient to have two names for the same plant. However, the name *M. baileyi* persisted in the catalogues of some nurserymen, because having sold the plant under that name they did not wish to change it; to do so might be bad for trade, however good for botany.

Moreover, botanists, after some consultation, agreed that the new plant did not *quite* match the old; there are slight and possibly constant differences between them. It was therefore considered that it might be convenient to retain *baileyi* as a varietal name and call the plant *M. betonicifolia baileyi* [sic *M. betonicifolia* Franch. var. *baileyi* (Prain) Edwards *nomen nudum*; *M. b.* forma *baileyi* (Prain) Cotton]' (KW, *Pilgrimage for Plants* (1960)).

Kingdon Ward's introductions brought the plant firmly into cultivation and to a highly appreciative gardening public. Under the name *Meconopsis baileyi* it became known as Bailey's blue poppy, the Tibetan blue poppy or even the Himalayan blue poppy (although this latter attribution could justifiably be applied to another and arguably finer species, *M. grandis*). Seed was widely distributed and the plant was soon well-established in Britain and Europe, as well as Canada, the USA and New Zealand. In the July 1929 monthly record of the Royal Geographical Society, a lengthy note was submitted by Arthur R. Hinks, FRS: "At the Chelsea Show of 1926 a superb 'blue poppy', *Meconopsis baileyi*, was shown by

Lady Aberconway and the Hon. H. D. McLaren, and then first attracted general notice, though it had received an Award of Merit of the R.H.S. at their fortnightly show of the preceding April 7. In early June of the present year [1929] a large bed of the plant flowered magnificently in Kensington Gardens, within three minutes of the Society's House, and the Director of the Royal Botanic Gardens at Kew declares that it grows like a weed — which is very unusual in the beautiful flowers which have come from the Tibetan border …".

The taxonomic position remained unchanged until George Taylor (later Sir George) published his significant monograph of the genus in 1934. Here *Meconopsis baileyi* is considered a synonym of the earlier species, *M. betonicifolia*. Taylor added, "That the two plants are conspecific seems evident, and I have expressed the opinion that they may be regarded as geographical forms of the same species. Stapf [at Kew] has since accepted this view by identifying the cultivated plant as *M. betonicifolia* forma *baileyi*, and for the other form, on which the species was founded, he has proposed the name *M. betonicifolia* forma *franchetii*".

The position has remained unchanged until the present day, although the two formas are not generally used. Indeed forma *franchetii* technically, if recognised and following botanical convention, should be forma *betonicifolia*. However, the fact is that the only plant produced in quantity and widely-cultivated is the Tibetan plant resulting from Kingdon Ward's introductions and later those of Ludlow, Sherriff and George Taylor. Although generally listed in catalogues and other publications as *Meconopsis betonicifolia*, a few still hark back to *M. baileyi*. Incidentally, the former has received various spellings over the years including *betonicaefolia* and *betonicifolius*.

Since the monograph was produced some 75 years ago further material has been collected in Tibet, notably during the various expeditions of Ludlow and Sherriff between 1933 and 1949, on occasion accompanied by Dr J. H. Hicks or George Taylor. This material has substantially increased our knowledge of the Tibetan plant and allowed further comparisons to be made with its Chinese counterpart. Accompanying this plethora of herbarium material has come a great deal of fresh field

Meconopsis betonicifolia; Laojunshan, Lijiang County, NW Yunnan, 3800 m. PHOTO: HARRY JANS

Table 5
A COMPARISON OF CHARACTERS BETWEEN TIBETAN AND YUNNANESE POPULATIONS OF *MECONOPSIS BAILEYI/BETONICIFOLIA*

CHARACTERS	Yunnan populations *Meconopsis betonicifolia*	Tibet-Myanmar-NE India populations *Meconopsis baileyi*
Plants	stoloniferous	clump-forming
Leaf (bract) whorl	absent	present
Leaf lamina base[21]	cordate to truncate	broad cuneate to subtruncate
Basal leaves — marginal teeth[21]	5–9 on each side	8–19 on each side
Basal leaves — dimensions[21] (mm)	65–135 × 28–67	152–280 × 54–116
Style length (mm)	5–9	± obsolete –5.5
Stigma length (mm)	3.5–5.5	3–4.5
Fruit capsule pubescence	Very rarely with a few scattered bristles along the sutures	moderately to densely bristly
Fruit capsule size (mm)	25–33 × 8–9	(26–)28–40 × 10–16

data and many excellent habitat and plant photographs of them in the wild, notably by Harry Jans, John Mitchell and Toshio Yoshida, and others, and this has given a very much clearer insight to these fascinating blue poppies. All this data has been an invaluable asset in re-evaluating the various taxa involved.

At a glance the Tibetan and Yunnanese populations look strikingly similar, both being tall and elegant with exquisite four-petalled blue flowers. However, in comparing the Tibetan-Burmese and Chinese material of *Meconopsis betonicifolia-baileyi* I have not found the merging of characters as stressed by Taylor in his monograph. Indeed his description bears some surprising oversights, most notably with regard to the size and shape of the leaves, the disposition of the stem (cauline) leaves and the shape of the fruit capsules. Although it is clear that the Tibetan and Yunnanese populations are inherently variable the differences between them amount to more than eight characters and these are outlined in Table 5.

---

[21] Mature basal leaves only recorded.

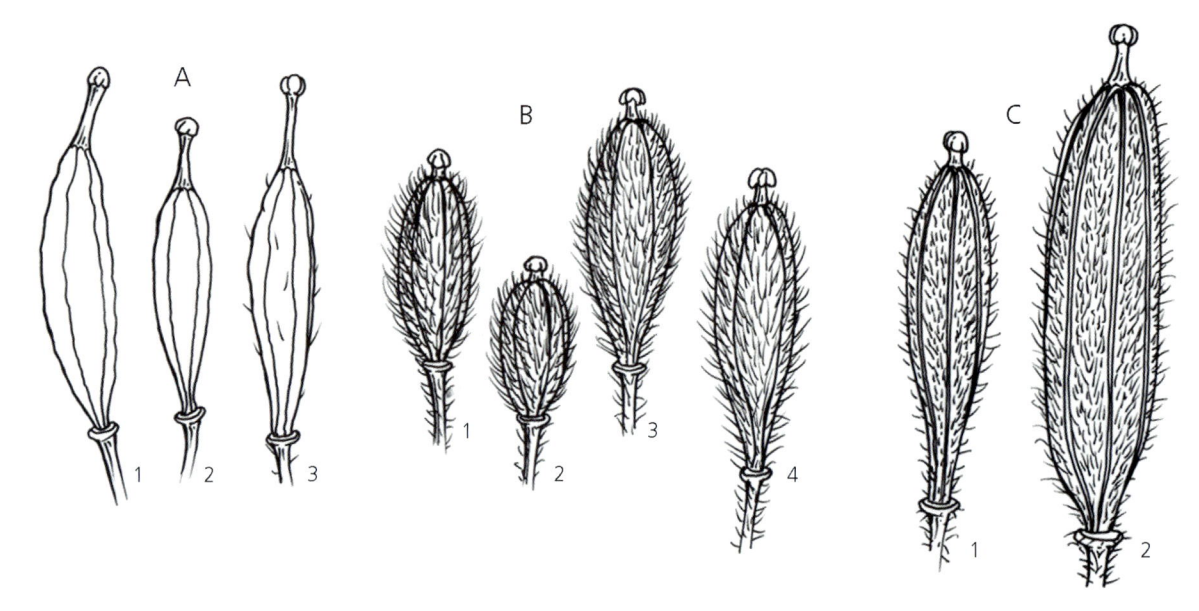

FIG. 12. Fruit capsules. **A1–3**, *Meconopsis betonicifolia*; **B1–3**, *M. baileyi* subsp. *baileyi*; **B4**, *M. b.* subsp. *pratensis*; **C1–2**, *M. simplicifolia*. **A** × 1¹/₃, **B–C** × 1. DRAWN BY CHRISTOPHER GREY-WILSON

Field observations by various people along with photographic evidence clearly show the Chinese plant to be stoloniferous. An extract from a letter penned by Stanley Ashmore (Coahuila, Mexico) to Geoff Hill (Member of the Meconopsis Group) comments as follows: "…..In the fall of 2000 I participated in a seed-hunting expedition headed by Dan Hinkley of Heronswood Nursery and made up of members of the Alaska Rock Garden Society. We visited a population of *Meconopsis betonicifolia* in north-west Yunnan in the area of Ninety-Nine Dragons Mountain [Jianchuang: This locality, by the main road between Dali and Lijiang (Dayan), is just west of the *locus classicus* for *M. betonicifolia* at Hoking (Heqing)]. I observed that, although the plants were similar in some respects to the *M. betonicifolia* that are in cultivation, there were some remarkable differences. Some of these have been noted by you and others. Growth was odd. The plants grew in colonies rather than clumps and appeared to be aquatic in some cases ….. Since then the two 'forms' — the Tibet and the Yunnan — have been growing side by side in the Blue Poppy Garden outside Palmer, Alaska. We have succeeded in making a fertile cross between the two. In addition, we have had the luxury of making observations of the Yunnan 'form' in the garden. The Yunnan form spreads by underground rhizomes somewhat like *M. quintuplinervia* ….. rhizomes are the size of a pencil and produce plants 6 inches [15 cm] from the mother plant….."

Unfortunately, these stolons have been overlooked by collectors in the past and are not present on any herbarium specimen that I have examined, very few of which have even a vestige of root; however, the stoloniferous habit can be clearly seen in some photographs of plants growing in the wild. I am unable to say whether or not the Tibetan populations are always clump-forming and non-stoloniferous and no one to my knowledge has been able to make such an investigation; however, judging by those in cultivation at the present time from various localities in south-eastern Tibet, NE Arunachal Pradesh (NE India) and N Myanmar I can state quite categorically that none are stoloniferous.

Right, from top:

Meconopsis betonicifolia; Laojunshan, Lijiang County, NW Yunnan, 3800 m. PHOTO: HARRY JANS

Meconopsis betonicifolia, flower details; Laojunshan, Lijiang County, NW Yunnan, 3800 m. PHOTO: HARRY JANS

Meconopsis betonicifolia; Laojunshan, Lijiang County, NW Yunnan, 4000 m. PHOTO: TOSHIO YOSHIDA

With these revelations in mind, I had no hesitation in reinstating the Tibet plant as a species in its own right, *Meconopsis baileyi*, and this was first presented in *The Alpine Gardener* (*Bull. Alpine Gard. Soc. Gr. Brit.*) in 2009. This species is clearly very closely allied to *M. betonicifolia* and along with *M. grandis, M. integrifolia, M. pseudointegrifolia* and *M. sherriffi*, fits snuggly into series *Grandes*. While some would undoubtedly interpret my findings rather differently (several have suggested that perhaps subspecies might have been more appropriate) the number of differences recordable along with the distinct geographical distribution of the two taxa have convinced me that separate specific status is warranted. This name-about change should not for once irritate gardeners who are quick to accuse taxonomists of "fiddling with names" for many knew the plant as *M. baileyi* over the years (indeed long after Taylor's monograph appeared) and even today some nurseries still list the plant under the now correct name of *M. baileyi*. Now the challenge is to persuade the others to let go of the name *M. betonicifolia* for the Tibetan plant. This should not be difficult as all the plants in general circulation at the present time are of the former while *M. betonicifolia* is very rare in cultivation and only in private hands. Even the recent edition of *The RHS Plant Finder* is out of date in this respect.

In hindsight it seems surprising that the Chinese plant, viz. *Meconopsis betonicifolia*, was not introduced into cultivation long before *M. baileyi*. After all, it had been discovered more than 30 years before *M. baileyi* was introduced into cultivation. In the first years of the twentieth century numerous plant collectors had been scouring western China for plants. Perhaps most notable, apart from Kingdon Ward himself, was George Forrest who brought many of Delavay's discoveries into cultivation. Forrest had systematically explored and collected in Yunnan between 1904 and 1932 and, apart from *Rhododendron* and *Primula*, *Meconopsis* was a genus to which Forrest would have paid particular attention (He found no less than 12 species in Yunnan!). Forrest did in fact search out and pinpoint Delavay's locality at Hoking and found the plants, but Kingdon Ward (1960, *Pilgrimage for Plants*) states that: "….. strangely enough the seeds he sent home did not do well — the little plants perished as seedlings. Nor did he find it more than once or twice. Later the American Joseph Rock collected it, but with no better result for horticulture". Kingdon Ward (1960) added, "…One inescapable conclusion, I think, is that this particular poppy has always been rare in Yunnan. Many of the most famous introductions from western China are widespread, being found in any suitable locality over an area of hundreds, or even thousands, of square miles, but this is not one of them. The fact that George Forrest, whose well-trained Chinese collectors covered a lot of ground and missed very little, only came across the plant once or twice is sufficient endorsement for the view that *M. betonicifolia* is a very rare plant in Yunnan. Yet, even while making every allowance for the difficulties and uncertainties; it seems extraordinary that forty years elapsed between the first discovery of this *Meconopsis* and its introduction to Europe."

The rarity of *Meconopsis betonicifolia*, stressed by Kingdon Ward, is perhaps rather overstated. The species is undoubtedly local, and although few modern collections exist (at least in European and American herbaria) there are almost 20 historical collections mostly gathered by George Forrest and Joseph Rock, but also several sheets of Delavay's original and earlier collections. Taken together these represent perhaps a dozen localities on the Lancangjiang (Mekong)-Jinshajiang (Yangtse) and the Lancanjiang-Chienchuan divides. On a map this forms a rough triangle from west of Weixi (Wei-hsi) in the north, Shigu and Heqing (Hoking) in the east and Chienchuan in the south, an area some 120 km north to south and 100 km west to east. Even today it can be expected to grow in additional localities within this area. However, it is suspected that many of its former locations will have deteriorated along with the mindless destruction of many forested areas in this part of China. Plants like *Meconopsis*, particularly the forest-dependent species, are very susceptible to disturbance.

In contrast, *Meconopsis baileyi* appears to be far commoner in the wild than its Yunnanese counterpart. It has a well-marked distribution in south-eastern Tibet (collections are recorded from Kongbo Province, Tsari and Zayul) from the Tsangpo valley extending south-eastwards for some 280 km to the Seinghku valley in northernmost Myanmar (Burma) and in recent years it has also been found in north-eastern Arunachal Pradesh (NE India).

It will undoubtedly prove irksome to some gardeners (and maybe others) to learn that virtually all the material in cultivation under the name *Meconopsis betonicifolia* should be redefined as *M. baileyi*. One could say things have gone full circle as, for many years after its initial introduction, this is the name under which it was revered by the gardening public. It is to be hoped that the material of the true *M. betonicifolia* at present in cultivation in America, and gathered in recent years in Yunnan, will become established and allow us to admire and grow this exquisite plant in European gardens alongside its more famous Tibetan cousin.

Stanley Ashmore who introduced seed of *Meconopsis betonicifolia* from the wild into Canada, has grown and observed the species both in the wild and in cultivation and has come to the same conclusion as I have as regards the separate identities of *M. baileyi* and *M. betonicifolia*. His observations and data has been extremely useful in this study. Ashmore has created hybrids between the two in his Alaska garden, reporting that they are intermediate in habit. Unfortunately, material sent to Britain by him as *M. betonicifolia* has so far proved to be this hybrid.

24. MECONOPSIS BAILEYI

Meconopsis baileyi Prain, *Kew Bull.* 1915: 161 (1915); Prain & Kingdon Ward, *Ann. Bot.* 1926, xl. 541, descr. emend. Type: *Bailey* 8, SE Tibet, Lunang, July 10, 1913 (K, holotype).
[syn. *M. betonicifolia* forma *baileyi* (Prain) Cotton, *Gard. Chron.*, Ser. 3, 85: 143 in obs. & Stapf, *Curtis's Bot. Mag.* 153, tab. 9185 (1930); *M. b.* var. *baileyi* (Prain) Edwards, *Gard. Chron.*, Ser. 3, 82: 506 in obs. (1929), nomen nudum].

DESCRIPTION. *Polycarpic, rarely monocarpic, plant*, generally clump-forming, without long underground stolons. *Stems* erect, one to several per plant (up to 6), to 1.2 m tall, generally rather less, with scattered sub-barbellate hairs to subglabrous. *Leaves* mid to dark green, paler beneath, the basal leaves petiolate, lamina oblong-ovate to oblong, oblong-elliptic or broad paddle-shaped, 15.2–32 × 5.4–11.6 cm, the base broad-cuneate (broad wedge-shaped) to more or less attenuate, more or less decurrent, generally rather asymmetric, the apex subobtuse to subacute, moderately to sparsely covered all over the upper surface, including the margin, with rather weak, rufous barbellate hairs (hairs 3–6 mm long), the hairs on the lower surface confined primarily to the midrib, the margin more or less shallowly crenate with 8–19 teeth on each side; petiole 12.5–24 cm long, occasionally longer, expanded and sheath-like at the base; cauline leaves ascending to suberect, sub-amplexicaul or

Meconopsis baileyi; Serkhyem La (west), SE Tibet, 3400 m. PHOTO: HARRY JANS

not, the uppermost 3–4 (the bracts) in a whorl at the stem top, the others alternate, otherwise details as for the basal leaves, although the margins can be rather jaggedly toothed. *Inflorescence* subumbellate with up to 6 flowers, sometimes with 1–3, occasionally more, extra flowers produced below from the uppermost alternate stem leaves. *Flowers* lateral-facing to nodding, broadly cupped to saucer-shaped, occasionally rather flattish, 7.5–14 cm across. *Pedicels* 5–17 cm long in flower, that of the terminal flower to 22.5 cm long, glabrous to sparsely patent-pubescent. *Buds* ovoid, the sepals densely patent-hairy. *Petals* 4 (rarely 5–6 in the terminal flower), sky blue to azure or dark sky blue, occasionally flushed with purple, rounded to ovate, 4.5–7 × 2.7–4.2 cm, obtuse to rounded at the apex. *Stamens* numerous, with filiform white filaments and yellow or orange-yellow anthers, 2–2.5 mm long. *Ovary* ovoid-ellipsoid, densely bristly, with a poorly-developed style not more than 5.5 mm long, often 2–4 mm, but sometimes ± obsolete; stigma pale green, brown eventually, 3–4.5 mm long, capitate, 4–7-lobed. *Fruit capsule* narrow ovoid (narrowly egg-shaped) to obovoid, (26–)28–42 × (9–)10–16 mm, beset with dense, spreading rufous sub-barbellate bristles, these sometimes breaking off at maturity to leave a stubble-like base, dehiscing at maturity by 4–7 (often 5–6) valves for about a quarter to one third of its length. 2n = 82. (Fig. 12B).

DISTRIBUTION. SE Tibet (Kongbo, Mainling, Nyingchi, Tsari, Zayul; particularly common in the Tsari valley and on the Serkhyem La, also in the La Chu, Lilung Chu, Lo La Chu, Lusha Chu, Rong Chu valleys and on the Lang La and Taku-pu La), NE India (NE Arunachal Pradesh), N Myanmar; 2896–3810 m (Map 9).

HABITAT. Mixed forests, *Rhododendron* forest, forest glades and margins, open shrubberies, grassy slopes and meadows, stream banks, rocky places.

FLOWERING. June–early August.

There is a full discussion regarding this species under the preceding, *Meconopsis betonicifolia*.

KEY TO SUBSPECIES OF MECONOPSIS BAILEYI

1a. Flowers nodding; upper leaves and bracts coarsely, incisely and jaggedly toothed. . . . .
 . . . . .subsp. **pratensis** (N Myanmar)
1b. Flowers lateral-facing, rarely semi-nodding; upper leaves and bracts toothed but neither incisely nor jaggedly toothed 2
2a. Basal leaves ovate to broad-oblong or paddle-shaped, with 8–13 pairs of marginal teeth; mature fruit capsules (9–)10–15 mm diameter . . . . subsp. **baileyi** (SE Tibet: excluding Nyima La)
2b. Basal leaves narrow elliptic-oblong, with 14–19 pairs of marginal teeth; mature fruit capsules 15–20 mm diameter . . . . subsp. **multidentata** (SE Tibet: Nyima La; NE Arunachal Pradesh)

subsp. baileyi

DESCRIPTION. *Basal leaves* ovate-oblong to paddle-shaped, with 8–13 crenate, marginal teeth on each side, the base broad-cuneate to somewhat attenuate; lower and middle cauline leaves similar to the basal but sessile, the upper semi-amplexicaul; upper cauline leaves and bracts increasingly smaller, not jaggedly toothed, sometimes subentire. *Flowers* lateral-facing to slightly nodding. *Style* variable, 1–4 mm long. *Fruit capsule* (26–)28–40 × 10–15 mm, patent-bristly at maturity, 4–6 (–7)-valved. (Figs 11A, 12B1–3).

DISTRIBUTION. SE Tibet: see above (Map 9).

SPECIMENS SEEN (all from SE Tibet). *Capt Bailey* 8 (K); *Kingdon Ward* 5683 (K), 5784 (BM, K), 11010 (BM); *Ludlow & Sherriff* 1865 (BM), 2110 (BM), 2219 (BM), 2636 (BM), 2641 (BM), 2675 (BM), 3869 (BM); *Ludlow, Sherriff & Elliot* 13742 (BM), 15025 (BM), 15747 (BM), 15807 (BM); *Ludlow, Sherriff & Taylor* 3621 (BM), 4636 (BM), 4636A (BM); *R. B. & Pradhan* 25 (K).

Although several seed collections were made, the majority of plants that came into cultivation in the 1920s–30s were from seed collected under the numbers *Kingdon Ward* 5784 and 11010 (Tsela Dzong, Doshong La, 1924–25 expedition). Other notable seed collections came into Britain under the numbers *Ludlow & Sherriff* 2641 and *Ludlow, Sherriff & Elliot* 15025 and 15807.

"The note accompanying Bailey's specimen says it was collected in flower on July 10, 1913, at the same spot at the same time, and there is no other species of *Meconopsis* in the neighbourhood with which it could be confused."
Kingdon Ward, *Annals of Botany* 40 (159), 1926.

Subsp. *baileyi* is widely cultivated. It is this plant that gardeners often refer to as 'the blue poppy' and it is one

THE GENUS MECONOPSIS
24. **MECONOPSIS BAILEYI**

Clockwise from top left:

Meconopsis baileyi subsp. *baileyi*; Serkhyem La (west), SE Tibet, 3400 m. PHOTO: HARRY JANS

Meconopsis baileyi subsp. *baileyi* in cultivation at Logan Botanic Garden, Scotland. PHOTO: CHRISTOPHER GREY-WILSON

Meconopsis baileyi subsp. *baileyi*; Serkhyem La (east), SE Tibet, 3450 m. PHOTO: JOHN MITCHELL

Meconopsis baileyi subsp. *baileyi* fruit capsules; Serkhyem La (west), SE Tibet, 3400 m. PHOTO: JOHN MITCHELL

of the easiest species to cultivate, tolerating somewhat drier conditions than some of its cousins. However, in these drier gardens, especially in the south of Britain and elsewhere it tends to behave as a biennial, or at least a short-lived perennial. In more agreeable climates it is often a long-lived perennial. For instance in her garden at Sherriffmuir, Scotland, Evelyn Stevens has plants in excess of 20 years old. Apart from the very beautiful blue forms, two distinct cultivars have been recognised in cultivation, var. *alba* (pure white flowers; several forms exist) and 'Hensol Violet' (violet-purple flowers).

subsp. pratensis Kingdon Ward ex Grey-Wilson stat. nov. Type: N Myanmar (Upper Burma), Advance base, Seinghku Wang, *Kingdon Ward* 6836 (K, holotype). [syn. *M. baileyi* var. *pratensis* Kingdon Ward, *Gard. Chron.*, Ser. 3, 82: 506 in obs. (1927); op. cit. 93: 367 (1933), *Ann. Bot.* 42: 856, tab. 16 fig. 1 (1928) and *Plant Hunting on the edge of the World*: tab. opp. 96 (1930)]

DESCRIPTION. *Basal leaves* broad paddle-shaped to oblong-ovate, 15–18 × 6.8–7.5 cm, with relatively few, generally 8–10, shallow, rather incised, crenate-dentate teeth on each side, the base broad-cuneate; lower cauline leaves similar to the basal but sessile; upper cauline leaves and bracts increasingly smaller and rather jaggedly toothed, with 3–7 teeth on each side. *Flowers* pendent, borne on arching pedicels, usually overtopped by the whorl of leaf-like bracts. *Style* short, 1–2 mm long. *Fruit capsule* 33–38 × 11–14 mm, the persistent style 1.5–2 mm long, 4–5-valved. (Fig. 11B, 12B4).
DISTRIBUTION. N Myanmar (Seinghku valley) (Map 9).
SPECIMENS SEEN. *Kingdon Ward* 6836 (K), 6862 (K), 7208 (K).

While it is clear that *Meconopsis baileyi* exists in the wild in several more or less distinct forms (for instance plants on the east and west flanks of the Serkhyem La (one of its best known localities) are rather different, those to the east growing in more open habitats than those to the west that are confined to rather dense forest) the puzzle remains as to the identity of Kingdon Ward's var. *pratensis*. The plant in question has a brief history:

This taxon was first noted by Kingdon Ward on his 1926 expedition to Upper Burma (northern Myanmar), to the Seinghku Valley and written-up initially in his book *Plant Hunting on the edge of the World* published in 1930. "The Burmese Alps are wonderfully rich in flowers of all kinds. Ascending the Seinghku valley, one emerged from the forest at an altitude of about 9,000 feet, into high meadow with scattered thickets of *Rhododendron*. This was not the end of the forest but the beginning of the alpine region …… In the next patch of meadow several tall Poppies are opening sky-blue flowers to display a shower of golden anthers in the centre. This Poppy is a lovely slender sea-green thing with dangling blue bubbles which swing to and fro in the breeze. It grows in great drifts and clumps all up the open valley, streaking the jade meadows with turquoise shadows. Most people now know *M. betonicifolia Baileyi*,

Meconopsis baileyi var. *alba*. PHOTO: CHRISTOPHER GREY-WILSON

Meconopsis baileyi 'Hensol Violet' at Ian Christie's garden, Kirriemuir, Angus, Scotland.
PHOTO: CHRISTOPHER GREY-WILSON

the famous blue Poppy of Tibet. This plant is very like it, only it grows in the open instead of in the woodland, coming out in hundreds and thousands in July, and ascending to 12,000 feet, though never growing in such serried ranks as do the primulas….. Amongst the hundreds of plants I saw in Tibet and in the Burmese Oberland, I never saw one with anything but azure blue flowers or with more than four petals". Kingdon Ward does not apply the epithet *pratensis* to the Burmese plant in his book; however, two years earlier, in 1928, in an article entitled 'Burmese species of *Meconopsis*' and published in the *Annals of Botany* it appears as *M. baileyi* Prain var. *pratensis* Ward and given a very brief description — "……The stem leaves are shorter and broader than those of *M. Baileyi* but in the present state of our knowledge the divergence is hardly sufficient to justify our taking an extreme view, though it may be remarked that year-old plants of the gathering are easily distinguished from year-old plants of *M. Baileyi*. The most striking difference, however, is horticultural rather than botanical. *M. Baileyi* is a woodland plant, found growing in thickets, under trees, or in shady places generally; the variety *pratensis* is a plant of the open meadows and steep rubble screes where it grows in dozens. It is a very rare plant on the Tibetan side of the range."

The field notes attached to Kingdon Ward collection 6862 (= *pratensis*) are worth presenting here:

"Advance Base Seinghku Wang, 9,500 ft" …..

"Rocky meadows and thickets. A fine plant 4–5 ft with sky blue flowers nearly 6 inches across, golden anthers and a slight fragrance of Shirley Poppy. Polycarpic. In rocky meadows and thickets. Unlike the Tibetan *M. baileyi* this plant does not grow in shady situations under trees and bushes, but out in the open. It is extremely abundant between 9,500 ft and about 12,500 ft."

Kingdon Ward made three gatherings of this plant under the numbers 6836, 6862 and 7208. The first two were gathered in flower in June and a fruiting specimen was added to a second sheet of number 6836 (a second sheet at K). This appears to be fruiting material from the previous rather than the present season. Kingdon Ward collected ample seed (presumably in the autumn under number 7208, for which I am unable to trace herbarium material, although it may have just been a seed collection. From this collection var. *pratensis* came into cultivation and is still in cultivation to this day, although scarce. The reason for this may be explained by F. C. Puddle writing from Bodnant in North Wales in 1930s 'Random Notes on Plants and Culture' in which he comments:

Meconopsis baileyi subsp. *pratensis* in cultivation. PHOTO: EVELYN STEVENS

"….. During the following expedition in 1926 he [Kingdon Ward] discovered *M. betonicifolia* [*baileyi*] var. *pratensis* ………It appears very evident that the collector himself did not realise that the variety *pratensis* was so superior to the variety *Baileyi* as it has proved under cultivation, and I am afraid that some of the original recipients of the seeds did not treat them seriously but merely considered them a good seed collection of the variety *Baileyi*, with the result that whereas *M. betonicifolia* var. *baileyi* has become popular and plentiful, the better form '*pratensis*' has remained in comparative obscurity. *M. betonicifolia* var. *pratensis* being, as its name implies, a plant of open meadow, it is the more suitable for the average garden. It is a very robust grower, flowering from a fortnight to three week earlier than the variety *Baileyi*. The flowers on well grown plants frequently measure six inches across, and the colour is more consistently sky-blue under comparatively sunny conditions than those of the woodland variety."

Meconopsis baileyi 'Inverewe', probably the same as *M. b.* subsp. *pratensis*. PHOTO: EVELYN STEVENS

Interestingly, the situation remains the same today. *Meconopsis baileyi* subsp. *baileyi* is widely cultivated and readily available from nurseries and good garden centres, while subsp. *pratensis* is scarcely available but certainly still in cultivation. Beryl McNaughton, Evelyn Stevens and Ian Christie all have it in their collections in Scotland and there is no doubt that the plant, which is a good robust herbaceous perennial is the same as the plant described by Kingdon Ward more than 80 years ago. Furthermore, the plant which goes under the cultivar name of 'Inverewe' appears to be identical, at least I can see no difference. In this case the cultivar name becomes both superfluous and synonymous.

The prime distinguishing characters of subsp. *pratensis* are:
1. Young emergent foliage strikingly flushed with purple-red, as often are the stems and pedicels.
2. Basal and lower stem leaves broad paddle-shaped, with relatively few shallow subcrenate teeth, becoming more deeply toothed at maturity.
3. Relatively short and broad middle and upper leaves and bracts, with incisely (rather jaggedly) and rather unevenly toothed margin.
4. Nodding broad-cupped flowers.
5. Sky-blue petals
6. A bristly style.

Interestingly, the large nodding flowers which clearly distinguish subsp. *pratensis* are not mentioned by Kingdon Ward directly, although he refers to them in his 1930 book as "dangling blue bubbles" which seems to infer a nodding character; however, the half-tone photograph taken in the wild that accompanies his article in the *Annals of Botany* (1928) clearly show the nodding flowers as well as the rather jagged bracts. The umbel of flowers commonly carries 3–5, occasionally 6 flowers, the upper stem with several additional later flowers borne in the axils of the uppermost leaves. Although, like subsp. *baileyi*, the flowers are normally 4-petalled they can on occasion (in cultivation at least) bear 5–6 petals.

It is a pity that the area of northern Burma (formerly Upper Burma) where subsp. *pratensis* is found has been out-of-bounds to foreigners for many years for it is the home of a number of species of *Meconopsis*, not least the beautiful yet little known *M. violacea*. With the improving political situation in Myanmar today one can perhaps foresee a time in the not too distant future when access to this region may once more be achieved and open to scientific study.

subsp. multidentata Grey-Wilson, **subsp. nov.** Differs from the typical plant, subsp. *baileyi*, and from subsp. *pratensis*, by its relatively longer and narrower basal leaves exhibiting a great number of marginal teeth, and by the broader fruit capsules; from subsp. *pratensis* it also differs in its lateral-facing rather than pendent flowers. Type: NE India, Arunachal Pradesh, near Mt Daphabum, 27°53'26"N, 97°04'03"E, *M. Wickenden* 03-224 (K, holotype).

DESCRIPTION. *Basal leaves* elliptic-oblong, 25–32 × 7.5–11 cm, with 14–19 marginal teeth on each side; lower cauline leaves similar to the basal but sessile or subsessile, with up to 15 marginal teeth on each side, the base somewhat attenuate to truncate-cordate. *Flowers* lateral-facing or somewhat inclined, the upper overtopping the whorl of leaf-like bracts. *Style* 3–4.5 mm long. *Fruit capsule* 3.7–4.2 × 1.5–2 cm, densely patent-bristly, but many of the bristles breaking off at maturity to leave a stubbly base, 6–7-valved. (Fig. 11C).
DISTRIBUTION. NE Arunachal Pradesh and SE Tibet (Nyima La); c. 3600 m.
HABITAT. Open woodland and shrubberies.
FLOWERING. June–July.
SPECIMENS SEEN. *Wickenden* 03-224 (K).

This is an interesting variant, very distinctive in its long narrow leaves with their multi-toothed margins. It was collected in 2003 by Michael Wickenden of Cally Gardens, Castle Douglas in Scotland, who travelled extensively in Arunachal Pradesh. He also collected a couple of other *Meconopsis*, *M. paniculata* and *M. grandis*, in the west of the state close to the Bhutanese border. Interestingly, the collection under discussion came from the other side of the state close to the Myanmar border. This locality is not so far from Kingdon Ward's var. *pratensis* which he described from the Seinghku valley in northern Burma (now Myanmar); however, the two are very different in their foliage and in the disposition of the flowers. A further development came upon examining the photographs of *M. baileyi* taken in Tibet by Toshio Yoshida to find that one of them recorded on the Nyima La in south-eastern Tibet matches the Wickenden collection in most respects, although details of the fruit cannot be deduced from the photograph (the plant while still in flower is also in very early fruit). These two localities are some 270 km apart and it is not inconceivable that subsp. *multidentata* will be found in other locations in the complex system of mountains that lie inbetween, especially as much remains unexplored botanically. Strangely, there appear to be no collections of the species gathered either by Kingdon Ward nor Ludlow & Sherriff on the Nyima La. They all visited the pass and collected other species of *Meconopsis*. Nor, as far as I can ascertain, is it mentioned in their various writings, although *M. baileyi* (or *M. betonicifolia* var. or forma *baileyi*) figures prominently from other localities in Tibet.

25. MECONOPSIS GRANDIS

Meconopsis grandis Prain, *J. Asiat. Soc. Bengal*, Pt. 2, *Nat. Hist.* 64: 320 (1895). Type: Sikkim 1892, *Gammie* (L, syntype); India 1887–8, Sikkim Himal, *Dr King's Collector* s.n. (BM, G, P, syntypes). Lectotype *Dr King's Collector* s.n., Sikkim (BM, lectotype; G, P isolectotypes) chosen here.

DESCRIPTION. *A robust, clump-forming, herbaceous perennial* to 1.2 m tall in flower, but as little as 35 cm on occasion, with an extensive fibrous root system, overwintering as a solitary, or cluster of, rufous-bristled buds at ground level; plant base invested with fibrous leaf-base remains beset with numerous rufous barbellate bristles; plants sometimes shortly rhizomatous (this applies particularly to subsp. *orientalis*). *Stem* erect, rather rigid, 10–40 cm long, sparsely to densely covered with spreading to somewhat deflexed barbellate bristles, gradually becoming glabrescent at maturity. *Basal and lower leaves* petiolate, the lamina elliptic to elliptic-oblong, lanceolate or ovate-lanceolate to oblanceolate, (4–)6.5–26 × 1.3–8.4 cm, attenuate into the petiole at the base, the apex acute to subacute, margin subentire to irregularly and coarsely serrate or crenate, especially in the lower half, covered on both sides by rufous barbellate bristles; upper cauline leaves sessile, the uppermost 3–5 aggregated into a false whorl from which the flowers arise[22], lanceolate to ovate or oblong-lanceolate, 4.2–7 × 0.8–1.2 cm, margined as the lower leaves, sometimes more coarsely

Meconopsis baileyi subsp. *multidentata*; S of Nyima La, Milin County SE Tibet, 3600 m.   PHOTO: TOSHIO YOSHIDA

---

[22] In specimens in which the stem is very short the false whorl of leaves may be hidden or partly hidden amongst the basal leaves. This gives the flowers the appearance of being scapose and in these instances the species is sometimes mistaken for *Meconopsis simplicifolia*. However, the two species are readily separated: *M. simplicifolia* has a larger number of petals, generally 6–9, while the filaments are the same colour if not somewhat darker (4–6 petals and white filaments in *M. grandis*), while at the same time the buds of *M. simplicifolia* are always upright or ascending, while in *M. grandis* and the related *M. betonicifolia* they are pendent or semi-pendent. The photograph in George Taylor's *The Genus Meconopsis* (Plate XVI) of *M. grandis* subsp. *grandis* clearly shows the apparently scapose condition of some plants, although others in the same group have a well-developed stem with a whorl of leafy bracts held well above the basal leaf-rosette.

25. MECONOPSIS GRANDIS

toothed, the base cuneate to cordate or subcordate; petiole of basal leaves 19–21.5 cm long, expanded and sheathing at the base. *Flowers* 1–3 normally, occasionally up to 5, sometimes with one or two superfluous buds (which may or may not open) borne in the axils of the lower cauline leaves, more rarely on pseudobasal scapes[22], nodding to half-nodding or lateral-facing, bowl- or saucer-shaped. *Pedicels* 7–32 cm long (to 40 cm in fruit), erect, covered in spreading to slightly deflexed barbellate bristles. *Buds* nodding to ascending, ovoid to narrow oblong-ovoid; sepals densely barbellate-bristly. *Petals* generally 4–6, occasionally up to 9, pale to deep blue, mauve-blue, purple, pinkish purple or wine-purple, obovate to suborbicular, 3.5–9 × 3–7.8 cm, slightly ruffled at the margin. *Stamens* numerous, the filaments filiform, whitish; anthers bright yellow or orange-yellow, occasionally purplish. *Ovary* ovoid to elliptic-ovoid, glabrous to densely bristly; style slender, 4–8 mm long, protruding well beyond the boss of stamens, capped by a clavate, 4–6-lobed, greenish or whitish stigma, the lobes generally decurrent onto the upper style. *Fruit capsule* narrow to broadly ellipsoidal to ellipsoid-oblong, 27–58 (–68) × 6–15 mm, splitting by 4–6 valves for a short distance from the apex, glabrous to densely bristly, the bristles ascending to patent, the persistent style 4–7 mm long. 2n = 164.

DISTRIBUTION. W & E Nepal, Sikkim, N Bhutan, NE India (Arunachal Pradesh), SE Tibet; 3000–4420 m (Map 10).

HABITAT. Rough meadows, rocky and grassy places, moist woodland (*Picea* usually) and woodland margins, shrubberies, streamsides.

FLOWERING. June–early August.

Meconopsis grandis belongs to a small group of species currently classified in subgenus *Grandes* series *Grandes* Prain. This series is distinguished by monocarpic or polycarpic species which die down to resting buds in the autumn and which bear simple, erect leafy stems carrying a distinctive whorl of leaf-like bracts (except in *M. betonicifolia*) near the apex from which one or several flowers arise. The stems may also, in some instances, support additional, generally smaller and sometimes aborted, flowers or buds in the axils of stem leaves below the whorl of bracts. All the species in the series, as in the closely allied series *Simplicifoliae*, bear an indumentum of barbellate hairs, as do the large monocarpic evergreen species of subgenus *Meconopsis*, which contains several widely cultivated species such as *M. paniculata*, *M. superba* and *M. wallichii*. However, whereas in the latter all the hairs (stems, leaves, pedicels, sepals and ovaries) are barbellate, in subgenus *Grandes* those of the stems

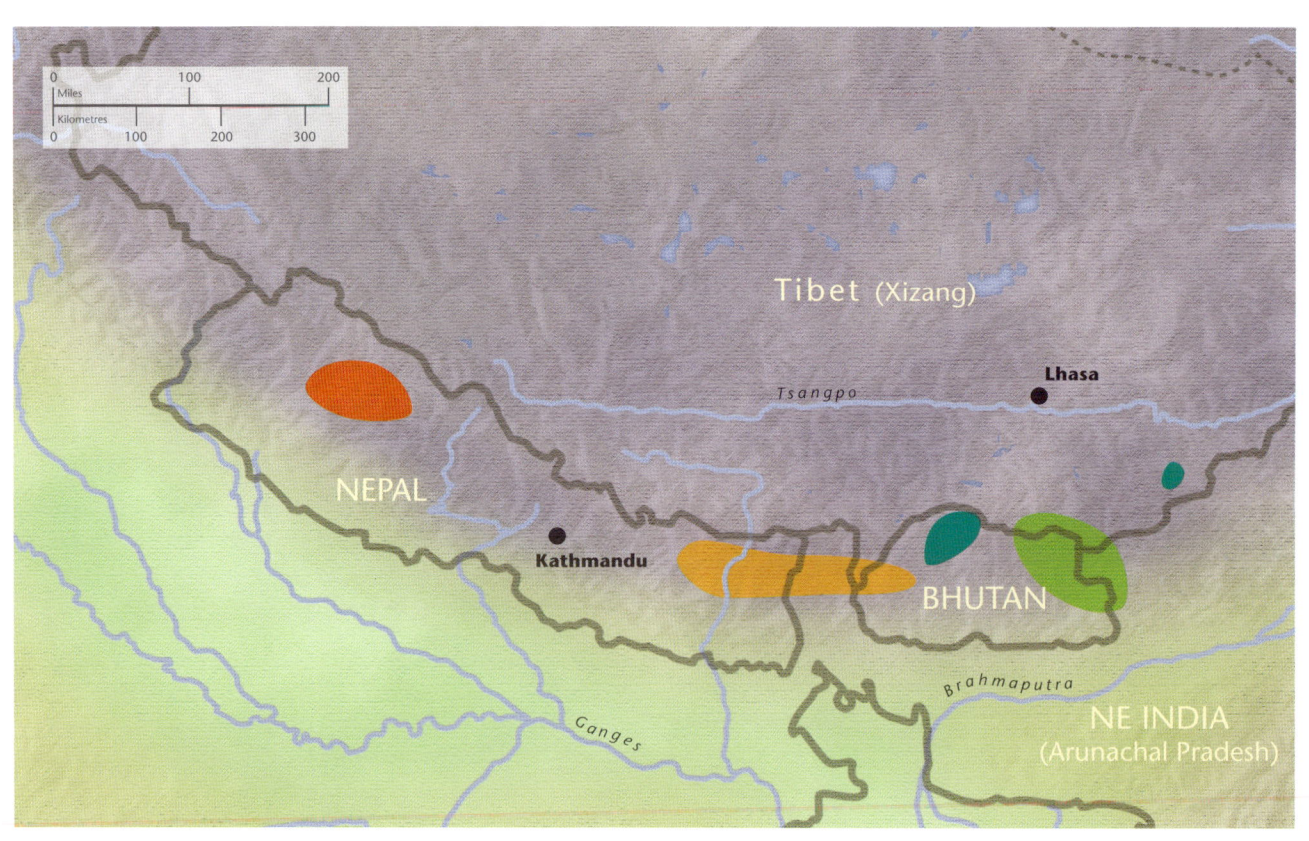

MAP 10. Species distribution: ■ *Meconopsis grandis* subsp. *grandis*, ■ *M. g.* subsp. *orientalis*, ■ *M. g.* subsp. *jumlaensis*, ■ *M. sherriffii*.

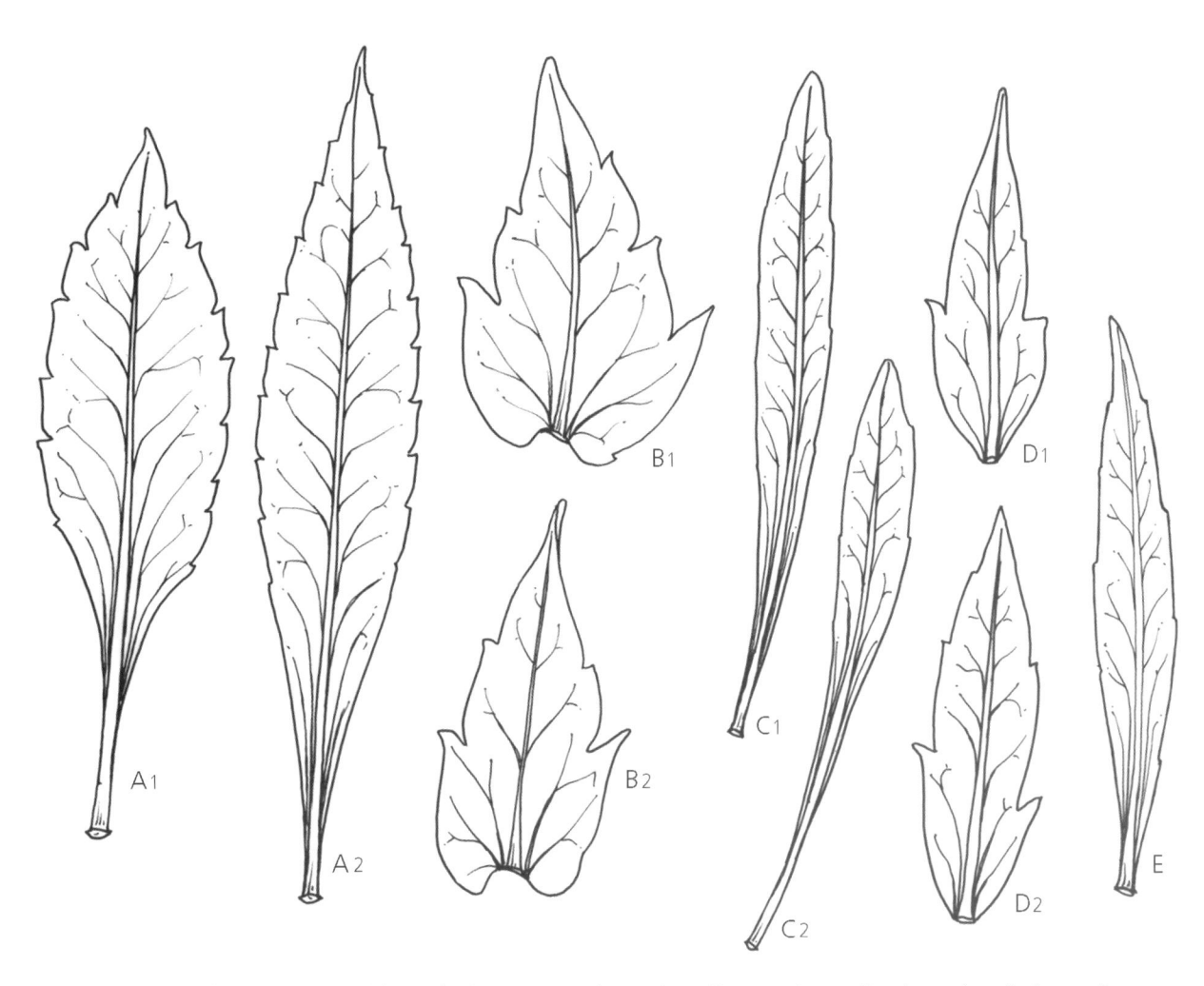

FIG. 13. Basal leaves and bracts, lower petioles and pubescence not shown. **A1–2**, *Meconopsis grandis* subsp. *orientalis*, leaves; **B1–2**, *M. g.* subsp. *orientalis*, bracts; **C1–2**, *M. g.* subsp. *grandis*, leaves; **D1–2**, *M. g.* subsp. *grandis*, bracts; **E**, *M. g.* subsp. *jumlaensis*, bract. All × ¹⁄₃. DRAWN BY CHRISTOPHER GREY-WILSON

and leaves are barbellate while those of the sepals and ovaries are often simple, although not invariably so.

Series *Grandes* contains 4 species as currently defined. While the three species already mentioned (*Meconopsis grandis*, along with *M. baileyi* and *M. betonicifolia*) bear basically blue or purple flowers, *M. sherriffii* has pink flowers. Of these species only *M. grandis* and *M. sherriffii* are truly Himalayan.

Meconopsis grandis was described in the late nineteenth century from several collections (primarily those of Gammie, Watt, and native collectors working for Dr King) made between 1881 and 1892 in the Jongri district of Sikkim. George Taylor (*An Account of the Genus Meconopis*, 1934) records that: "According to Gammie, *M. grandis* is found in this area only as a cultivated plant about the dwellings used in summer by shepherds. They are not primarily concerned with the aesthetic qualities of the plant, but grow it rather for the extraction of oil from the seeds, although the properties of this oil are not recorded. It has also been stated that the species is not truly native to Sikkim, but was originally introduced there from Nepal".

Along with *Meconopsis baileyi*, *M. grandis* forms a duo of species commonly referred to as the Himalayan blue poppies. These startlingly beautiful plants have an aura that gardeners find irresistible yet, in Britain at least, they can only be satisfactorily grown in the cooler damper parts of the country. While *M. grandis* is truly a Himalayan species, found primarily from east Nepal eastwards to NE India (Arunachal Pradesh) and just creeping into southern Tibet, the related *M. baileyi* is found to the north of the main Himalayan divide in the monsoon-fed regions of south-eastern Tibet. The third in the trio, *M. betonicifolia*, which finds its closest ally in *M. baileyi*, is confined to NW Yunnan in China.

Meconopsis grandis is without doubt one of the most beautiful members of this extraordinary genus, particularly in those forms from the north-east of Bhutan

25. MECOONOPSIS GRANDIS

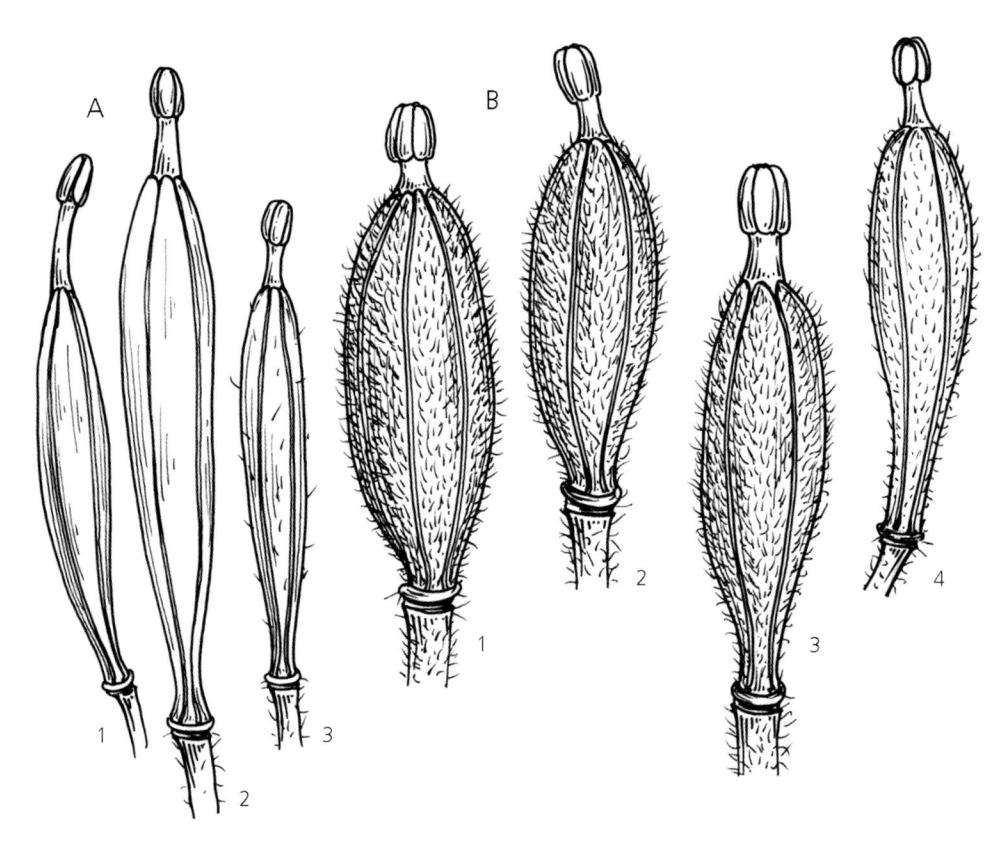

FIG. 14. Fruit capsules. A1–3, *Meconopsis grandis* subsp. *grandis*; B1–4, *M. g.* subsp. *orientalis*. All × 1¹/₃. DRAWN BY CHRISTOPHER GREY-WILSON

(represented by subsp. *orientalis*). Its introduction into cultivation, initially from Nepalese material (subsp. *grandis*) and later Bhutan (subsp. *orientalis*) led to the establishment in gardens of 'big perennial blue poppies' whose origins are discusssed below.

David Prain (see *Annals of Botany* 20: 364 (1906) was apparently responsible for introducing *Meconopsis grandis* into cultivation: it first flowered at the Royal Botanic Garden Edinburgh in 1895. The Nepalese form (subsp. *grandis*, see below) was not introduced until later, flowering in cultivation for the first time in 1932. However, it appears that it was the introductions from north-eastern Bhutan and the neighbouring region of NE India (Arunachal Pradesh) by Ludlow & Sherriff (sometimes accompanied by others) that introduced into cultivation the most robust and splendid forms of the species (here defined as subsp. *orientalis*). Today, with hindsight, proper cataloguing and record keeping of all these collections would have been extremely helpful in our overall understanding of this species and its evolution in cultivation: unfortunately, the majority of the introductions became muddled in gardens and their authenticity lost. As a result a lot of time can be wasted trying to track the origin of a particular plant and whether or not it can be traced back to an original collection.

To gardeners these were novel and beautiful plants and to most their origins were of little consequence. In any event, the species were soon overcome by hybrids in favoured gardens, vigorous hybrids in the main that were more adaptable and less temperamental in the garden environment and, furthermore, reliably perennial.

In studying *Meconopsis grandis* it has been necessary to re-examine the historical and more recent herbarium collections at the Royal Botanic Garden Edinburgh (E), at the Royal Botanic Gardens Kew (K) and at the British Museum, Natural History (BM). In his monograph of the genus of 1934, George Taylor makes no attempt at subdivision of the species, although he discusses the Sikkim and Nepalese forms. This is scarcely surprising since much of the material of the species observable today was collected after that date. Indeed, the species was not discovered in Bhutan until 1933 and the first seed introduction from that country was of *Ludlow & Sherriff* 600 collected the year of publication of the monograph, but plants from this seed introduction would not have flowered at the very earliest before 1935, more probably in the following year.

Having examined all the available material of *Meconopsis grandis*, it is clear that the western and eastern populations, separated as they are by the vast mountainous expanses of western and central Bhutan, are rather different in their overall characters and these are outlined below. These differences alone are not sufficient to warrant species distinction but they are, when combined with their geographic isolation, enough for subspecific division and this is the course which I think it is reasonable to adopt. At the same time, there is a small and far more isolated population of *M. grandis* in western Nepal and this has also been taken into account in this revision.

KEY TO SUBSPECIES OF MECONOPSIS GRANDIS

1a. Basal leaves 4.0–8.4 cm wide; stem leaves broad-cuneate to cordate at base; petals 6.2–9.0 cm long; fruit capsule 12–15 mm wide, densely bristly at maturity. . . subsp. **orientalis** (E Bhutan, NE India, SE Tibet)
1b. Basal leaves 1.3–4.8 cm wide; stem leaves with a narrow-cuneate base; petals 3.5–6.8 cm long; fruit capsule 6–11 mm wide, glabrous to slightly bristly at maturity 2

2a. Small plant not more than 35 cm tall in flower; basal leaves not more than 2.7 cm wide . . . .
. . . . subsp. **jumlaensis** (W Nepal)
2b. Plants larger, 40–60 cm tall in flower normally; basal leaves 2.0–4.8 cm wide . . . .
. . . . .subsp. **grandis** (E Nepal, Sikkim)

subsp. grandis

[syn. *M. grandis* subsp. *occidentalis* Grey-Wilson mss!]

DESCRIPTION. Basal leaves 10–21 × 2–4.8 (–5.8) cm, the margin entire to remotely and shallowly toothed or sinuate. *Inflorescence* 1–3 (–4)-flowered. *Bracts* 1–3.0 cm wide, with a long-cuneate base, entire to remotely and shallowly toothed. *Petals* 4 (–6), 3.5–6.8 (–7.8) × 3–6.4 (–7.2) cm. *Fruit capsule* 3.7–6.8 × 0.6–1.1 cm, glabrous or sparsely to moderately furnished with slender bristles most of which have fallen away by the time the fruit is mature, the persistent style 6–7 mm long, 2.5–3 mm diameter. *Pollen* in tetrads. 2n = 118, 120. (Figs 13C1,2, D1,2, 14A1–3).

DISTRIBUTION. E Nepal (Ramechhap, Sankhuwasabha, Solukhumbu & Taplejung districts; particularly common in the upper Arun Khola and Barun Khola area, especially to the south and east of Everest, Jaljale Himal and Topke Gola), Sikkim (Jongri region), W Bhutan (E of Chomolhari); (3100–)3353–5487 m (Map 10).

Subsp. *grandis* is also recorded from Mustang District (C Nepal) but probably as a result of introduction; the plant is an important medicinal herb used in the treatment of fevers as a result of liver and lung disease. The *Flora of Nepal* also records that "Seeds are roasted and pickled by Sherpas and Tamangs." It is also recorded that in Sikkim shepherds grow or tolerate *Meconopsis grandis* in the vicinity of their huts for the extraction of oil from the seeds.

HABITAT. Rocky meadows, stream and river banks, low scrub, forest margins, vicinity of shepherd's huts and shelters.

FLOWERING. June–July (–early August)
SPECIMENS SEEN. *AGSES* 207 (K); *J. J. Barnet* 16 (BM); *R. Bedi* 88 (K); *Beer* 8245 (BM), 8309 (BM); *Cave* s.n. Jongri (BM E); *Dhwoj* 248[23] (BM E); *B. J. Gould* 1150 (K); *Kanai et al.* 720713 (BM), 721251 (BM E), 723043 (BM E), 723044 (BM E); *Dr King's collector* s.n. Sikkim (BM, K); *Long et al.* 615 (E); *McCosh* 244 (BM); *Wollaston, Mt Everest 1922 Expedition* (Kew No. 402, K); *Noshiro et al.* 9261221 (BM E), 9760385 (BM E); *Ohba et al.* 8580421 (E); *Polunin, Sykes & Williams* 5506 (BM); *E Sharma* 515 (BM); *Shrestha & Joshi* 348 (BM); *Stainton* 367 (BM), 382 (BM E), 561 (BM E), 1668 (BM E), 1704 (BM); *G. Watt* 5485 (K); *Williams* 869 (BM); *Zimmermann* 808 (BM).

Subspecies *grandis* is locally common from eastern Nepal to western Bhutan, being commonests in rocky alpine meadows of the Arun valley and its tributaries where it often forms substantial clumps. Although the flowers can be almost pure blue they are more often purplish or pinkish and, in some instances, ruddy purple. In contrast, the more easterly subsp. *orientalis* tends to have bluer and deeper flowers, although they can open purple before changing to blue.

Plants from around Thudam in the Lumbasamba Himal, just east of the Arun Valley are particularly interesting.

---

[23] *Dhwoj* 248 (BM) typifies the fleeting pubescent nature (when present) of the fruit in subsp. *grandis* for, on the same specimen, the young fruit is moderately bristly but the near-mature fruits are glabrescent, without a sign of a bristle.

25. MECONOPSIS GRANDIS

Meconopsis grandis subsp. *grandis*; N of Topke Gola, Jaljale Himal, E Nepal, 4000 m. PHOTO: TOSHIO YOSHIDA

Meconopsis grandis subsp. *grandis* hybrid, possibly introgressed with *M. simplicifolia*; S of Thudam, Lumbasamba Himal, E Nepal, 4200 m.
PHOTO: TOSHIO YOSHIDA

subsp. orientalis Grey-Wilson, *Sibbaldia* 8: 81 (2010). Type: S Tibet, Cho La, *Ludlow, Sherriff & Hicks* 20801 (BM, holotype). The collection at the British Museum (Natural History) is also accompanied under the same number by a fine black and white photograph.

DESCRIPTION. *Basal leaves* 13–26 × 4.6–9.4 cm. *Inflorescence* 2–5-flowered. *Bracts* 3.2–6.1 cm wide, rounded to auriculate-cordate or subcordate at the base, subentire, to sinuately lobed to coarsely toothed. *Pedicels* in fruit 23–30 cm long. *Flowers* 12–18.8 cm across. *Petals* 4, rarely 5–6, bright sky blue to deep blue, mauve-blue or deep purple-lilac, (5.8–)6.2–9 × 4.8–7.8 cm. *Fruit capsule* 2.8–4.6 × 1.2–1.5(–1.9) cm, very bristly, the bristles stout and persistent in the mature fruit, the style 4–6 mm long, the stigma 3.5–5 mm diameter. (Fig. 13A1,2, B1,2, 14B1–4).

DISTRIBUTION. N & NE Bhutan, NE India (NW Arunachal Pradesh), adjacent parts of Tibet (Cho La, Po La); 3658–4268 m (Map 10).

FLOWERING. June–July.

SPECIMENS SEEN. *Kingdon Ward* 13711 (BM); *Ludlow & Sherriff* 387 (BM, E), 600 (BM, E), 875 (BM), 1021 (BM), 1315 (BM); *Ludlow, Sherriff & Hicks* 20671 (BM, E), 20801 (BM), 21069 (BM), 21431 (BM); *Ludlow, Sherriff & Taylor* 6416 (BM); *Rohmoo* 927 (BM, E).

Meconopsis grandis subsp. *orientalis* appears to be rather scarce in Tibet (Xizang). It has been recorded from two passes just to the north of the NE Bhutan-Arunachal Pradesh area (see *Ludlow & Sherriff* 1315 and *Ludlow, Sherriff & Taylor* 6416).

Kingdon-Ward 13711, collected in Arunachal Pradesh (then Upper Assam) is located just east of the Bhutanese border not far from the Bhutanese collections of subsp. *orientalis*. It was collected on the Orka La and the specimen at the British Museum (Natural History) under that number fits well within the circumscription of that subspecies. In recent years Peter Cox and others have also collected and photographed the plant in the same general vicinity in Arunachal Pradesh and plants from seed collected then are at present in cultivation in Britain.

Subspecies *orientalis* is the most robust of the three subspecies of *Meconopsis grandis*, a leafy plant with broader leaves and bracts than in the other subspecies. The flowers are by far the largest of those of the three subspecies and, according to field observations and recent photographic evidence, are often a good deep blue. In general terms it is similar in general habit to subsp. *grandis* but the broader, more coarsely toothed leaves and bracts clearly distinguish it. In addition, the fruit capsules are relatively broader and shorter, persistently bristly and with a shorter style. Other differences can be observed in the living plants. Most prominent is the shape and disposition of the flowers, those of subsp. *grandis* being lateral-facing to half-nodding or nodding and saucer-shaped to shallowly bowl-shaped, while those of subsp. *orientalis* are generally nodding and are a deep bowl shape. The third subspecies, subsp. *jumlaensis* bears lateral-facing to somewhat ascending, occasionally

Above: *Meconopsis grandis* subsp. *orientalis*; Chomjuk-Mg, Sakden, NE Bhutan.
PHOTOS: DAVID & MARGARET THORNE

slightly nodding, flowers that are widely open, flattish saucer-shaped. This latter subspecies also tends to have rather narrower, less rounded, less overlapping petals that are generally ruffled towards the margins.

subsp. jumlaensis Grey-Wilson, *Sibbaldia* 8: 82 (2010). Type. W Nepal, Ghurchi Lagna, 11,500 ft, June, *Polunin, Sykes & Williams* 4371 (BM, holotype; E, isotype).

DESCRIPTION. *Small tufted herbaceous perennial* to 35 cm tall in flower, occasionally more, the stems and leaves adorned with pale brown barbellate hairs. *Leaves* 6.5–29 × 1.3–2.7 cm, margin remotely toothed; petiole 19–21.5 cm long. *Scapes* 1-flowered, 12–24 cm long (to 63 cm long in fruit). *Bracts* 4.2–12.5 × 0.8–1.8 (–2.5) cm, narrowed to a cuneate base. *Petals* 4, occasionally 5–7, sky blue to bright or dark blue, occasionally purplish red, 4.5–5.7 × 3.4–5.0 cm. *Stamens* with white filaments and bright yellow to orange-yellow anthers. *Ovary* glabrous. *Fruiting pedicels* to 63 cm. *Fruit capsules* 19–21.5 mm long, glabrous. *Pollen* in monads. (Fig. 13E).

DISTRIBUTION. W Nepal, Jumla region (Dolpo & Jumla districts; Barbaria Lekh, Bhurchula Lekh, Chaudhabise Khola, Ghurchi Lagna, Maharigaon, Sialgarhi, Sisne Himal, Suli Gad); 3353–4420 m (Map 10).

HABITAT. Grassy places on rocky slopes, forest margins with rhododendrons.

FLOWERING. May–July.

SPECIMENS SEEN. *J. M. Bailey* s.n. 1936, Guchi (E); *J. J. Burnet* 16 (BM); *Pendry et al.* JRS A21 (E); *Polunin, Sykes & Williams* 270 (BM), 2101 (BM, E), 2286 (BM, E), 4371 (BM, E, L), 4613 (BM, E), 5423 (BM), 5437 (BM), 5506 (BM).

This is an interesting subspecies collected just a handful of times in the Jumla region of western Nepal. This region appears to represent a western outlier where a number of otherwise eastern Himalayan species are found in isolation. Apart from *Meconopsis grandis* one can cite *M. paniculata* and *Incarvillea himalayensis*. *M. grandis* is a splendid plant, more dwarf than the other subspecies and for that reason it would be a fine plant to get into cultivation. There is a fine photograph of this new subspecies in Polunin & Stainton's *Flowers of the Himalaya* (103, pl. 13) published in 1984 that shows a broad clump in perfect flower.

While the differences with the other subspecies are primarily quantitative the significance of the pollen (monads rather than tetrads) seems unclear and would perhaps reward further investigation.

Within Series *Grandes*, *Meconopsis grandis* probably finds its closest ally in *M. baileyi*. While the two do

Above: *Meconopsis grandis* subsp. *orientalis*; Chomjuk-Mg, Sakden, NE Bhutan.
PHOTOS: DAVID & MARGARET THORNE

not overlap in distribution in the wild, the latter being generally more easterly, they do hybridise readily in cultivation and this has lead to a plethora of bold hybrids popularly called 'the big blue poppies' or 'the perennial blue poppies' and quite a few of these have been given cultivar names. The fact that the majority of these hybrids are sterile is not surprising since the parent species have disparate chromosome numbers: in the case of *M. baileyi* 2n = 82, *M. grandis* 2n = 164 (or apparently 118–120 according to Ratter 1968; however, the material upon which this was based may not have authenticated wild-source material and the number is very close to that of some observed hybrids), so that hybrids have a count of 2n = 123 or thereabouts. Interestingly, fertile hybrids have appeared amongst these poppies in recent years (e.g. 'Lingholm' being perhaps the best known example) and such plants are known to be allotetraploids with a count of 6n = 246.

While the exact hybrid origin of nearly all the big blue poppies (or the big blue perennial poppies) is shrouded in doubt and mystery, it is also uncertain whether or not other species were involved in this cultivated hybrid complex. Over the years other species have been suggested that were known to be in cultivation at the time the first hybrids appeared, especially *M. simplicifolia*, which was far more widely cultivated then than now, and possibly *M. integrifolia*, although the involvement of the latter in this hybrid process seems to me to be highly doubtful, especially as known garden hybrids between blue- and yellow-flowered species invariably result in cream or pale yellow hybrids: the vast majority of 'big blue poppies' are however blue. It has been suggested that simple DNA tests would prove the parentage of the various poppies involved in this hybrid complex which arose entirely within cultivation. However, in many instances it is probable that complex crossing between hybrids and one or other parent or between hybrids may have occurred making the picture very complicated. Added to this must be the time-consuming task of DNA analysis of a large number of plants and the sheer cost involved.

MECONOPSIS GRANDIS IN CULTIVATION

As has already been pointed out, the history of this fine species in cultivation is rather confused and somewhat fragmentary, especially compared to that of the widely grown *Meconopsis baileyi* (*M. betonicifolia* of gardens). It is at least known that its introduction to cultivation in Britain was in the late nineteenth century, for it first flowered at the Royal Botanic Garden Edinburgh in 1895. The introduction is credited to David Prain (*see Annals of Botany* 20: 364, 1906). Prain had in the same year described the species, basing his description on various collections made by native collectors (1887), by George Watt (1881) and Gammie (1892). All these collections had been made in the Jongri district of Sikkim and observations made at the time indicate that the plant was in all probability cultivated around summer enclosures used by the local shepherds and herdsmen. This plant is attributable to subsp. *grandis*, restricted to

Meconopsis grandis subsp. *jumlaensis*; below Khalichaur, Jumla District, W Nepal, 3480 m. PHOTO: SHUICHI NOSHIRO

Meconopsis grandis subsp. *jumlaensis*; below Khalichaur, Jumla District, W Nepal, 3480 m. PHOTO: SHUICHI NOSHIRO

eastern Nepal and perhaps Sikkim (if it is considered to be native there).

According to George Taylor (*An Account of the Genus Meconopsis*: 70, 1934) "The Nepal form flowered in cultivation for the first time in 1932, and from a horticultural point of view was regarded as distinctly disappointing, being of much smaller stature than is widely associated with *Meconopsis grandis* and with flowers of a very inferior wine-purple. There are indications that this form will improve in cultivation and opinions regarding its horticultural merit may require modification. It may be observed that in the natural state the occurrence of both colour forms [blue or purple] is confirmed by the statements of native collectors".

Unfortunately, the persistence in cultivation of these early introductions is not recorded. We do know that it was still in cultivation in the early 1930s because a fine plate of it, painted by Lilian Snelling, was reproduced in *Curtis's Botanical Magazine* (146, tab. 9304) in 1933 with an accompanying text by Otto Stapf. Incidentally, the fine black and white plate (Pl. 16) of the species in Taylor's (1934) monograph shows an excellent stand of *Meconopsis grandis* (subsp. *grandis*) in a bed at the Royal Botanic Garden Edinburgh. This photograph had in fact been taken early on in the twentieth century and the original print is still in the archives at Edinburgh.

However, it was the introductions of the Bhutanese forms (subsp. *orientalis*) that firmly established the species in cultivation. It was discovered in that country in 1933 by the famous collecting partnership of Frank Ludlow & George Sherriff and over the next fifteen years they made repeated forays into that country and also southern Tibet (Xizang). Their first gathering of *M. grandis* was made on the Me La in the north-east of the country under the number 387, the flowers described as 'bright sky blue'. However, of their collections of *M. grandis* several stand out; *Ludlow & Sherriff* 600 (often referred to as 'Sherriff 600' or *grandis* GS600, as Sherriff was collecting alone at the time in the company of Danon, a local Lepcha) gathered on the Nyuksang La just south of Sakden in 1934; *Ludlow, Sherriff & Hicks* 20671★ ('Betty Sherriff's dream poppy'), found in NE Bhutan close to Shingbe in 1949, which Harold Fletcher (*A Quest for Plants*: 344, 1975) describes as "…..certainly one of the supreme gems of the Ludlow and Sherriff collections"; *Ludlow, Sherriff & Hicks* 20801 collected on the Cho La, NE Bhutan, also in 1949; *Ludlow, Sherriff & Hicks* 21069, also from the Me La near Shingbe. There were other collections made by Ludlow & Sherriff which are represented in herbarium collections, but it is not known whether or not these were accompanied by seed collections; they were all gathered in NE Bhutan in the vicinity of the Cho La (on the Bhutan-Tibet border), Me La and around Sakden and Shingbe in what is a relatively small area of the Himalaya. Ludlow, Sherriff & Taylor collected the species (6416) on the Po La just north of the Bhutan border in southern Tibet in 1938, while Kingdon Ward collected the species on the Orka La east of Sakden under the number 13711 on the Bhutan-Assam (Arunachal Pradesh in NE India) border in 1938.

Returning to GS600 (known widely by aficionados of *Meconopsis* cultivation affectionately as 'Sherriff 600' or as *M. grandis* GS600), the story of its discovery makes for interesting reading and is related in Harold Fletcher's account of Frank Ludlow & George Sherriff's explorations, *A Quest of Flowers*:

"On the Nyuksang La, Sherriff found a plant that surpassed in beauty all the primulas and every other plant on the pass — a most magnificent form of *Meconopsis grandis* (600) which he and Ludlow had recorded from Bhutan for the first time in 1933. It was occupying open stony ground beside *Primula waltonii*. Well might Sherriff have echoed the remarks of Kingdon Ward when he collected Bailey's Blue poppy, *Meconopsis betonicifolia* [*baileyi*], near Tumbatse in SE Tibet in 1924; 'Among a paradise of primulas the flowers flutter out from amongst the sea green leaves like blue and golden butterflies'.

Meconopsis grandis subsp. *grandis*; N of Topke Gola, Jaljale Himal, E Nepal, 4100 m. PHOTO: TOSHIO YOSHIDA

Perhaps the collection of *M. betonicifolia* [*baileyi*] is the achievement best associated with Ward's name, for it is now firmly established in cultivation both in Britain and overseas. And it could well be that the discovery, and the introduction to cultivation, of 'Sherriff 600', as this marvellous plant is now known to horticulture, will be ranked as Sherriff's greatest achievement. It is a finer plant by far than *M. betoncifolia* [*baileyi*]; and it is a finer plant by far than the form *M. grandis* from Sikkim which grew for many years in the rock garden at the Botanic Garden, Edinburgh, and which carried only a solitary nodding flower on a 12 or 18-inch flower stem. Sherriff 600 grows to twice that height, sometimes higher, and bears several glorious deep blue flowers often as much as 6 inches [15 cm] across".

The second collection, *Ludlow, Sherriff & Hicks* 20671, is the sort of story that legends are made of and, although well known to *Meconopsis* aficionados, I relate its discovery here for completeness quoting from Harold Fletcher's account of the plant explorations of Ludlow & Sherriff. In May 1949 Mrs Betty Sherriff, along with Hicks and a Bhutanese, Tsonpen, were encamped in north-east Bhutan. Harold Fletcher takes up the account:

"Their collecting at Shingbe began auspiciously — and in somewhat surprising circumstances. During the night of 25 May, Betty Sherriff dreamt that her husband walked into her tent, stood beside her camp bed, and gave her instructions for collecting on the following day. She was to seek out below the camp a small track leading to the Me La; to follow the track for about three miles until it bifurcated; to take the right hand fork and walk some 300 yards to a large rock mass. On the far side of the rock she would see a poppy she hadn't seen before. As Sherriff left the tent he turned, shook a finger at her, and said 'Be sure you go'. The next morning, at their usual 5.00 a.m. breakfast, when she told Hicks and Tsongpen the substance of her dream, they were both very sceptical and urged her to keep to their original plan of collecting in a particular valley they called the Glacier Valley. But the dream, and especially the shaking finger, had been so vivid and the instructions so clear and positive, that she determined to leave the rest of the party to seek her dream poppy. She had no problems; she found the track easily; she found the mass of rock easily; and behind it she found a glorious blue meconopsis which she hadn't collected before, a form of *Meconopsis grandis* (20761). Hicks, at first unbelieving, returned to the spot the next day and took several photographs. More important, in the autumn he succeeded in collecting seed of what is certainly one of the most supreme gems of the Ludlow & Sherriff collections; when plants from these seeds flowered in some Scottish gardens they became known as 'Betty's Dream Poppy'......"

Meconopsis grandis subsp. *grandis*; S of Dudhkund, Solu, E Nepal, 3950 m. PHOTO: TOSHIO YOSHIDA

While this is a good story it has been disputed. Those who have been lucky enough to read through Betty Sherriff's diaries of the period report that on the days in question no mention is made of the fabled 'dream poppy'!

However, it was Dr J. H. Hicks who returned in the autumn to the site and collected seed, having joined Ludlow and the Sherriffs as Medical Officer. He was also responsible for collecting seed under the number 21069 and probably also 20801. Full credit should be given to this stalwart doctor who almost alone should be credited with the fine forms of *Meconopsis grandis* introduced into gardens and included under the Ludlow & Sherriff 'umbrella'. However, as with so many of these introductions, little is recorded as to their success or indeed permanence in cultivation. Although we can be reasonably certain that GS600 played a significant and lasting part in the unwitting development of the 'big perennial blue poppies', much detail has been lost. The fact is that plants have been grown in diverse gardens and passed on to fellow gardeners under the number GS600 for many years now; however, it is now very uncertain that these plants all share the same origins. Certainly

25. MECONOPSIS GRANDIS

Meconopsis grandis subsp. *grandis*; S of Dudhkund, Solu, E Nepal, 3950 m. PHOTO: TOSHIO YOSHIDA

quite a few different plants are to be found under that number today, although some have been singled out and given distinct cultivar names. The doubts creep in when it is realised that the original plant was fertile (otherwise how would all the hybrids arrive) and that plants grown under that number today are sterile. This surely leads one to suspect that hybridity has tainted the collection sometime over the years. Only detailed DNA analysis might be able to solve this dilemma, although it has already been pointed out that an in-depth analysis is both time-consuming and costly. To add to this is the very real possibility that the original seed collection of GS600 gave rise to quite a number of similar but not identical offspring. This of course can apply to any seed collection and which makes them inherently tricky to maintain in cultivation under a single number, although it does, at the same time, give rise to the possibility that particularly interesting offspring may arise.

Shortly after completing the manuscript for this monograph I received an exciting communication from David and Margaret Thorne (Oct. 2013) who, during the summer, had managed to enter eastern Bhutan from Arunachal Pradesh in north-eastern India. Furthermore, they had trekked into some of the areas around Sakden (the Orka La for instance) that had been so profitable to Ludlow & Sherriff many years previously. Even more exciting was the arrival some days later of a disk containing numerous photographs of meconopsis, particularly some rivetting shots of *Meconopsis grandis* subsp. *orientalis*. This was exciting for a number of reasons, not least the fact that they were the first really good colour photographs of this particular subspecies from the wild that I had seen and also that the photographs feature all aspects of the subspecies, thus helping to fill in missing details of the plant's morphology. What the photographs show quite clearly is that subspecies *orientalis* is a most sumptuous and wonderful thing in its finest forms, a rival to any in the genus. The nodding blue lanterns, like globose lampshades, drift across the misty moors of eastern Bhutan, at places in profusion. The Thornes photographed a number of forms, some to match the fabled GS600 (collected in the same general area) in the size, poise and colour of their blooms. In another form the unfurling petals are a deep purple but quickly change to blue as the flowers mature. This feature is sometimes seen in some meconopsis cultivars, most notably 'Barney's Blue'.

NOTE. *Ludlow, Sherriff & Hicks* 20801 from the Chola La in NE Tibet has exceptionally large flowers c. 16 cm across. The sheets at the British Museum (Natural History) are accompanied by several half-tone photographs of a colony of plants in the wild, from which this particular number was collected.

In summary the picture is extremely complicated. While some of the hybrid poppies will be simple crosses between two distinct species, others would appear to have more complex origins. Hybrids between *Meconopsis baileyi* and *M. grandis* will probably always be triploid because of the discrepancy in the chromosome numbers between the two species. Such plants are sterile so can only be reproduced by vegetative means. However, fertile hybrids have arisen within the hybrid group by a process of chromosome doubling giving rise to allohexaploid plants from the triploid stock. Such a process will have the effect of isolating the different hybrids from one another, either because of reduced fertility or chromosome incompatibility.

Meconopsis grandis subsp. *orientalis* is certainly one of the most fabled flowers of the Himalaya, particularly in the finest forms found in Bhutan and the adjacent regions of Tibet. Little wonder then that it has been chosen as the national flower of Bhutan.

The big perennial blue poppies have arisen in gardens since the collections of Ludlow & Sherriff and their accompanying collectors. It is reasonable to assume that they have all arisen by chance, for few deliberate recordable crosses were made during the period (primarily the 1940s to the 1980s). That we have so many different ones in cultivation today is the result of three main factors: first the ease with which the parent species hybridised in gardens; second the robust nature of the majority of these hybrids (many are infertile) allowing them to be readily vegetatively propagated and distributed amongst growers in favoured parts of the country (the phenomenon is almost exclusively a Scottish and Irish one!); third the dedication of the Meconopsis Group in analysing and cataloguing the clones, giving the finest and most distinct cultivar names, although some clones like the Irish 'Slieve Donard', dating back to 1935, were known long before the Group was initiated.

As has been said, it is generally assumed that all these big blue poppies, beautiful and enticing as they are, are hybrids between two prime species, *Meconopsis grandis* and *M. baileyi* (*M. betonicifolia* of gardens): *M. betoncifolia* sensu stricto is restricted to south-western China and was not in cultivation at the time, so it can be safely removed from the hybrid equation). However, as has already been pointed out, other species could have been involved: most notably *M. simplicifolia* which was present in many collections during the critical period. Another species, *M. integrifolia*, was also in many *Meconopsis* collections at the time and has often been suggested as another parent in the 'hybrid pot'; however, I see no evidence in any of the big perennial blue poppies to suggest this, besides which most hybrids involving this yellow-flowered species and blue species such as *M. simplicifolia* (*M. ×harleyana* G. Taylor) or *M. betonicifolia* (*M. ×sarsonsii* Sarsons ex *Gard. Chron.*) produce pale yellow, cream or white flowers, not blue. The known hybrid involving *M. grandis* with a yellow species is with *M. integrifolia* (*M. ×beamishii* Prain) and George Taylor points out in his 1934 monograph of the genus that 'Of the known hybrids known in the genus between blue-flowered and yellow-flowered species this is the only one which shows any indication of blue in the flower colour [they are yellow throughout or with a purple blotch towards the base of each petal]'. This strongly suggests that yellow is dominant over blue in these species.

The case for excluding *Meconopsis simplicifolia* from the 'hybrid pot' is far less sure. Certainly, chromosome studies (unpublished data) undertaken on behalf of the Meconopsis Group in the last few years indicate the probable hybrid nature of many of the perennial big blue poppies. *M. baileyi* has 82 chromosomes, it seems likely that *M. grandis* has double that number at 164, while the

Meconopsis grandis subsp. *orientalis*; Chomjuk-Mg, Sakden, NE Bhutan. PHOTO: DAVID & MARGARET THORNE

hybrids have intermediate numbers, around 110–120 or 123. However, *M. simplicifolia* also has 82 chromosomes so it cannot be eliminated at this stage. Further work, especially more chromosome counts and DNA analysis of the species involved, is required before more positive pronouncements can be made. It is perhaps also important to note that *M. simplicifolia* is a stemless species, with all the flowers being strictly scapose, while the other species bear flowers in a well-developed inflorescence. While the scapose habit may of course be recessive, and this is by no means certain, it does not manifest itself in any of the 'big perennial blue poppies'.

While many of these fine garden hybrids are presumed to have resulted primarily from crosses between *Meconopsis baileyi* and *M. grandis*, or complex backcrosses involving hybrids and either parent, the possibility remains that some may have originated between subspecies of *M. grandis* itself, i.e. subsp. *grandis* × subsp. *orientalis*.

Interestingly, while all these big perennial blue poppy hybrids were arising inadvertently in gardens, one of their progenitors, *Meconopsis grandis*, was gradually disappearing and, despite more recent introductions, it is today rather rare in cultivation. This only emphasises the vigour of the hybrids and their sustainability in gardens. The impact of this species on the big blue poppies is paramount; without its input these wonderful plants would not exist.

The first formal hybrid recorded between *Meconopsis grandis* (2n = 164) and *M. baileyi* (2n = 82; *M. betonicifolia* of gardens), two species firmly established in cultivation by the early 1930s, was named *M.* ×*sheldonii* by George Taylor in the *Gardener's Chronicle* in 1936. It is safe to assume that this hybrid was between the Sikkim or Nepalese form of the former (i.e. subsp. *grandis*) and *M. baileyi*. This is borne out by the leaf characters and flower characters which are almost intermediate between those of the parent species. *M.* ×*sheldonii* was a sterile hybrid (2n = 123) and could only be propagated by vegetative division of the parent clump. *M.* ×*sheldonii* had first been raised by W. G. Sheldon (Oxted, Surrey) in 1934. In fact it is recorded that this interesting hybrid arose on several occasions in different gardens, both unintentionally and deliberately so it would seem.

A number of similar plants, placed in the recent classification of cultivated blue poppies in the infertile blue group, have been named: 'Bobby Masteron', 'Bryan Conway' and 'Maggie Sharp' and clearly should be placed under *Meconopsis* ×*sheldonii*, but it is not yet clear whether or not other cultivars collected into this Group should also be referred to *M.* ×*sheldonii*. The important point is that plants in this group would appear to all be descended from an original hybrid or hybrids

Meconopsis grandis subsp. *orientalis*; Chomjuk-Orka La, Sakden, NE Bhutan. PHOTO: DAVID & MARGARET THORNE

between *M. grandis* subsp. *grandis* and *M. baileyi*, but until cytologial studies reveal otherwise, *M. simplicifolia* cannot be ruled out of the equation, especially as concerns the other cultivars named within the group.

The famous and widely popular 'Slieve Donard' requires mentioning here, it also being a cross between *Meconopsis baileyi* and *M. grandis* sensu stricto. It is recorded (see James L. S. Cobb *Meconopsis*, 1989) that the cross was also made by Alec Curle of Edinburgh and that a plant given to Edrom Nursery was named 'Ormswell'. Other plants from the same cross were transferred via Hugo Patten and Marjorie Dickie to the Slieve Donard Nursery in Ireland. One of these was at first invalidly called *M. grandis* 'Prain's Variety' but this was later changed appropriately to 'Slieve Donard'. This stresses the importance of seedling variation in any batch of hybrid seedlings. It appears that from *M.* ×*sheldonii*, in all likelihood, fertile allohexaploids (counts show that these plants have 246 chromosomes or near that) arose spontaneously in gardens and have been named 'Lingholm', a cultivar readily raised from seed and now widely available and, as expected, it shows some seedling variation. This fine cultivar is classified in the fertile blue group. 'Kingsbarns' is similar and, along with several other cultivars, which may or may not be directly related, are classed in the same group.

This has an important bearing on the input of *Meconopsis grandis* in these hybrids. Of course further cytological studies will undoubtedly tell us more but the picture as I see it at the present time is thus:

Given the date and timing of introductions it is impossible that *Meconopsis grandis* subsp. *orientalis* was involved in *M.* ×*sheldonii* and that perhaps implies that it has no input into the infertile blue group. However, the introduction of this fine subspecies from Bhutan between 1935 and 1949 almost certainly gave rise to what is classified today as the George Sherriff group. Hybrids between *M. grandis* subsp. *orientalis* and *M. baileyi* should not be placed under *M.* ×*sheldonii*. The two subspecies have given rise to very different groups of plants. In the case of the George Sherriff group (which are sterile or practically so; very occasionally a few seeds are produced) many of the plants are closer to the *M. grandis* parent in their leaf and flower characteristics than they are to *M. baileyi*. It has been assumed that these stemmed solely from the introductions of *Ludlow & Sherriff* 600 (GS600) collected in Bhutan in 1934; however, I see no real evidence that all have arisen from this number although, having said that, it is quite clear that this collection has played a very significant and dominant role in the evolution of the George Sherriff group. In the broad range of characters that the George Sherriff group presents many have more features in common with the latter introductions and it would be certainly wise to look at the seed collections of Dr J. H. Hicks under the *Ludlow, Sherriff & Hicks* numbers 20671, 29801 and 21069 (see above) as potential parent material.

What of *Meconopsis grandis* in cultivation today? It is regrettable that many fine introductions, especially those from Bhutan, have disappeared from cultivation, overwhelmed no doubt by their more versatile hybrid offspring. Considering the number of seed collections made in eastern Nepal in the past 30 years it is perhaps surprising how few plants survive from these introductions. One of the earliest was *Polunin, Sykes & Williams* 5423 (or PSW5423). This was collected in west Nepal at Barbare Lekh, north-east of Jumla and was cultivated for a number of years, so is referable to subsp. *jumlaensis*. The herbarium specimen located at the British Museum (Natural History) clearly records that this was a seed collection. Plants raised from this source were available for quite a few years and then it apparently disappeared. However, The Meconopsis

Above left: *Meconopsis grandis* subsp. *orientalis*, showing colour dimorphism on the same individual; Surkatna-Chomjuk, Sakden, NE Bhutan.
PHOTO: DAVID & MARGARET THORNE

Above right: *Meconopsis grandis* subsp. *orientalis*; recently fertilised ovary expanding. PHOTO: DAVID & MARGARET THORNE

Group records that since it was last listed in the *RHS Plant Finder* in 1998 a plant has come to light. In brief, it came via Dr Jim Gauld of Aberdeen who had in turn purchased it from Jack Drake's nursery at Inshriach in the Cairngorms. Plants were handed to The Meconopsis Group for study and analysis. The plant in question, a vigorous clone, possesses the prime features of the Nepalese plant in its narrow leaves and glabrous fruits and it can bear up to three flowers per stem. This is a fine plant, however, some doubt must remain as to its authenticity, as the clone is apparently sterile and some hybrid contamination cannot be ruled out.

Another collection, this time from eastern Nepal, is singled out simply as 'Single-headed Blue' and this may be the same as the plant going around, probably incorrectly, as 'Sikkim Form'. This is another very persistent clone and is fertile: seed has been distributed from it via seed exchanges for the last few years as true *Meconopsis grandis* (now of course subsp. *grandis*). Even here there seem to be at least two forms: quoting Evelyn Stevens "I think it is the same as one of the two fertile true *grandis* that I have ... one has frilly, gappy flowers ('Single-headed Blue') and still present at Cluny, and the other has rounded petals and looks like the plant of *M. grandis* illustrated in Alf Evan's 'The Peat Garden'".

In addition, mention has already been made of recent seed collections (of subsp. *orientalis*) made in Arunachal Pradesh by Peter Cox and his party in 2004 (NAPE; Nagaland Arunachal Pradesh Expedition) and a number of plants from this source are still in cultivation. Interestingly, there are marked similarities between the wild Arunachal Pradesh plants and those in cultivation under the George Sherriff group. These include the broad-lanceolate, toothed, basal leaves and tall leafy stems which carry often 3–4 flowers. However, there are some minor differences, the two prime ones being the lack of red-purple pigmentation in the young foliage of the Arunachal Pradesh plants which is such a marked feature in the George Sherriff group, and the less bristly fruit capsules of the Arunachal Pradesh wild-collected plants. However, taking the overall variation that is apparent, then the Arunachal Pradesh plants fit comfortably within *M. grandis* subsp. *orientalis*. It has to be pointed out, at the same time, that plants of the George Sherriff group, although almost certainly involving subsp. *orientalis*, are probably of hybrid origin (in cultivation) and there is some chromosomal evidence to back this up. The Arunachal Pradesh plants are shortly rhizomatous in nature and this is also revealed in cultivated plants.

As well as these there are plants in cultivation of subsp. *grandis* derived from the Kew Edinburgh Kangchenjunga Expedition (KEKE) which were collected in the Kangchenjunga region of eastern Nepal in 1989. Plants derived from this seed collection have persisted in cultivation, although it is not widely available and plants tend to be rather short-lived; however, it does set seed reliably. Plants from this source are notable for their narrow cauline leaves and extremely long pedicels, generally carrying one half-nodding, campanulate flower per stem.

One problem facing anyone trying to describe these beautiful and tantalising plants is defining the flower colour. While pure tones are the exception most of the blue poppies have petals flushed or infused with colours other than blue: hints of pink, mauve, purple, lilac and lavender can creep into the blooms. To add to this, flowers can change colour from first opening to maturity and those growing in stronger light may well fade long before the petals fall. To add to this the same plant can look a different colour at different times of day, while divisions of a single clone planted in different gardens, or indeed in different parts of the same garden, can look different. It is not surprising that blue poppies prove such fickle plants to photograph. In addition to this, the same plant in the same location can change colour subtly from one year to another. All this seems to indicate that it is perhaps not the soil type or pH that causes these fluxes in colour. It has been suggested that the pH of the pigments that control flower colour can be affected by temperature and unstable weather conditions during bud development, causing the variation in colour that we observe. Clearly to understand this, a detailed scientific laboratory study is required.

POSTSCRIPT. Whilst I realise that insufficient recording, coupled with insufficient data, regarding the origin of the big perennial blue poppies reads like poor science, it is important to realise that in the last 20 years or so our understanding of this group of plants has made major strides forward thanks to the work of The Meconopsis Group and others. Many problems remain unresolved. Foremost there is the need to establish a firm scientific basis and for this it is necessary to look at the chromosomes and DNA of wild plants or, at least plants of known wild origin. Several samples need to be taken from a number of different populations in the wild to gauge the amount of variability in a particular taxon. As far as *Meconopsis grandis* is concerned, samples are required from all three subspecies. Samples of populations of *M. baileyi*, *M. betonicifolia* and possibly *M. simplicifolia* will also be required. Only then can a proper analysis be made of the role played by these species in the origins of the big perennial blue poppies. While some work has already been done, much remains: it would be an excellent research subject for a doctorate! In the meantime we have many fine poppies to admire. Yet one question above all remains in my mind which I feel only these studies will tell us, that is are some of these so-called hybrid poppies actually the true wild species in disguise in cultivation?

26. MECONOPSIS SHERRIFFII

Meconopsis sherriffii G. Taylor, *New Fl. & Silva*, 9: 155 (1937). Type: S Tibet, Drichung La, Nr Charme, 15–16,000 ft, July 1936, *Ludlow & Sherriff* 2309 (BM, holotype; E, isotype).

DESCRIPTION. *A clump-forming herbaceous perennial*, with one to many tightly clustered leaf-rosettes, forming rather dense patches up to 60 cm across, occasionally more, with a fibrous rootstock, the base of the plant thickly enveloped with a conspicuous mat of marcescent leaves and leaf-bases (this sometimes up to 25 cm deep) beset with pale rufous, barbellate hairs. *Stems* up to 30 in mature plants, erect, fairly stout, 7–44 cm long, green, often flushed brown, covered in somewhat deflexed, pale rufous, barbellate hairs. *Leaves* mostly in basal rosettes, the lamina lanceolate to elliptic-lanceolate, (3.5–)4.8–12.5 × (1.1–)1.8–2.4 cm, mid, soft, bluish-green above, paler beneath, often rather channelled, the base attenuate into the petiole, the apex acute to subacute, often adorned with a tuft of hairs, three-veined in the lower two-thirds, the margin entire, covered on both surfaces with pale rufous or golden, somewhat deflexed, barbellate hairs giving the leaves, especially the undersurfaces, a rather hoary appearance; cauline leaves 3–5 in a single whorl forming a shuttlecock-like shape, more or less sessile, linear-lanceolate, to 12 × 1.6 cm, acute, with a similar pubescence to the basal leaves; petiole linear, 6.2–22 cm long, expanded towards the base. *Flowers* solitary, 7–13 cm diam., facing laterally, cupped to saucer-shaped. *Pedicels* (scapes) brownish, rather stout, erect, 10.5–23 cm long (to 60 cm long in fruit), pubescent as for the stems. *Buds* ovoid-ellipsoid, nodding to half-nodding. *Sepals* narrow-ovate, covered with patent rufous hairs on the outside. *Petals* bright pink to rosy red, (5–)6–8(–9), elliptic-oval to oval-obovate, 3.4–6.5 × 3–4.4 cm, somewhat fluted and with a wavy margin. *Stamens* numerous in a rather tight boss, about a quarter the length of the petals, the filaments linear, whitish or straw-coloured, the anthers pale yellow to golden. *Ovary* greenish or blackish purple, sometimes striped, subglobose, covered in appressed, fawn-colour bristles; style obsolete or to 1.5 mm long, blackish purple,

Meconopsis sherriffii; en route to the Jaze La, C Bhutan. PHOTO: MARTIN WALSH

Meconopsis sherriffii; Danji, C Bhutan, 4712 m. PHOTO: MARTIN WALSH

bearing a large 5–7-lobed, capitate, cream, greenish or brownish-purple stigma. *Fruit capsule* ellipsoid, 24–30 × 12–18 mm, (4–)5–7-valved, covered in ascending to semi-appressed barbellate bristles, with a style, when present, up to 2 mm long.

DISTRIBUTION. N Bhutan; Upper Pho Chu (Chojo Dzong and Gafoo La), Upper Mangde Chu (Passu Sefu) districts; SE Tibet (Drichung La, Charme); 4024–5100 m (Map 10).

HABITAT. Moraines, boulder fields, cliffs and rock ledges, open shrubberies (often of juniper and dwarf rhododendron and shrubby honeysuckles) or amongst coarse herbage, rich yak pastures.

FLOWERING. Late June to early August.

SPECIMENS SEEN. Bowes-Lyon 3376 (BM); *Ludlow & Sherriff* 2309 (BM, E), 16704 (BM, E), 2723 (E); *Ludlow, Sherriff & Hicks* 17231 (BM, E); *Ludlow, Sherriff & Taylor* 6389 (E).

Meconopsis sherriffi was discovered by George Sherriff on the Drichung La above the Chayul Valley in SE Tibet, north of the NE India (Arunachal Pradesh) border, in June 1936. His collecting partner Frank Ludlow considered it to be the gem of the whole 1936 collecting expedition, describing its flowers as 'rose-pink like the first flush of dawn on the snows'. The discoverer was less enthusiastic about the find ….. "Came across a nice *Meconopsis*, 2309, which I do not know, pinkish wine-red, one on a scape and I don't think more than two to a plant. Leaves basal and cauline, all very thickly covered with bristles' (from Sherriff's diary, June 1936).

Strangely, a few years later, in June 1949, the pair were in NW Bhutan and Ludlow collecting in the company of Danon (a Lepcha who had previously worked for the Botanical Survey of India) and Ramzana (a Kashmiri; Ludlow's personal servant, whose real name was Ramzan Mir) paused for a while at Foomay having crossed the Kangla Karchu La. Whilst resting Danon and Ramzana brought a *Meconopsis* (16704) into camp which Ludlow thought must be new, not linking it to the Tibetan gem collected some 13 years previously. 'It is a plant with a very large rose-pink flower borne singly on a scape which varies in height from 18–30 inches. Half way up the stem a whorl of 3 or 4 cauline leaves more or less sessile. The basal leaves are lanceolate and petiolate. The stem and leaves are covered with soft fawn-coloured hairs and each leaf at its apex has a little tuft of these hairs. The most important fact connected with this plant, apart from its loveliness, is that it is perennial and gives off lateral shoots ………It occurs abundantly, locally, on cliffs and in boulder scree …….It is a most attractive plant and, if new, will create quite a sensation in horticultural circles. Danon and Ramzana went to an adjoining valley and found the pink meconopsis in even greater numbers than in the original valley I went up today. They say that the valley for great distances was just pink with the blooms. Danon reported that in an area of two feet in diameter he counted twenty-six flowers….".

Shortly after this the three left Chojo Dzong and on July 9th came upon the meconopsis once again while crossing the Gafoo La (5152 m): "On a boulder strewn moraine the new pink poppy was growing to perfection. Some of the blooms were enormous. The flowers look you straight in the face and do not hold their heads down like *M. simplicifolia* as do many other meconopses. I noticed that this poppy always seems to choose a position facing north ….. This poppy inhabited this boulder strewn moraine for about half a mile and then vanished…..".

Five days later, while crossing the Saga La they came upon it again ….. "We suddenly came upon the new pink meconopsis in great abundance, and here it grew not at the base of boulders, or on cliffs, but just mixed up with *Meconopsis paniculata* in the grassy stony shrubby valley-bed….." (from Ludlow's diary, July 1949).

The meconopsis was not a new species but it was a new record for Bhutan, extending its range well to the west of its previous Tibetan haunts. Ludlow & Sherriff, accompanied by Hicks, recorded that *Meconopsis sherriffii* grew in the company of three other species, *M. horridula*, *M. paniculata* and *M. simplicifolia* in the valley of the Mangde Chu, Passu Sefu.

As it is known at present, *Meconopsis sherriffii* has a disjunct distribution with a distance, as the crow flies, of about 300 km between the northern Bhutanese sites and the type locality in south-eastern Tibet; however, it can be reasonably well be expected to occur in some of the remote mountains between the two, especially close to the Tibet (Xizang) frontier, a region little-explored botanically.

Sometimes, as in the perennial *Meconopsis grandis*, the whorl of cauline leaves is held close to the basal leaf-rosettes so that the plant appears to be stemless and the flowers scapose, but close examination will always reveal a stem, however short.

Meconopsis sherriffii is a very beautiful and distinctive plant. In the wild plants can be quite substantial, forming quite dense patches and, while the flowers are always solitary, they are produced in sufficient numbers and are large enough to make a considerable impact. Some variability can be seen in plant height and flower size and while the flower colour is more often than not a bright pink it does tend to rosy-red on occasion. In some plants the indumentum is sufficiently dense and shaggy to give the leaves and stems a distinctly hoary appearance which is especially accentuated when plants are enveloped in mist.

Meconopsis sherriffii, flower details; near Danji, Lunana, C Bhutan, 4700 m.
PHOTO: TOSHIO YOSHIDA

The species was discovered in 1936 and described a year later by George Taylor just three years after the publication of his monograph. It is thanks almost entirely to Ludlow & Sherriff and their collaborators that we know as much about the species as we do: most of the herbarium specimens are of their collections! In more recent times *Meconopsis sherriffii* has been seen and photographed in the wild by several individuals in particular Martin Walsh and Tim Lever in the company of David and Margaret Thorne and others, and separately by Toshio Yoshida, and there is a fine photograph taken by the latter in his book *Himalayan Plant Illustrated*, the plant in question being photographed near Dangey, Lunana.

Meconopsis sherriffii undoubtedly finds its closest ally in *M. grandis*. There is of course the obvious difference in flower colour and whereas the former normally has flowers with 6–8 petals, in the latter there are 4–6. A more notable difference can be seen in details of the fruit capsule, those of *M. sherriffii* being relatively shorter and broader and with a very short style that is sometimes more or less obsolete.

Ludlow & Sherriff introduced seed into cultivation from a collection made in September 1938 on the Drichung La (the type locality in Tibet) but it did not persist for long from this introduction. A further introduction was made by them in 1949, this time from Bhutan, north of Bumthang. After a depressingly wet and unproductive seed harvest Sherriff was anxious to gather a record harvest but became increasingly depressed until his friend Ludlow sent news that he had collected five pounds of *Meconopsis sherriffii* seeds: "That in itself is a pretty good collection to take back from this trip because it will almost certainly grow well from this low altitude seed. On the Drichung La [Tibet] it was found at from 15,500–17,000 feet as far as I remember, whereas this is mixed up with *Meconopsis paniculata*, *simplicifolia* and *horridula* all of which are cultivated at home". Although the second introduction faired far better than the first, *M. sherriffii* has never proved anything but tricky in cultivation. However, it is well recorded in the *Bulletin of the Alpine Garden Society* (19 (4): 373) in December 1951, along with a fine half-tone photograph of three plants in a pot each sporting a solitary flower. The potful was exhibited by Mrs Knox Finlay of Keillour in Perthshire and received an Award of Merit, with the comment that "The flowers measure three and a half inches across with six to eight overlapping frilly petals. They are an exquisite shade of deep rose pink set off by golden stamens and a jade green stigma". This and all subsequent material in cultivation stem from Ludlow & Sherriff's 1949 collection.

Writing in *The Alpine Gardener* (72 (3): 341) in September 2004, Alan Furness who lives in Northumberland wrote:

"My final plant is a species that has given me periods of great satisfaction, alternating with phases of despair and frustration. My first two plants of *Meconopsis sherriffii* were raised from the few seeds distributed by the SRGC [Scottish Rock garden Club] 1984/85 list ...... Where it came from I have no idea, but I am sure it was growing well in several Scottish gardens at the time; it was certainly commercially available then from Jack Drake's Inshriach nursery at Aviemore. Since those first two seedlings twenty years ago I have managed to hold on to it through periods of plenty and periods of famine. The hot summer of 2003 almost finished it off completely; my last two mature flowering plants in the garden died and I lost all ninety of my pricked-out seedlings. Fortunately, both flowering plants did not succumb

Meconopsis sherriffii; Danji, C Bhutan, 4712 m. PHOTO: MARTIN WALSH

until August, several weeks after I had collected another batch of seed. That seed germinated well this spring, as I hope did the seed I sent to Aberdeen and Tromso as an insurance, and the seedlings are growing on well, so far. In terms of flowering plants my graph has hit rock bottom compared to ten years ago when I had thirty or more flowering plants and was giving away the spares!"

"Out of flower *Meconopsis sherriffii* forms a rosette of narrow, strap-like leaves, similar to many forms of *M. integrifolia*. In autumn the rosette dies back, with the growth point visible amongst the dead leaves; this rarely forms the tight resting bud typical of *M. integrifolia*. For that reason I have also found it best to protect the dormant plants with a cloche and to infill over the plant with some crumpled-up horticultural fleece. *M. sherriffii* is an early starter, especially in mild spells of winter weather and, once the flower stem and bud begin to develop, spells of hard frost can do serious damage, usually resulting in the death of the plant. However, assuming that plants avoid such damage, my own experience has been that 60% of all plants that I grew died after producing their first flowering scape (i.e. they were monocarpic), with the rest lasting up to five years, and the best producing eight 35–40 cm long , single-flowered scapes ..... Those plants that developed a perennial habit grew well in my gritty, loamy soil, heavily enriched with well-rotted beech leafmould; the preferred site being one that offered some dappled shade at midday. It was said that Betty Sherriff would divide perennial plants of *M. sherriffii* in her garden at Ascreavie. This was something that I never achieved. All the dead perennial plants that I examined showed that the multiple crowns developed laterally from one root system, making division impossible."

Prior to its demise in cultivation James Cobb (1989) commented that "Seed is usually scarce even on good plants and much of what is offered is no more than aborted seed. The seeds are slightly tear-shaped and obviously plump when viable. They store over winter quite satisfactorily and usually germinate readily, though such is the scarcity of seed it is unnecessary to warn against sowing it too thickly! The seedlings grow on quite rapidly but are prone to damping off and care is required to keep seed pots moist not soggy ....They should be pricked on at the usual stage [generally 6–8 weeks after germination] and then potted on into 10 cm (4 in) pots .... They are very much a plant that insists on summer humidity. This is easily achieved in a shaded frame and as long as the plants are reasonably cared for in pots they will grow on .... sooner or later they will have to go out and as they are greedy feeders sooner is better than later ....The plants require drying off in winter and small tent cloches with added dried leaf protection is good ....The following spring the plants

Meconopsis sherriffii; Danji, Upper Pho Chu, C Bhutan. PHOTO: TIM LEVER

will become multi-rosetted and should flower. Many will also flower even if they have not become multi-rosetted and this is invariably fatal ..... There is also a major frost hazard since the species is not frost hardy to flower bud damage .....".

Plants lingered on in cultivation right up until the turn of the present century, with seed being made available from cultivated plants from time to time; however, most growers lost all their plants to unseasonally hot summers during that period. Seedlings are especially vulnerable to increased temperatures and despite careful watering and heavy shade none survived. Increased summer temperatures, whether or not caused by man-made global warming, are the greatest threat to maintaining meconopsis in cultivation, especially the smaller species.

series Integrifoliae

This series is characterised by its monocarpic habit, young rosettes (those before flowering stage is reached) that die away during the autumn to leave a large overwintering bud at ground level, and by the presence of yellow flowers. *Meconopsis integrifolia* is often referred to as the yellow poppywort or Farrer's lampshade poppy, but it represents a complex of species widely distributed in the mountains of central and southern Tibet, western and north-western China, as well as northern Myanmar. The complex is exclusively one of the higher mountains, being found at altitudes from about 2750 to 5200 m. Whilst those found in central and eastern Tibet and north-western China are plants of rather dry regions away from the prime influence of the summer monsoon, those to the south of the range (extreme southern and south-eastern Tibet and SW China and Myanmar) are plants of the summer monsoon belt and receive a considerably higher rainfall. This as we shall see has markedly affected their evolution: this applies also to the *M. horridula* complex which covers much of the same area as well as the greater part of the Himalayan range.

As stated, the plants of the *Integrifoliae* are monocarpic, probably taking two to four years in the wild to reach flowering stage. As the overwintering buds expand in the spring the emerging leaves can be very attractive, being adorned with golden, russet or reddish hairs, the blades sometimes with a flush of pink; this is just as noticeable in the wild as it is in cultivated specimens and seems to be a common feature of this complex. Once plants have reached maturity the leaf-rosette quickly expands to reveal the emerging inflorescence and the plump flower buds that can be erect or pendent. The larger members of the complex can attain a height of more than a metre and carry in excess of a dozen large flowers; they are majestic plants being rivalled only by some forms of *Meconopsis grandis*. *M. integrifolia* and its closest allies have the largest flowers in the genus and flowers in excess of 20 cm diameter are not unusual.

Forrest, Monbeig, Kingdon Ward and Joseph Rock collected extensively in the north-west of Yunnan, particularly along the great divides separating the Salween, Mekong and Yangtse rivers, and made numerous collections of the large yellow meconopsis or poppyworts. The majority of their specimens show a stout plant with a distinct stem with a whorl of leaf-like bracts at the top supporting several flowers, with often several scattered leaves below, these sometimes supporting additional flowers. While plants to the north bear globular or bowl-shaped ascending to upright flowers (*Meconopsis integrifolia*), many of those to the south are less stout plants with nodding or semi-nodding flowers and have, in the past, been assigned to *M. pseudointegrifolia*.

Kingdon Ward (*Gardener's Chronicle* 1926) accepted both *Meconopsis integrifolia* and *M. pseudointegrifolia*, backed by observations that he was able to make in the field, suggesting at the same time that the 'intermediate form' (i.e. '*brevistyla*': see Hort. ex Prain in *Country Life* 1923: 112 (1923) in obs.) should be given specific status as well. However, George Taylor in re-examining all the available herbarium material while preparing his monograph of the genus in the early 1930s concluded that:

> "Careful examination of the copious field material and observation of cultivated specimens has shown that there is no justification for retaining specific or varietal status for these forms. It is true that the extremes of variation afford a very striking contrast, but they are so closely intergraded by intermediate types that it is impossible to recognise any clearly defined units within the aggregate ......".

Meconopsis integrifolia subsp. *integrifolia*; W of Kangding, W Sichuan, 4200 m. PHOTO: TOSHIO YOSHIDA

The taxonomist is left with a dilemma in such instance: whether to lump everything into one hugely variable species, which is often highly unsatisfactory, or to try and separate the complex on the evidence available from field and herbarium data, whilst at the same time trying not to be unduly swayed by the evidence from cultivated material which often looks very distinctive and easily separable.

Having travelled extensively in western and south-western China, particularly in the 1980s and 1990s, I have had the opportunity of seeing this species and its allies on numerous occasions. It is quite a feature of some high alpine meadows and moorland, often growing as large, dispersed colonies, sometimes of hundreds of individuals. In bloom, with its large and conspicuous yellow poppy flowers, it is unmistakeable, even at a distance, being often the tallest plant in the herb communities in which it grows, except when it ventures into the open montane shrubberies. Both the Chinese and Tibetans collect the plant for medicinal purposes and it is quite common to see groups of 'locals' coming down from the passes with bunches of the plant in flower, with roots attached.

I have always been struck, as have others, by the differences between the populations that I have seen in the field and these differences can also be noted in herbarium specimens (it and its allies, along with *Meconopsis horridula* sensu lato, are the most widely collected meconopsis). In northern Sichuan and southern Gansu where plant collectors such as Reginald Farrer and Ernest Henry ('Chinese') Wilson observed them, the plants are often large with upright, generally globular flowers of rich golden yellow, reminiscent of large globeflowers (*Trollius eurpoaeus*). Superficially similar plants with paler more cupped flowers can be found in the Lijiang mountains (Yulongxueshan) in NW Yunnan and in the adjacent part of SW Sichuan, areas much explored by the French missionary and collector Abbé Delavay at the end of the nineteenth century and more comprehensively a few years later by George Forrest and others. To the west of this latter locality, along the great divides of the Salween, Mekong and Yangtse rivers, and extending well into southern Tibet, the yellow poppies are paler, often cream, with nodding or semi-nodding flowers with wide-spreading petals. Such plants were observed and collected in the past by various well-known plant hunters including Frank Kingdon Ward, George Forrest, Ludlow & Sherriff and Joseph Rock. In this region other populations can be found with more or less scapose flowers, while populations to the south (around Dali), from the evidence to hand, comprise consistently dwarf plants with markedly hairy foliage. Other dwarf plants can be found on several of the high passes in south-eastern Tibet.

From field observations it struck me that all these plants represented either one extremely variable species (as interpreted by George Taylor) or a series of closely related taxa. As I began to examine the large number of herbarium specimens available more thoroughly, other features soon became apparent some of which had been commented upon by Taylor in his 1934 monograph.

While others may disagree with me I have long felt that the blanket treatment of these various elements was wrong and in 1996 I published a revision in *The New Plantsman* under the title '*Meconopsis integrifolia*, the yellow poppywort and its allies' in which I recognised two prime species, *M. integrifolia* and *M. pseudointegrifolia*, as well as various subordinate taxa. Since that date a great deal more material, especially digital photographs, has come to light and field observations by me and others has focused on this interesting complex. What is clear from all of this is that, although the complex has

Meconopsis integrifolia subsp. *souliei*, semi-mature fruits; Xuemengkeng pass, Moxi to Kanding, 3800 m. PHOTO: HARRY JANS

an extensive range covering much of central, southern and eastern Tibet, western and south-western China, northwards as far as Qinghai and Gansu, the variation has very much a geographical bias and, as a result, I have been able to modify and expand by original findings and these are presented here. First, however, it is necessary to examine variation within the complex:

1. The presence of a well-developed stem is to be found throughout the distributional range. At high altitudes plants are often more dwarf and with a reduced stem and these can be considered as high altitude forms or ecotypical variants; however, in the southern part of the range (in NW Yunnan and SE Tibet) plants are found that are naturally short-stemmed with the stem and inflorescence whorl of bracts wholly or partly concealed by the basal leaf-rosette.

2. The presence or absence of three-veined leaves (i.e. with 3 prime veins running the length or most of the length of the leaf-lamina) also has a geographical bias. All the plants in Gansu, western and northern Sichuan, Qinghai and E & NE Tibet correspond to this type. In the south and south-west of the range (S & SE Tibet, NW Yunnan, SW Sichuan) the majority of plants are weakly three-veined only towards the base, otherwise they are pinnately veined. This is with the exception of plants growing on the Daxueshan and Yulongxueshan in NW Yunnan and in the vicinity of Muli in SW Sichuan. In these latter places the leaves have well-developed three veins but not as strongly so as populations to the north. In these southern localities the three-veined character is exclusive to these locations as far as it is known. Furthermore the character links clearly to 5 below.

3. Whether or not the flower buds are nodding or erect. On the whole plants to the north and north-east of the range have erect buds while those to the south and south-west have nodding or semi-nodding buds and this affects the poise of newly opened flowers, those of the former have erect or ascending flowers, those of the latter nodding or lateral-facing flowers. Again there is an exception in some of those southerly populations with semi-scapose flowers; in these the buds are ascending. In all four species the fruits are held on stiffly erect pedicels.

One point that needs further clarification is the disposition of the flowers in the *Meconopsis integrifolia* alliance. In all taxa involved the fruit pods are erect at maturity, that is whether or not the flowers are nodding or erect when they are in bloom. In *M. integrifolia* subsp. *integrifolia* and *M. pseudointegrifolia* the flowers are always erect or ascending, in *M. integrifolia* subsp. *souliei* and *M. lijiangensis*, lateral-facing, while in *M. sulphurea* they are nodding or half-nodding. In this latter species as the flowers age and are pollinated the pedicels begin to assume an upright position and this is often achieved before all the petals have faded and fallen, thus giving the impression at times that the flowers were erect in the first instance. In plants coming into flower the buds of *M. integrifolia* and *M. pseudointegrifolia* are erect while those of *M. sulphurea* are nodding. In the remaining species they are generally half-nodding.

4. The shape and disposition of the flowers. The problem perhaps with this complex is that they all bear large yellow flowers and at a glance they can seem very similar; however, there are significant differences in the flower shape, governed by both the disposition of the flowers and in the actual petal shape. The most northerly populations bear large, erect or ascending globose or semi-globose flowers with the petals markedly incurved. Plants

Meconopsis lijiangensis; Daxueshan, NW Yunnan. PHOTO: JOHN & HILARY BIRKS

in western Sichuan and on the Daxueshan and Yulongxueshan in NW Yunnan and in the vicinity of Muli in SW Sichuan, which generally have lateral-facing flowers, have deeply cupped flowers, the petals curved but not incurving. The majority of populations in the remainder of north-western Yunnan, in southern and south-eastern Tibet and in northern Myanmar have nodding saucer-shaped flowers with wide spreading petals.

5. The presence of a style. Numerous authors, significantly David Prain, Kingdon Ward and George Taylor, have written and commented on the stylar features in the *Meconopsis integrifolia* complex. The differences in the extreme forms, from a broad and sessile stigma, to a small stigma set atop a distinct slender style is very marked. Yet again these differences have a more or less regional basis with the lack of a style and an elaborated broad stigma very much a feature of northern populations, a well-developed slender style and small stigma characteristic of the majority of population in south-western China and southern and south-eastern Tibet.

6. The shape and details of the fruit capsule are partly governed by the stylar characters (5 above) so that fruit capsules lacking a style or with a very short one, tend to be broad and barrel-shaped, those with a distinct style more ellipsoid or fusiform in general shape.

If these variations are looked at in a geographical context then there is a more or less clear demarcation between the various features, in isolation or in combination with one or other. As a result it is possible to recognise 4 distinct though clearly allied species as the key below demonstrates. In this account the geographical sort of the *Meconopsis integrifolia* complex becomes clearer and the circumscription of each entity more readily defined. Thus *M. integrifolia* in the strict sense is more or less confined to the north (Gansu, Qinghai, W & NW Sichuan and the bordering regions of Tibet), *M. sulphurea* has a decidedly southern distribution (SW Sichuan, C & S Tibet just reaching SW Qinghai, NW Yunnan), *M. pseudointegrifolia* is restricted to scattered localities in NW Yunnan (not the Lijiang area) and SE Tibet), while *M. lijiangensis* is concentrated in NW Yunnan (Lijiang and Yunnan-Sichuan border area) and SW Sichuan (Muli area).

Meconopsis sulphurea; Litiping, near Weixi, NW Yunnan, 4100 m.
PHOTO: CHRISTOPHER GREY-WILSON

KEY TO SPECIES OF SERIES INTEGRIFOLIAE

1. Stigma broad and sessile, the stylar arms markedly decurrent on to the top of the ovary/fruit capsule; flowers and buds ascending to upright............. 27. **M. integrifolia** (W & NW China, E Tibet)
+ Stigma narrow and with a short to long style, the stylar arms not decurrent on to the top of the ovary/fruit capsule; flowers and buds generally nodding or half-nodding 2

2. Plants subscapose, often with solitary flowers; style short-bristly ....
.... 29. **M. pseudointegrifolia** (NW Yunnan, SE Tibet)
+ Plants with a well-defined stem, bearing 2 or more flowers; style glabrous 3

3. Flowers generally nodding to half-nodding with usually 6–9 petals; leaves pinnately-veined (only 3-veined at the base of the lamina); style distinct, 3–11 mm long, the base of the stigma lobes not recurrent onto the style .... 30. **M. sulphurea** (SW China, SE Tibet)
+ Flowers lateral-facing to ascending with 5–6 petals; leaves 3-veined in the lower half to two-thirds; style short, not more than 4 mm long, the base of the stigma lobes decurrent onto the style ....
.... 28. **M. lijiangensis** (W China, E Tibet)

27. MECONOPSIS INTEGRIFOLIA

Meconopsis integrifolia (Maxim.) Franch., *Bull. Soc. Bot. France* 33: 389 (1886); and *Pl. Delav.* 1: 41 (1889). [syn. *Cathcartia integrifolia* Maxim., *Bull. Acad. Imp. Sci. Saint-Petersbourg* Sér. 3, 23: 310 (1877). Type: W China, Gansu 1872, *Przewalski* (LE); *M. integrifolia* sensu Taylor, *The Genus Meconopsis*: 58 (1934), pro parte]

DESCRIPTION. *Monocarpic herb* 20–100 cm tall, covered for the most part in rather soft golden-yellow or rufous, barbellate hairs, with a narrow dauciform taproot to 10 cm long, supported by numerous fibrous roots. *Stem* simple, erect, often stout, to 35 mm diameter near the base, striate, moderately to densely patent to somewhat deflexed hairy. *Leaves* mostly aggregated into a basal rosette, the outer spreading, the inner ascending, the lamina oblanceolate to obovate or more or less elliptical, to 37 × 5 cm, the base tapered gradually into the petiole, the apex acute to rounded, margin entire, the lamina prominently 3-veined, especially in the lower two thirds, the central one (midrib) keeled beneath, densely hairy on both surfaces at first but ultimately glabrescent; petiole broad-linear, with longitudinal veins, much expanded at the base. *Cauline leaves* few, spirally arranged, ascending to erect, similar to the basal but generally slightly smaller, the lowermost with a short, less expanded petiole, the upper cauline leaves (generally 3–7 (–10) in number); bracts erect, aggregated into a false whorl, more or less sessile, narrow-elliptic to lanceolate or linear-oblanceolate; other details as for the basal leaves. *Inflorescence* subumbellate with up to 7 (rarely as many as 10) flowers, sometimes with an additional flower or two below the umbel borne from an upper cauline leaf-axil, very rarely solitary and apparently scapose; flowers bowl-shaped to subglobular, to 16 cm diameter, upright to ascending or sometimes lateral-facing. *Pedicels* 8–32 cm (to 47 cm long in fruit), patent-pubescent. *Buds* semi-nutant but ascending as the flowers reach anthesis, ovoid to elliptic-ovoid. *Sepals* ovate to oval, 20–34 mm long, patent pubescent. *Petals* 6–8 (–9) usually, butter yellow to deep gamboge, oval to obovate or suborbicular, 5.8–7.8 cm long and broad, broadly overlapping, strongly incurving to spreading. *Stamens* numerous with linear filaments the same colour as the petals or whitish; anthers yellow, orange, brownish or blackish depending on their maturity. *Ovary* broad ovoid to narrow-ellipsoid, densely beset with rufous or golden ascending to appressed barbellate bristles, 4–6 (–7)-valved; style broad, to 4 mm long or obsolete; stigma broad, capitate, 4–7-rayed, included within the boss of

Meconopsis integrifolia subsp. *integrifolia*; Huanglong, Songpan County, Xueshanliang, NW Sichuan, 3900 m.   PHOTO: HARRY JANS

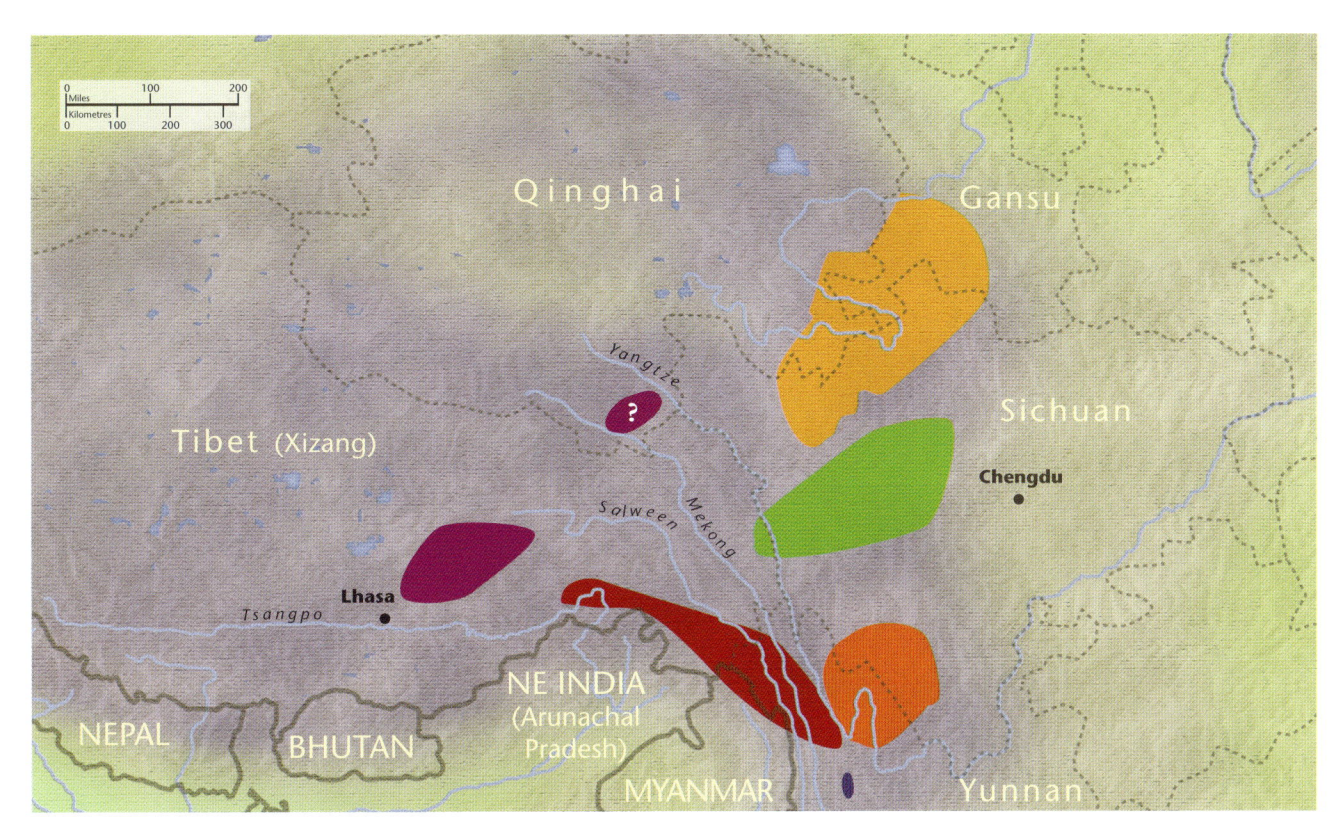

MAP 11. Species distribution: ■ *Meconopsis integrifolia* subsp. *integrifolia*, ■ *M. i.* subsp. *souliei*, ■ *M. i.* subsp. *gracilifolia*, ■ *M. lijiangensis*, ■ *M. sulphurea* subsp. *sulphurea* (*M. pseudointegrifolia* follows the same distribution as *M. sulphurea* subsp. *sulphurea*), ■ *M. pseudointegrifolia* subsp. *daliensis*.

stamens, in estylar plants covering the entire top of the ovary and more or less to pronouncedly decurrent onto the upper third of the ovary. *Fruit capsule* ellipsoid to ellipsoid-oblong, 35–46 mm, splitting for about one-third of its length from the top at maturity into 4–7 valves. *Seeds* subreniform, the testa longitudinally pitted. 2n = 74.

DISTRIBUTION. W & NW China: Gansu, Qinghai, W & NW Sichuan and neighbouring parts of Tibet (Xizang); 3100–5300 m (Map 11).

HABITAT. Mountain moorland, rocky slopes, open shrubberies, alpine meadows.

FLOWERING. May–July.

The name *Meconopsis integrifolia* was first proposed in 1889 by Franchet, although the species had been previously described as *Cathcartia integrifolia* by Maximowicz in 1877, based upon specimens collected by Przewalski in Kansu (now Gansu) five years previously. Franchet's opinion that the plant was indeed a species of *Meconopsis* and not *Cathcartia* has been widely accepted ever since.

Przewalski's plant is very typical of many other specimens from the northern part of the range of the complex: the plants have a stout stem bearing two or more erect and globular flowers usually, the leaves are clearly three-veined and the stigma broad and sessile with wide radiating rays that run flange-like onto the top of the ovary. The globular yellow flowers are very impressive, especially when large numbers of plants adorn a mountain slope, and they remind me of giant globeflowers (*Trollius europaeus*). Undoubtedly plants that conform to these characteristics can be regarded as the true *Meconopsis integrifolia* sensu stricto. This plant was brought into cultivation in Britain, flowering for the first time in 1904 from seed collected in western Sichuan in the vicinity of Tatsienlu (now Kangding) by E. H. Wilson (referable to subsp. *souliei*), a locality well to the south of the original gathering of Przewalski. A plant from this source was featured in the *Flora and Silva* in 1905 (3: 80, tab. opp. 80).

In the same year (1904) further seed purporting to be *Meconopsis integrifolia* was sent to Britain from the Koslov expedition to south-eastern Tibet, from "the valleys and headwaters of the Mekong". Capt. Koslov collected seed on behalf of the Imperial Botanic Garden at St Petersburg, who in turn sent a small consignment to A. K. Bulley in Cheshire. A flowering plant from this accession was featured alongside that of the previous in the *Flora and Silva*. Koslov's plant differed significantly by having solitary flowers borne on basal scapes and flowers with a prominent style and narrow stigma. In 1906 David Prain re-examined all the material available at the time and decided that the Koslov plant with its scapose habit and prominent style should be referred to a new species which he appropriately called *M. pseudointegrifolia*. Re-examination of this material with similar material from the region of south-eastern Tibet and north-western Yunnan shows that the leaves are consistently pinnately veined and the flowers nodding or half-nodding.

Although Koslov is generally credited with introducing *Meconopsis integrifolia* into British gardens it was from an isolated plant that flowered in 1904 and then died out; however, as we have seen, the Koslov plant is now regarded as a different species, *M. pseudointegrifolia*. The broader introduction into European gardens was due almost certainly to Abbé Farges who sent seed from Sichuan to Paris where plants flowered as early as 1896 (probably *M. integrifolia* subsp. *souliei*). However, its permanence in our gardens was due almost certainly to Ernest Henry Wilson who sent back a number of collections (primarily of subsp. *integrifolia*) of seed from Sichuan, especially in the early 1900s. In more recent times seed has been introduced from a number of sources, both of the true *M. integrifolia* and, more importantly, *M. sulphurea*.

Meconopsis integrifolia is undoubtedly a majestic species and a spectacular plant both in the wild and in cultivation, although it has to be admitted that it is not the most elegant species, handsome rather than pretty. The size of the blooms as we have already seen can be most impressive. Reginald Farrer called this bold species the 'Lampshade Poppy' and he was clearly enraptured by its presence in the wild:

> "*Meconopsis integrifolia* is very magnificent and portly in the highest turf of Thundercrown [Siku Alps; Lei-go-shan], standing stiffly up in early June with its huge lemon-pale globes in sumptuous but rather graceless and gawky candelabra of colour. Here, as I say, it loves the long high alpine hay at some 12,000 to 13,000 feet, and is found in no other situation but over all the open flanks of the grassy slopes, where it bloom is at its height before the herbage is well up, while still the Alps are sere and brown. No meadow, however, is too coarse for it; and, at its lowest limit, at some 7500 to 8000 feet, it luxuriates amid the coarsest tangle of tall Asters and Berberids, the Asters, in September, enclosing the huge upstanding pods of the Poppy in a lush jungle of leafage and blossom. (The Lampshade Poppy)"
> [Reginald Farrer *The English Rock Garden* 2: 506, 1928]

The distribution of *Meconopsis integrifolia* is somewhat uncertain, especially as it has been widely confused with *M. pseudointegrifolia* and *M. sulphurea* in the literature; however, photographic evidence along with herbarium records would suggest that this familiar species is in all probability restricted to central and southern Qinghai, Gansu and western and north-western Sichuan and the neighbouring regions of Tibet (Xizang).

KEY TO SUBSPECIES OF MECONOPSIS INTEGRIFOLIA

1a. Stigma broad, generally 10–14 mm diameter, markedly decurrent onto the upper part of the ovary; the flowers upright to ascending with the petals markedly incurving . . . . subsp. **integrifolia** (Gansu, Qinghai, NW Sichuan)
1b. Style present, generally 2–4 mm long, the stigma narrower, not more than 7 mm diameter, not or only somewhat decurrent onto the ovary; the flowers slightly nodding to lateral-facing with the petals curved to spreading. . . . . subsp. **souliei** (W Sichuan, E Tibet)

subsp. integrifolia

DESCRIPTION. *Plants* mostly 50–100 cm tall in flower with a prominent stem which often bears scattered leaves and superfluous flowers below the whorl of leaf-like bracts. *Flowers* erect, deeply cupped or goblet-shaped, the petals mid-yellow to deep gamboge, often markedly incurved. *Ovary* barrel-shaped, densely bristly; stigma sessile, with 4–6 broad radiating arms, forming a cap 10–14 mm diameter. *Fruit capsule* 5–6 (–7)-valved, with the stigmatic lobes decurrent in the top $1/5$. 2n = 74. (Figs 15A, B, 16A)

DISTRIBUTION. S & SW Gansu (Huichuan, Xiahe), E, SE & S Qinghai, W & NW Sichuan; 3100–5300 m (Map 11).

HABITAT. Mountain moorland, stabilised screes, open shrubberies, rocky places.

FLOWERING. From late May–July.

SPECIMENS SEEN. *Bartholomew & Gilbert* 34 (E); *Boufford et al.* 29471 (E); *R. C. Ching* 555 (E); *Farrer* 92 (E); *Ho et al.* 34 (BM); *Licent* 4352 (BM, K), 4670 (K); *McLaren* 86 (E); *Potanin* s.n. 1886 (E); *Przewalski* s.n. April 1873 (E), s.n. regio Tangut 1880 (K); *Purdom* 92 (E), 592 (K), 692 (E, K); *Rock* 12416 (E, K), 13073 (K), 13651 (K), 14086 (E, K); *SBQE* 51 (E), 394 (E), 416 (E), 692 (E); *SQAE* 477 (E); *T. T. Yü* 15127 (E).

NOTE. A Chinese collection (CG 81-0009 (BM) collected in the NE part of the Qinghai-Xizang Plateau represents

FIG. 15. Basal leaves, pubescence not shown. **A–B**, *Meconopsis integrifolia* subsp. *integrifolia*; **C**, *M. i.* subsp. *souliei*; **D–E**, *M. lijiangensis*; **F–G**, *M. pseudointegrifolia*; **H–J**, *M. sulphurea* subsp. *sulphurea*. All × $^1/_3$. DRAWN BY CHRISTOPHER GREY-WILSON

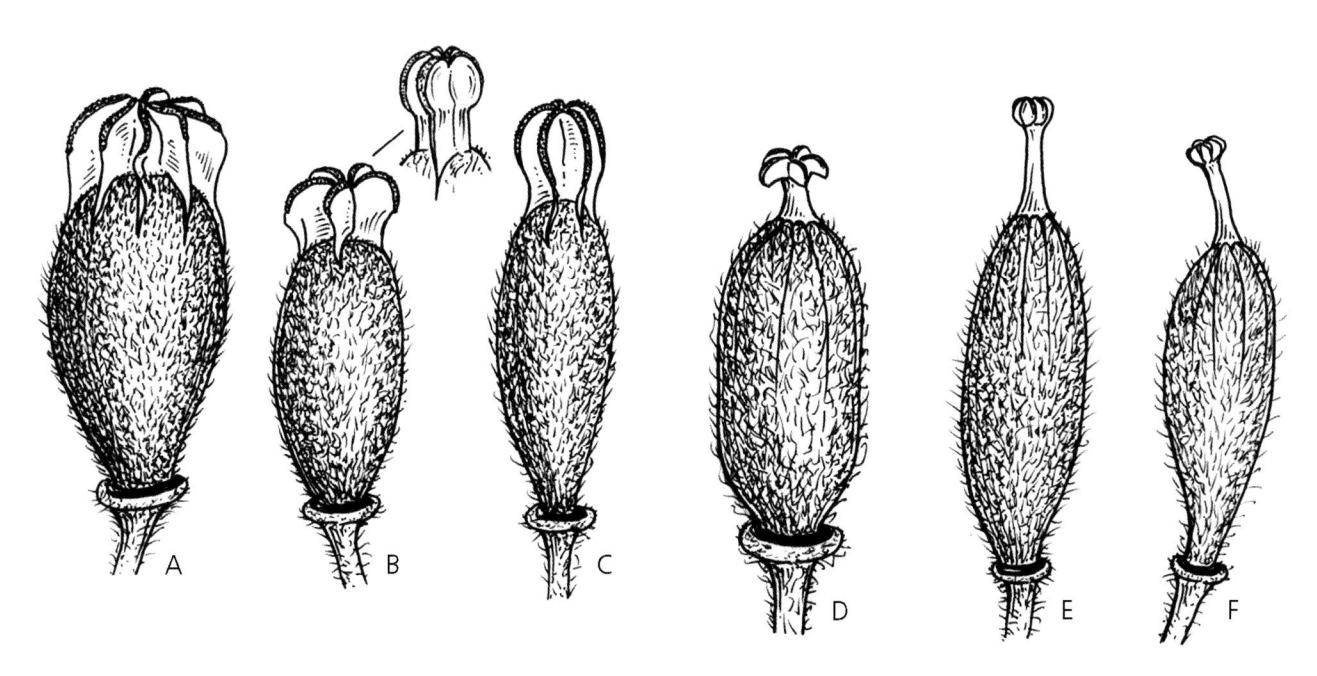

FIG. 16. Fruit capsules. **A**, *Meconopsis integrifolia* subsp. *integrifolia*; **B**, *M. i.* subsp. *souliei*; **C**, *M. lijiangensis*; **D**, *M. pseudointegrifolia*; **E–F**, *M. sulphurea* subsp. *sulphurea*. All × $1^1/_3$. DRAWN BY CHRISTOPHER GREY-WILSON

Meconopsis integrifolia subsp. *integrifolia*: Huanglong, Songpan County, NW Sichuan, 4000 m. PHOTO: CHRISTOPHER GREY-WILSON

Meconopsis integrifolia subsp. *integrifolia*, semi-mature fruits; near Jigzhi, SE Qinghai, 4100 m. PHOTO: CHRISTOPHER GREY-WILSON

an unusually dwarf, densely hairy plant just 15 cm tall. Unfortunately, it is not backed up by any other specimens from the area and it is difficult without further better material to make a proper evaluation. The specimen in question is just in flower, but carries several additional buds, partly concealed within the basal leaf-rosette.

subsp. souliei (Fedde) Grey-Wilson, stat. nov.
[syn. var. *souliei* Fedde in Engl., *Pflanzenr.* 4, 104: 262 (1909). Type: W China, W Sichuan, near Tatsienlu (now Kangding) Soulié 2435, China 1894 (BM, E, P, syntypes); Soulié 2047, China 1894 (P, syntypes); *M. souliei* (Fedde) Farrer, *Engl. Rock Garden* 1: 482 (1919), *nomen provisorium*; *M. integrifolia* sensu Taylor, *The Genus Meconopsis*: 58 (1934), pro parte].

DESCRIPTION. *Plants* rarely more than 60 cm tall, the stem often less prominent, often shorter than the basal longest leaves. *Flowers* lateral-facing to somewhat ascending, open bowl to deeply cupped, the petals primrose to pale gamboge, spreading, curved but not markedly incurved. *Ovary* ellipsoid, densely bristly; stigma sessile or subsessile, forming a cap 7–10 mm diameter. *Fruit capsule* 4–5-valved, with the stigmatic lobes decurrent in the upper $^1/_5$–$^1/_3$. (Figs 15C, 16B).

DISTRIBUTION. W Sichuan (Baoxing, Balangshan, Kanding (Minya Konka and other mountains), Siguniang, Litang mountains (including Geniexueshan), Batang mountains) and neighbouring mountains of eastern Tibet; 4000–4500 m (Map 11).

HABITAT. Rocky alpine meadows, mountain moorland, cliff ledges.

FLOWERING. June–July.

SPECIMENS SEEN. *R. Cunningham* 208; *McLaren* s.n., 1938 (E); *Limpricht* 1709 (K); *D. P. Mussot* 39 (K); *Pratt* 37 (K), 756 (BM, E), 869 (BM, E, K); *Rock* 17717 (E); *H. Smith* 3833 (BM), 10569 (BM); *Soulié* 2435 (E); *E. H. Wilson* 3167 (BM, K), 3029 (K), s.n. W China (BM, K).

Subsp. *souliei* has more open, less globular flowers than subsp. *integrifolia* and, while the flowers of the latter are always upright, those of the former, like the buds, are slightly nodding to somewhat ascending. The smaller stigmatic cap also seems to be a consistent and distinctive feature.

On the Balangshan, above Wolong, where I have observed it on a number of occasions, subsp. *souliei* is a common plant, scattered across the mountain slopes but never in the large colonies sometimes seen of subsp. *integrifolia*. Here it grows on occasions in association with both *Meconopsis punicea* (here at the southern end of its range) and the recently described blue-flowered local endemic *M. balangensis* (formerly included within

Above: *Meconopsis integrifolia* subsp. *souliei*; Balangshan, Wolong, W Sichuan. PHOTO: TONY KIRKHAM

Right, from top:

Meconopsis integrifolia subsp. *souliei*, semi-mature fruits; Miyalou, Zhegushan, W Sichuan, 4050 m. PHOTO: HARRY JANS

Meconopsis integrifolia subsp. *souliei* with *Primula longipetiolata*; Balangshan, Wolong, W Sichuan. PHOTO: HARRY JANS

Meconopsis integrifolia subsp. *souliei*; Balangshan, Wolong, W Sichuan. PHOTO: CHRISTOPHER GREY-WILSON

M. rudis). Subsp. *souliei* is, in its various guises, certainly a more elegant plant than subsp. *integrifolia*, yet it is still a relatively rare plant in cultivation: unfortunately, it is difficult to tell how widely it is cultivated as it harbours under the umbrella of *M. integrifolia* without distinction. The cultivar 'Wolong' (given an Award of Merit (AM) in 1993 by the Royal Horticultural Society) came from seed collected by the American Reuben Hatch in 1983 in the Wolong area west of Chengdu and where the famous Giant Panda reserve is located. Although this plants fits well within the circumscription of the species it is distinctive in cultivation from other forms with its reddish haired emerging leaf-rosette and its pale-lemon yellow, half-nodding to semi-erect flowers.

28. MECONOPSIS LIJIANGENSIS

Meconopsis lijiangensis (Grey-Wilson) Grey-Wilson, stat. nov.

[syn. *M. integrifolia* sensu Taylor, *The Genus Meconopsis*: 58 (1934), pro parte; *M. integrifolia* subsp. *lijiangensis* Grey-Wilson, *The New Plantsman* 3 (1): 33 (1996). Type: SW China: NW Yunnan, Lijiang, Yulongxueshan, 'Satseto', 12,500 ft, *Rock* 24851 (E, holotype; K, isotype)].

DESCRIPTION. *Monocarpic plant* mostly 45–75 cm tall with a rather slender taproot to 10 mm diameter at the top, the base of the plant enveloped with old petiole remains accompanied by a felt of barbellate hairs. *Stem* well developed, 11–30 cm long, somewhat ridged, especially below the bracts, patent with brownish barbellate hairs. *Leaves* petiolate, bluish green, mostly borne in a rather lax basal rosette, the basal elliptic to elliptic-lanceolate or elliptic-oblanceolate, 6.5–21 × 1.4–3.2 cm, the base gradually attenuate into the petiole, the apex subobtuse to subacute, three-veined from the base for about three-quarters the length, the margin entire, the lamina sparsely to moderately covered with brownish barbellate hairs, especially along the margin and along the midrib beneath; petiole strap-like, 7–16 cm long, expanded and sheath-like at the base, pubescent like the leaves, more densely so towards the base; cauline leaves few, 1–3 normally, scattered, similar to the basal but rather smaller and transitional with bracts, short-petiolate. *Inflorescence* a 3–5-flowered pseudoumbellate raceme, sometimes with one or two additional flowers borne in the axils of the cauline leaves; flowers lateral-facing to somewhat ascending, cup-shaped, 9–12 cm across. *Bracts* 4–6, similar to the leaves but often narrower, linear-elliptic to linear-lanceolate, 8–18.3 × 0.6–1.1 cm, attenuate into a short petiole or subsessile, pubescent as for the leaves. *Buds* ovoid to ellipsoid, half-nodding, 20–32 mm long, the sepals oval to ovate, patent-hairy. *Pedicels* ascending to erect, 5–29 cm long (to 35 cm long in fruit), moderately to densely patent with brownish barbellate hairs, more densely so just below the flowers. *Petals* 5–6, pale to mid-lemon or sulphur yellow, oval to ovate-oval, 4–7 × 3.3–4.8 cm, half-spreading, not markedly incurving, the margin entire. *Stamens* numerous with linear filaments the same colour as the petals or whitish, the anthers orange, browning on ageing, 2–3 mm long. *Ovary* ellipsoid, 14–19 × 8–11 mm, adorned with dense

Meconopsis lijiangensis; Daxueshan, NW Yunnan, 4150 m. PHOTO: CHRISTOPHER GREY-WILSON

ascending to subappressed dark rufous hairs; style short but distinct, 1–4 mm long, glabrous, the stigma subclavate, (4–)5–6-rayed, 5–7 mm long, 6–8 mm diameter, the base or the rays sometimes shortly decurrent onto the top of the ovary. *Fruit capsule* broad-ellipsoid, 35–45 × 15–19 mm, narrowed at the base and there 5–6 mm diameter, generally 5–6-valved, glabrescent or with some hair bases still present at maturity, the persistent style 3–5 mm long. (Figs 15D, E, 16C).

DISTRIBUTION. SW Sichuan (Muli southwards), NW Yunnan (especially Daxueshan, Hongshan and Yulongxueshan); 2745–3810 m (Map 11).

HABITAT. Mountain moorland, open shrubberies, rocky meadows.

FLOWERING. June–July.

SPECIMENS SEEN. *Delavay* 40 (BM, K), s.n. 'Likiang' (K); *Forrest* 2284 (E), 5661 (BM, E), 10181 (BM, E), 10470 (BM), 16373 (BM, E, K), 16892 (E, K), 20439 (BM, E), 28226 (E), 28394 (BM, E), 28512 (E), 30076 (BM), 30081 (BM), 30084 (E), 30087 (BM, E), 30091 (E); *Handel-Mazzetti* 3542 (K); *KEY* (Kunming Edinburgh Yulongxueshan Exped.) 273 (E); *Kingdon Ward* 5134 (E), 5153 (E); *Kunming Edinburgh Gothenburg Exped.* KEG 307 (E); *McLaren* 8 (BM), 35 (BM, E), 46 (BM), 132D (BM); *Maire* s.n. (E); *Rock* 3358 (E), 3385 (E), 3989 (E), 5564 (E), 5565 (E), 6114 (E), 14615 (E), 16020 (E), 18241 (K), 23395 (E, K), 23960 (E), 24462 (E), 24797 (E, K), 24851 (E, K), 25344 (E); *SBLE* 450 (K); *T. T. Yü* 7642 (E), 14615 (BM), 15127 (BM); *Schneider* 1830 (K).

In 1899 Abbé Delavay collected specimens of a large yellow-flowered *Meconopsis* on the 'Lichiang' Range (now Yulongxueshan) in north-western Yunnan where seed was subsequently collected by a number of individuals, particularly George Forrest. The Lijiang plant was assigned to *M. integrifolia*, despite the fact that it differed from the typical plant in having a narrower, more elliptical, ovary and the presence of a short though distinct style. All the material subsequently collected in the vicinity of Lijiang, right up to the modern era, conform to these characteristics: all this material, including those collections of Delavay and Forrest, as well as those of more recent times (especially that of the Sino-British Lijiang Expedition, SBLE) show that the plant in question is remarkably uniform and, furthermore, such plants have been cultivated over the years, especially from the earlier Forrest introductions. In addition, herbarium collections show that specimens collected on the Big Snow Mountain (Daxueshan) on the Yunnan-Sichuan border, and those on the mountains around Muli in south-western Sichuan, conform to

Meconopsis lijiangensis; Hongshan, Xianggelila County, NW Yunnan, 4535 m. PHOTO: HARRY JANS

the same characteristics, although these populations are rather more variable than the Lijiang ones. Two collections, *Kingdon Ward* 5134 and 5153 are marked 'Litang-Yalung Divide' without precise co-ordinates, but presumably not far removed from the Muli area.

This taxon was treated by me in *The Plantsman* in 1996 as a subspecies of *Meconopsis integrifolia* and equated, at the same time, with David Prain's var. *brevistyla*. Unfortunately, as elaborated under *M. pseudointegrifolia* I had followed the generally accepted view that all the short-styled plants were referable to this variety while in truth there are two distinct taxa with short styles involved. The position of subsp. *lijiangensis* seems to me to be very unsatisfactory, especially in the light of more recent research and accordingly I have here raised it to specific level.

Meconopsis lijiangensis has a distinctive look with its deeply cupped, rather floppy flowers and narrow, darkly

Meconopsis lijiangensis; Yulongxueshan, Wu-to-di, NW Yunnan, 3950 m. PHOTO: CHRISTOPHER GREY-WILSON

bristled fruits with their short, yet stout, styles. Like *M. integrifolia* the leaves are three-veined, but in the latter the veins travel to the apex of the leaf-lamina, while in the former they stop well short. The fruit capsules are very different: those of *M. integrifolia* are barrel-shaped and capped by a broad sessile stigma, those of *M. lijiangensis* are more ellipsoidal, narrowed at the top and with a short style and small stigma. *M. sulphurea* differs primarily in its pinnately veined leaves, in the nodding or half-nodding flowers, which generally carry more petals, in the 7–10-valved (not 5–6) fruit capsules that bear a longer and relatively narrower persistent style. *M. pseudointegrifolia* is clearly differentiated by its upright, often solitary, subscapose flowers and its tubby, densely 'furry' fruit capsules. Other differences are also observable.

29. MECONOPSIS PSEUDOINTEGRIFOLIA

Meconopsis pseudointegrifolia Prain, *Ann. Bot.* 20: 353 (1906). Type: E Tibet (Xizang), *Koslov* (K holotype?). [syn. *M. brevistyla* Hort. ex Prain, *Country Life* 54: 112 (1923) in obs., nomen provisiorum; *M. pseudointegrifolia* var. *brevistyla* Prain ex Kingdon Ward, *Gard. Chron.*, Ser. 3, 79: 438 (1926) in obs.; *M. integrifolia* var. *microstigma* Prain ex Kingdon Ward op. cit.; *M. integrifolia* sensu Taylor, *The Genus Meconopsis*: 58 (1934), pro parte; *M. pseudointegrifolia* sensu Grey-Wilson pro parte, non Prain, *The New Plantsmen* 3(1): 33 (1996); *M. integrifolia* var. *uniflora* C. Y. Wu & H. Chuang, *Fl. Yunnanica* 2: 28 (1979), pl. 8: 4. Type: China, NW Yunnan, Zhongdian, 25 Aug. 1937, *K. M. Feng* 2195 (KUN, holotype)].

DESCRIPTION. *Monocarpic plant* to 35 cm tall in flower, often only 15–22 cm (to 38 cm tall in fruit), with a long slender taproot 6–13 mm diameter. Stem 4–11 cm long, to 13 mm diameter, sparsely to moderately villous. *Leaves* mostly in a basal rosette, often numerous and dense, moderately to densely villous, the outer generally smaller than the middle and inner leaves, with a tendency to dimorphism, the lamina elliptic to elliptic-oblanceolate, the innermost often linear-elliptic, 3.2–9.2 (–15) × 0.8–2.2 cm, the base gradually attenuate into the petiole, the apex subacute to subobtuse, moderately to densely villous, especially when young, 3-veined in the basal one to two thirds, margin entire; petiole 3–8 cm long, often rather indeterminate, villous, especially towards the base. *Inflorescence* subumbellate, with 1–3 (–5) flowers, often solitary, generally appearing to be scapose or subscapose due to the fact that the whorl of bracts is hidden or partially hidden within the basal leaf-rosette; flowers erect to ascending, bowl-shaped, 10–12.5 cm across. Bracts 4–6, ascending, borne in a close pseudowhorl, linear to linear-elliptic, 50–90 × 3–8 mm, pubescent as the leaves, sessile or subsessile. *Pedicels* (pseudoscapes) erect, 7.5–17 cm long (to 36 cm long in fruit), 4.5–8 mm diameter, moderately to densely villous, especially just below the flower, becoming somewhat glabrescent in fruit. *Buds* erect, patent-villous. *Petals* 6–8, sulphur yellow, cream or white, oval to obovate, 4.5–6 × 3–5.5 cm, with an entire, sometimes undulate margin. *Stamens* numerous, with linear filaments 17–20 mm long and yellow anthers 2.5–3.5 mm long. *Ovary* barrel-shaped, densely lanate, 10–24 × 13–15 mm, with a short, rather thick, hairy style, 1–4 mm long; stigma 7–9 mm diameter. *Ovary* densely hairy, the ridged style and stigmatic rays short-bristly, the style 3–6 mm long. *Fruit capsule* broad-ellipsoid, 28–45 × 16–23 mm, 9–11 mm diameter at the base, densely lanate-sericeous at first but increasingly

29. MECONOPSIS PSEUDOINTEGRIFOLIA

glabrescent at maturity, 6–9-valved, the persistent style to 4–6 mm long. (Figs 15F, G, 16D).

DISTRIBUTION. NW Yunnan (Salween-Mekong-Yangtse watersheds, Yongning), neighbouring regions of Tibet (probably west as far as the Pa La and Temo La, c. 94°E); 3960–5180 m (Map 11).

HABITAT. Rocky meadows, cliff ledges, boulder slopes, on limestone and sandstone.

FLOWERING. Late May–July.

SPECIMENS SEEN. *Aldén et al. KEG* 947 (E); *Delavay* 40 (P); *Forrest* 13169 (BM, E), 14678 (BM, E, K), 17250 (E), 20202 (E), 20394 (BM, E), 30091 (BM), 30602 (BM, E); *Kingdon Ward* 1009 (E), 5393 (E); *McLaren* 86 (BM, E, K), s.n. 'N' collection 1933 (BM); *Rock* 8277 (BM, E, K), 10348 (E, K), 22887 (BM).

Meconopsis pseudointegrifolia is an attractive plant bearing relatively large pale yellow, cream or whitish flowers which at a distance often appear to be scapose; however, close examination of both living specimens in the field and herbarium specimens shows that there is a distinct yet short stem, revealing the presence of a reduced inflorescence. While the solitary flower state seems to be common in the wild, occasionally plants may bear several subordinate blooms. The small stigma and relatively short, clearly ridged style are very characteristic, as are the rather broad-based, narrow, barrel-shaped fruit capsules.

The species is found in scattered localities in north-western Yunnan, especially on the Baimashan and in the mountains close to Weihsi on the Yangtse-Mekong divide, but it is also found further to the north around Yongning (formerly Yung Ning) close to the Yunnan-Sichuan border and in various localities in south-eastern Tibet, mostly above 4300 m altitude.

Meconopsis pseudointegrifolia was described from a plant presented to the Royal Botanic Gardens Kew by A. K. Bulley of Ness, Cheshire, and which flowered in the rock garden in May 1906. Bulley had in return received seed from the Imperial Botanic Garden in St Petersburg that had been collected in China by one Captain Koslov. The details given were "SE Tibet, from the valley of one of the headwaters of the Mekong". This locality, still within Tibet today, lies at the north-western end of the Hengduanshan which stretches well down into north-western Yunnan. David Prain (1906) relates this episode more fully:

Below, from left:
Meconopsis pseudointegrifolia; Mi La (Kongbo Pa La), SE Tibet. PHOTO: ANNE CHAMBERS
Meconopsis pseudointegrifolia; Ganden-Samye, SE Tibet, 5000 m. PHOTO: TOSHIO YOSHIDA
Meconopsis pseudointegrifolia; Nyima La, SE Tibet. PHOTO: ANNE CHAMBERS

29. MECONOPSIS PSEUDOINTEGRIFOLIA

Meconopsis pseudointegrifolia: Baimashan, NW Yunnan, 4900 m.
PHOTO: CHRISTOPHER GREY-WILSON

Meconopsis pseudointegrifolia, in an unusual pale-coloured form; Baimashan, NW Yunnan, 5000 m. PHOTO: JOE ATKIN

"During 1899–1901 a journey was made in Tibet by Capt. P.K. Koslov. Circumstances led to his wintering just below the level of the Tibetan plateau in the valley of the Ra-chu, one of the headwaters of the Mekong. In this locality, situated approximately in lon. 97°30'E. and lat. 29°30'N., some of the members of Koslov's party made a collection of seeds for the Imperial Botanic Garden, St. Petersburg. A number of the seeds thus obtained were sent by Dr. Fischer de Waldheim to Mr. A.K. Bulley, Neston. Among the plants raised by Mr. Bulley was a *Meconopsis* (group *Grandes*) with sanguinaroid yellow flowers. This plant was figured by Mr. Bulley (*Flora and Silva*, iii.80) in 1905 as *M. integrifolia*, a species which it indeed resembles and to which it is nearly related, but from which is differs as *M. simplicifolia* differs from *M. grandis*. Another example of this species, presented by Mr. Bulley to Kew, flowered in the Alpine Garden in May, 1906. having thus had an opportunity of studying it and *M. integrifolia* side by side in the living state, the writer has been able to confirm the opinion already expressed by Bulley (*Flora and Silva*, iii.191) and by Hemsley (*Curtis's Bot. Mag.*, t. 8027) that this species is distinct from *M. integrifolia*. In the correspondence which has taken place regarding this plant, Dr. Fischer de Waldheim has invited the writer to name and describe the species. It is dealt with below as *M. pseudointegrifolia* (Pl. XXV)".

This plant was subsequently figured and described by David Prain in the *Annals of Botany* in the same year that it first flowered at Kew. The accompanying plate shows a scapose plant with partly three-veined leaves and large upward-facing, saucer-shaped flowers, while the ovary clearly bears a style that is expanded towards the tip. Unfortunately, there does not appear to be an accompanying herbarium sheet so that, in its absence, the colour plate becomes the type of the name *Meconopsis pseudointegrifolia*. In my revision of these yellow-flowered species '*Meconopsis integrifolia*, the yellow poppywort and its allies' published in 1996, I recognised three subspecies of *M. pseudointegrifolia*. Besides subsp. *pseudointegrifolia*, subsp. *robusta* and subsp. *daliensis*. Since then a lot material and photographs have come to light and I am now able to reassess these taxa. It is clear that subsp. *robusta*, with its large semi-nutant flowers and prominent style, along with the fact that it possesses a prominent subumbellate, sometimes semi-racemose, inflorescence, is best considered as a separate species and to this end I have in this work raised it to specific level under the epithet *sulphurea*, as *robusta* would become a later homonym (i.e. *M. robusta* sensu Hook. f. & Thomson non Grey-Wilson). Subsp. *daliensis*, on the other hand, presents more of a problem because no further material has come to light (it exists only from two herbarium collections!) and it has never been photographed in the wild to my knowledge. For the moment it is probably still best regarded as a subspecies of *M. pseudointegrifolia*.

David Prain was in no doubt that several taxa were involved in the *Meconopsis integrifolia* complex as this extract reveals:

> "....... but it is in his identification that Franchet is wrong and his description, as Forrest's specimens show, is rigidly accurate. Now, we must definitely lift the form represented by Delavay n. 40, Forrest nos. 13169, 14678, 17250, out of *M. integrifolia* Franch. (*Cathcartia integrifolia* Maxim.). So this form clearly belongs to the plant which Elwes flowered in March 1918 and sent to me to be named with the indication that he could not fit it into '*integrifolia*' as grown by us from the Kansu and Szechwan seed. In sorting the sheets today I find that Elwes has already sent me the same thing in Decr. 1915 — not from his own garden but from that of Mr. Dimsdale who was growing it and was lost in wonder at the time of year it selected for flowering .... What I particularly want you to observe however is that *M. integrifolia* Franch. quoad descr. sed syn. *Cathcartia integrifolia* encl. is not *M. pseudointegrifolia* Prain, the species to which all the rest of Forrest's specimens of the 'Integrifolia' group belong. What we are to do with Franchet's plant as opposed to Maximowicz' one I hardly know. You must try and study the three as they grow. There are two possibilities, either the Yunnan '*integrifolia*' is a distinct species or a variety. If the latter it may be regarded as *M. integrifolia* var. *microstigma* (as its stigma is less than $^1/_3$ the size of the Kansu plant; or it may be regarded as *M. pseudointegrifolia* var. *brevistyla* (as the style is so much shorter).
>
> My own inclination at present is towards the latter view and I have so named it tentatively in our herbarium. But I should not be surprised if you should be able to prove it a good species, as Forrest's thinks."
>
> [Extract from a letter dated 24th August 1919 from Sir David Prain]

While it is clear that both Forrest and Prain were referring to the subscapose plants as presented in the description above, a great deal of ambiguity surrounds short-styled and small stigma plants found in the wild, primarily in NW Yunnan, but also in south-western Sichuan and south-eastern Tibet, and the tendency by Kingdon Ward and others has been to place all such plants under the '*brevistyla*' umbrella'. However, it is clear that there are two distinct taxa involved, leaving aside typical representations of both *Meconopsis integrifolia* and *M. pseudointegrifolia*. On the one hand are plants with subscapose flowers and fruits and weakly 3-veined leaves and a short yet slender style while, on the other hand, are those plants with a well-developed stem carrying two or more flowers in a subumbellate inflorescence, with a short but thick style along which the base of the stigmatic rays are often decurrent, and with well-marked 3-veined leaves. There are ample specimens of both these conditions represented in herbaria, and while the former is found in NW Yunnan and SE Tibet (*M. pseudointegrifolia*) it does not overlap with the latter which in restricted to the Lijiang mountains and those to the north of Zhongdian, as well as those in the vicinity of Muli across the border in SW Sichuan (See *M. lijiangensis*).

subsp. daliensis Grey-Wilson, *The New Plantsman* 3 (1): 36 (1996). Type: SW China, NW Yunnan, Dali, Talishan (Cangshan Mountains), *Forrest* 29125 (E, holotype)].

DESCRIPTION. *Plants* not more than 25 cm tall in flower, often more dwarf, densely sericeous-bristly on leaves, stems and pedicels. *Stem* short, not more than 10 cm long with a ruff of petiole base remains and barbellate bristles at the base. *Basal leaves* few, generally 5–7, ascending, not graded, elliptic to oblanceolate, 3-veined only towards the base (in the lower quarter), otherwise pinnately veined. *Bracts* 3–5 leaf-like, borne in a single whorl held just above the basal leaf-rosette. *Flowers* 2–5, erect to ascending, yellow, up to 9 cm diameter. *Ovary* densely sericeous, the style short but distinct, 1–3 mm long. *Fruit capsule* barrel-shaped, densely hairy.
DISTRIBUTION. NW Yunnan (Dali — N Cangshan, mountains S of Lijiang — Sungkwe Hochin range); altitude not recorded (Map 11).
HABITAT. Exposed mountain moorlands, rocky slopes, moraines.
FLOWERING. Late June–August.
SPECIMENS SEEN. *Forrest* 29125 (BM, E); *McLaren* 62 (BM, K).

Subsp. *daliensis* is very little known and probably rare in the wild, being restricted as far as it is known to the northern part of the Cangshan (Dali mountains). In herbaria I have only been able to trace two specimens, both dating back to the early part of the previous century. As far as I am aware it has not been seen in recent times, although there may be more modern specimens tucked away in one of the Chinese herbaria. The status of the plant remains questionable until such time as more material, and it is to be hoped photographs, become available. At present it is best harboured under *Meconopsis pseudointegrifolia*. Interestingly, the McLaren sheet cited indicates that the specimens (which are in flower; there are five plants on the Edinburgh sheet and two on the British Museum sheet) was collected in May, which is particularly early for a meconopsis, mid-June to early August being the prime flowering time for them in western and south-western China. It is distinct in its dwarf and densely sericeous habit but perhaps more especially in the basal leaf rosette: in subsp. *pseudointegrifolia* the basal rosette consists of numerous leaves which are graded from the outer small ones to the inner larger ones; in subsp. *daliensis* there are relatively few leaves which are all similar and not graded. I have resisted the temptation to raise this plant to specific level until such time that more data becomes available.

30. MECONOPSIS SULPHUREA

Meconopsis sulphurea Grey-Wilson, sp. nov.
[syn. *M. integrifolia* sensu Taylor, *The Genus Meconopsis*: 58 (1934), pro parte; *M. pseudointegrifolia* subsp. *robusta*[24] Grey-Wilson, *The New Plantsman* 3 (1): 35 (1996). Type: SW China, NW Yunnan, Judian District (Mekong-Yangtse Divide), Dacaoba, 3350 m, June 1987, *SBLE* 450 (E, holotype; K, isotype)].

DESCRIPTION. *Monocarpic herb* 25–120 cm tall covered for the most part in soft golden or rufous barbellate hairs, with a stout napiform taproot, 40 cm long, 12–35 cm diameter. *Stem* distinct, erect, stout in large plants, to 3.2 cm diameter near the base, striate, generally with several scattered leaves below the whorl of leaf-like bracts, moderately to densely patent hairy. *Leaves* mostly in a basal rosette, which in unflowered plants die away to a large central, overwintering resting bud in the autumn; lamina oblanceolate to elliptic to 40 × 5 cm, the base attenuate into the petiole, the apex subobtuse to subacute, margin entire to somewhat toothed, pinnately-veined but the base usually somewhat 3-veined, moderately to densely pubescent when young, but becoming gradually pubescent, particularly above, at maturity, especially in flowering specimens, the petiole equalling or longer than the lamina; cauline leaves few, spirally arranged, similar to the basal but gradually smaller and narrower up the stem, short-petiolate or sessile, the base generally decurrent onto the stem, the uppermost leaves (bracts) erect and aggregated into a false whorl, generally numbering (1–)5–9(–12). *Inflorescence* a subumbellate raceme with up to 18 flowers opening in succession, often supported by several lateral flowers borne from the axils of the lower cauline leaves; flowers bowl-shaped to broad saucer-shaped, nodding to half-nodding at maturity but becoming erect as the flowers fade and the fruit capsules develop. *Pedicels* to 45 cm long, subglabrous to patent-pubescent. *Buds* nodding, oval to obovoid. *Sepals* elliptic-oval to obovate, patent-hairy on the exterior. *Petals* (5–)6–9, pale lemon or sulphur yellow, elliptic to oval or obovate, 5.5–10.5 cm long, curved, not incurved, but often spreading widely apart. *Stamens* numerous, with the filiform filaments the same colour as the petals, the anthers yellow or orange-yellow, but browning on ageing. *Ovary* obovoid to ellipsoidal, densely to sparsely ascending to appressed, soft-bristly; stigma rather narrow and with a distinct slender style 4–9 mm long; stigma capitate, protruding beyond the boss of stamens, the stigmatic rays usually 7–10, somewhat decurrent onto the upper part of the style. *Fruit capsule* ellipsoid to narrow-obovoid, 40–48 × 16–19 mm, densely patent-bristly to almost glabrous, splitting at maturity by (6–)7–10 valves, the valves up to one third the length of the capsule, the persistent style 6–14 mm long. 2n = 76[25] (Fig. 3E).

DISTRIBUTION. Extreme SW Sichuan, NW Yunnan (Salween-Mekong-Yangtse watersheds, but not to the east of the Yangtse-Mekong watershed) S & SE Tibet (north of Lhasa eastwards and south-eastwards, extending north-east to Yushu County) and NE Myanmar (Upper Burma); (2700) 3360–4575 m (Map 11).

Meconopsis sulphurea; Laojunshan, Lijiang County, NW Yunnan, 3800 m. PHOTO: TOSHIO YOSHIDA

---

[24] It is not possible to raise subsp. *robusta* to specific status as a species under that name already exists: see *Meconopsis robusta* Hook. f. & Thomson.

[25] Ying *et al.* (2006) as *Meconopsis integrifolia*.

HABITAT. Open woodland, woodland margins, mountain moorland, swamp margins, rocky slopes, rhododendron scrub, stabilised screes and moraines.
FLOWERING. Late May–July (– early August).

Meconopsis sulphurea is one of the boldest and handsomest of the yellow-flowered species that has long hidden within its more northerly ally *M. integrifolia*. Both David Prain (1906) and Frank Kingdon Ward (1924 & 1926) considered it to be a distinct species along with another which they referred to as *M. brevistyla*. However, upon the publication of George Taylor's monograph in 1934 these and other variants were all included within *M. integrifolia* without qualification. Taylor comments that "It is unfortunate that this well-known plant should have been the source of so much misunderstanding and, as a consequence, should be burdened with such a cumbrous synonymy. In dealing with this species most authors have apparently concentrated on finding differences between the various forms, neglecting the intermediate conditions and thus failing to recognise the very close resemblances which they show".

The history of *Meconopsis pseudointegrifolia* (formerly including material now classed as *M. sulphurea*) needs to be examined and I can do no better than to again quote from Taylor's monograph:

"In 1904 plants which were at first referred to *M. integrifolia* and figured under that name flowered simultaneously in two gardens in this country. The seed from which these had been raised were derived from independent sources: the one set of seeds was from the Koslov expedition to south-eastern Tibet and the other from Wilson collected at Tatsienlu in western Sichuan. Both plants were figured in 'Flora and Silva' and comparison suggested that they belonged to different species. The Koslov plant as grown had the flowers borne singly on basal scapes and characterised by the presence of a prominent style; that of Wilson had the flowers raised on a stem and had sessile stigmas with conspicuous radiating lobes. The question of the identity of these plants was referred to Prain who decided that the latter was typical *M. integrifolia* while the former, in virtue of its habit and conspicuous style, he described as a new species under the name *M. pseudointegrifolia*. Forrest and Monbeig subsequently collected forms in Yunnan in which the flowers in addition to having conspicuous styles were borne in a cluster on the stem. It was then realised that the separate identity of *M. pseudointegrifolia* depended only on the presence of a long style."

Despite Taylor's analysis, detailed examination of the entities involved neatly show that a combination of characters along with geographical provenance clearly separate *Meconopsis integrifolia* from *M. sulphurea*. The presence of a distinct style in the latter, coupled with pinnately veined leaves, the nodding or half-nodding poise of the flowers and the shape of the mature fruit capsule are all significant. These differences can be readily observed in the wild and also in cultivated plants. Those accessions from the wild that are in cultivation also reveal interesting differences at the seedling stage. Peter Cox of Glendoick Gardens in Scotland tells me that the two can be readily identified at the seedling stage: while the leaf-lamina vein details are revealed at an early stage in the plant's life, the young leaves of *M. integrifolia* are narrow-elliptical with an entire margin, the counterpart in *M. sulphurea* (*M. pseudointegrifolia* Hort.) are broader and relatively shorter with a few remote, though shallow, teeth along the margin. This is an extremely useful way of separating the species at an early stage, especially in due of the fact that nearly all the material collected in recent years, both seed and herbarium specimens, has come in under the name *M. integrifolia* without qualification.

Meconopsis sulphurea; Tianchi, Zhongdian, NW Yunnan, 3400 m.
PHOTO: CHRISTOPHER GREY-WILSON

30. MECONOPSIS SULPHUREA

In its stylar characters and disposition of the flowers the pink-flowered *Meconopsis sherriffii* most closely resembles *M. sulphurea*. Yet the former is a markedly perennial species in the wild, forming substantial clumps in time, while the leaves are strongly 3-veined in the lower half. It is interesting to note that in the wild the former occupies territory to the west and south-west of the latter, but there is no evidence that the two grow together anywhere in the wild.

As we have seen (see p. 177) *Meconopsis integrifolia* was originally described as a species of *Cathcartia* by Maximowicz in 1877 based upon specimens collected by Przewalski in Gansu in 1872 and 1872. A few years later in 1886, Franchet in Paris realised that this plant was in fact a species of *Meconopsis* and published the transference from the one to the other. To complicate matters he identified a number of collections made by Abbé Delavay from the Lijiang (then Lichiang) mountains as being the same taxon, despite that material bearing the presence of a style, although a short one. Franchet clearly felt that the presence or lack of a style was of no taxonomic importance and this is the position adopted by George Taylor in 1934.

Meconopsis sulphurea leaf-rosette showing overwintering bud; Serkhyem La, 4700 m.
PHOTO: HARRY JANS

KEY TO SUBPECIES OF MECONOPSIS SULPHUREA

1. Flowers and buds nodding to half-nodding, generally with 6–9 petals, these 5.5 cm long or more; basal leaves elliptic to oblanceolate, the lamina generally more than 2.5 cm wide .... subsp. **sulphurea** (SW China — SW Sichuan, S & SE Tibet, NW Yunnan but not Dali-Cangshan; N Myanmar)

+ Flowers and buds ascending to erect, the flowers deeply cupped, generally with 6 petals, these rarely more than 5.2 cm long; basal leaves linear-elliptic to narrow-elliptic, the lamina not more than 2.6 cm wide .. .. subsp. **gracilifolia** (W China from C Tibet to SW Qinghai)

subsp. sulphurea

DESCRIPTION. Plant very variable in height, often 70–90 cm, occasionally to 1.2 m. *Basal leaves* elliptic to oblanceolate, the lamina mainly 2.5–5 cm wide, the innermost similar to the outer, though often a bit smaller; cauline leaves present usually. *Bracts* similar to the cauline leaves, often rather narrower. *Buds* nodding to semi-nodding. *Flowers* nodding to semi-nodding, saucer-shaped. *Petals* normally 6–9, cream to lemon-yellow or pale gamboge, elliptic to oval or obovate, 3.8–10.5 × 2.6–7.8 cm. *Ovary* sparsely to densely appressed to subappressed bristly; style 4–12 mm long in fruit. (Figs 15H, J, 16E, F).

DISTRIBUTION. Throughout the range of the species except for the area to the north and east of Lhasa, extending north-eastwards to the Kyushu Xian; (2700) 3360–4575 m (Map 11).

SPECIMENS SEEN. *Aufschnaiter* s.n. 1951–52 (BM); *Farrer* 1735 (E); *Forrest* 10452 (BM, E), 10470 (E), 12522 (BM, E, K), 13169 (BM), 13311 (K), 14118 (BM, E), 16551 (E, K), 17250 (E), 19660 (BM, E), 20727 (BM, E), 20754 (BM, E), 22033 (K), 22308 (E, K), 23419 (BM, E), 23424 (BM, E), 25933 (E, K), 28512 (BM), 28753 (E), 30015 (E), 30083 (BM, E), 30091 (BM), 30094 (BM, E), 30102 (BM, E), 30115 (BM), 30108 (BM, E), 30602 (E); *Hanbury-Tracy* 15 (BM), 201 (BM); *KEG* (Kunming Edinburgh Gothenburg Exped.) 1017 (E), 1464 (E); *Kingdon Ward* 174 (E), 513 (E), 555 (E), 4169 (E), 5627 (BM, E, K), 5749 (BM, K), 5766 (E, K), 5910 (K), 5984 (K), 6125 (K), 6213 (K), 7098 (K), 7099 (K), 10512 (BM); *Ludlow & Sherriff* 9668 (BM); *Ludlow, Sherriff & Elliot* 13153 (BM, E), 13748 (BM), 13853 (BM), 14458 (BM), 15042 (BM, E), 15323

(BM, E), 15657 (BM, E), 15778 (BM); *Ludlow, Sherriff & Taylor* 5043 (BM, E), 5043A (BM), 5043B (BM, E), 5058 (BM, E), 5058B (BM), 5146 (BM, E), 5891 (BM), 6041 (BM), 6082 (BM, E), 6082A (BM), 5058A (BM); *McLaren* 35 (E), 62 (E), 132 (E, K); *Monbeig* 21 (E), s.n. (K); *Rock* 8753 (E, K), 9227 (E), 9386 (E), 10348 (E), 22314 (E), 22316 (E), 22428 (E), 22633 (E), 22692 (E), 22753 (E), 22828 (K), 22850 (E), 22887 (E, K), 23100 (E), 23254 (E), 23334 (E, K), 23432 (E), 24797 (BM), 25298 (E), 25408 (E), 25096 (E); *Sino-American Botanical Exped. Yunnan SSY* 254 (E); *Sino-British Lijiang Exped. SBLE* 339 (K), 450 (E, K); *Sino-Scottish Exped. NW Yunnan* 535 (E); *T. T. Yü* 7913 (E), 10894 (BM), 14660 (E), 19644 (E), 22151 (E), 22520 (E).

NOTE. Very dwarf plants are found on occasion at the higher altitudinal range for the species, generally growing in exposed moorland or rocky terrain. *Hanbury-Tracy* 15 from the Kharu La in eastern Tibet, which is recorded as bring fragrant, is one such collection, as is *Ludlow, Sherriff & Taylor* 5043.

Depauperate specimens of *Meconopsis sulphurea* are sometimes observed in the wild, sometimes growing at no great altitude nor necessarily in an unsuitable habitat. One such plant is ACE 705 (NW Yunnan, above Benzilan, E Baimashan, 4050 m): just 17.5 cm tall and with linear-lanceolate leaves, 40–44 × 6–7 mm, the pale primrose flowers just 5.5–5.8 cm across with 5 petals and a style c. 5.5 mm long. Frank Kingdon Ward described var. *gracilis* based on two collections (numbers 7098 and 7099, both at K) from the Seinghku Valley in northern Myanmar (Burma) stating (*Annals of Botany* 42: 861 (1928) that: "This plant differs from typical *M. pseudointegrifolia* [sulphurea sic] in its altogether slighter build, smaller and narrower leaves, fewer and paler flowers, smaller in all their parts, and in being less hairy. I recorded the flowers of 7098 as 'Ivory white', those of 7099 as 'sulphur' ..... It resembles the type in its inflorescence, elongated style, and large stigma, in the number of petals (6–8), entire leaves and narrow linear bracts; it differs from it only in the minor points mentioned above". He goes on to

Meconopsis sulphurea; Serkhyem La, SE Tibet, 4400 m. PHOTOS: HARRY JANS

Meconopsis sulphurea, mature fruit capsules; Serkhyem La, SE Tibet, 4700 m. PHOTO: HARRY JANS

say that "..... var. *gracilis* is a rare plant in the Seinghku valley. About six plants were discovered, scattered on the steep flank of the range, growing up amongst the dwarf 'Lapponicum' rhododendrons, at 13,000 feet in one locality only."

Prominent collections of this subspecies were made by Delavay at the end of the nineteenth century and later George Forrest and, in more recent times, by the Sino-British Lijiang Expedition (SBLE) and the CLD expedition.

subsp. gracilifolia Grey-Wilson **subsp. nov.** Like the typical plant (subsp. *sulphurea*) in stature but basal leaves linear-elliptic to narrow-elliptic and flowers and buds ascending to erect, the flowers deeply cupped, generally with 6 petals. Type: Tibet, Reting, 60 miles N of Lhasa, *Ludlow & Sherriff* 8924 (E, holotype; BM, E, isotypes).
[syn. *M. integrifolia* sensu Taylor, *The Genus Meconopsis*: 58 (1934), pro parte]

DESCRIPTION. Like subsp. *sulphurea* in stature, often 35–70 cm tall, but sometimes as little as 20 cm tall in flower. *Basal leaves* narrow-elliptic to linear-elliptic, the lamina mainly 5.5–11 × 1–2.6 cm, the innermost sometimes very narrow, 4–10 mm wide, 3-veined in the lower half; cauline leaves absent or 1–2. *Bracts* up to 11, similar to the inner basal leaves, 7.5–12 × 0.35–0.9 cm. *Buds* ascending to upright. *Flowers* 3–7, ascending to upright, deep bowl-shaped, 8–10 cm across. *Petals* normally 6, pale lemon-yellow, oval to obovate, 3.8–5.2 × 2.4–3.6 cm. *Ovary* ovoid, densely appressed bristly; style 3–7 mm long. *Fruit capsule* 35 × 16 mm, with a persistent style 6–9 mm long.

DISTRIBUTION. Central Tibet, N & E of Lhasa, extending north-eastwards into SW Qinghai; 4270–4725 m (Map 11).

HABITAT. Semi consolidated screes, rocky ground with other herbage, ditches or close to running water.

FLOWERING. July–August.

SPECIMENS SEEN. *Guthrie* 2039 (K); *Ho, Bartholomew & Gilbert* 913 (BM); *Ho, Bartholomew, Watson & Gilbert* 2235 (BM, E); *Littledale* s.n. (K); *Ludlow & Sherriff* 8924 (BM, E), 9085 (BM), 9668 (BM), 9945 (E), 9981 (BM, E).

NOTE. *Ho et al.* 2235, represented in Britain by single sheets in the Edinburgh and Kew Herbaria, is unusual in having a basal rosette of rather upright very narrow, grass-like leaves; however, in other respects it matches specimens gathered by Ludlow & Sherriff to the north of Lhasa, in the vicinity of Reting. The number in question was gathered in the Yushu Xian, mountains in Qinghai province, close to the north-western border with Sichuan. Another, *Ho et al.* 913, also from Qinghai (Gande Xian, 35°55'N, 99°44'E) also has very narrow basal leaves, not more than 10 mm wide. Although widely separated by the vast mountainous plateau expanses between Lhasa and the south-western Qinghai localities, this is very much an under-explored region botanically and it is more than possible that subsp. *gracilifolia* occurs in other localities in these mountains.

Meconopsis sulphurea subsp. *gracilifolia*; Ganden-Samye, E of Lhasa, S Tibet, 5000 m, early July. PHOTO: TOSHIO YOSHIDA

section SIMPLICIFOLIAE

This section is undoubtedly very closely related to the previous, differing most obviously in the strictly scapose nature of the flowers. Like the members of section *Grandes* plants, although often polycarpic, can be monocarpic with the rosette leaves withering away in the autumn to leave the plant as an overwintering bud, or a collection of buds at ground level, generally with the lower half just below the soil surface. All the species are beset with varying amounts of barbellate hairs, especially on the leaves and scapes. Three of the four members of this section (*Meconopsis punicea*, *M. quintuplinervia* and *M. simplicifolia*) are well known horticulturally, with the former two well-established in cultivation today. The fourth member of this association, *M. nyingchiensis*, is more problematic and certainly very little known or understood.

KEY TO SERIES OF SECTION SIMPLIFICOLIAE

1. Leaf-lamina pinnately veined or weakly 3-veined towards the base or in the lower half; flowers saucer-shaped to cupped or flattish; petals normally 5 or more .... series **Simplicifoliae**
+ Leaf-lamina with 3 or 5 well marked longitudinal veins reaching towards the leaf tip; flowers campanulate to campanulate-infundibuliform; petals normally 4 .... series **Puniceae**

series Simplicifoliae

The series is primarily characterised by having blue, purple, lavender or reddish purple, relatively large, flowers that are lateral-facing to half-nodding. Series *Simplicifoliae* is confined to the central and eastern Himalaya and south-eastern Tibet.

KEY TO SPECIES OF SERIES SIMPLIFICOLIAE

1. Flowers saucer-shaped or shallowly cupped, horizontal to ascending, normally with 5–9 petals; scapes and fruits with reflexed bristles; plants usually monocarpic; stigma clavate .... 31. **M. simplicifolia** (C Nepal to Sikkim, Bhutan and SE Tibet)
+ Flowers campanulate, ascending to nodding, normally with 4–6 petals; scapes and fruits glabrous or with spreading to ascending bristles; plants polycarpic; stigma cupuliform .... 32. **M. nyingchiensis**

31. MECONOPSIS SIMPLICIFOLIA

Meconopsis simplicifolia (D. Don) Walp., *Repert. Bot. Syst*. 1: 110 (1842).
[syn. *Papaver simplicifolium* D. Don, *Prodr. Fl. Nepal*.: 197 (1825). Type. C Nepal, Gossain Than [sic Gosainkund], *Wallich* 8125 (K, holotype, isotype); *Stylophorum simplicifolium* (D. Don) Spreng in L., *Syst. Veg*. Ed. 16 (4) 2: 203 (1827); *Polychaetia scapigera* Wall., *Numer. List*: 277, sub. n. 8125 ★1832); *Meconopsis uniflora* Gumbl., *Garden* 22: 90 in obs. (1882), nom. illegit.; *M. simplicicaulis* Wood, *Garden* 24: 115 (1883), laps. pro simplicifolia; *M. simplicifolia baileyi* Farrer, *Engl. Rock Gard*. 1: 481 in obs (1919), nom. illigit.; *M. simplicifolia* var. *baileyi* Kingdon Ward, *Gard. Chron*., Ser. 3, 79:340 in obs. (1926)]

DESCRIPTION. *Monocarpic, occasionally polycarpic, plant to 60 cm tall in flower, often less, sometimes as little as 20 cm, with a narrow-dauciform taproot to 13 cm long, or more or less fibrous-rooted*; base of the solitary leaf-rosette beset with old membranous petiole remains and numerous golden or golden-brown, barbellate bristles. *Leaves* all basal, ascending to erect, occasionally spreading, dying down to an overwintering bud at ground level, borne in a lax to rather dense rosette, soft green, somewhat paler beneath, the lamina lanceolate to oblanceolate or ovate-lanceolate, 4.5–16 × 0.9–4.2 cm, the base attenuate into the petiole, the apex acute to subobtuse, not or weakly to moderately 3-veined from the base, the margin entire to remotely serrate or crenate, sometimes coarsely so, rarely somewhat lobed, sometimes pronouncedly undulate, covered on both surfaces with pale rufous or golden barbellate hairs; petiole linear, 2–16 (–20) cm long, somewhat expanded at the base, leaves rarely subsessile. *Flowers* solitary, scapose, up to 6 (–9) per leaf-rosette, nodding to half-nodding, occasionally

Meconopsis simplicifolia subsp. *grandiflora*; S of Yumtang, NE Sikkim, 3550 m. PHOTO: TOSHIO YOSHIDA

ascending, deeply bowl- to saucer-shaped, 4–12 cm diameter. *Pedicels* (scapes) often rather stout, erect, 16–67 cm long (to 85 cm long in fruit), (3–)5–11 mm diameter, covered in dense, golden to pale rufous, horizontal to deflexed, barbellate bristles. *Flower buds* ascending to erect, narrow- to broad-ovoid, the sepals covered with dense, spreading, barbellate hairs. *Petals* (4–)6–9, sky to deep blue, lavender, mauve-purple, dark purple, purple-blue, rosy purple or reddish blue, often with rather darker veins, especially towards the base, rarely pure white, obovate to oval, 20–48 × 12–44 cm, the margin entire to minutely erose. *Stamens* a third to half the length of the petals; filaments linear, the same colour as the petals, although often darker; anthers orange to orange-yellow, oblong, c. 2 mm long. *Ovary* ellipsoid to oblong-ellipsoid, usually densely appressed setose, rarely glabrous; style protruding well beyond the boss of stamens, (1–)4–9 mm long, the stigma pale green to pale brown or whitish, capitate to subclavate, 4–9-lobed, 3.5–5.5 mm long, the lobes more or less decurrent on the upper part of the style; style 1–7 mm long, occasionally subobsolete. *Fruit capsule* narrow-ellipsoid to oblong-ellipsoid, 33–67 × 8–23 mm, 4–9-valvate, prominently ridged, generally half-nodding at first but becoming erect as the fruit develops, covered in sparse to moderately dense, deflexed, barbellate bristles, very occasionally glabrous, splitting for about one third of its length from the apex when ripe, the persistent style 3–12 (–14) mm long. *Seeds* elliptic to reniform in outline, the testa densely papillose. $2n = 82, 84$.

DISTRIBUTION. C & E Nepal, Sikkim, Bhutan, NE India (Arunachal Pradesh), SC & SE Tibet; 3050–4575 m (Map 12).

HABITAT. Margins of forests and forest glades, dwarf *Rhododendron* scrub, shrubberies (*Berberis* and *Salix* in particular), streamsides, amongst boulders, damp meadows.

FLOWERING. Late May–early August.

This species was first described as a *Papaver* (*P. simplicifolium*) by David Don in 1825 based upon Wallich collections cited from Gossain Than (today Gossainkund), a locality well known to todays' trekkers in central Nepal and located close to the famous Langtang Valley. It was transferred to *Meconopsis* about 18 years later, almost 30 years after the genus had been established.

Meconopsis simplicifolia is often said to be unsurpassed among the blue poppies, although in truth the flower colour can vary from pure sky blue to muddy purple or reddish purple, but many shades of blue, lilac and lavender can be found in the wild. In many ways the larger flowered forms of *M. simplicifolia* come close to *M. grandis* in general appearance and the two species are sometimes confused, especially in photographs; while the former is strictly scapose in habit, the flowers of the latter are always borne on a bracteate stem, however short that might be. In addition, *M. grandis* normally has few petals (usually 4–6) while the filaments are white, whereas the latter has flowers with up to 9 petals and filaments of a similar colour to the petals, often darker. In addition, the two are quite different in the details of the fruit capsules, those of *M. simplicifolia* being heavily ridged. One interesting and rather unusual feature of *M. simplicifolia* can be observed in the often erect flower buds: in most species of *Meconopsis* the flowerbuds are nodding, at least at first, this whether or not the flowers open in an upright or nodding position.

Meconopsis simplicifolia is remarkably variable in the wild and the differences between the extreme forms are

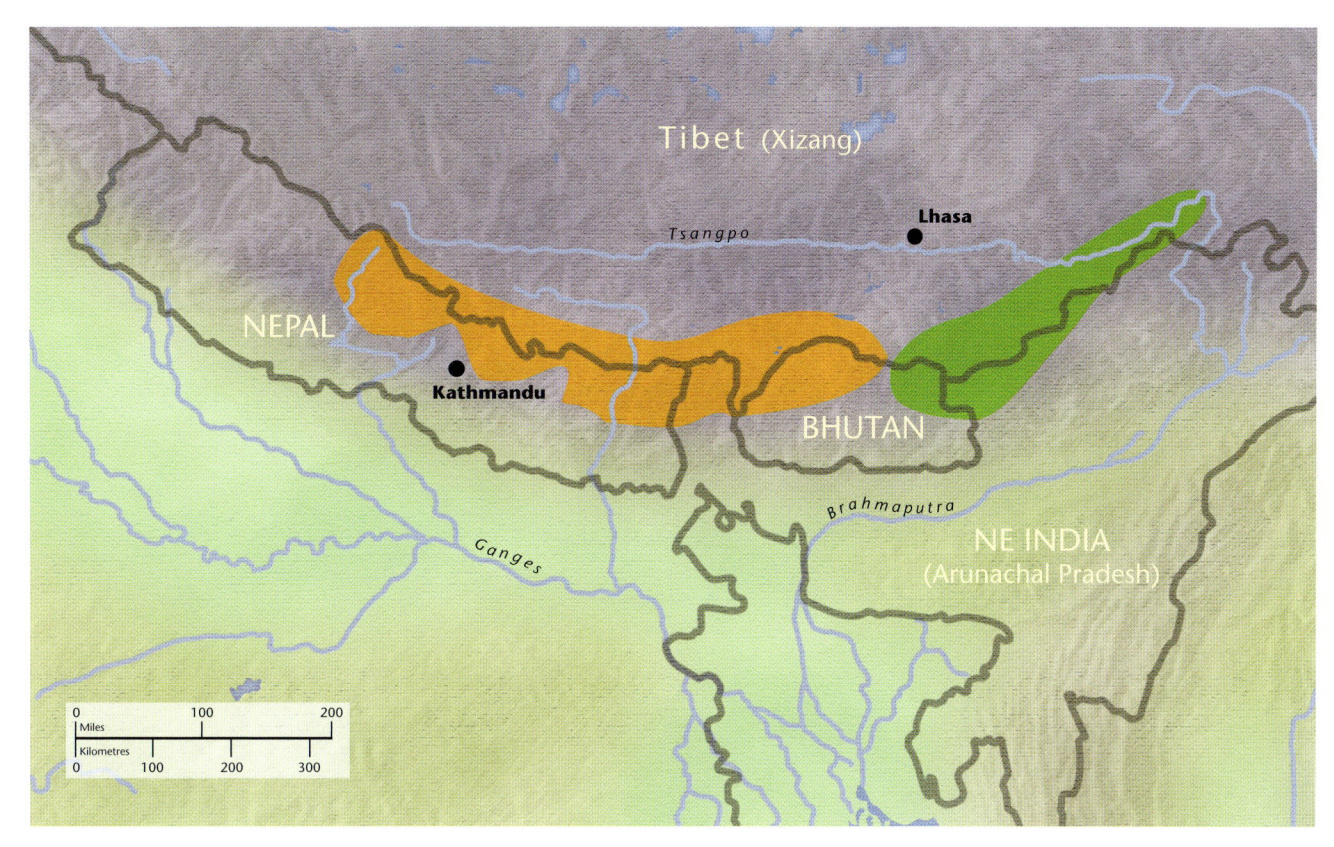

MAP 12. Distribution of *Meconopsis simplicifolia*: ■ *M. simplicifolia* subsp. *simplicifolia*, ■ *M. s.* subsp. *grandiflora*.

quite marked. Slight plants with thin leaves and long, very slender scapes carrying nodding pale campanulate flowers contrast markedly with robust forms with, thicker, tougher leaves and stout scapes bearing larger deeply coloured, flatter, lateral facing flowers. While it is quite clear that the species requires further detailed field analysis, these two extremes, accompanied by the vast percentage of herbarium specimens, fall into two groups, the one concentrated in the central and eastern Himalaya from central Nepal to western Bhutan and the neighbouring parts of southern Tibet, the other primarily in north and north-eastern Bhutan and south-eastern Tibet.

This interesting and variable species can behave as a monocarpic or polycarpic plant, especially in cultivation. That it is primarily a polycarpic species can be ascertained readily by examining the numerous herbarium specimens, many of which show the presence of offsets or lateral shoots but, furthermore, quite a few bear the remains of pedicels (scapes) from the previous season (or seasons).

KEY TO SUBPECIES OF MECONOPSIS SIMPLICIFOLIA

1. Flowers small, often shallowly cupped and half-nodding, the petals usually (4–)5–6, not more than 30 × 17 mm; leaf-lamina up to 10.5 × 4.2 cm, not 3-veined except at the very base, the lamina base generally rather abruptly narrowed into the petiole, the margin with a few, sometimes coarse, teeth, rarely subentire; fruit capsule (33–)42–67 mm long, (6–)8–14 mm diameter, 3–5 mm diameter at the base, 4–6-valved, the persistent style 6–12 (–14) mm long in fruit. . . . . subsp. **simplicifolia** (Nepal, N India (Sikkim), W & C Bhutan, S Tibet)

+ Flowers relatively large, generally flat or saucer-shaped, and lateral-facing, the petals usually 6–9, 33–48 × 24–44 mm; leaf-lamina not exceeding 8.2 × 2.9 cm, weakly to strongly 3-veined, the lamina base generally gradually attenuate into the petiole, the margin entire, rarely somewhat toothed; fruit capsule 33–47 mm long, 16–23 mm diameter, 6–10 mm, diameter at the base, 6–9-valved, the persistent style 3–7 mm long . . . . subsp. **grandiflora** (N & NE Bhutan, SE Tibet)

subsp. simplicifolia

DESCRIPTION. *Plants* often monocarpic, sometimes polycarpic. *Leaves* relatively few, generally less than 20 in the basal rosette, the lamina thin, to 10.5 × 4.2 cm, not 3-veined or only in the basal quarter, the lamina base generally rather abruptly narrowed into the petiole, the margin subentire to remotely toothed, often undulate. *Flowers* relatively small, broad-campanulate, nodding to half-nodding. *Petals* usually (4–)5–6, not more than 30 × 17 mm. *Fruit capsule* (33–)42–67 mm long, (6–)8–14 mm diameter, 3–5 mm diameter at base, 4–6-valved, the persistent style 6–12(–14) mm long in fruit. (Fig. 12C1–2).

DISTRIBUTION. Nepal (Mustang and Myagdi districts east to Taplejung and Panchthar districts, but apparently absent from Gorkha, Manang Nuwakot & Sindhupalchok districts), eastwards to W & C Bhutan and adjacent parts of S Tibet; 3300–4570 m (Map 12).

HABITAT. Rocky and grassy slopes, low scrub, open forest and forest margins.

FLOWERING. May–July.

SPECIMENS SEEN. *AGSES* 47 (K); *Bailey* s.n. 26 June 1937 (E), s.n. 12 June 1937 (E); *R. Bedi* 250 (K); *Beer* 8267 (BM), 25420 (BM), 25521 (E); *Botanical Exped. Hiamalaya* (1985) 8580251 (E); *Bowes Lyon* 3245 (BM); *Cave* 153 (E), 421 (E), 1803 (E), s.n. 27 Oct. 1916 (E), s.n. 6 Nov. 1917 (E), s.n. 26 June 1919 (E), s.n. 1.07.1922 (E); *S. Chapman* 1936 (K); *C. B. Clarke* 46389 (BM); *R. E. Cooper* 23 (E), 88 (E), 311 (E), 516 (E), 812 (E), 909 (E), 938 (E), 999 (E); *Dhwoj* 20 (E), 138 (BM, E), 233 (E); *Fifth Botanical Exped. Himalaya* (1972) 723040 (E); *First Darwin Nepal Fieldwork Training Exped.* (DNEP1) 149 (E); *Flora of East Nepal* (1995) 9596115 (E); *Ganesh Himal Central Nepal Exped.* (1994) 9400075 (E), 9420218 (E); *Gould* 158 (K); *Hara et al.* 514 (BM); *Hooker* s.n. Sikkim, 12-14,000 ft (BM, E, K); *Hara et al.* 514 (BM); *Jaffrey* s.n., 1885 (BM); *Kanai et al.* 320360 (BM), 720708 (BM), 723040 (BM, E), 723042 (BM); *King's collector* s.n. 1887 (E, K); *Lachinpinga* 8 (K); *Long, McBeath, Noltie & Watson* 252 (E), 330 (E), 505 (E); *Ludlow & Sherriff* 122 (BM), 672 (E); *Ludlow, Sherriff & Hicks* 16471 (BM, E), 17301

Below, from left:

Meconopsis simplicifolia subsp. *simplicifolia*; S of Nyima La, Milin County, SE Tibet, 4100 m. PHOTO: TOSHIO YOSHIDA

Meconopsis simplicifolia subsp. *simplicifolia*; W of Topke Gola, Jaljale Himal, E Nepal, 4200 m. PHOTO: TOSHIO YOSHIDA

Meconopsis simplicifolia subsp. *simplicifolia*; Chelei La, Thimphu district, W Bhutan, 3800 m. PHOTO: TIM LEVER

(BM), 19854 (BM); *Miyamoto et al.* 9400075, 9420218 (BM), 9596115 (E); *Ohba, Kikuchi, Wakabayashi, Suzuki, Kurosaki, Rajbhandari & Wu* 8580251 (E); *Polunin* 1076 (BM); *Ribu & Rohmoo* 5334 (K), 5736 (E), 6005 (E, K); *Rohmoo Lepcha* 3 (BM, E), 960 (E), 1012 (E), 1066 (BM); *K. N. Sharma* 29 (BM), 95 (E), E400 (BM), E456 (BM); *T. B. Shrestha* 21 (BM); *Stainton* 743 (E), *Stainton, Sykes & Williams* 3054 (BM, E), 6073 (BM), 6113 (BM, E), 6630 (BM, E), 8643 (BM); *Starling, Upward, Brickell & Mathew* AGSES 47 (E); *E. H. Walsh* 146 (BM); *Third Darwin Nepal Fieldwork Training Exped.* AY126 (E); *G. Watt* 5438 (E); *Zimmermann* 594 (BM).

NOTE. *Miyamoto et al.* 9400075 & 9420218 (both E!), collected from central Nepal (Bagmati Zone, Rasuwa District, around Seto Kund at 3930, in August 1994) are unusual in their relative broad leaves and accompanying small flowers. *Leaves* 9–11.5 × 3.8–4.2 cm, with 6–15 cm long petioles. *Scapes* very thin, 2.5–3 mm diameter, to 36 cm long. *Petals* 22–30 × 22 mm. *Dhwoj* 138 from Tangba, Nepal, in contrast has very long slender leaves, 23–26 × 2–2.5 cm. These require further detailed investigation.

Meconopsis simplicifolia subsp. *simplicifolia*, leaf-rosette; Maroothang, Tongsa district, C Bhutan, 3800 m. PHOTO: TIM LEVER

Meconopsis simplicifolia subsp. *simplicifolia*; south side Rupina La, Gorkha Himal, C Nepal, 3600 m., flowering in mid July. PHOTO: ALAN DUNKLEY

subsp. grandiflora Grey-Wilson, subsp. nov.

Differs from the typical plant (subsp. *simplicifolia*) by its weakly to strongly 3-veined leaves that are gradually attenuate at the base, by its relatively large 6–9-petalled flowers, and by the relatively broader 7–9-valved, not 4–6-valved, fruit capsules. Type: SE Tibet, Takpo Province, Chubumba La, Langong, 28°51'N, 93°47'E, 12,500 ft, 2 June 1938, *Ludlow, Sherriff & Taylor* 3944 (holotype, E).

DESCRIPTION. *Plants* robust, often polycarpic. *Leaves* numerous, generally 20 or more, the lamina relatively thick, not exceeding 8.2 × 2.9 cm, 3-veined at least in the lower half, gradually attenuate into the petiole, the margin entire, rarely somewhat toothed. *Flowers* large, generally flat or saucer-shaped, lateral-facing. *Petals* usually 6–9, 33–48 × 24–44 mm. *Fruit capsule* 33–47 mm long, 16–23 mm diameter, 6–10 mm diameter at the base, 6–9-valved, the persistent style 3–7 mm long.

DISTRIBUTION. N & NE Bhutan (Upper Mo Chu to Upper Kulong Chu districts, especially to the north of Bumthang, Lhuntsi and Tashigang), NE India (Arunachal Pradesh)[26] and SE Tibet (particularly common in Tsari and on many of the major passes such as the Bimbi La, Chubumba La, Doshong La, Kechen La, Mira La, Nyima La and Temo La); 3200–4575 m (Map 12).

---

[26] The species is almost certainly present in Arunachal Pradesh (NE India). Although there are no herbarium specimens to substantiate this, I have recently seen photographs of it taken in the west close to the Bhutan border by David and Margaret Thorne, Tim Lever and others.

31. MECONOPSIS SIMPLICIFOLIA

SPECIMENS SEEN. *Bailey* 8/06/1928 (E), 14/06/1928 (E); *R. E. Cooper* 311 (E), 1300 (E), 1694 (E, K), 1833 (BM, E, K), 2178 (E, K), 2975 (E, K), 3226 (E); *B. J. Gould* 2055 (K); *H. E. Hobson* s.n. Yatung, Tibet (K); *Kingdon Ward* 5716 (K), 5737 (K), 5855 (BM, E, K), 6245 (K); *Ludlow & Sherriff* 38 (E), 92 (BM, E), 122 (BM, E), 672 (BM), 1029 (BM, E), 1613 (BM, E), 1637 (BM, E), 1687 (BM, E), 1735 (BM), 1790 (BM), 1845 (BM, E), 1877 (BM), 2088 (BM), 2093 (BM, E), 2176 (BM), 2202 (BM), 2203 (BM), 2569 (BM), 2662 (BM), 2655 (BM), 3068 (BM, E), 3223 (E), 3263 (BM, E), 3386 (BM), 3909 (BM), 3944 (BM); *Ludlow, Sherriff & Elliott* 13821 (BM), 14311 (BM, E), 14383 (BM, E), 14445 (BM, E), 15149 (BM, E); *Ludlow, Sherriff & Hicks* 16313 (BM), 16348 (BM, E), 16710 (BM, E), 16775 (BM), 18995 (BM, E), 19205 (BM, E), 19802 (BM), 20285 (BM), 21162 (BM, E); *Ludlow, Sherriff & Taylor* 3906 (E), 3909 (BM, E), 3944 (E), 4789 (BM), 4789a (BM, E), 4789b (BM), 4789c (BM, E), 6083 (BM, E), 6416 (E); *Ribu & Rhomoo* 5736 (E); *E. H. Walsh* 146 (BM); *Sinclair & Long* 5260 (E).

NOTE. *Ludlow, Sherriff & Elliott* 15152 (BM) from SE Tibet, 'Deyang La, 13,500 ft' appears to be a hybrid with the following details: Leaves quite toothed. Calyx green; Corolla light blue, petals 4. Stamens with white filaments and orange anthers. Style and stigma green. See also under *Meconopsis nyingchiensis*.

Meconopsis simplicifolia subsp. *grandiflora*, newly emerged leaf-rosettes; Bumthang valley, C Bhutan. PHOTO: CHRISTOPHER GREY-WILSON

Below, from left:

Meconopsis simplicifolia subsp. *grandiflora*, flowers and young fruit capsules; W Bhutan.
PHOTO: ROSIE STEELE

Meconopsis simplicifolia subsp. *grandiflora*; SE Tibet. PHOTO: KENNETH COX

Meconopsis simplicifolia subsp. *grandiflora*, immature fruit capsules; W Bhutan.
PHOTO: ROSIE STEELE

Above: *Meconopsis simplicifolia* subsp. *grandiflora*, flower details; Yaksa, Thimphu district, W Bhutan, 4300 m. PHOTO: TIM LEVER

Meconopsis simplicifolia, in one form or another, has been in cultivation for more than one hundred and fifty years, although it has faltered in cultivation a number of times during this period. Its initial introduction was from seed collected by Joseph Dalton Hooker in Sikkim in 1848 and so represents subsp. *simplicifolia*. Seed was supplied to Kew from the Calcutta Botanic Garden in the early years of the twentieth century and a flowering plant from this source was featured in *Curtis's Botanical Magazine* in 1911 (vol. 137, tab. 8364). R. E. Cooper collected seed in Sikkim and Bhutan in 1913 and 1914 respectively and many of the plants in gardens during the first half of that century derived from that source. Further collections were made in Nepal during the years up to the publication of George Taylor's monograph in 1934. Certainly the species was more common in cultivation then than today, despite fresh introductions of seed from time to time.

At least two forms were recorded in cultivation. One, generally attributed to a Capt. Bailey (subsp. *grandiflora*) introduction from SE Tibet, has sky-blue flowers and is generally monocarpic and is often referred to in the literature as 'Bailey's Form'. The other is generally more robust and has deep blue or purple-blue flowers and is usually polycarpic; this probably stems from more than one introduction. In recent years further seed has been introduced into cultivation, particularly from the Alpine Garden Society expedition to Sikkim in 1983 and from various Bhutan collections, particularly those of Sinclair & Long from the Royal Botanic Garden Edinburgh. Plants from these sources are still in cultivation, although by no means common. The more robust forms of subsp. *grandiflora* with their large and enticing deep blue or purple flowers are as beautiful as any of the Himalayan blue poppies. The slighter forms encompassed within subsp. *simplicifolia* often have paler, smaller and more nodding blooms: at their best they are elegant and delightful plants, but certainly more demure and less brazen than some forms of subsp. *grandiflora*.

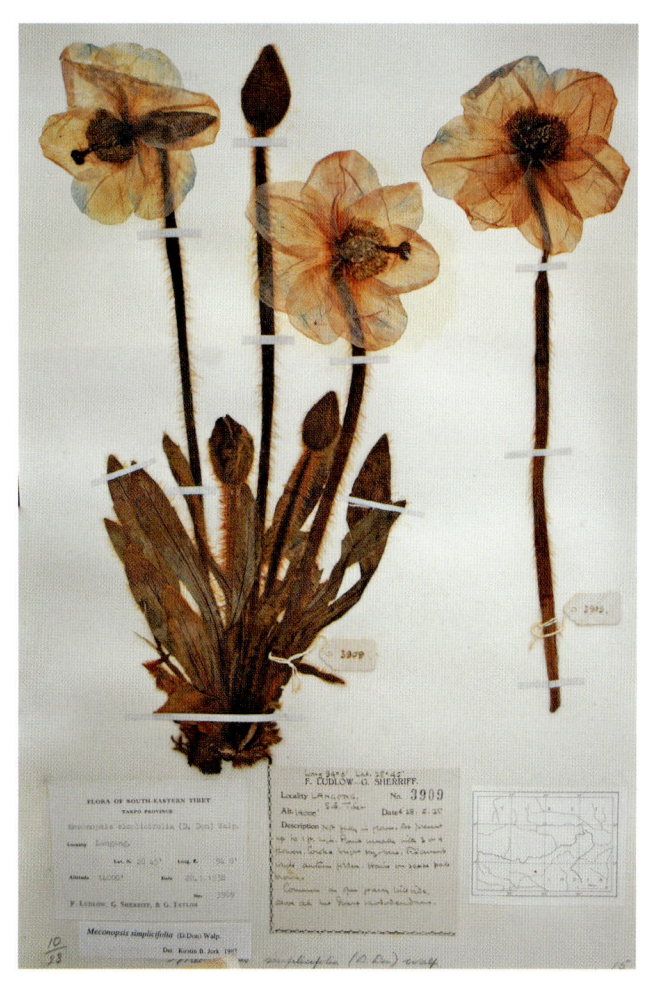

Meconopsis simplicifolia subsp. *grandiflora*: herbarium sheet of *Ludlow & Sherriff* 3909 (BM), Langong, SE Tibet.

Meconopsis simplicifolia subsp. *grandiflora*; half-tone photo of a colony from which *Ludlow, Sherriff & Taylor* 3909 (Langong, SE Tibet, 4270 m) was derived. PHOTO: FRANK LUDLOW

As in all the blue poppies, flower colour can vary considerably and this is especially noticeable in cultivation. However, it is clear that it can also vary a lot in the wild and the flower colour may vary according to aspect and soil conditions but perhaps most importantly by the age of the flowers. Several collections remark on this fact, e.g. 'It sometimes opens red …. but when fully open is a fine dark rich blue; filaments purple-blue; anthers golden.' From field notes of *Ludlow, Sherriff & Hicks* 19205.

In cultivation *Meconopsis simplicifolia* has been crossed with *M. grandis* giving rise to the plant described as *M. ×hybrida*. This was raised at Bodnant Garden in North Wales by F. C. Puddle (see *Gardener's Chronicle*, Ser. 3, 77: 406 in obs.) and the plant was said to be robust ("over four feet") with large 6 in (15 cm) flowers of brilliant blue. Unfortunately, no full description was drawn up of this interesting hybrid, which soon disappeared from cultivation despite being polycarpic. Furthermore, herbarium specimens were not prepared of it; however, it should be possible to recreate this interesting hybrid.

32. MECONOPSIS NYINGCHIENSIS

Meconopsis nyingchiensis L. H. Zhou, *Bull. Bot. Lab. N.E. Forest. Inst., Harbin* 1980 (8): 98 (1980) f. 2. Type: SE Xizang, Nyingchi (Shejilashan), 3600–3800 m, 1972, *Xizang Medic. Herb. Exp.* 3446A (NWBI, holotype).

DESCRIPTION. *Perennial herb* to 50 cm tall in flower, with a short conical taproot branching beneath, beset at the base by withered leaf-base remains and barbellate bristles. *Leaves* in basal rosettes, green above, paler beneath, the outermost spreading, the inner ascending to erect with the apex somewhat recurved, the lamina oblancolate to elliptic-oblanceolate or more or less spathulate, 3–8 × 0.8–2 cm, papery, the base cuneate and decurrent onto the petiole, the apex obtuse to subacute, margin entire to remotely and shallowly crenate, with sparse whitish or pale tawny-coloured, barbellate hairs overall, more densely so on the dorsal surface towards the apex, 3-veined from base to more than halfway, sometimes almost to the apex, especially prominent beneath; petiole greenish or purplish, 2–10 mm long, 2–3 mm wide, broadening at the base, beset with tawny-coloured barbellate bristles, especially towards the base. *Flowers* broad-campanulate, half-nodding (semi-nutant), 1–2 per leaf-rosette, scapose, 8–10 cm diameter (when flattened); buds erect, becoming semi-nutant as they reach maturity, the sepals densely patent-bristly. *Pedicels* (scapes) fairly robust, erect, to 50 cm long, 5–6 mm diameter, covered with sparse, somewhat deflexed barbellate hairs, denser towards the top. *Buds* ovoidal, the 2 sepals obovate to subrounded, densely yellow-setose. *Petals* 6, pale blue to pale purple, broad-oval to obovate, 4–5 × 3–4 cm. *Stamens* numerous, half the length of the ovary, the filaments filiform, c. 10 mm long, whitish; anthers yellow, narrow-elliptic, 1.5–2 mm long. *Ovary* broad-ellipsoidal (flask-shaped), pale green, with sparse to rather dense yellowish bristles, subappressed at first, narrowed at the top to 3–4 mm diameter but without an obvious style, the green stigma consisting of c. 5 broad oval, infolded flanges, later apparently cupuliform, pleated and waved towards the perimeter. *Fruit capsule* unknown at maturity, but immature capsule pale green, patent-pubescent, apparently 5-valvate, the persistent stigma browning on ageing. (Fig. 17).

DISTRIBUTION. SE Tibet (Nyingchi[27]; Shejilashan); 3600–3800 m.

---

[27] The town of Nyingchi lies on the Nyang Chu close to and a tributary of the mighty Tsangpo river, to the west of Namcha Barwa. However, Nyingchi is also the name of the large accompanying territory that stretches west of the Serkhyem La towards the Mi La, well to the east of Lhasa. The photographs of *Meconopsis nyingchiensis* were taken close to the main Chengdu-Lhasa highway that passes through the town of Nyingchi.

HABITAT. banks, open shrubberies, woodland margins.
FLOWERING. July.

Meconopsis nyingchiensis is probably the least known of any species of *Meconopsis*, existing as far as I am aware only as the original type collection which was gathered by a Chinese team close to Nyingchi in 1972. The primary aim of the expedition, from the North-west Botanical Institute, was to collect medicinal herbs. Since then few further sightings have been made; however, it has been photographed in the Nyingchi area on at least two occasions in the intervening years, first by the Canadian Pam Eveleigh and second by the German Dieter Zshummel. Unfortunately, both photographs are of plants in the late flowering phase, but they do show the developing fruit capsules with their telltale flanged stigma. The stigma is unique in the genus, consisting as far as one can observe of 5 oval, slightly overlapping flange-like lobes. There is no obvious style. However, the two plants photographed differ in a number of respects: in one the leaves are ascending and clearly 3-veined for much of their length and the flowers are deeply cupped, while in the other the leaves spread out almost horizontally and are only 3-veined towards the base, and the flowers are rather flat. This begs the question as to

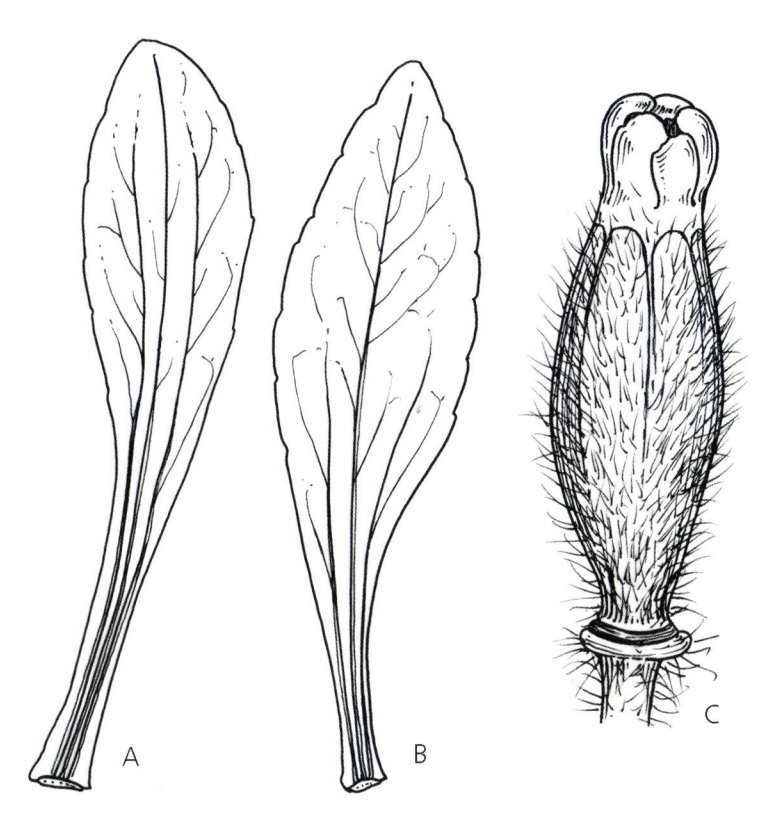

FIG. 17. *Meconopsis nyingchiensis*: **A–B**, basal leaves from separate individuals; **C**, fruit capsule. Leaves × 2; capsule × 3. DRAWN BY CHRISTOPHER GREY-WILSON

Above left: *Meconopsis nyingchiensis*, flower; Serkhyem La, SE Tibet, 4550 m.   PHOTO: PAM EVELEIGH

Above right: *Meconopsis* cf. *nyingchiensis*; SE Tibet.   PHOTO: DIETER ZSCHUMMEL the origin of these plants. While it is impossible at the present time to prove otherwise, I strongly suspect that this taxon is of hybrid origin. It is difficult to speculate on putative parents although *M. simplicifolia*, with which *M. nyingchiensis* clearly affiliates, is certainly one and is well known from many localities in south-eastern Tibet. The other possible parent is far more circumspect: *M. baileyi*, *M. sulphurea* and *M. speciosa* are all found in the surrounding mountains.

At present, without detailed field observation, it is impossible to say more about this interesting plant and until such time that it can be proved otherwise it must remain as a rare and little known species.

33. MECONOPSIS ×HARLEYANA

Meconopsis ×harleyana G. Taylor, *New Fl. & Silva* 6: 48 (1933). Type: Cult. Edinburgh Bot. Gard. 1933 (BM, holotype).
[syn. *M. simplicifolia* Bailey's var. × *M. integrifolia* Brooker, *Gard. Chron.*, Ser. 3, 83: 353, fig. 167 (1928); *M. simplicifolia* var. *eburnea* Hort. ex R. L. Harley, *New Fl. & Silva* 4: 199, fig. 74 (1932)]

Meconopsis ×*harleyana*, a hybrid between *M. integrifolia* and *M. simplicifolia*, first arose spontaneously in the garden of R. L. Harley in 1926 in England. It was at first regarded as a yellow form of *M. simplicifolia*, but its hybrid origin soon became suspected, especially since both *M. integrifolia* and *M. simplicifolia* were growing in close proximity in his garden. It is upon this plant that George Taylor based his description, noting that 'it was raised from seed saved from Bailey's form of *M. simplicifolia* and flowered for the first time in 1925' and 'Seeds have been produced [of the hybrid sic], though sparingly, and the hybrid appears to be so fixed as to breed true.'

Taylor also associates the hybrid with Kingdon Ward's 'Ivory Poppy' collected on the Temo La in south-eastern Tibet in 1924. This assertion seems to have been backed-up by additional collections made by Ludlow, Sherriff & Taylor on the Sang La, also in south-eastern Tibet, in June 1938. George Taylor told the story (see *A Quest for Flowers* p. 202):

> "The most conspicuous plant of the moorland was *Meconopsis integrifolia* whose fountain of yellow flowers rose elegantly through the carpet of *Rhododendron laudandum* and *Potentilla fruitcosa* var. *grandiflora*. At this elevation, about 13,000 feet, the plants were up to four feet in height and very homogeneous in character. All had prominent, slender, cylindrical styles and the ovaries were densely covered with golden-brown appressed hairs. *Meconopsis simplicifolia* grew in association but was not so prominent, as its flowers barely showed above the hummocky rhododendrons. Colonies of the species grew in small clearings. But the most exciting plant was one [5053, a b & c] bearing pure white to pale yellow flowers, which occurred sporadically in association with *M. integrifolia* and *M. simplicifolia*. At a glance this was recognised as Kingdon Ward's Ivory poppy, which was discovered in 1924 on the nearby Temo La. In my 'Account of the Genus Meconopsis' I tentatively assigned this plant to *M.* ×*harleyana*, but little did I imagine then that I would have the opportunity of confirming my opinion in the field. The hybrid, in habit and flower colour, was easily picked out and we counted about twenty individuals in the area where the plants overlapped. The Ivory poppy was not found isolated from both parents. In the majority of the specimens the flowers were borne on basal scapes and the leaves were usually notched, characters which have been derived from *M. simplicifolia*. The flowers were up to 5 inches in diameter, and in texture, shape and colour showed the influence of *M. integrifolia* though occasionally the petals had a faint flush of mauve from *M. simplicifolia*. In its ovary characters, the plant was intermediate between the parents. The capsule of *M.* ×*harleyana*, in contrast to the turgid state of those of the parents, were narrow and spindly and when open they showed rows of abortive ovules. All the plants examined were monocarpic and by September had completely withered with the gaping capsules containing powdery undeveloped seeds."

The name *Meconopsis* ×*harleyana* applies only to hybrids between *M. integrifolia* and *M. simplicifolia*. This creates a bit of dilemma for, while it is almost certain that these two species were involved in the first instance, the parentage of the hybrid in the wild is not. Since George Taylor's 1934 monograph of the genus, *M. integrifolia* has been divided into several species, with the re-instatement of David Prain's *M. pseudointegrifolia* and the addition of *M. sulphurea*. While *M. integrifolia* sensu stricto was well-established in cultivation early in the twentieth century primarily from Sichuan collections of E. H. Wilson, neither *M. pseudointegrifolia* nor *M. sulphurea* were in cultivation, in fact the latter was probably not introduced until the 1960s. That being so, then the parentage of *M.* ×*harleyana*, a garden hybrid, cannot be disputed and the characters of the hybrid, midway between the parent species, bear this out. The wild plants assigned to the same hybrid by George Taylor do not match the garden hybrid in some respects, especially in the extraordinary length and features of the fruit capsules. Taylor in his description of the Ivory Poppy in the wild clearly states that the ovaries of *M. integrifolia* all had 'prominent, slender cylindrical styles' a particular feature of *M. sulphurea*. Research has in fact shown that only *M. sulphurea*, not *M. integrifolia*, is to be found on the Sang La, with *M. pseudointegrifolia* found much further east and in NW Yunnan. The parentage of the Ivory poppy is therefore *M. simplicifolia* × *M. sulphurea*. The name *M. ×harleyana* cannot therefore be applied to it and a new name must be created as follows:

34. MECONOPSIS ×KONGBOENSIS

Meconopsis ×kongboensis Grey-Wilson non Hort. ex R. L. Harley, **hybr. nov.** Plant intermediate between *M. sulphurea* and *M. simplicifolia*, the leaves 3, occasionally 5-veined, the flowers scapose with cream petals; fruit capsule developing, fusiform-ellipsoid, sterile. Type: SE Tibet, Kongbo Province, Sang La, 29°35'N, 94°43'E, *Ludlow, Sherriff & Taylor* 5053 (BM, holotype & isotypes; E, isotype).

DESCRIPTION. *Polycarpic plant*, forming dense tufts, the crown invested with a felted mass of old petiole remains accompanied by numerous barbellate rufous bristles. *Leaves* ascending to erect, narrow elliptic-oblanceolate to oblanceolate, 4.5–18 × 0.9–2.6 cm, gradually attenuate at the base into the petiole, subacute at the apex, the margin subentire to slightly and remotely sinuate or toothed, 3- occasionally 5-veined towards the base; petioles 5–12 cm long, narrow strap-like, 5–8 mm wide. *Flowers* broad bowl-shaped, semi-nutant, 8.5–13.5 cm across. *Scapes* (pedicels) up to 6 per plant, 22–50 cm long (to 69 cm long in fruit), moderately to densely, patent, slender-setose, densest immediately beneath the flowers, the bristles barbellate, rufous. *Buds* subnutant, broad-ellipsoidal, c. 3.2 × 1.9 cm, the sepals densely bristly. *Petals* 5–7, pale creamy yellow, obovate, 4.7–6.3 × 2.8–4.2 cm. *Stamens* numerous, the filaments the same colour as the petals, linear, 13–16 mm long, the anthers orange-yellow, 2–2.5 mm long. *Ovary* oblong-ovoid, densely ascending-setose, the bristles rather short, semi-patent, yellowish brown, the style 9–10 mm long, the stigma oblong-clavate, 6–10 mm long. *Fruit capsules* fusiform-ellipsoidal, 2.6–2.8 cm long, 0.8–1.1 cm diameter, 5–7-valved, ascending-setose, becoming semi-glabrescent at maturity, the persistent style (including the stigma) to 15 cm long, the capsule apparently not developing fertile seeds, or only sparingly so.
DISTRIBUTION. SE Tibet (Kongbo, Nyima La, Rong Chu, Sang La, Temo La); c. 4210 m.
HABITAT. Scrub and boulder fields.
FLOWERING. Late June–July.
SPECIMENS SEEN. *Ludlow, Sherriff & Taylor* 5053 (BM, E), 5053a (BM, E), 5053b (BM, E), 5053c (BM, E).

Meconopsis ×kongboensis (right) with *M. simplicifolia* subsp. *simplicifolia* (left); SE Tibet. PHOTO: KENNETH COX

Ludlow & Sherriff added a full note to their herbarium sheet number 5053, the type specimen, as follows; "With both parents, growing mostly in potentilla and dwarf rhododendron scrub, frequent. Leaves noticeably notched. Scapes basal, sometimes leafy. Petals texture of *M. integrifolia*, 5–7, pure white to sulphur-yellow. Scapes all basal. Filaments same colour as petals, anthers orange-yellow. Ovary light green with spreading light brown hairs — not so densely as *M. integrifolia*, and more densely than *M. simplicifolia*. Style cylindrical at base, ridged below clavate stigmas; stigmatic rays green at first, then brown."

In referring to *Meconopsis integrifolia* the authors imply one of the parent species. This as has been explained (see p. 202) is in fact *M. sulphurea*. That *M. sulphurea* is the second parent is easy to establish as it is recorded from the sites where the hybrid was collected. In fact *Ludlow, Sherriff & Taylor* 5043 was collected on the Sang La, the type locality of *M. ×kongboensis*, while 5043a was collected on the Nyima La, close by.

Meconopsis ×kongboensis is one of the few substantiated *Meconopsis* hybrids found in the wild. The hybrid comes more or less midway in character between the parent species. It is clearly a very beautiful plant with its large ivory flowers borne on substantial scapes. This together with the fact that the plants are polycarpic make it a handsome and desirable plant for the garden. To my knowledge it has rarely been seen in recent years.

series Puniceae

This series is one of the most highly distinctive in the genus, containing as it does two of the most familiar species known to gardeners, *Meconopsis punicea* and *M. quintuplinervia*. The series is characterised primarily by polycarpic, occasionally (at least in the wild) monocarpic, tufted plants with spear-shaped entire leaves with 3- or 5 principal longitudinal veins. The flowers are solitary and scapose, being borne high above the foliage on slender, often rather wispy stalks. The flowers are generally 4-petalled, nodding and more or less campanulate. Both species are centred on the area where Gansu, Szechuan and Qinghai meet, although *M. punicea* tends to have a rather more southerly distribution. *M. quintuplinervia* is the only species of *Meconopsis* to be found in both Shanxi and Shaanxi provinces.

KEY TO SPECIES OF SERIES PUNICEAE

1. Petals crimson, rarely white, considerably longer than broad . . . . 35. **M. punicea** (W & NW China and E Tibet)
+ Petals blue, purple-blue, lavender, lilac or violet, rarely much longer than broad . . . . 36. **M. quintuplinervia** (NW China, NE Tibet)

35. MECONOPSIS PUNICEA

Meconopsis punicea Maxim., *Fl. Tangut.* 1: 34 (1889). Type: China, NE Tibet (Qinghai today!), 1884, *Przewalski s.n.* (K); N Sichuan, 1885, *Potanin s.n.*, China (E, K, P, W). Lectotype chosen here: *Potanin s.n.*, Szetschuan septentr. 1885 (K; isolectotypes E, P, W). [syn. *Cathcartia punicea* Maxim., *Fl. Tangut.* 1:35, iii, tab. 23, figs. 12–21 (1889), nom. ambig.; *Meconopsis punicea* var. *elliptica* Z. J. Cui & Y. S. Lian; *M. punicea* var. *glabra* M. Z. Lu & Y. S. Lian.]

DESCRIPTION. A tufted perennial, sometimes monocarpic, to 58 cm tall in flower (to 75 cm in fruit), but often only half that height, with a slender taproot to 5 cm long, bearing a mass of fibrous roots, the neck of the plant invested with persistent fibrous leaf bases beset with yellowish or brown, occasionally ginger, barbellate bristles. *Leaves* in solitary or clustered basal rosettes, dying away in the autumn to small resting buds at ground level, yellowish or bluish green; lamina narrow-obovate to oblanceolate, (1.5–)3–21 × 0.9–4 cm, with 3 (–5) longitudinal veins, attenuate at the base and decurrent onto to the petiole, the apex acute to subobtuse, margin entire, covered overall in sparse to dense yellow-brown barbellate, patent bristles; petiole (1.7–)2.2–18(–24) cm long, somewhat expanded and sheathing at the base. *Flowers* solitary, 1–6 per leaf rosette, borne on simple scapes, nodding, campanulate, 6–10 cm long. *Pedicels* (scapes) erect, 20–55 cm long, 4–5 mm diameter, somewhat ribbed, rather slender, the same colour as the leaves but suffused with red towards the top, covered in yellow-brown barbellate, rather soft, slightly deflexed bristles. *Buds* nodding, ovoid to oblong, 18–28 mm long, densely yellow-brown pubescent, with barbellate hairs. *Sepals* ovate, 1.5–4 cm long, barbellate on the outside. *Petals* 4 (occasionally 6–8), deep carmine, occasionally dull scarlet, carmine-pink, often darker towards the base, rarely white, satiny, rhombic-elliptic to ovate-elliptic, occasionally linear-elliptic, (4.5–)5.6–10.4 × (1.5–)2.1–5 cm, often somewhat crumpled, especially shortly after the flowers open, diverging towards the apex, sometimes spreading widely apart, the apex subacute to obtuse or rounded, margin entire. *Filaments* the same colour as the petals, filiform, flattened, 10–30 mm long, the anthers yellow or brownish yellow, 3–4 mm long. *Ovary* ovoid or oblong, 10–30 mm long, densely covered in ascending to appressed yellow-

Meconopsis punicea in habitat; Gongan Len Pass, Huanglong, NW Sichuan, 3600 m.
PHOTO: CHRISTOPHER GREY-WILSON

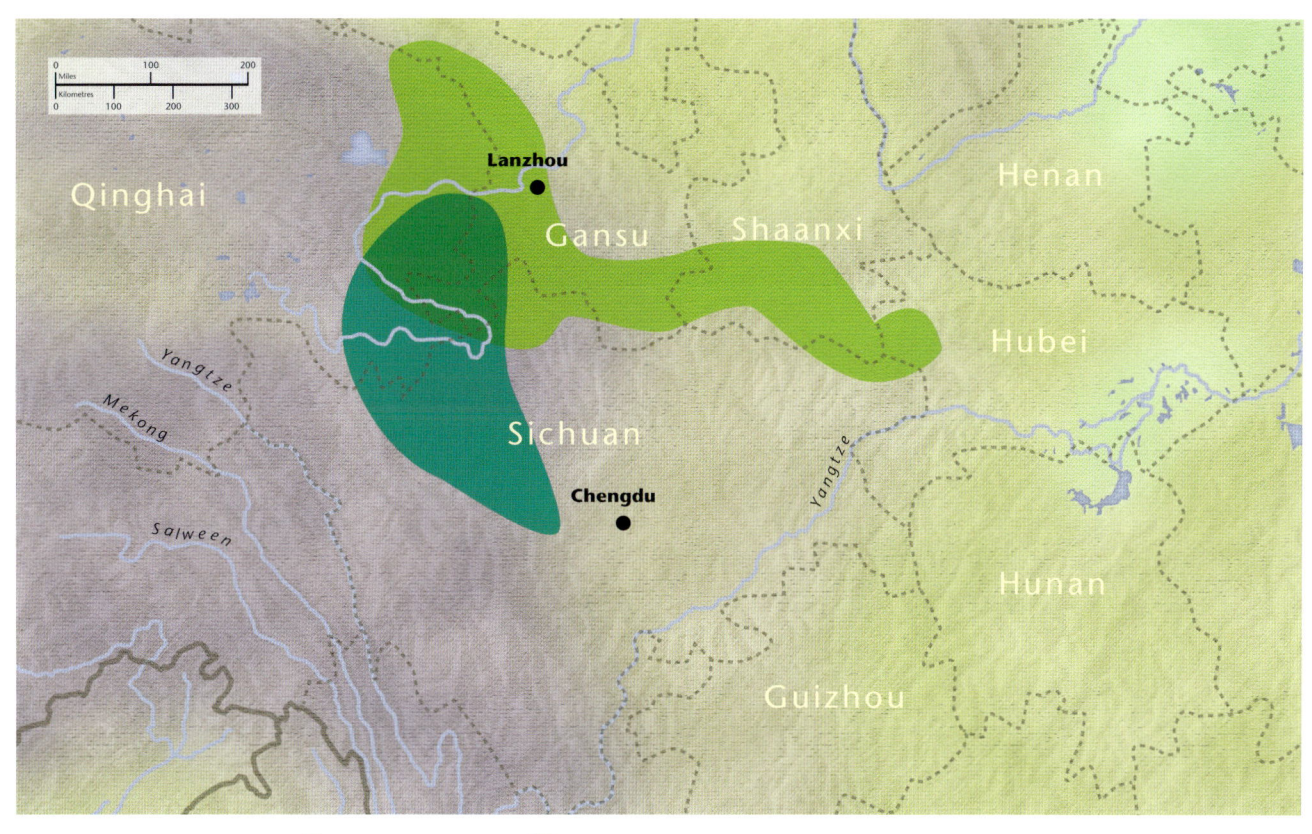

MAP 13. Species distribution: ■ *Meconopsis punicea*, ■ *M. quintuplinervia*.

brown barbellate bristles; style very short, not more than 1 mm long, or obsolete, the stigma capitate, 4–5.5 mm diameter, white, with 3–6 radiating lobes. *Fruit capsule* oblong-ellipsoidal, 18–25 × 8–14 mm, densely yellow-brown ascending barbellate-setose, 3–6 (often 4)-valved, splitting for a short distance from the apex, at first nodding but soon becoming erect as the fruit develops. *Seeds* densely papillose, later appearing pitted.

DISTRIBUTION. N & W China (SW Gansu, S & SE Qinghai, W & NW Sichuan); (2800–)3400–4600 m (Map 13).

HABITAT. Mountain moorland, shrubberies, margins of woodland, amongst low rhododendrons or willows (*Salix*), drier parts of marshes.

FLOWERING. June–August.

SPECIMENS SEEN. *Bartholomew & Gilbert* 375 (E), 1216 (BM, E); *Botanical and Forestry Department Hong Kong* 1379 (K); *Boufford et al.* 33523 (E); *J. W. Brooke* s.n. (K) *R. C. Ching* 818 (E); *W. P. Fang* 4021 (E, K); *Farrer* 175 (E); *Ho, Bartholomew & Gilbert* 375 (BM), 1216 (BM); *Kunming Edinburgh Exped. Sichuan* 100 (E); *Long et al.* SBQE 415, 691 (E); *Potanin* s.n. Szetschuan septentrionale 1885 (E, K); *Przewalski* s.n. Tibet borealis (K); *Purdom* 698 (E, K); *Rock* 12195 (BM), 12423 (E, K), 12840 (K), 12942 (BM), 13053 (K), 13642 (E, K),

Meconopsis punicea, with unusually broad petals; near Hanguan, NW Sichuan, 3950 m. PHOTO: CHRISTOPHER GREY-WILSON

35. MECONOPSIS PUNICEA

Meconopsis punicea, flower at bud burst stage; Balangshan, above Wolong, W Sichuan, 4110 m. PHOTO: HARRY JANS

Meconopsis punicea, petals fully developed before sepal drop; Mengbi Pass, Barkham, W Sichuan, 3990 m. PHOTO: HARRY JANS

14506 (E, K); *SBQE* 64 (E), 65 (E), 216 (E), 244 (E), 370 (E), 415 (E), 476 (E), 570 (E), 691 (E), 724 (E); *H. Smith* 3837 (BM), 4280 (BM); *E. H. Wilson* 1378 (K), 3031 (BM, E, K), 3168 (BM, K), 9103 (K).

Meconopsis punicea was discovered in north-eastern Tibet (now Qinghai province) in 1884 by the Russian Grigori Przewalski and in the following year further material was collected by another Russian, Nicolai Potanin in north-western Sichuan. These combined collections laid the basis for the description of *M. punicea* penned by Maximowicz. However, he was clearly uncertain as to where the species truly belonged, for it was described as a species of *Meconopsis* but the accompanying plate clearly refers it to the genus *Cathcartia*.

This species is clearly closely allied morphologically to *Meconopsis quintuplinervia*. The two are obviously similar in leaf and general habit but differ significantly in flower colour and the size and proportions of the petals. In addition, both species bear a very short or more or less obsolete style, and while the filaments of *M. punicea* are linear and flattened, those of *M. quintuplinervia* are filiform throughout. One important difference is observable in the rootstock, that of the former has a taproot surmounted by a crown from which the shoots arise, that of the latter is fibrous with each leaf-rosette arising from a tuft of fibrous roots.

Meconopsis punicea is certainly one of the most distinctive members of the genus with its sumptuous satiny red pendent blooms held high above the basal leaf-rosettes on sinuous stalks. It has, as far as records go, a rather narrow north-south distribution, being found north as far as the mountains south of Amne Machin (Anyémaqén) in Qinghai Province and the mountains of south-western Gansu, southwards to western Sichuan, its southernmost locality being the Balangshan, west of Chengdu, close to the famous giant panda reserve at Wolong. It is primarily a plant or open, exposed mountain moorland where its drooping flowers flutter enticingly in the slightest breeze.

There is no doubt that this striking species has awed all those who have seen it, especially growing on its bleak moorland habitats in those remote regions close to the Tibetan frontier.

"No species has drawn such superlative and almost extravagant epithets as *M. punicea*. Those who have had the privilege of seeing the species in its natural habitat, where it brightens the copses and meadows with splashes of the most vivid colour, are unanimous in their eulogies, and one feels from the description that the ecstatic joy of a first acquaintance with the plant is warranted."

[George Taylor, *Genus Meconopsis*; 55, 1934]

"In cultivation the species is commonly more floriferous, bearing up to thirty blooms on each plant, and these sometimes show multiplication in the number of petals. Farrer has stated that *M. punicea* is monocarpic, but it is recorded that plants have flowered in two successive years in a Scottish garden and have all the appearance of continuing to do so." [GT: *op. cit.*: 55]

"*Meconopsis punicea* far surpasses all English description and all English effort, as you begin to see it, bloodily flaunting in and out of the scantier coppice in the Tibetan valley-bottoms opposite Jô-ni, first appearing at some 10,000 feet and thence ascending to the high grass-ridges, haunting the glade-edges and light bushery of the glen, until in the open hayfields it rages furiously over all the hill, between 11,000 and 13,000 feet, dappling the distances with blood like any poppy in an English field; and, in the little grassy hollows along the crests, hovering in flapping flags of vermilion above the rippling sea of golden-eyed purple Asters. For in England those dim flags of scarlet flop; on the Tibetan Alps they blaze and flap — huge expanded stigg goblets of wave-winged butterflies of incandescent blood, that compel from me a palinode to my previous rather cold description of *M. punicea*, as alone I had hitherto known it, showing no trace of its own true sinuous and serpentine magnificence ...." (The Blood Poppy)
[Reginald Farrer, *The English Rock Garden*: 507, 1928]

"Mixed with this scrub are herbs in great variety, the Poppyworts (*Meconopsis*) being particularly abundant. Possibly the commonest herb between 12,500 feet and 14,000 feet is *Meconopsis punicea*, a lovely species having large dark scarlet nodding flowers. (It was from near this vicinity that I succeeded in introducing this plant in 1903.) The violet-blue flowered *M. henrici* [almost certainly *M. psilonomma*] is common between 13,000 feet and 14,000 feet. From 11,500 feet to 13,000 feet the gorgeous *M. integrifolia*, growing 3 feet tall, with its peony-like, clear yellow flowers 8 to 11 inches across, occurs, but is not plentiful."
[E. H. Wilson, *A Naturalist in Western China*: 138, 1986 edition]

Meconopsis punicea came into cultivation very early in the nineteenth century, seed having been collected by E. H. Wilson in 1903. Wilson had made the journey on this occasion especially to gather seed of this species. However, although plenty of seed was introduced and a plant flowered in a private garden the following year, *M. punicea* quickly died out of cultivation. However, some years later Joseph Rock re-introduced it from seed collected in Gansu and on this occasion it was far more successful. Thereafter it has had a rather chequered history, disappearing from cultivation in the middle of the last century. However, further introductions in more recent times, notably that of Cox & Hutchinson in 1986 (under the number *C.H. & M* 2586) has led to its better permanence in gardens. From this seed two forms appeared, one with ginger hairs and rather open leaf-rosettes, the other plainer, with pale hairs and more upright leaves. Although *M. punicea* is not particularly difficult to grow it is tricky to maintain. Many plants

Meconopsis punicea; Deng Zhanzwa, NW Sichuan, 3500 m. PHOTO: HARRY JANS

Meconopsis punicea; S of Aba, NW Sichuan, 4000 m.
PHOTO: CHRISTOPHER GREY-WILSON

prove not to be polycarpic (perennial): this is partly because they tend to flower over a long season, often producing their last flowers in the late autumn, with the result that plants become exhausted and unable to survive the following winter. Added to this is the fact that cultivated plants tend to have low fertility and like many *Meconopsis* plants are wholly or partly self-incompatible. The best means of ensuring seed set, even in a favourable summer, is to grow plants in groups. Division of multi-headed clumps is not possible as all spring from a single taproot as has been noted above. Should this be damaged or rot then the entire plant is lost. This is very different to the situation found in *M. quintuplinervia* in which vegetative division is straight forward in many instances because of the fibrous nature of the rootstock.

Producing seed is a hit and miss affair in cultivation and after wet cool summers little if any is produced even if the fruit capsules develop and this requires having a number of mature plants in close proximity. In the wild the species enjoys ample moisture during the summer months when it is in flower, even though many grow on the periphery of the monsoon belt. In addition, during these months when flowering and seed set is taking place the air is generally buoyant and warm, even hot at times. In cultivation lack of fertility can be put down to a reduction in pollen produced, in a lack of suitable pollinating insects and often less than ideal weather conditions at the vital period from flowering to fertilisation. In a bad year no seed is produced: this has been noted in the past and is the prime reason why the species disappeared from cultivation in the mid twentieth century. The answer today is surely to grow plants where there is a greater expectation of seed set (e.g. Tromso Botanic Garden or in favoured gardens in the Scottish highlands) and to ensure that some seed is stored under proper conditions in a seed bank. The other hope is to develop some truly perennial forms that will last in the garden for a number of years. A number exist already but are difficult to propagate and seed tends to produce mainly monocarpic offspring.

Those lucky enough to have seed are advised to sow it in the autumn or early winter for germination the following spring. As always thin sowing is essential, especially so as to avoid any problems with damping-off. The seed appears to have a built in dormancy factor and spring-sown seed will fail to germinate until the following spring in most instances. Pricked on and well fed, the young plants will grow away rapidly, some even starting to flower the same autumn; however, most will wait until the second year. Fortunately, the species is quite frost-hardy and early-produced flower buds are not easily damaged. *Meconopsis punicea* has the propensity to flower all through the summer, continuing to throw up additional flowers into the autumn, even in the early winter on occasions. Plants grow slower on a more meagre diet and will probably not flower until their third year. However grown, this is a superb plant and one well worth persevering with. It is perhaps a pity that no white form has appeared in cultivation, although they are occasionally seen in the wild and these have been assigned to forma *albiflora*, from a plant collected in south-eastern Qinghai.

There is a tendency for plants to produce flowers with multiple petals in cultivated specimens and one such plant has been developed by Ian Christie in his Scottish nursery at Kirriemuir.

forma albiflora L. H. Zhou, *Bull. Bot. N.E. Forest Inst., Harbin* 8: 98 (1980). Type: Qinghai, Jigzhi, Golog steppe station, *L. H. Zhou* 105 (NWBI, holotype).

Plant with relatively small flowers, the petals white, 5–6 cm long, 2–3 cm wide; filaments 8–13 mm long.

36. MECONOPSIS QUINTUPLINERVIA

Meconopsis quintuplinervia Regel, *Gartenflora* 25: 291, tab. 880, figs. b, c, d (1876). Type: N China, Gansu, *Przewalski*, cult. St Petersburg Botanic Garden (LE, holotype).

[syn. *M. punicea* var. *limprichtii* Fedde, *Repert. Spec. Nov. Regni Veg.* 17: 197 (1921). Type: Shanxi, Tsing-ling-schan, Malten am Wang-sso-nai, obchalt des Passes Da-tsing-ling, sudlich Hsi-an-fu, 2300 m. Fl. Fr 20.VIII, 1916, *Limpricht* 2665 (LE); *M. biloba* L. Z. An, Shu-Y. Chen & Y. S. Lian, *Novon* 19 (3): 286–288 (2009). Type: Gansu, Lingtan County, Lianhuoshan, 3050 m, July 2004, *L. Z. An* 2047243 (LZU, holotype).]

DESCRIPTION. *A patch-forming, stoloniferous perennial* to 36 cm tall in flower, with few to many leaf-rosettes, densely covered at the base by a collar of yellowish or brownish barbellate bristles and petiole remains. *Leaves* all basal, to 25 cm long overall, the lamina elliptic to lanceolate, oblanceolate or more or less obovate, 2.2–14.8 (–18) × 0.6–3 cm, with 3–5 longitudinal veins, gradually attenuate into the petiole, acute to obtuse at the apex, margin entire, sometimes slightly undulate, covered overall in straw-coloured bristles, sometimes rather sparsely so; petiole 7–9.8 cm long, somewhat broadened and flattened at the base and with membranous margins. *Flowers* solitary, generally only one per leaf-rosette, scapose, nodding, narrow to broad bells to 4.8 cm long. *Pedicels* (scapes) erect, rather wiry, to 32 cm long (to 50 cm long in fruit), 1–2 mm diameter, covered in patent to slightly reflexed bristles, sometimes densely so, especially towards the top. *Buds* nodding, broad-ovoid. *Sepals* oval, c. 20 × 15 mm, sparsely bristly on the exterior. *Petals* 4, occasionally 5–6, lavender-blue, purplish, mauve, mauve-purple, azure blue, amethyst, rosy-lilac, rarely white or white with a lavender base, rounded to oval or elliptic-oblanceolate, 28–50 × 16–42 mm, slightly fluted or ruffled, with a rounded apex,

Above: *Meconopsis quintuplinervia*; S of Jigzhi, SE Qinghai, 3950 m. PHOTOS: CHRISTOPHER GREY-WILSON

occasionally notched or shallowly bilobed. *Stamens* with filiform filaments, 8–9 mm long, the same colour as the petals or whitish; anthers 1–1.5 mm long, buff, yellowish or white. *Ovary* ovoid to oblong, covered in dense, appressed fulvous bristles, rarely glabrous; style 1–6 mm long, sometimes more or less obsolete; stigma capitate with 3–6 radiating lobes. *Fruit capsule* erect, oblong-ellipsoidal, 15–25 × 9–12 mm, 3–6 (often 4)-valved, sparsely to densely covered with ascending to appressed barbellate bristles, occasionally glabrous or subglabrous, the bristles sometimes pronouncedly swollen at the base, occasionally glabrous. 2n = 76, 84.

DISTRIBUTION. N China: S & SW Gansu, W Hubei (Shennongjia), NE, E & SE Qinghai, W Shanxi, S Shaanxi, N & NW Sichuan, NE Tibet; 2300–4600 m. 12,000–14,000 ft (Map 13).

HABITAT. Alpine meadows, grassy slopes, open forest, often of birch (*Betula*) or larch (*Larix*), open shrubberies, occasionally on moist rocky slopes.

FLOWERING. June–September.

var. quintuplinervia

DESCRIPTION. Ovary and fruit capsule densely appressed-setose.

DISTRIBUTION. As above except for Ningshan, Shaanxi province.

SPECIMENS SEEN. *R. C. Ching* 519 (E); *Farrer* 118 (BM, E, K); *Rev. Jos Giraldi* s.n., Shensi septentr. (K), 766 (K); *Ho, Bartholomew & Gilbert* 36 (BM); *Rev. Fr. Hugh* s.n. 1898–1900 (BM); *Licent* 4353 (BM, K), 4655 (BM, K); *Long et al.* SBQE 690 (E); *Potanin* s.n. 1885 (E); *Przewalski* s.n. Tangut, Kansu (K); *Purdom* s.n. Minchow, W Kansu (K), 118 (E), 414 (E, K), 709 (E, K); *H. J. Ridley* 100 (E); *Rock* 12608 (BM), 12611 (E, K), 12742 (E, K), 13046 (E, K), 13620 (E), 14144 (BM), 14224 (E, K), 14309 (E, K), 14462 (K); SBQE 25 (E), 68 (E), 69 (E), 414 (E), 678 (E), 690 (E); *H. Smith* 3828 (BM), 3845 (BM).

var. glabra M. C. Wang & P. H. Yang, *Bull. Bot. Res.*, Harbin 10 (4): 43 (1990). Type: Shaanxi, Ningshan Xian, Cai zi-ping, 2400–2900 m, 21 June 1989, *Wang ming-chong & Yang ping-hou* 300 and same locality, 16 Sept. 1989, *Yang ping-hou* 301 (Herbarium North-western College of Forestry, syntypes)

DESCRIPTION. Ovary and fruit capsule glabrous.
DISTRIBUTION. Shaanxi (Ningshan).
SPECIMENS SEEN. No authenticated specimens seen.

Meconopsis quintuplinervia is one of the most exquisite members of this alluring genus. With its neat mats of lance-shaped leaves and its dainty nodding bells of blue, mauve, rose-lavender or lilac flowers held high above the foliage on wispy scapes, it is unmistakeable. Caught in the slightest breeze the flowers flutter like exotic butterflies. Amongst the truly perennial species cultivated in gardens it is perhaps the greatest gem. The species comes from the northern range of the genus in the mountains of northern Sichuan, south-eastern Qinghai, southern Gansu and the adjacent provinces to the east, a vast region where the summer monsoon is far lighter than it is in the high mountains of Yunnan and much or western and southern Sichuan. It is often found in the wild growing in association with *M. integrifolia* and *M. punicea* and with the latter it does on occasions hybridise (*M. ×cookei*) in the wild (see p. 214).

The species was discovered by Przewalski in Gansu in the 1870s who sent seed to St Petersburg Botanic Gardens. It was described there from cultivated specimens in 1876. Its introduction to Britain stems primarily from several collections made by Reginald Farrer, also from Gansu, during the 1914–15 period. Besides the normal lavender or purplish forms, Farrer noted albino forms in the wild. Sadly, these have never been seen in cultivation.

Meconopsis quintuplinervia; N of Jigzhi, SE Qinghai, 4050 m.
PHOTO: CHRISTOPHER GREY-WILSON

Interestingly, albino forms of *Meconopsis punicea* have also been noted in the wild but these too have not been seen in cultivation to my knowledge (this latter was described in 1980 by L. H. Zhou as forma *albiflora*).

In the wild *Meconopsis quintuplinervia* stretches over a considerable altitudinal range, being found below the treeline occasionally in grassy glades or above in open shrubberies, but it is most often seen in rocky alpine moorland where it can make substantial clumps. It is the most northerly distributed of any species of *Meconopsis*. Interestingly, the flower colour in the wild, as has been described, ranges from quite pale tones to deep rich ones, sometimes within the same colony, with even an occasional albino turning up. In cultivation, on the other hand, plants bear attractive rather pale lavender flowers and as it is such a good garden plant, one wonders why some of the richer coloured variants have not found their way into our gardens from the wild? Reginald Farrer explored the Gansu mountains during his 1914–15 expedition finding *M. quintuplinervia* to be common in what he called the Da-Tung Alps not far south of the high dry desert lands of Inner Mongolia and this must be as far north as it is found, although it could conceivably be found in the ranges to the north of the Datong He, a tributary of the Huang Shui, the upper reaches of the Yellow River or Huanh Ho. Today, province changes place the Da-Tung Alps (the Dabanshan on modern maps) not in Gansu but in the neighbouring province of Qinghai, the Dabanshan being the high ranges to the north of Xining, the capital.

Reginald Farrer (1928) comments that:

"It is indeed a gracious and lovely thing, with its single bell-shaped flowers of the softest lavender-blue swinging high upon their bare stems above the group of pale-haired, greyish foliage crowded in the turf below. The supremely important point about *M. quintuplinervia*, however, is that it is undoubtedly perennial, and thus forms a grand addition to the garden …….. This beautiful treasure inhabits the fine (as a rule) alpine turf of Kansu [Gansu]-Tibet border between 9000 and 13,000 feet. We first met it, still dormant, amid the snows of Chago-ling; on Thundercrown and all up the Min S'an [Minshan] it abounds, as also, in amazing profusion, in the northerly ranges of the Da-Tung. In the Minchow district it trenches on cultivated land, and there, at the edges of culture-patches on the rounded green hills, it becomes quite unrecognisably splendid in the steep enriched embankment down the slope, waxing into masses of foliage a foot across and almost as deep, with 40 or 50 great swaying vases of lavender all hovering at once, on 1 or 2 feet stems, above the tangle of leaves below ……. on Thundercrown there

Meconopsis quintuplinervia; Jigzhi Shen Shan, SE Qinghai, 4150 m.
PHOTO: HARRY JANS

Meconopsis quintuplinervia; Jigzhi Shen Shan, SE Qinghai, 4150 m.
PHOTO: HARRY JANS

36. MECONOPSIS QUINTUPLINERVIA

Meconopsis quintuplinervia; Jigzhi, Stone Mountain, SE Qinghai, 4100 m. PHOTO: JOHN & HILARY BIRKS

"Will anyone wonder that even I, hating as I do the Wardour Street popular names with which Ruskin tried to 'affubler' such known beauties as Saxifrage and Campanula, should now yield to the same weakness, and try to give my beloved Tibetan Poppy an English name to which she has no right at all? But I hope it is only proleptically that I forge the name Harebell Poppy. I hope that the plants beauty and its charm and its permanence will so ensure its popularity in gardens as to make a popular name inevitable. And, that being so, there will be 'Harebell Poppy' ready made. For indeed, to cherish or even to purchase, a plant called *Meconopsis quintuplinervia* is as impossible as to love a woman called Georgiana: mitigating substitutes inevitably have to be invented, Some there are, indeed, who misprise these, and find them insufficiently 'showy.' Alas, the prevailing fault of *Meconopsis* is not modesty; but it is in modesty that *M. quintuplinervia*, alone of her kind, excels. Not for her the blatant crude enormousness of *M. integrifolia*, the sinister and snaky splendour of *M. punicea*, the hard clear glory of *M. simplicifolia*: her fluttering butterflies aim at a more quiet charm, that only the dull-eyed and deboshed in taste could dream of calling dull."

(Reginald Farrer, *Rainbow Bridge*: 224, 1986 edition)

"So much for the gentle blue-purple of the Harebell as she normally lives, and so deliriously abounds over that highland lawn in the high heart of the Alps [S Gansu]; but what of her varieties? *M. quintuplinervia*, indeed, is a stable species in her variable race, and not one in a million flowers shows any difference. But in the very millions that possess the upper fells of the Da-Tung, varieties at last do arise. Further on towards the Clear Lake there is a broad triangular stretch of the moor, between two becks, on which the poppy shifts to rosy and amethyst tones and, very rarely, you come on forms in which the delicate loveliness of the bells is of real azure blue, as it were muted and veiled with a hinted reminiscence of their normal lavender so as to have an opalescent effect of shot silk. And then, more rarely still, there are albinos and albinoids, almost unimaginably beautiful, like living drops of purity or incarnate snowflakes — exactly what the Snowdrop ought to be, and isn't ....."

(R. Farrer, *Rainbow Bridge*: 224, 1986 edition)

was a notable little rocky grot which in June was filled with a rose-scented jungle or rose-pink Peonies, above and amid which floated the innumerable expanded blue butterflies of the Poppy. The number of petals, though usually 4, can often be 6; and it may be noted that the original diagnosis seems to have been made from specimens smaller than the usual type of *M. quintuplinervia*, and far inferior to the best."

(*The English Rock Garden*: 506)

"...... And here the whole mountain-side was curtained with cushions of the Harebell Poppy, coagulated so abundantly in every moss-fall as to show me not only that here, in the Da-Tung [Dabanshan sic], we have the very central point of the plant, but also that, above all other Meconopsids, it thirsts for moisture and abundantly repays it." ".......Then as the valley climbs and narrows, the beck has to be crossed and recrossed: and on its right flank, among the boscage, which now begins to be of Rhododendron, the Harebell Poppy takes charge, and all the slope flickers with its lavender butterflies. coyly peering." Farrer, *Rainbow Bridge*

Meconopsis quintuplinervia clearly finds its closest ally in the startling red-flowered *M. punicea*. Both species bear basal tufts of spear-shaped leaves with 3 or 5 longitudinal veins. Apart from the obvious difference in colour, those flowers of *M. punicea* are considerably larger, the petals up to nearly 10.5 cm long on occasions. Other differences can be noted in the ovary and fruit characters: in *M. quintuplinervia* there is nearly always an obvious style and the capsules are sparsely to moderately ornamented with bristles, while in *M. punicea* the style is obsolete or almost so, and the fruit capsules are densely beset with bristles. From a horticultural point of view the important difference between the two species (neither of which set seed readily in cultivation) is seen in the taprooted nature of *M. punicea* which is strongly contrasted with the fibrous-rooted nature of *M. quintuplinervia*, which makes the latter very much easier to propagate. The closeness of the two species is emphasised by the readiness with which they can cross in the wild. The hybrid, *M. ×cookei*, which is sterile, can often be found wherever the two parent species grow in close proximity with one another in the wild.

In cultivation seed is sometimes available but it is never plentiful. Like *Meconopsis punicea* it is subject to problematic dormancy: seed needs to be sown in the autumn for germinating the following spring. Sown in the spring, the seed will remain ungerminated until the following spring. Unfortunately, although fruit pods are often produced many contain no or very little viable seed, although they may contain numerous undeveloped ovules ('dust'). On the plus side, plants are readily propagated from careful divisions as each piece will carry its own portion of fibrous root. In the garden a sunny or part-shaded, but not hot, site suits its best. Although plants will thrive in both mildly acid to calcareous soils, they will only do so given ample humus, well-rotted friable compost or leafmould being ideal. Dead-heading plants the moment they have flowered encourages repeat blooming, although at the sacrifice of potential seed production.

Meconopsis quintuplinervia, the name literally means five-veined, referring to the leaves, is without question one of the gems of the genus, especially in gardens where its dainty bells cannot fail to attract attention. In cultivation several forms are recognised from a tall rather lax-growing one up to 50–60 cm tall in flower and a more compact one to 30 cm. The latter, 'Kaye's Compact' form, originated from Reginald Kaye's nursery, Silverdale in Lancashire who in turn had found it growing in a local garden. Its origin beyond that is unknown. The plant forms a close mat of numerous leaf-rosettes 30 cm or more across and producing numerous pale lavender flowers in the early summer. This attractive form does set some seed, so can be classed as partly self-compatible at least, although the seedlings do not come true to type. *M. quintuplinervia* is certainly frost hardy and coming into growth rather late in the spring avoids late spring frost damage to the developing buds. Despite this, some growers place a cloche over their plants during the winter months for added protection. This at the same

Meconopsis quintuplinervia 'Kaye's Compact' form in cultivation at Sheriffmuir, Scotland. PHOTO: CHRISTOPHER GREY-WILSON

time may ward off the attentions of birds that can pull tufts out of the plants in their search for grubs; plants can be wrecked in such a manner and a piece of wire netting may prove a simple solution to their attention.

Meconopsis biloba, described as recently as 2004, is based on a collection in which the petals are bilobed to almost halfway; however, in all other details of leaf, flowers and fruits, as well as pubescence and geography, this species fits comfortably within *M. quintuplinervia* and I see no reason to recognise it on present evidence, even at varietal level. Although lobing of the petals is rare in *Meconopsis* it is not unique to *M. quintuplinervia*. For instance both doubling of the petals, notching and shallow lobing have been encountered in the closely related *M. punicea* and in some forms of *M. betonicifolia* and *M. integrifolia* the petal margin can be deeply erose, cut or shallowly lobed. The authors of *M. biloba* point out the similarity of their species to *M. quintuplinervia*, stressing that it "differs from all known *Meconopsis* species in China in its bilobed petals".

37. MECONOPSIS ×COOKEI

Meconopsis ×cookei G. Taylor, *J. Roy. Hort. Soc.* 76: 231 (1951). Type: based on a cultivated specimen from Devonhall, Perthshire from A. Harley, 27 May 1947, from a plant supplied the previous year by R. B. Cooke, Kilbryde, Corbridge, Northumberland (BM, holotype).

DESCRIPTION.[28] *Polycarpic or occasionally monocarpic plant to 40 cm tall, the stem base beset with old leaf base and petiole remains as well as dense barbellate bristles. Leaves petiolate, all borne in basal rosettes, solitary or several clustered together, the lamina elliptic to elliptic-oblanceolate or oblanceolate, 4–10 × 1.3–1.8 cm, 3- or 5-veined, with 3 or 5 longitudinal veins, the base attenuate into the petiole, the apex subobtuse to subacute, the margin entire, the lamina covered on both sides with straw-coloured softish bristles; petiole 2.2–6.5 cm long. Flowers solitary, nodding, one or several per leaf-rosette, shaped like a flared skirt, 6.8–7.9 cm long. Scapes (pedicels) erect, somewhat sinuous, 35–39 cm long, beset with somewhat deflexed straw-coloured hairs. Buds oblong in outline, c. 27 × 15 mm, the sepals with dense patent, softish bristles on the exterior. Petals 4, pale plum-coloured, often a rather muddy hue, or occasionally rich plum-purple, satiny, elliptic-oblong to elliptic-oblanceolate, 6.8–7.9 × 3.2–3.4 mm. Stamens with linear filaments similar in colour to the petals, and yellow or brownish anthers. Ovary suborbicular,*

---

[28] The description is of the species in the wild.

shrouded in dense, semi-appressed, pale bristles; style 2 mm long, the stigma 5–7 mm diameter, dark purplish. Fruit capsules not developed.

DISTRIBUTION. SW Gansu, SE Qinghai, NW Sichuan; 3780–4215 m.

HABITAT. Open hillsides, steep rocky and grassy slopes, moorland, sometimes in open low scrub.

FLOWERING. June–July.

SPECIMENS SEEN. *Long, Brickell, McBeath, Strangman, Steele, Lu, Deng & Zhang* (SBQE) 425 (E), 686 (E), 696 (E); *H. Smith* 3174 (BM).

Along with *Meconopsis* ×*kongboensis*, *M.* ×*cookei* is the only authenticated natural *Meconopsis* hybrid that has been found in the wild. Of the two *M.* ×*cookei* is by far the best known and probably the most numerous in the wild, and wherever *M. punicea* and *M. quintuplinervia* grow in close proximity to one another this hybrid is likely to arise. Perhaps surprisingly, *M.* ×*cookei* was first described from a cultivated plant created by Randall Cooke in Corbridge, but later found in the wild. That the hybrid should occur at all is not unsurprising as the two parent species are very closely related. While there

Meconopsis ×*cookei* (*M. punicea* × *quintuplinervia*); Bama Zhiqin valley, SE Qinghai, 3573 m. PHOTO: HARRY JANS

Above left: *Meconopsis* ×*cookei* (*M. punicea* × *quintuplinervia*); S of Jigzhi, SE Qinghai. PHOTO: CHRISTOPHER GREY-WILSON
Above right: *Meconopsis* ×*cookei* (*M. punicea* × *quintuplinervia*); Bama Zhiqin valley, SE Qinghai, NW China, 3575 m. PHOTO: HARRY JANS

is clearly no incompatibility mechanism that prevent cross pollination between the two species, the success of the hybrid is severely limited by the fact that it is sterile and has not been known to set seed. This means that plants are generally scattered in ones and twos through populations of the parents, never to my knowledge being found in more substantial numbers. Their colour, often a wishy-washy plum or peuce makes them easy to identify in the field. Occasionally more strikingly coloured hybrids with rich plum or plum-purple flowers are to be seen, though these are not the norm. This may be a reflection on the seed parent which is generally assumed to be *M. punicea*.

In general terms *M.* ×*cookei* comes midway between the parent species. Plants tend to be a short-lived perennial, with foliage about the size of that of *M. quintuplinervia*. The flowers are about midway in size between the parent species but are close to the flared skirt shape of *M. punicea* rather than the bell shape of *M. quintuplinervia*. In its best coloured forms *M.* ×*cookei* can be a most attractive plant and eagerly sought by gardeners, although it is not common in cultivation and few plants exist today. Since the original hybrid was made it has been recreated in cultivation, notably by Andrew Harley of Devonhall and has been described by James Cobb (1989) as "a muddy compromise between the two" parents which will hardly endear it to gardeners, yet in its deeper richer plum-coloured forms it is a worthy and striking plant; sadly these have never been in cultivation. However, Harold Fletcher writing in the *Alpine Garden Society Bulletin* (29: 36, 'A Decade at Edinburgh') in 1961 has an opposing view of the hybrid stating that "*M. cookei* that splendid hybrid between *M. punicea* and *M. quintuplinervia* which, though known well before 1951, has only been in fairly general cultivation during the past ten years. Though the plants lacks the striking colour of *M. punicea* yet many regard it as an improvement on *M. quintuplinervia* from which it has inherited the perennial habit".

The fact that *Meconopsis* ×*cookei* is relatively easy to create in cultivation has tempted growers to try to produce a robust genuinely perennial hybrid with a good colour. This goal has been achieved by Lesley Drummond who gardens near Forfar in Scotland, who made a deliberate *M. punicea* × *M. quintuplinervia* cross. The resultant seedlings (half a dozen in all) produced plants of intermediate form but variable in colour. One handed on to Evelyn Stevens proved to be particularly floriferous and she suggested the name 'Old Rose' for it. This is a fine plant with pretty deep rose flowers but, more importantly for gardeners, it is soundly perennial and easy to propagate by division; Irishman's cuttings (pre-rooted portions). It is now obtainable from several growers, particularly Ian Christie's nursery, Kirriemuir, also in Scotland. Drummond went on to backcross 'Old Rose' with *M. punicea*, this time to attempt to produce a reliably perennial *M. punicea* look-a-like. Repeated backcrossing and careful selection eventually gave rise to several good perennial plants with flowers more or less indistinguishable from *M. punicea*. One seedling, again in the hands of Evelyn Stevens, proved to be both multi-crowned and double-flowered (i.e. with twice the normal compliment of petals).

Subgenus CUMMINSIA

Subgenus *Cumminsia* is by far the largest of the four subgenera in this account, comprising more species than the other three put together. Apart from two species, *Meconopsis bella* and *M. delavayi*, all are monocarpic plants taking, in the wild at least, several years to reach flowering size. The leaf-rosettes wither back in the autumn and the plants overwinter as resting buds at or just below the soil surface. Unlike the other subgenera which all possess barbellate hairs, in subgenus *Cumminsia* the hairs are simple and non-barbellate, although looked at under a microscope they possess minute nodes at intervals. The hairs vary a great deal from species to species, being very fine, sometimes brittle, or more robust and bristle like, especially on the stems (when present), pedicels and sepals. In the most robust types such as *M. horridula* and *M. speciosa*, the hairs are stiff and 'spine'-like, often pungent and prickly to the touch, sparse or dense. In addition, the hairs, especially in the pungent kinds, can vary in length considerably even on the same leaf. Plants can have a well-developed stem and inflorescence in which some or all of the flowers are bracteate, or they may bear all their flowers singly on basal scapes. Partial agglutination of the central scapes in the rosette is common in the subgenus amongst the scapose species, this giving rise to a pseudoraceme which is nearly always accompanied by separate basal scapes. The fruit capsules nearly always bear a persistent style and vary from smooth to very prickly.

section ACULEATAE

The four species of this section produce relatively dense rosettes of leaves in which the lamina is shallowly to deeply pinnatifid, sometimes the lower leaf-lobes bipinnatifid. All parts of the plant are covered moderately to rather densely in stout bristles and the flowers are commonly 4-petalled, although not invariably so. Although the section is primarily western Himalayan in origin, one species, *Meconopsis speciosa*, is found north of the main Himalayan divide in south-eastern Tibet and the neighbouring parts of north-western Yunnan.

KEY TO SPECIES OF SECTION ACULEATAE

1. Leaf-lamina undulate to shallowly lobed.. . . . . 40. **M. latifolia** (N Kashmir)
+ Leaf-lamina shallowly to deeply pinnatifid. 2

2. Flowers all borne on basal scapes. . . . . 39. **M. neglecta** (NW Pakistan; Chitral)
+ Flowers borne in part or wholly bracteate inflorescences 3

3. Leaves toothed to shallowly pinnatifid, 7–9.2 cm wide; only the uppermost flowers ebracteate; style of fruit capsule not exceeding 6 mm long . . . . 40. **M. latifolia** (N Kashmir)
+ Leaves deeply pinnatifid, not more than 5 cm wide; only the lowermost flowers in the inflorescence bracteate; style of fruit generally more than 6 mm long 4

4. Style stout; ovary bristles reddish-brown . . . . 41. **M. speciosa** (SE Tibet, NW Yunnan)
+ Style slender; ovary bristles straw-coloured to golden . . . .
 . . . . .38. **M. aculeata** (N Pakistan, Kashmir, NW India)

38. MECONOPSIS ACULEATA

Meconopsis aculeata Royle, *Illustr. Bot. Himal.*: 67 (1834) tab. 15. Type: NW Himalaya (Kashmir or Kumaon), *Royle* (native collectors), India 1828–31 (LIV, holotype).

[Syn. *M. guilelmi-waldemarii* Klotzsch & Garcke, *Bot. Ergebn. Reise. Prinz Waldem.*: 129, tab. 36 (1862); *M. racemosa* sensu Silva Tarouca non Maxim., *Freiland-Staud.*: 150, Fig. 201 (1910); *M. aculeata* var. *typica* Prain, *Bull. Misc. Inform.*, Kew 1915: 144 (1915); *M. aculeata* var. *nana* Prain *loc. cit.* (1915) & 1916, *App.*: 63 (1916); *M. aculeata* forma *normalis* Prain *loc. cit.*: 145 (1915); *M. aculeata* forma *acutiloba* Prain *loc. cit.*: 145 (1915); *M. latifolia* sensu Silva Tarouca & Schneider, *Freiland-Staud.* 1927: 249, fig. 248 (1933)].

DESCRIPTION. *A rather variable monocapic plant 22–75 cm tall in flower with a stout, cylindrical-tapering to fusiform taproot, to 12.5 cm long, 7–13 mm diameter near the top. Stem beset with scattered, patent, fine pale brown or tawny-coloured bristles, these generally 4–7 mm long. Leaves greyish or bluish-green, borne in a lax rosette in the year(s) before flowering, to 27 cm long including the petiole, the lamina oblong to elliptic oblong, 4.4–12.5 × 2.3–5 cm, pinnatifid, with 2–5 pairs of oblong,* occasionally unevenly bilobed, lobes, generally with broad sinuses between the pairs, the lowermost pair sometimes distant from those above, the lobes very variable in shape from broad-oblong to elliptic-oblong, obtuse to rounded, or subtriangular and acute or subacute, the upper part of the leaf lamina sometimes pinnatisect, sparsely covered on both surfaces with straw-coloured, more or less pungent, bristles; cauline leaves similar to the basal but decreasing in size upwards, the uppermost less lobed, sometimes scarcely lobed at all, sessile and merging with the bracts; petiole to 15 cm long, expanded towards the base. *Inflorescence* simple racemose, occupying most if not all of the stem, sometimes confined to the upper half only, the flowers 7–15 normally, 4.7–7.4 cm across, semi-nodding, all but the uppermost bracteate. *Pedicels* 6.3–22 cm long. *Buds* nodding, subglobose, 13–15 × 14–17 mm. *Petals* 4, rarely 5 or 6, blue (pale sky blue to azure, rarely dark blue), violet-blue, violet-purple, purple-blue, reddish blue occasionally pink, reddish or white, rounded to obovate or obovate-oblong, 27–40 × 24–33 mm. *Stamens* numerous, the filaments the same colour as the petals although generally much darker, the anthers golden-yellow. *Ovary* subglobose to ellipsoidal, covered in appressed or semi-appressed straw-coloured bristles, the style 3.5–8 (–12) mm long, expanded at the

Meconopsis aculeata; Balugere, Himachal Pradesh, NW India, 3400 m. PHOTO: DAVID & MARGARET THORNE

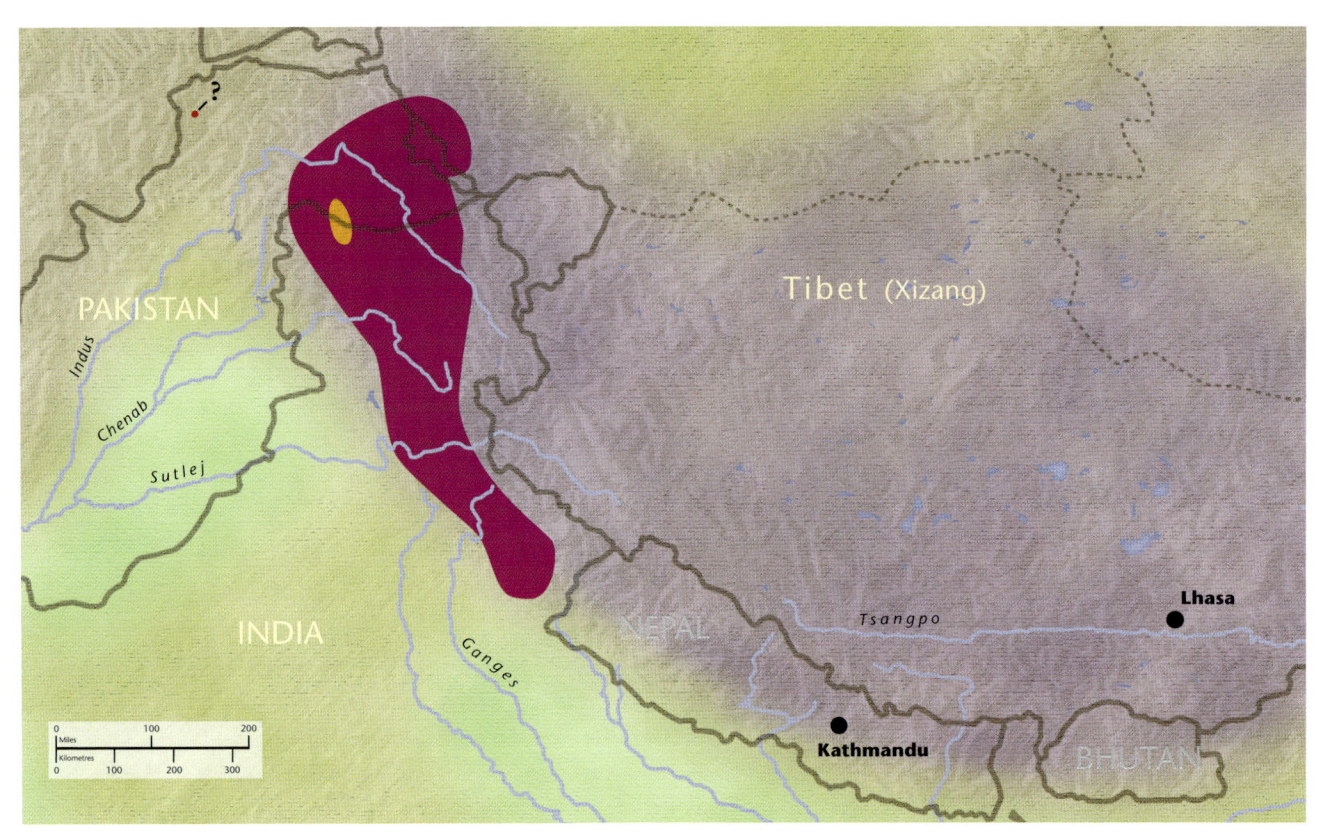

MAP 14. Species distribution: ■ *Meconopsis aculeata*, ■ *M. neglecta* (possible location), ■ *M. latifolia*.

base in fruit, the stigma cream or yellowish, occasionally purplish. *Fruit capsule* oblong-ellipsoidal to subglobose, 13–19 × 5.5–11 mm, moderately to densely patent bristly, 4–6 (–8)-valved, splitting for a short distance from the apex at maturity. *Seeds* subreniform, adorned with longitudinal rows of pits. 2n = 28, 56.

DISTRIBUTION. W Himalaya from North-west Frontier Province (Hazara District) to N India (Kumaon), especially in Kashmir, extreme SW Tibet (Xizang); (2450–)2895–4650 m (Map 14).

HABITAT. Damp rocky places, open scrub, streamsides.

FLOWERING. Late May–July.

SPECIMENS SEEN. *A. Barclay* 160 (BM); *F. Billiet & J. Leonard* 6633 (K); *N. L. Bor* 9724 (E), 15021 (E, K); *C. B. Clarke* 24634 (K), 31057 (K), 31291 (K); *Dr Cleghorn* May 1864 (K); *Clements* 189 (K); *H. Collett* 812 (K); *R. E. Cooper* 5445 (E), 5649 (E); *J. P. Crowden* s.n. Zanskar (K); *J. R. Drummond* 1237 (K), 1331 (BM), 4359 (BM), 4365 (K), 6297 (K), 8339 (K), 8341 (K), 8342 (BM, K), 14153 (K), 14825 (K), 14837 (BM, K); *Duthie* s.n. 25 July 1893, Kashmir (BM), s.n. 3.07.1899 (K), s.n. 17 July 1899, Chamba State (K), 1051 (BM), 1913c (K), 13083 (E), 19136 (K), 24220 (K), 25483 (K); *Edgeworth* s.n. 1844 (K); *R. Ellis* 274 (K), 336 (K); *M. A. Evershed* s.n., Kashmir (BM); *Falconer* 117 (K); *Gamble* s.n. (K), 20 (K), 23508 (K); *Grace, Bailey, Bull & Carr* 162 (K); *G. S. Hart* s.n. Dharmsala Forest (E!); *P. P. Huggins* 35 (BM); *Inayat* s.n. Hazara 1899 (K); *Jacquemont* 699 (K), *H. Jaeschke* 136 (K); *Jamshed* 18 (BM); *F. E. W. Kenning* 107 (K); *J. H. Lace* s.n. Kashmir (K), 405 (BM, E), 1238 (E), s.n., Drati Pass, Chamba State (E); *D. Lance* s.n. (K); *Ludlow* 126 (BM, E); *Ludlow & Sherriff* 12 (BM), 7981 (BM), 8058 (BM), 8260 (BM), 9139 (BM, E); *Lieut. Maclagan* 623 (E); *Margaret Munro* s.n. Thirot Nulla (E); *M. Nath* 23514 (K); *Parmanand* 278 (E); *Polunin* 215 (E), 593 (BM); *Range Officer* s.n., Oct. 1933, Kumaon (BM); *C. D. Sayers* 3622 (BM), 3721 (BM); *E. Schelpe* 3484 (BM), 3500 (BM, E); *Sherriff* 7363 (BM), 7442 (BM, E); Southampton University Botanical Exped. 162 (BM); *J. L. Stewart* 100 (E); *R. R. Stewart* 1554 (K), 2537 (K); *Strachey & Winterbotoom* 2 (BM, K); *T. Thomson* s.n. (K); *M. K. Timmins* 200 (BM); *F. E. W. Venning* 107 (K); *Wallich* 8122 (K); *Mrs G. Ward* s.n., Kashmir (BM); *G. Watt* 2184 (E), 2497 (E), 3301 (E), 8627 (E), s.n. 1878 (K); *Winterbottom* s.n. (K).

NOTE (on herb sheet): useful as a purgative. Two sheets (E), *G. Watt* 8627 and *Polunin* 215 include both a caulous as well as acaulous specimen. See *Meconopsis neglecta* below.

38. MECONOPSIS ACULEATA

Meconopsis aculeata is a distinct species restricted to the western Himalaya, centred on Kashmir, along with two other members of series *Aculeatae*, *M. latifolia* and *M. neglecta*, although the three do not overlap in distribution. It is a pretty plant with, in its finest forms, delightful azure blue flowers, although others are equally attractive with shot-silk purple-blues. In the wild it is particularly variable, often in the same colony, particularly in height and in the degree of dissection of the leaves. Plants sometimes appear to be scapose but this is often due to damage to the young inflorescence as it emerges: this can happen through trampling or grazing and has been noted in other species e.g. *M. lancifolia*. It has as its closest ally *M. latifolia* which has a much more restricted distribution in northern Kashmir; the latter is readily distinguished by coarsely toothed to shallowly lobed, significantly broader leaves, by the fact that most of the inflorescence is bracteate, and by its longer, shorter-styled, fruit capsules. The other member of the series, *M. speciosa* has an easterly Sino-Himalayan distribution and is an altogether less leafy plant with narrow racemes of flowers and stout-styled fruit capsules.

Reginald Farrer, who was clearly enraptured by the species in cultivation comments in *The English Rock Garden* (p. 474, 1918) that "*M. aculeata* can be perennial or biennial or monocarpic. But, whatever its powers of

Above left: *Meconopsis aculeata*; Rupin, Himachal Pradesh, NW India, 4000 m.   PHOTO: JOHN & HILARY BIRKS

Above right: *Meconopsis aculeata*; Rohtang, Himachal Pradesh, NW India, 3750 m.   PHOTO: DAVID & MARGARET THORNE

38. MECONOPSIS ACULEATA

endurance may be, its powers of beauty take the breath. It stands about a foot to 16 inches high at the most, and the whole plant is of blue-grey tone, horrid everywhere with bristling hairs and spininesses. The leaves are irregularly feathered into notably deep rounded or pointed lobes (note this), and are on long footstalks, and bristly almost invariably on both sides. The flowers are borne on long footstalks, also, in a most graceful, loose spike; and they are of noble size, four-petalled, of a crumpled pale blue silk, filmed with a diaphanous irridescence of violet — against their golden spray of stamens a colour indescribable in its pure beauty, like a sky of dawn remembering very faintly the first touch of amethyst in which it died the night before. Nor do they seem to vary into worse tones with the copiousness of other blue Meconopsids ...... and *M. latifolia* beats it, both for beauty and constitution."

The wide variation seen between and within populations of *Meconopsis aculeata* has led in the past to various varieties and forms being described by David Prain (1915), these based primarily on height and leaf division, but also on the fact that plants occasionally produce a fibrous root system, a feature that has been noted occasionally in cultivated plants. However, after close examination I am unable to maintain any of these subordinate taxa: George Taylor (1934) came to the same conclusion.

Meconopsis aculeata has been in cultivation intermittently since 1864, the original seed sent from NW India by a Dr Cleghorn. It was a flowering specimen from this source that was the subject of the plate in *Curtis's Botanical Magazine* (tab. 5456), published in 1864. The species is grown as a medicinal herb in both NW India and in neighbouring Tibet.

Above left: *Meconopsis aculeata*; Balugere, Himachal Pradesh, NW India, 3750 m. PHOTO: DAVID & MARGARET THORNE

Above right: *Meconopsis aculeata*; Ranglati, Himachal Pradesh, NW India, 3620 m. PHOTO: DAVID & MARGARET THORNE

One intriguing possibility is that *Meconopsis aculeata* has been in cultivation even longer than supposed and I am grateful to Molly Mahood for passing this information onto me. Molly Mahood is compiling a *John Clare Flora* which will contain the 370 or so plants that the poet mentions in his writings. In this there are two allusions to a blue poppy. It is recorded that when Clare was a gardening apprentice at Burghley House in Northamptonshire in 1810, or thereabouts, he found a yellow and a "dark blue" poppy, both "very beautiful" and "not wild" that had seeded themselves among the vegetables. The blue one, he adds "is rare it is as deep tinged as the 'corn bottle' (i.e. cornflower). He took seed home and discovered that the yellow one was *Papaver cambricum* (syn. *Meconopsis cambrica*) and perennial, but the blue one to be an annual. It has been suggested that this blue poppy was in fact *M. aculeata* and although the suggestion that the plant in question was an annual (it can behave, and often does, as a biennial in cultivation!) the possibility remains that its introduction into gardens was earlier than previously recorded. Molly Mahood continues [via email] that "At first I thought this impossible, since Himalayan blue poppies were unknown in Britain until the 1880s and not brought into cultivation here until 1926. But I also see that *M. aculeata* was and is highly valued for medicinal properties (hence the present scarcity), and so wonder if its seeds could have been imported in the 18th century by the East India Company. If the flowers turned out to be beautiful, the Marquis of Exeter would certainly have got hold of them for his garden."

Although *Meconopsis aculeata* has been in cultivation for many years it has, nonetheless, never been common in gardens. It is a pretty plant and well worth persevering with. It requires a damp spot in order to succeed but not one that becomes waterlogged in the winter. It will also tolerate dappled shade to some extent. In the wild it inhabits damp places, often amongst rocks and frequently by streams or rivers. In the western Himalayan areas it inhabits, the summer monsoon is much lighter than it is farther east and damp habitats are at a premium. For this and similar moisture-loving species (primulas in particular) gardeners devise special humus beds or containers in which the moisture level can be controlled through the seasons. I have seen fine stands of this plant

Right, from top:

Meconopsis aculeata; Marhi, Himachal Pradesh, NW India, 3260 m.
PHOTO: DAVID & MARGARET THORNE

Meconopsis aculeata, flower detail; Rohtang La, Himachal Pradesh, NW India, 3950 m.   PHOTO: JOHN & HILARY BIRKS

in various Scottish gardens, notably Peter Cox's garden at Glendoick, where it grows in glades in the woodland garden along with other meconopsis, primulas, lilies and other delights.

Plants are prone to slug-damage in the early spring as the fresh new growth begins to emerge. This can be alleviated by placing a wide collar of sharp grit around the neck of the plants or by using safe slug baits if that is your method of control.

James Cobb (1989) reports that "A strikingly beautiful white form with purple anthers and stigma turned up from wild seed and such plants are worth much effort to establish". White-flowered forms have been reported from the wild on a number of occasions over the years but seedlings from them invariably produce normal blue-flowered plants. In recent times a number of seed collections have come in from the wild and with luck these might produce a greater range of variability and even forms that may be slightly more dry-tolerant and therefore more adaptable to the garden environment, at least in the drier parts of Britain and elsewhere.

Meconopsis aculeata; Suru Valley, Zanskar, NW India, 3000 m, flowering in mid August. PHOTO: ALAN DUNKLEY

39. MECONOPSIS NEGLECTA

Meconopsis neglecta G. Taylor, *The Genus Meconopsis*: 102 (1934). Type: Pakistan, Chitral, Kafiristan, *Maj. S. M. Toppin* 761 (K, holotype).

DESCRIPTION. *Monocarpic herb* forming a small lax rosette (to 20 cm across) in the years before flowering, with a dauciform taproot 10–15 cm long, 8–15 mm diameter at the widest, the top beset with some fibrous remains of old leaf bases. *Leaves* all basal, the lamina lanceolate-elliptic, 3.7–7 × 1.2–1.4 cm, deeply pinnatifid, with 3–4 pairs of oblong, obtuse, lobes, covered on both surfaces and along the margin with rather weak yet pungent bristles; petiole to 6.5 cm long, expanded somewhat at the base. *Flowers* solitary, scapose, up to 12 per plant, saucer-shaped, to 4 cm across. *Pedicels* (scapes) ascending to erect, 11.5–15 cm long, sparsely bristly. *Buds* ovoidal, 11–12 mm long, patent bristly. *Petals* 4, almost certainly blue, rounded to broad-obovate, to 21 mm long and almost as broad. Stamens numerous, the same colour as the petals, with yellow anthers. *Ovary* ellipsoid, covered with appressed pungent bristles, with a distinct but short style 0.5–1.2 mm long, bearing a capitate stigma. *Fruit* unknown.

DISTRIBUTION. W Pakistan, Chitral (Kafiristan), close to the Afghanistan border. Precise locality and altitude unrecorded (Map 14).

SPECIMENS SEEN. *S. M. Toppin* 761 (K).

Meconopsis neglecta is closely related to the western Himalayan *M. aculeata* and is the most westerly distributed member of the genus *Meconopsis*. It differs from the latter in two prime features, the acaulous habit with all the flowers borne on basal scapes, and in the details of the ovary, especially the very short style. Other differences can also be noted including the narrow leaves and small flowers. If this plant were found within the general distribution of *M. aculeata* then it would probably be regarded as a depauperate specimen; however, it is found well to the north-west close to the border of Afghanistan, while *M. aculeata* is found from Hazara eastwards. George Taylor remarks in his 1934 monograph that "Previous to this record of *M. neglecta* no species (apart from *M. cambrica* [now excluded from the genus!]) has been found west of the River Indus". Growing as it does in the eastern Hindu Kush one might perhaps expect it to be found on the Afghan side of the frontier, although I have no evidence to date that it has. While it is true that *M. neglecta* is known only from a single gathering it differs significantly in overall characters and for the time being is best regarded as a distinct yet little known species.

40. MECONOPSIS LATIFOLIA

Meconopsis latifolia (Prain) Prain, *Kew Bull.* 1915: 146 (1915).

[Syn. *M. sinuata* var. *latifolia* Prain, *Curtis's Bot. Mag.* 134, t. 8223 (1908) & *Bull. Misc. Inform., Kew* 1909, app.: 95 (1909). Type: based on a cultivated plant from seed collected in Kashmir by Appleton in 1906 (K, holotype); northern Kashmir, *Falconer* (K, paratype); *M. sinuata* Irving non Prain, *Gard. Chron.*, Ser. 3, 44: 91, fig. 88 (1908); *M. aculeata* T. Smith non Royle, *Gard. Chron.*, Ser. 3, 44:91, figs. 38–39 (1909); *M. psilonomma* Wehrh. non Farrer, *Gartenstaud.* 1: 468, fig. (1930)]

DESCRIPTION. *A rather variable monocarpic plant* to 1.2 m tall in flower but often only half that height or as little as 30 cm, with a stout cylindrical to subfusiform taproot, to 30 cm long and 2.2 cm diameter. *Stem* erect, pale glaucous green, sometimes with a purplish flush, covered in spreading, rather pale or straw-coloured bristles, sometimes densely so. *Leaves* in a lax basal rosette in the years before flowering, then spirally arranged along the stem, merging gradually with the bracts, glaucous-green, paler beneath, sparsely adorned with straw-coloured bristles: basal and lower cauline leaves to 27 cm long, including the petiole, the lamina oblong-ovate to oblong or lanceolate, 13–25.5 × 7–9.2 cm, cuneate to rounded at the base, subacute to obtuse at the apex, the margin often markedly undulate, coarsely, often rather unevenly, serrate, to subpinnately lobed; middle and upper leaves similar to the basal but decreasing in size up the stem and merging with the bracts, the petiole gradually shorter with the uppermost sessile; petiole to 20 cm long, broadening towards the base. *Inflorescence* a narrow or conical raceme, with numerous flowers (up to 56 have been counted), these sometimes occupying the entire length of the stem, only the uppermost flowers ebracteate, the bracts indistinguishable and merging with the upper leaves, gradually reduced up the inflorescence; flowers 6.6–7.6 cm across, cupped to saucer-shaped, facing sideways or somewhat nodding. *Pedicels* slender, 2.2–9 cm long, covered with spreading bristles, especially just beneath the flowers. *Buds* nodding, ovoid, sparsely to moderately bristly. *Petals* 4, pale to mid blue or violet-blue, sometimes pale milky blue or white, obovate to suborbicular, 32–35 × 25–33 cm. *Stamens* numerous, the filaments pale to deep blue or violet, 7–8 mm long, the anthers orange-yellow, 1.5–2 mm long. *Ovary* ovoid, beset with dense more or less appressed straw-coloured bristles; style purple-black or greenish, 3–4 mm long bearing a rounded (capitate) to oblong, purple-black stigma, glabrous. *Fruit capsule* ellipsoid-oblong to obconic, 21–30 mm long, 6–10 mm diameter, covered in dense, spreading, spines, 4–7-valved, splitting for a short length from the apex at maturity, the persistent style 4–6 mm long. *Seeds* subreniform, the testa with longitudinal rows of pits. 2n = 56.

DISTRIBUTION. Confined to northern Kashmir in the western Himalaya; 3350–3940 m (Map 14).

HABITAT. Screes, rock crevices, cliff ledges and bases, rocky meadows.

FLOWERING. June–July.

Meconopsis latifolia; N Kashmir, NW India. PHOTO: HEATHER ANGEL

40. MECONOPSIS LATIFOLIA

SPECIMENS SEEN. *C. B. Clark* 29299B (B, K); *Drummond* 4359 (BM); *Duthie* s.n. (K); *G. L. Fuller* 248 (K); *Ludlow & Sherriff* 7818 (BM, E), 9139 (BM), 1459 (E); *Pinfold* 266 (BM); *Polunin* 745 (BM), 9610 (BM); *Range Officer* Oct. 1933, Almora Div., Kumaon (BM); *Col. D. F. Sanders*, s.n. Srinagar-Gilgit road (BM).

NOTE. The three sheets of *Ludlow & Sherriff* 7818 at the Natural History Museum are accompanied by a fine black and white photograph taken in the field in the Erin Valley, near Bandapur in Kashmir.

This beautiful species was introduced into Britain in 1906 as seed collected in Kashmir by Ltn Appleton and it was originally described from a cultivated plant from this source. The collector noted that it "is always found growing in the crevices of rocks or among loose piles of stone debris on stone slides and below cliffs. It likes the full sun, and springs to full growth after the snow melts off, while the ground is still damp".

Meconopsis latifolia was initially allied to *M. sinuata* and first described as a subspecies of that taxon and as such it was figured in 1908 in *Curtis's Botanical Magazine*. However, the leaf and fruit capsule characters and the nature of the inflorescence were clearly different and just seven years later the plant was raised to specific status. *M. latifolia* finds its closest allies in *M. aculeata*, also restricted to the western Himalaya, and in other members of the *M. horridula* complex, especially *M. rudis*, with which it bears a superficial similarity. It is at once distinguished from them by its very leafy character, the leaves with broad, shallowly lobed or toothed margins. *M. aculeata*, which is also endemic to the western Himalaya, has far more dissected foliage and noticeably longer pedicels. The half-tone plate in George Taylor's monograph (Pl. XXV, 1934) shows a fine group in flower in cultivation and exhibits the leafy character very well. Another character that drew many admirers was its tendency to bear its flowers in dense multi-flowered inflorescences involving much of the height of the plant with a number (often 20 or more) of flowers open at a time giving a conical effect, which is especially prominent as the lower flowers open.

Judging by the relatively few herbarium collections that exist of this fine species, compared for instance with *Meconopsis aculeata*, the conclusion must be drawn that *M. latifolia* is scarce in the wild and with a very limited distribution. It is known from a small area to the north and north-west of Bandipur (Bandapur) in Jammu and Kashmir, straddling the disputed frontier between India and Pakistan, today a highly sensitive area which is unlikely to entice the plant hunter. Although it comes within the distributional range of *M. aculeata* the two species have never been recorded growing together, not even in close proximity in the wild.

R. Bevan in his article entitled 'Kashmir Diary' published in the *Bulletin of the Alpine Garden Society* in 1948 comments that on the Viji Gali pass which lies to the south of the Gurais valley "The flowers were few, but I was delighted to find scattered plants of *Meconopsis latifolia* in full bloom at 12,600 feet, its sky-blue torches contrasting well with the dark granite blocks". It is interesting that in the half-tone photograph of the plant in the wild that accompanied his article and the more recent colour photograph published in this work, the inflorescences are rather narrower and more columnar than those from cultivated plants. This may be accounted for by the growing conditions or the fact that seed was

Meconopsis latifolia, part of inflorescence; cultivated.
PHOTO: JAMES COBB

introduced from a plant or plants with more conical inflorescences. In both, the flowers span more or less the full height of the plant.

George Taylor in his 1934 monograph remarks that: 'By many horticulturists this is regarded as the finest species in cultivation, and it is certainly a plant of high merit which has not become as common in gardens as one would have expected.' In its finest forms it makes a column or cone with many flowers opening in succession. Unfortunately, it is now a very rare plant in cultivation, if it hasn't disappeared altogether, which is a shame as I would certainly concur with Taylor that it deserves its garden status. Strangely enough, although it comes from southern Kashmir which was well explored during British colonial times and later by trekkers and tourists, many of whom accompanied flower tours, it has been relatively little photographed in the wild: Heather Angel's photograph, which is included in this work, is one of very few of it from the wild that I have come across! Today this whole area is one of political unrest and it is unlikely that new material (seed, herbarium specimens or photographs) can be obtained in the foreseeable future.

In cultivation, plants are tolerant of rather drier conditions than most *Meconopsis*. Plants often behave as biennials, emerging in their second spring as a silvery pink expanding rosette. The inflorescence that follows can attain one metre in height and carry beautifully formed, cup-shaped blooms of a glorious duck-egg blue, like the finest shot silk, although darker blue forms are also known. In gardens it thrives best in a moist soil and dappled shade, although in cooler regions full sun is preferable. Ample water during long dry periods and regular feeding ensures strong plants. As with *M. aculeata*, the emerging spring growth is very prone to damage from marauding slugs.

There is a suspicion that some of the plants cultivated in the past were in fact hybrids between *Meconopsis aculeata* and *M. latifolia* although this has not been substantiated. However, James Cobb (1989) has observed them in his Fife garden: "A darker blue form that has set viable seed has appeared in the garden in the last two years but it must be at least possible that it is a hybrid. In general, although *M. latifolia* grows mixed up with *M. horridula* [probably *M. prattii* or *M. zhongdianensis*], they do not appear to interbred. There is however a puzzling race of hybrids with variable coloured stigmas that may have one parent *M. latifolia* and the other *M. aculeata*. These hybrids are very attractive because, although the plants are all mauvy-blue, it is that wonderful rippling colour of shot silk. In general these plants produce very little seed which is further evidence for hybridity".

41. MECONOPSIS SPECIOSA

Meconopsis speciosa Prain, *Trans. & Proc. Bot. Soc. Edinburgh* 23: 258, tab. 2 (1907). Type: NW Yunnan, Mekong-Salween Divide, 27°28'N, 12–13,000 ft, 1906, *Forrest* 468 (E, syntype); NW Yunnan, 1917, *Forrest* 14455 (BM, E, K, W syntypes). Lectotype: *Forrest* 14455 (E), chosen here; (isolectotypes BM, K, W).

[syn. *Meconopsis ouvrardiana* Hand.-Mazz., *Anz. Akad. Wiss. Wien, Math.-Naturwiss. Kl.* 59: 247 (1922). Type: NW Yunnan, Mekong-Salween Divide, Landsang-dijang to Ludjiang, 28°9'N, 4300–4400 m, July 1916, *Handel-Mazzettii* 9556, China 1916 (W, syntype; B, E, P, isosyntypes)]

DESCRIPTION. *A monocarpic plant* (8–)12–70 cm tall in flower, often only 20–35 cm, with a rather stout narrow napiform to fusiform taproot to 31 cm long and 0.8–2.5 cm diameter, sometimes branched. *Stem*, when present, thick, 12–24 mm diameter near the base, terete,

Meconopsis speciosa subsp. *speciosa*; Baimashan, NW Yunnan, 4100 m. PHOTO: JOHN & HILARY BIRKS

somewhat striate, bracteate, glabrous to sparsely bristly, the bristles often brownish, generally with a raised reddish or blackish brown base. *Leaves* in a rather lax, spreading basal rosette, mid- to deep glossy green, sometimes with a slightly purplish flush, paler or somewhat glaucous beneath, oblong to lanceolate-oblong or ovate-oblong, occasionally linear-elliptic, 4.4–14 × (0.6–)1.1–4.8 cm, attenuate at the base into a broad petiole, the apex rounded to subacute, pinnatifid, with up to 8 pairs (commonly 3–5) of rounded to oblong pinnae with rounded sinuses in between, sometimes more shallowly or more incisely lobed with narrow, more acute sinuses, the pinnae rounded to oblong or obovate, sometimes triangular, the lower often asymmetrically bilobed or trilobed, 5–20 × 4–12 mm, the margin entire, the midrib broad, adorned with scattered, rather sparse, reddish brown or blackish brown, patent bristles, this colour often extending onto the raised lamina surface at the base of each bristle; petiole 2–14 cm long, greenish, linear-strap-shaped, thinly winged, expanded towards the base. *Bracts* present in caulous forms, borne on the lower half or third of the stem, the lowermost very similar to the lower leaves, short-petiolate, the upper sessile, increasingly smaller and more bract-like, lobed or unlobed, the uppermost sometimes linear-lanceolate and entire, shorter than the accompanying pedicel. *Flowers* 4–8.6 cm across, cupped to rather flat, laterally facing or slightly nodding, commonly 6–22 (but in racemose plants as many as 48), scapose or, more commonly, in a well-defined racemose inflorescence, fragrant (hyacinth-like). *Scapes* stout, erect, 5–19.5 cm long (to 36 cm long in fruit), free or the central ones partly agglutinated, green flushed with reddish or purplish brown, scattered with rather sparse reddish brown or straw-coloured, patent bristles, more congested towards the expanded top, often glabrous near the base; pedicels in racemose individuals ascending, 2–8 cm long, pubescent as per scapes. *Buds* nodding, ovoid to obovoid, the sepals greenish or variously flushed or marbled with purple or reddish brown, adorned with patent, rather sparse reddish or golden-brown bristles. *Petals* often 4, occasionally 5–8, pale to mid azure or Cambridge blue to turquoise, sometimes with a ruddy purplish or mauve flush, occasionally maroon, or white with a bluish flush, translucent, rounded to obovate, 20–45 × 20–40 mm, the margin somewhat fluted and undulate. *Stamens* numerous, about one-third to half the length of the petals, the filaments the same colour as the petals, although much darker, the anthers yellow or orange-yellow, greying on maturity, oblong, c. 1 mm long. *Ovary* ellipsoid to ovoid, greenish to dark purplish brown, 5–8 mm long, densely covered with dark reddish brown semi-appressed to semi-patent bristles; style linear,

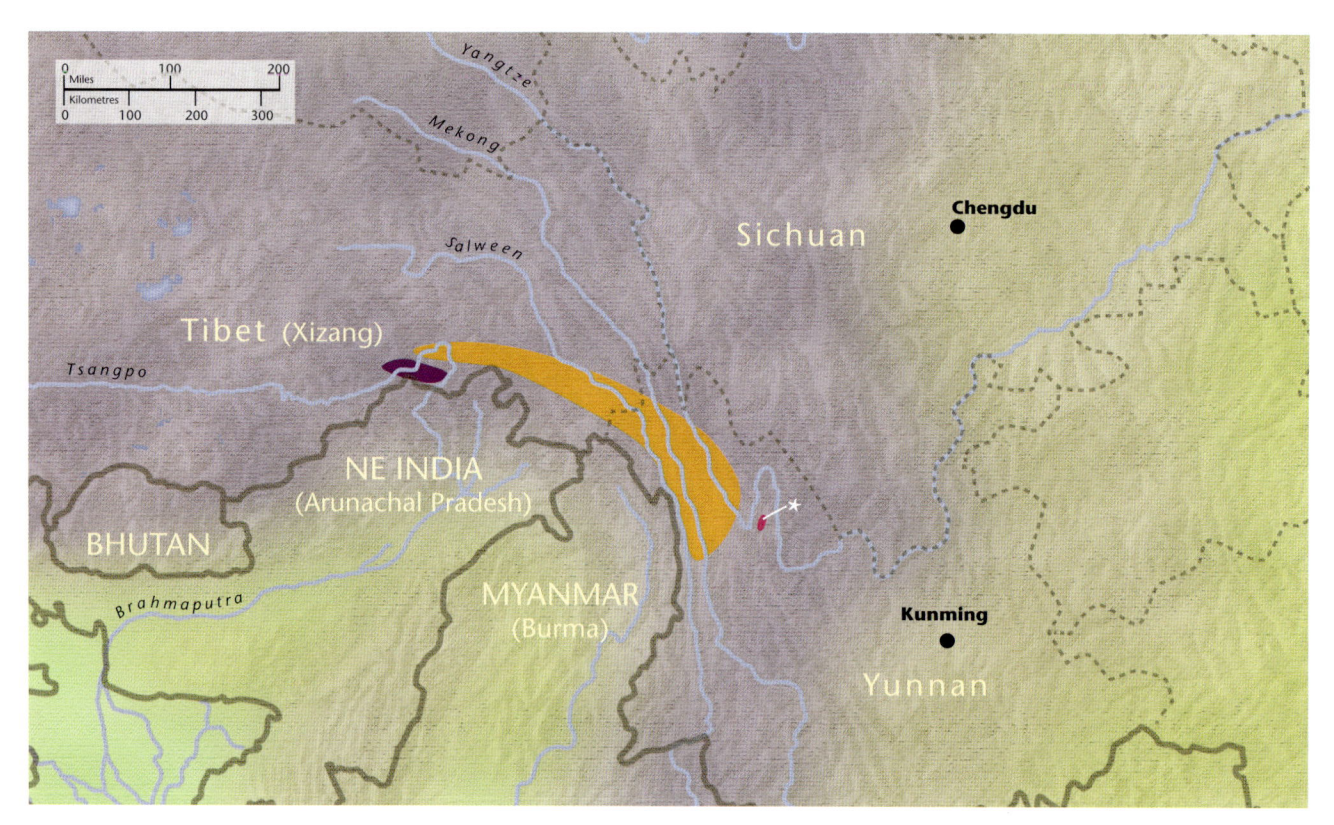

MAP 15. Distribution of *Meconopsis speciosa*: ■ *M. speciosa* subsp. *speciosa*, ■ *M. s.* subsp. *yulongxueshanensis*, ■ *M. s.* subsp. *cawdoriana*.

THE GENUS MECONOPSIS
41. MECONOPSIS SPECIOSA

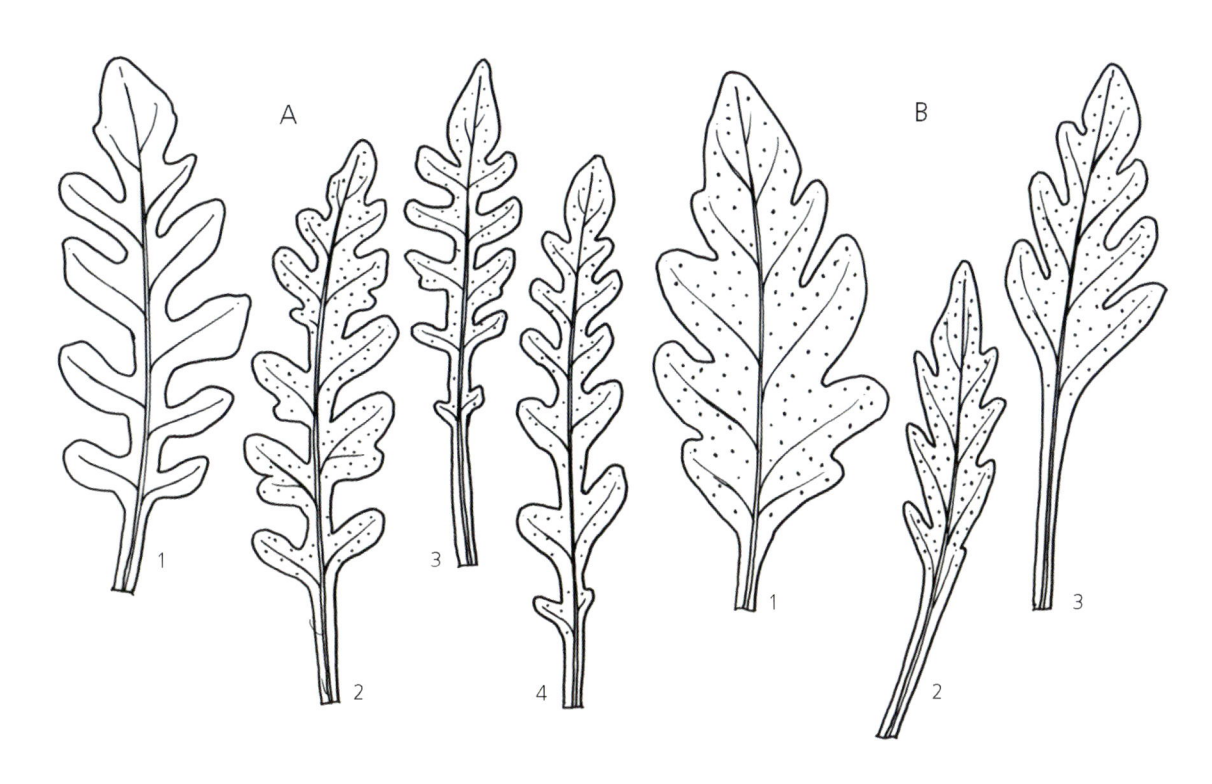

FIG. 18. *Meconopsis speciosa* basal leaves, lower petioles and pubescence not shown: **A1–4**, subsp. *speciosa*; **B1–3**, subsp. *cawdoriana*. All × 1.
DRAWN BY CHRISTOPHER GREY-WILSON

3–7.5 mm long, slightly ridged, often deep purple-brown, often with a few bristles at the base, the stigma capitate, 2–2.5 mm diameter, cream, greenish yellow to blackish purple, shallowly 4–8-lobed, subclavate, the lobes closely adhering to free and patent. *Fruit capsule* blackish green, oblong-ellipsoid, 15–30 × 6–13 mm, 2–9 mm diameter at the base, 4–8-valved (commonly 5–6-valved), covered in persistent, patent purplish or brownish purple bristles, splitting for a short distance (up to one third), from the apex when mature, with a persistent style 3–10 mm long, 2–2.5 mm diameter at the base. *Seeds* black, subreniform, the testa shallowly pitted in longitudinal rows. (Fig. 18).

DISTRIBUTION. SE Tibet and NW Yunnan (Dêqên (Baimashan), Gongshan, Weixi); 3658–5182 m (Map 15).
HABITAT. Rocky alpine meadows, screes (primarily limestone, occasionally granitic), boulder areas, rock crevices and ledges, open alpine scrub, humusy cliff ledges.
FLOWERING. July–August.

Meconopsis speciosa is one of the most beautiful high alpine members of the genus. It is readily recognised on account of its pinnately lobed leaves and generally rather pale, yet striking, flowers, with their dark ovaries. The fact that the flowers are fragrant adds greatly to its overall appeal, as does the fact that strong plants can boast a dozen or more flowers open at the same time. The species was discovered in NW Yunnan in 1905 by George Forrest

Meconopsis speciosa subsp. *cawdoriana*; Serkhyem La, SE Tibet, 4500 m. PHOTO: HARRY JANS

and seed was sent by him to Britain; however, there is no record of plants persisting in cultivation for very long, or whether or not they ever reached flowering status. George Taylor (1934) comments that — "All the plants which I have seen growing under the name *M. speciosa* have been either *M. aculeata* or *M. latifolia*". Subsequently in the early years of the twentieth century further seed introductions were made by Forrest followed by Handel-Mazzetti, Kingdon Ward and Joseph Rock and on several occasions since those early years but the results have been no more encouraging. Like *M. horridula* sensu stricto, another high alpine species, *M. speciosa* has proved to be a fickle plant in cultivation and is likely to remain so, especially considering the extreme environment that the two inhabit in the wild.

The species probably finds its closest allies in *Meconopsis aculeata* and *M. latifolia*, two species restricted to the western Himalaya centred on Kashmir, sharing with them the pinnately lobed leaves, the multi-flowered racemes and the dark ovaries, although of the three only *M. speciosa* appears to have fragrant flowers. It differs from *M. latifolia* most obviously in the sparser, narrower, more leathery looking leaves with their deeper divisions. *M. aculeata* differs in its more papery leaves with their sharper divisions, in the laxer inflorescence with long slender pedicels, and in the generally rather smaller fruit capsules.

It is noted that in the Tibetan-speaking regions where it is found, local people gather the roots of *Meconopsis speciosa* for medicinal purposes.

KEY TO SUBSPECIES OF MECONOPSIS SPECIOSA

1a. Plant stemless and bractless, the flowers scapose, the scapes generally separate or occasionally partly agglutinosed; leaves rather incisely, often shallowly, lobed with 3–4 pairs of lobes generally, with narrow sinuses in between; fruit capsule not more than 9.5 mm diameter, narrowed to 2–4 mm at the base, the persistent style 3–5 mm long . . . . subsp. **cawdoriana** (SE Tibet)
1b. Plant with a well-developed stem, bracteate in the lower part of the inflorescence, the flowers all borne in a simple raceme; leaves deeply lobed with 5–8 pairs of lobes generally, with broad sinuses inbetween; fruit capsule 6–13 mm diameter, narrowed to 4–9 mm at the base, the persistent style 5–10 mm long **2**

2a. Fruit capsule oblong-ellipsoid, 10–13 mm diameter, narrowed to 5–9 mm at the base . . . . subsp. **speciosa** (SE Tibet, NW Yunnan)
2b. Fruit capsule narrow-cylindric, 6–9 mm diameter narrowed to 4–5 mm at the base. . . . . subsp. **yulongxueshanensis** (NW Yunnan; Yulongxueshan)

subsp. speciosa

DESCRIPTION. *Plants* to 60 cm tall in flower, with a well-defined stem. *Basal leaves* pinnatisect with 4–8 pairs of obtuse, rounded to oblong pinnae generally more or less at right angles to the midrib, the sinuses between each rounded and obtuse, the lowermost pinnae often unequally bi- or trilobed. *Bracts* present in the lower half of the inflorescence, the lowermost like the basal leaves, the upper increasingly smaller and more simple. *Inflorescence* racemose, the pedicels not more than 8 cm long, generally green, sometimes purple-flushed; flowers patent to semi-nodding. *Ovary* and style purplish-brown or greenish, the stigma blackish purple to deep green. *Fruit capsule* 19–27 × (9–)10–13 mm, (4–)5–8-valvate, the persistent style 5–10 mm long. 2n = 56. (Fig. 18A1–4).
DISTRIBUTION. SW China: SE Tibet (Bome, Ninjingshan, Pero La, Kucha La, Sang La, Tsarung) and NW Yunnan (Baimashan, Doker La — Meilixueshan, Fuchuanshan, Mekong-Salween, Mekong-Yangtse Divide, Salween-Kiuchiang Divide); (2440–)3505–4575 m (Map 15).
HABITAT. Rocky alpine meadows, stony slopes, screes and ridges, boulder-moorland, generally on limestone.
FLOWERING. June–early August.
SPECIMENS SEEN. ACE (*Alpine Garden Society China Exped.*) 802 (E), 1312 (E); *Forrest* 468 (E), 9824 (BM), 13240 (BM, E, K), 14235 (BM, E, K), 14455 (BM, E, K), 14615 (BM, E), 17384 (BM, E, K), 18725 (BM, E, K), 19059 (BM, E, K), 19082 (BM, E, K), 19099 (BM, E, K), 19791 (BM, E), 20086 (BM, E), 20168 (BM, E), 20328 (BM, E), 20779 (E), 30103 (BM, E); *Handel-Mazzetti* 9556 (E); *Kingdon Ward* 765 (E), 817 (E, K), 1031 (E), 12121 (BM); *Ludlow, Sherriff & Elliot* 13206 (BM, E), 13967 (BM, E), 14110 (BM), 15809 (BM); *McLaren* 318a (E, K); *Rock* 9991 (E), 18364 (E), 22164 (E), 22315 (E, K), 22379 (E, K), 22430 (E), 22541 (E), 22695 (E), 22894 (BM, E, K), 23098 (E), 23288 (E,

Above, from left:

Meconopsis speciosa subsp. *speciosa*; Baimashan, NW Yunnan, 4100 m. PHOTO: JOHN & HILARY BIRKS
Meconopsis speciosa subsp. *speciosa*; Baimashan, NW Yunnan, 4100 m. PHOTO: JOHN & HILARY BIRKS
Meconopsis speciosa subsp. *speciosa*; Meilixueshan, NW Yunnan, 4100 m. PHOTO: MARGARET NORTH

K), 23430 (E), 33098 (E); *Soulié* s.n., Tsekou, Tibet Orientale (BM, E); *T.T. Yü* 22150 (E), 22741 (E).

Subsp. *speciosa* has a relatively limited distribution from NW Yunnan, north-westwards into south-eastern Tibet. It is most common on the great divides between the Mekong and Yangtse and the Mekong and Salween rivers (the Hengduanshan and Nushan) and it is in this region, especially in and around the Fuchuanshan that the largest and boldest specimens are to be found: for instance *Forrest* 14235 and 19791.

Frank Kingdon Ward, who came across it on a number of occasions, was enraptured by this species:

"Up here, at 17,000 feet, springing from amongst huge blocks of grey stone, I found the Cambridge blue poppywort (*Meconopsis speciosa*), one of the most beautiful flowers in existence .....".

[Kingdon Ward, *The Land of the Blue Poppy*: 109, 1913]

"One specimen I noted was 20 inches in height, crowned with 29 flowers and 14 ripened capsules, with 5 buds visible below — 48 flowers in all ..... each flower is between 3 and 4 inches in diameter, coloured brilliant azure blue, with the texture of Japanese silk, massed with old gold in the centre ..... Indeed the plant seems to go on throughout the summer unfurling flower after flower out of nowhere — like a Japanese pith blossom thrown into water — for the stem is hollow and the roots shallow. Another bore 8 fruits, 15 flowers and 5 buds, and a third, 15 inches high, had 6 flowers, each 3½ inches across, besides 14 buds. But for a certain perkiness of the stiff prickly stem, which refuses any gracefulness of arrangement to the crowded raceme, and the absence of foliage amongst the blooms, these great azure-blue flowers, massed with gold in the centre, would be the most beautiful I have ever seen. The Cambridge blue poppy is, moreover, unique amongst the dozen species of poppywort known to me from this region, in being

41. MECONOPSIS SPECIOSA

sweetly scented. On one scree I counted no less than forty of these magnificent plants within a space of a few square yards but, scattered as they were amongst big boulders, they only peered up here and there and did not look so numerous….. ..The natural history of this *Meconopsis* is also interesting. It is a plant which produces an enormous number of seeds, very few of which ever germinate, for the capsules are attacked by a grub before the seeds ripen, and cruelly decimated"

[Kingdon Ward, *Mystery Rivers of Tibet*, 1923].

The late George Taylor, the *Meconopsis* authority for much of the early and mid twentieth century, wrote on the flora of the Sang La (see abbreviated account in *A Quest of Flowers*: 203 (1975)) which lies to the west of Namche Barwa in south-eastern Tibet, just north of the Indian frontier (Arunachal Pradesh). Taylor comments that:

"The other species seen on the Sang La were *M. horridula* [5061] — the racemose form just coming into flower on the scree slopes: *M. impedita* [5024] on earthy banks under *Rhododendron* and bearing intensely violet-purple satiny flowers on spiny basal scapes, and *M. speciosa*. I remember hearing the late George Forrest extolling the virtues of *M. speciosa* and he was full of regret that the species had never become established in gardens from his expeditions. Having seen the plant in its native habitat — on boulder scree or in crevices of dry rocks with a southern exposure — I can well understand Forrest's sentiments. The flowers are usually a beautiful silky azure-blue, though I saw some plants with rich maroon petals. We collected a quantity of seeds of these forms but apparently no success has attended our attempted introduction."

Despite the accolades heaped upon this highly attractive species it has failed to respond well to cultivation. The failure of this species from early introductions of seed has already been noted, a pity because *M. speciosa* is a very beautiful plant. In the modern era several seed collections have been obtained from NW Yunnan and there are a few plants lingering in specialist collections. Like *M. horridula* sensu stricto, this is a high altitude species that inhabits bleak rocky places and, while some gardeners may find a way of conquering it in cultivation, it is never going to be easy. For those who have been lucky enough to see it in the wild there is the memory of graceful spires of flowers set against a harsh background and a delightful fragrance that few other *Meconopsis* are able to offer.

Above left: *Meconopsis speciosa* subsp. *speciosa*; Galung La, SE Tibet, 4135 m. PHOTO: PAM EVELEIGH
Above right: *Meconopsis speciosa* subsp. *speciosa*, inflorescence and bud detail; Galung La, SE Tibet, 4135 m. PHOTO: PAM EVELEIGH

Meconopsis speciosa subsp. *cawdoriana*; Serkhyem La, SE Tibet, 4500 m. PHOTO: JOHN MITCHELL

Meconopsis speciosa subsp. *cawdoriana*; Serkhyem La, Linzhi (Nyingtri) County, SE Tibet, 4500 m. PHOTO: TOSHIO YOSHIDA

subsp. yulongxueshanensis Grey-Wilson **subsp. nov.** differs from the other two subspecies in its narrow cylindric rather than oblong-ellipsoid, few-valved fruit capsules. Type: NW Yunnan, 'Lichiang Range', 'Ma K'a Ho Mt', *McLaren* 318d (BM, holotype; E, isotype).

DESCRIPTION. *Plant* to 40 cm in flower. *Basal leaves* with 2–4 pairs of broad round to oval, obtuse pinnae, these not further lobed. *Inflorescence* racemose. *Fruit capsule* oblong-ellipsoid 17–30 × 6–9 mm, 4–5 mm diameter at the base, 4-valvate, the persistent style 7–10 mm long.
DISTRIBUTION. NW Yunnan (Yulongxueshan); altitude not recorded (Map 15).
SPECIMENS SEEN. *McLaren* 318a (BM, E, K), 318d (BM).

This is a little known taxon represented by just two collections, *McLaren* 318a in flower and *McLaren* 318d in fruit, and is distinguished primarily by its unusually long and slender fruit capsules. In addition, the basal leaves have fewer and broader lobes.

subsp. cawdoriana (Kingdon Ward) Grey-Wilson **comb. & stat. nov.**
[Syn. *M. cawdoriana* Kingdon Ward, *Gard. Chron.* 1926, Ser. 3, 79: 308 (1926), *Ann. Bot.* 40: 536 (1926) and *J. Roy. Hort. Soc.* 52: 21 (1927). Type: SE Tibet, Temo La, 15,000 ft, *Kingdon Ward* 5751[29] (BM, holotype; E, isotype); *M. pseudohorridula* C. Y. Wu & H. Chuang, *Fl. Xizang* 2: 234 (1985). Type: SE Tibet, Nyingchi (She-ji-la-shan), Qinghai-Xizang Exped. 75-1249 (KUN, holotype)].

DESCRIPTION. *Plant dwarf*, rarely more than 30 cm tall in flower, practically acaulous. *Leaves* yellow-green, like subsp. *speciosa* but the lobes often incised at an acute angle to the midrib, with a narrow, subacute sinus between each. *Bracts* absent. *Flowers* solitary, up to 12 per plant, nodding to semi-nodding, borne on basal scapes, these sometimes partly agglutinated, especially those of the first flowers to open; scapes erect to ascending, 5 × 20.5 cm long, generally reddish-brown. *Ovary* and style greenish; stigma cream. *Fruit capsule* 15–17 × 7–9.5 mm, the persistent style 3–5 mm long. (Fig. 18B1–3).

---

[29] The sheet of *Kingdon Ward* 5751, an isotype of *Meconopsis cawdoriana* Kingdon Ward, is annotated *M. morshediana*. This latter name has never been validly published to my knowledge and resides only as a name on a herbarium sheet in the author's hand-writing. However, David Prain, following information from Kingdon Ward described *M. impedita* var. *morsheadii* in 1915 based upon a Capt. F. M. Bailey collection (number 9) on the Nyima La in SE Tibet, which was later upgraded to species status by him in 1928. This particular taxon is generally regarded as a synonym of *M. impedita*.

DISTRIBUTION. SE Tibet (Buri Tsepo La, Deyang La, Doshong La (Kongbo), Mira La (Nyang Chu), Nyima La, Peru La, Sang La, Serkhyem La, Temo La); 3505–4420 m (Map 15).
HABITAT. Similar to subsp. *speciosa*.
FLOWERING. Late June–mid August.
SPECIMENS SEEN. *Kingdon Ward* 171 (E), 5751 (BM, E, K); *Ludlow & Sherriff* 6235 (BM); *Ludlow, Sherriff & Elliot* 12257 (BM), 13257 (BM), 13967 (E), 14110 (BM), 14404 (BM), 15197 (BM), 15260 (BM, E), 15267 (BM), 15268 (BM), 15444 (BM); *Ludlow, Sherriff & Taylor* 4792 (BM), 4792A (BM), 4792B (BM, E), 4792C (BM), 5213 (BM, E), 5940 (E); 6084 (BM, E); 6235 (BM).

FIELD NOTE. (*Ludlow, Sherriff & Taylor* 6084) collected on the Mira La, Nyang Chu at 16,000 ft on August 16th 1938 "….Loose granitic scree. Leaves and stem with light brown hairs, which are swollen to purple bulbous base. Capsule livid green with black-purple ribs and light brown hairs, swollen to a conspicuous black-purple bulbous base. Style black-purple. Stigma brown ……".

This interesting subspecies has a very limited range on a few mountains in SE Tibet, generally to the west of the subsp. *speciosa* locations, the latter being confined more to NW Yunnan and the adjacent Tibetan border area. It is a generally smaller and more delicate-looking plant than subsp. *speciosa*, due mainly to the fact that the plants are stemless, with all the flowers borne from the basal leaf-rosette on separate scapes (pedicels). In addition, the leaves are characteristically more incisely lobed, the lobe tips generally more acute. A similar situation is found in *Meconopsis horridula* sensu stricto in which scapose and racemose forms can be commonly found in the wild. In some specimens two or several of the central scapes are partly agglutinated, giving rise to a pseudoraceme and in these types one or several leaves (bracts) may be carried irregularly up the 'stem'. However, even in this latter state subspecies *cawdoriana* can be readily separated from the other two subspecies by means of leaf and fruits characters.

Commenting on collection number 5751 as "*M. cawdoriana*" Kingdon Ward states that it "Resembles *M. horridula* in habit, but the plant is less prickly and the scapes fewer; also flowers are fragrant, the petals 4, instead of 6–8."

Meconopsis pseudohorridula C. Y. Wu & H. Chuang was described in the *Flora of Tibet* (*Florae Xizangica*) from the mountains in the vicinity of Nyingchi (She-ji-la-shan) which lies close to the Chengdu-Lhasa highway to the west of the Serkhyem La in SE Tibet. Interestingly, this is also the same general type locality of the little known and unrelated *M. nyingchiensis* L. H. Zhou. The plant has been seen and photographed in recent times on the Serkhyem La by various individuals including Harry Jans and John Mitchell, confirming its close allegiance to *M. speciosa* and not *M. horridula*.

section RACEMOSAE

This section contains a complex of closely inter-related, strictly monocarpic, species, mostly inhabiting bleak high altitude habitats, some descending below the treeline. They are characterised by often possessing a fearsome array of stiff, often stout, pungent bristles which adorn nearly all parts of the plant apart from the petals. The leaves are primarily elliptic or lanceolate in outline with an entire, sinuous margin, or with a few, rather irregular, and generally uneven lobes or teeth. The inflorescence is a simple raceme or the flowers are borne separately on basal scapes. There is a strong tendency in the latter condition for the central scapes in the leaf-rosette to become partly agglutinated to produce a pseudo- or false raceme.

KEY TO SERIES OF SECTION RACEMOSAE

1. Filaments all linear, spreading, not incurving . . . . series **Racemosae**
+ Filaments of inner stamens strongly incurved and dilated . . . . series **Heterandrae**

series Racemosae

The *Meconopsis horridula* aggregate

This series, along with series *Cumminsia*, is the largest in subgenus *Cumminsia* and has a wide distribution that includes much of the central and eastern Himalaya, western China and much of Tibet except for the north, west and north-west. Apart from their prickly nature, the species bear linear, non-dilated filaments to the stamens and flowers with 4–9 petals, mostly in the blue-purple-lavender range but occasionally pink, yellow or white.

There is no doubt that the *Meconopsis horridula* complex presents the botanist with a huge dilemma and, while it is certain that the complex can never be resolved in a manner that will satisfy all those interested in the genus, some attempt has to be made create some logical order out of what is a wide-ranging and unbelievably complicated picture. Of course the easiest, and feeblest, solution is to 'lump' everything into one gloriously variable taxon leaving aside inherent variation and regional differences. The alternative is to try and create some sort of order that will at the least satisfy the perpetrator. Of course it is one of the jobs of the taxonomist and botanist to order things in a systematic and logical manner by way of making sense out of the living world. They (perhaps like most humans) prefer order to chaos, prefer things to fit into neat compartments, and above all prefer clear unmuddled classification; however, Nature does not always oblige. In general we see things at one point in time, whereas evolution generally takes places gradually over a vast time scale. In the case of the *M. horridula* complex many of the elements have not yet resolved themselves into clear and well-defined species — the process is still ongoing. The reader might well say, and with some justification, that cytological/DNA studies might be the answer; however, a large sampling right across the complex, taking into account the wide morphological as well as geographical span that it represents, would be required and to date that has not been undertaken. In fact in the 2011 study by Kadereit *et al.*, 33 taxa of *Meconopsis* were listed and used in a molecular analysis including *M. horridula*, this based on a single Gene Bank accession. Admittedly this study did not set out to look at the *M. horridula* complex, rather to ascertain the relationship between the Welsh Poppy, *M. cambrica*, and the Asiatic representatives of the genus.

First it is necessary to look into the history of *Meconopsis horridula* and its allies. *M. horridula* was described by Hooker & Thomson in 1855 in the *Flora of India* based upon specimens collected by them in the Sikkim Himalaya in 1848. These specimens are typical of many in the western and central Himalaya in which the flowers are borne on basal scapes or in which some of the scapes are partly agglutinated. Then in 1877 Maximowicz described briefly *M. racemosa* based upon specimens collected by Przewalski in Gansu in which the plants bear flowers in a distinct raceme. This taxon was later made a variety of *M. horridula* by David Prain in 1896, describing at the same time var. *rudis*, based upon a specimen collected by Abbé Delavay in the Likiang [Lijiang — Yulongxueshan] mountains of north-western Yunnan. At the same time Prain also described *M. sinuata* var. *prattii* which was later sunk into synonymy under *M. horridula*. Perhaps significantly, var. *racemosa*, var. *rudis* and var. *prattii* were later (1914) all recognised by Prain as species in their own right. Such was the uncertainty that few of the accounts of the time dealt with these various taxa in the same way. It should be added that before George Taylor's 1934 monograph was published two further species of significance were added by Frank Kingdon Ward to the complex, viz. *M. prainiana* in 1926 and *M. calciphila* the following year, the first based on a collection from SE Tibet, the second from 'Upper Burma' [northern Myanmar]. When Taylor's monograph appeared in 1934 all these various elements were consumed within *M. horridula* without recognition even at varietal level.

Taylor declared that: "The polymorphic character of this species is reflected in the somewhat extensive synonymy, several of the forms having been given specific or varietal rank. It is in the disposition of the flowers

Meconopsis horridula subsp. *horridula*; above Manang, Marsyandi valley, C Nepal, 3650 m. PHOTO: CHRISTOPHER GREY-WILSON

that the most evident variation is displayed. These may be borne on simple basal scapes or on a central flowering axis which is bracteate towards the base but ebracteate towards the apex, but all transitions are found between these extremes......... Marked variation is also shown in the number and colour of the petals, in the texture of the leaves and colour of their spines, and also in the shape, pubescence, and dehiscence of the capsule........"

About *Meconopsis racemosa* he states "Prain in 1896 included *M. racemosa* under *M. horridula*, but recognised it as a variety on account of the presence of a central flowering-stem, although he expressly stated that a complete transition could be traced to the typical condition with basal scapes only. It is, however, impossible to regard Maximowicz's species even as a variety of *M. horridula*, as the forms are so intimately intergraded."

About *Meconopsis rudis* he states "While it must be admitted that field specimens from the Likiang [Lijiang mountain or Yulongxueshan], the type-locality of *M. rudis*, have a distinctive facies due to the association of several characters, I have been unable to segregate the plant as a definable taxonomic unit. In these Likiang plants the leaves are usually conspicuously spotted with purple or bear purple spines, are often lobed at the margin and in addition have a glaucous appearance. It appears that they must be regarded as belonging to a localised strain whose separate recognition as a taxonomic entity is undesirable."

Of the latter two species described, Taylor comments that "A particularly robust form collected by Kingdon-Ward in 1924 at Temo La was described by him as *M. prainiana*, but I have been unable to find any structural character of this plant to merit it even at varietal rank, and I believe that it is merely a luxuriant state of *M. horridula*. The same author has described one of his collections from Upper Burma [N Myanmar] as *M. calciphila*, but here again I have been unable to observe any satisfactory character sufficient to sustain separate taxonomic identity for this plant."

As defined by George Taylor the very polymorphic *Meconopsis horridula* "....... in addition to being the most widely distributed member of the genus, occurs throughout the greatest altitudinal range and indeed reaches the upper limit of flowering-plant vegetation. It was collected by the Mount Everest Expedition in 1921 at an elevation of 19,000 feet [5790 m], and under the severe conditions prevailing at that height is much reduced in stature, consisting of a small basal rosette and several basal scapes hardly an inch in height."

Thus in this 1934 account scapose plants scarcely 5–10 cm tall were allied to plants exceeding (at times) more than 1 m in height and carrying multi-flowered racemes, while other details such as leaf shape, pubescence characteristics and colour, petals, shape and number and fruit-capsule details are all absorbed into a single taxon without division. This situation remained thus until well into the 1990s when I and others began to question the circumscription of *Meconopsis horridula* (See Grey-Wilson *Poppies — the poppy family in the wild and in cultivation*, 1993 & 2000).

To be fair, George Taylor was working on far less material than that available to the modern researcher and a great deal more herbarium material has been added, especially in the years since the 1950s from the Himalaya, and since the 1970s from China and Tibet. In addition, the advent of modern photography, particularly the digital camera, has opened up a whole new world to the taxonomist, allowing the study of numerous plants in the wild,

Meconopsis lhasaensis; Guo La, N of Lhasa, S Tibet, 4200 m.
PHOTO: TOSHIO YOSHIDA

the colour and poise of flowers, leaves and other details that are often masked or less readily observable in dried specimens. Of course, after the publication of his monograph George Taylor had a chance to see *Meconopsis* in the wild for himself, accompanying Ludlow & Sherriff in south-eastern Tibet in 1938, but this did not affect his opinion on the polymorphic character of *M. horridula* sensu lato.

On the other hand, Reginald Farrer writing in 1926 (the third imprint of *The Rainbow Bridge*), who had observed plants in the wild, especially in southern Gansu and north-western Sichuan, was clearly of an opposing opinion:

" I am having an aside with such experts as are ambitious to perpend upon Poppies. In particular the Horrid Group of *Meconopsis* is especially entangled, confused, and difficult to decipher. All its members have bristles, all are monocarpous, and all have flowers in varying tones of blue and azure. The Indian species may have to be left out of account, but the Tibetan ranges up the Border give us *M. Prattii*, *M. racemosa*, *M. rudis*, *M. horridula* — and the disappointing *M. eximia* — all differentiable, but not always easy to differentiate. Yet, after much experience in the field during 1914 and 1915, with many hundreds and thousands of specimens, I remain convinced that the Horridulous Blue Poppy of the Min S'an and the Da-Tung is always the same species, *M. prattii* [*M. racemosa sic.*], with the unvarying ash-grey anthers (I ought to say that Professor Balfour sometimes hovers on the verge of wanting to separate them). But in almost every other respect — shape of leaf, shape of pod, length of style, the species appears to me to fluctuate despairingly, and I grow more and more struck by the temerity with which Herbarium botanists build up a species on half a dozen dried specimens on a sheet, whereas in many a detail the species would give them the lie, if they could see it by the thousand blossoming on the hill-side. In point of fact, a specific difference is merely the least common denominator in a vast generalisation of individuals, all of whom vary this way and that to their own sweet will, but all of whom possess the one or two quite unvarying details that by their universality and invariability constitute a species.

And, to arrive at these bedrock differences, it is really imperative to see the plant not only living, and wild, but abundant. It is only on a vast number of living individuals that you can satisfactorily establish a general formula for a 'species', since only a vast number can yield you its one common factor, by giving full perception of the plant's range of variation, so that you can arrive at the one point (or more) in which it never varies at all......"

[Reginald Farrer *The Rainbow Bridge*: 232, 1986 edition (following the third impression of 1926]

Having also seen numerous plants of the *Meconopsis horridula* complex in the wild in both Nepal and western and north-western China I find that I very much share Farrer's opinion on the subject. It is certainly true that if all the characters are taken en bloc over the entire range where they are found then there is a considerable overlap; however, taken on a regional basis many of these characters begin to separate out and it is quite possible to identify distinct taxa, generally based on a subtle combination of characters. Of these the following characters, presented in no particular hierarchy, are of prime consideration:

Meconopsis zhongdianensis; Napahai, N of Zhongdian, NW Yunnan, 3250 m. PHOTO: HARRY JANS

1. Leaves green or glaucous
2. Leaf-margin entire, sinuate or lobed
3. Bristles on leaves and stems concolorous or with a dark base
4. Bristles uniform or a mixture of stout-long and fine-short
5. Flowers scapose, partly agglutinosed or borne in distinct racemes[30]
6. Bracts present or absent
7. Disposition of the flowers i.e. nodding, half-nodding or lateral-facing, occasionally ascending
8. Fruit capsule characteristics

Other important factors taken into account apart from geographical distribution, include habitat and altitudinal range. While some will be found at or below the treeline, others will be found high above on bleak exposed mountain ridges, screes and moraines.

Taking all this into account the *Meconopsis horridula* complex can be reconciled into the following:

M. horridula: plants from the Himalaya and the Tibetan Plateau, north-eastwards as far as Qinghai, well above the treeline.

M. lhasaensis: plants of the Tibetan Plateau centred on Lhasa, above the treeline.

M. prainiana: plants from a restricted area of south-eastern Tibet, N Myanmar and NE India (Arunachal Pradesh), below the treeline.

M. prattii: plants from south-eastern Tibet, western, south-western Sichuan and north-western Yunnan, at or above the treeline.

M. racemosa: plants from western and north-western Sichuan, Qinghai, southern Gansu and north-eastern Tibet, above the treeline.

M. rudis: plants restricted to north-western Yunnan and south-western Sichuan, above the treeline.

M. zhongdianensis: plants restricted to NW Yunnan, Zhongdian area, below the treeline.

In addition, plants from western Sichuan (Balangshan in particular) formerly included in George Taylor's circumscription of *Meconopsis horridula* have been removed entirely from the complex because of the peculiar and consistent

---

[30] The floral characters are an important feature and have been much misunderstood in the past. In scapose plants each flower is borne on an independent pedicel directly from the basal leaf-rosette, one flower issued from each of the upper axils in the leaf-rosette. On occasions, two or more scapes become fused together or agglutinated in their lower part. Such fusing does not correspond to a true inflorescence, especially as the scape traces remain distinct. True racemes are commonly found in the *M. horridula* complex. These have a central axis bearing relatively short-pedicelled flowers that are spirally arranged (with the uppermost flower in the inflorescence opening first, followed in succession with those lower down), with all the traces confluent within the inflorescence axis. It is perhaps rather unfortunate that every plant within the complex found in the wild that possesses a raceme is often automatically called *M. racemosa*. This has caused a great deal of ambiguity in the past and probably will continue to do so in the future, especially in horticulture and amongst gardeners. *M. racemosa* sensu stricto is not known from the main Himalayan ranges, while racemose forms of the *M. horridula* aggregate are frequent, especially in the east of its range.

Meconopsis prainiana var. *lutea*; Bangajang, W Arunachal Pradesh, NE India. PHOTO: DAVID & MARGARET THORNE

occurrence of the inflated filaments to the inner whorls of stamens. This plant has recently (2011) been described under the appropriate name *M. balangensis* Tosh. Yoshida, H. Sun & Boufford, which is allied to the closely related *M. heterandra* described by the same authors the previous year. These two species can be found in the following series, *Heterandrae*. In addition, another new species, *M. bijiangensis*, also a member of the *M. horridula* complex was described by H. Ohba, T. Yoshida & H. Sun in 2009.

The prime morphological differences are outlined in the key presented below.

This scheme only partly resolves analysis of the complex and even after the removal of these various taxa from *Meconopsis horridula* there remains a great deal of variation to consider. For this reason *M. horridula* is now referred to as the *M. horridula* aggregate.

KEY TO SPECIES OF SERIES RACEMOSAE

1. Dwarf plants 6–20 (–27 cm tall in fruit); flowers scapose to agglutinated 2
+ Medium to tall plants 14–95 (–120) cm tall; flowers primarily borne in a racemose inflorescence, occasionally on basal scapes on the same plant 3

2. Leaves often with dark dots (base of bristles), the bristle generally stout; flowers lateral-facing to nodding; style broadened at base in fruit, sometimes markedly so . . . . 42. **M. horridula** (C & E Himalaya & S & SE Tibet)
+ Leaves thin, papery, without dark dots, the bristles thin and weak; flowers lateral-facing to ascending; style linear in fruit . . . . 43. **M. lhasaensis** (S Tibet centred on Lhasa)

3. Leaves and stems dotted with raised dark bases to the bristles 4
+ Leaves and stems without dark dots, the bristles concolorous 5

4. Leaves glaucous above, the basal 15–42 mm wide; fruit capsule not more than 15 mm long . . . . 48. **M. rudis** (SW China; NW Yunnan, SW Sichuan)
+ Leaves deep green or yellowish green, sometimes purple-flushed, the basal 7–14 mm wide; fruit capsule 25–38 mm long . . . . 49. **M. bijiangensis** (NW Yunnan; Biluoxueshan, Gaoligongshan)

5. Petals 4 (rarely the terminal flower in the inflorescence 5–6-petalled), blue, pale yellow or white . . . . 47. **M. prainiana** (NE India (Arunachal Pradesh), N. Myanmar, SE Tibet)
+ Petals 5–9, blue, blue-purple, ruddy purple, wine-purple, crimson-purple or crimson, more rarely pale yellow or white 6

6. Fruit capsule (minus style) 15–34 mm long; stigma green, greenish yellow or 7
+ Fruit capsule (minus style) 9–17 mm long; stigma white, occasionally greyish 8

7. Inflorescence generally with more than 20 crowded flowers; anthers cream or white; petals blue, blue-purple, purple or claret. . . . . 46. **M. zhongdianensis** (NW Yunnan)
+ Inflorescence with 18 flowers or fewer; anthers orange-yellow; petals blue-purple, purple-crimson, dark red or yellow. . . . . 50. **M. georgei** (NW Yunnan)

8. Inflorescence not secund; stigma white, protruding beyond the boss of stamens; fruit capsule with erect to ascending spines; anthers white or cream. . . . . 44. **M. racemosa** (W China, C, E & SE Tibet)
+ Inflorescence secund or semi-secund; stigma greenish or yellowish-green, included within the boss of stamens; fruit capsule with spreading spines; anthers yellow or grey-yellow. . . . . 45. **M. prattii** (SW China & SE Tibet)

42. MECONOPSIS HORRIDULA

Meconopsis horridula Hook. f. & Thomson, *Fl. Ind.* 1: 252 (1855). Type: *Hooker* s.n., Sikkim 1849 (syntypes, G, K, L, LD, M, P, UPS). Lectotype chosen here, *Hooker* s.n. Sikkim 14–17,000 ft (K).

[syn. *Meconopsis horridula* var. *typica* Prain, *J. Asiat. Soc. Bengal*, Pt. 2, *Nat. Hist.* 64 (2): 313 (1896); *M. h.* var. *abnormis* Fedde in Engl., *Pflanzenr.* 4, 104: 258 (1909). Type: *King's Collector* 522 (P); Dungboo, India 1879 (G, P); *M. horridula* Kingdon Ward, *Field Notes Pl. Shrubs & Trees*, 1924–25, 34 [1925]. [sphalm. norridula]; *M. racemosa* forma *horridula* Farrer, *Engl. Rock Gard.* 1: 480, in obs.(1919); *M. duriuscula* Prain, *Country Life* 54: 111, in obs. (1923), nomen nudum; *M. rigidiuscula* Kingdon Ward, *Gard. Chron.* Ser. 3, 79: 308, in obs. (1926)].

DESCRIPTION. *A dwarf monocarpic plant* to 28 cm tall in flower but as little 12 cm with a dauciform to narrowly elongate-fusiform taproot to 8 mm diameter at the top. *Leaves* all borne in a basal rosette, spreading to ascending, moderately to densely covered on both surfaces with stout subpungent to pungent, patent bristles, often mixed with numerous smaller weaker bristles, varying in colour from pale straw to yellowish or purplish, with or without a purple-black base which, on the stoutest bristles is often raised and wart-like; leaf-lamina elliptic to elliptic-oblanceolate or linear-oblanceolate to linear-oblong, 5–11.5 × 1.2–3.6 cm, cuneate to attenuate at the base, the apex acute to subobtuse, the margin entire or with one or several uneven jagged teeth on either side, often undulate, sometimes with the margin somewhat revolute; petiole 1–11 cm long. *Flowers* up to 29 (generally 4–17) borne on basal pedicels (scapes), the central ones sometimes partly agglutinosed (for up to two-thirds of their length), 4.5–8.2 cm across, generally cupped, nodding to lateral-facing, sometimes ascending. *Pedicels* erect to ascending, 11–27 cm long (to 33 cm long in fruit), mostly 2–3 mm diameter, patent-bristly, often densely so, especially just beneath the flowers. *Buds* ovoid to ellipsoid, pendent or semi-pendent, the sepals patent-bristly, the bristles similar to those on the leaves and pedicels. *Petals* 6–10, pale to deep blue, mauve-blue, purple, reddish purple or wine-red, occasionally lilac, rose or pink, very variable in shape from elliptical to obovate, oblanceolate or almost rounded, 19–38 × 15–34 mm, partly overlapping or not, the margin often somewhat fluted or frilled. *Stamens* numerous, the filaments similar in colour to the petals, although generally darker, rarely whitish, the anthers yellow or golden, 1.5–2 mm long. *Ovary* subglobose to ovoid, densely covered with appressed or obliquely ascending, stout pungent bristles, the style 4–7 mm long, generally broadening towards the base, often striate, the stigma subclavate, whitish, creamish or greyish, 2.5–4 mm long, slightly exceeding the boss of stamens, 4–7 (–9) lobed. *Fruit capsule* ovoid to oblong-ellipsoid, 12–20 × 6–11 mm, adorned with stout, patent to ascending, pungent whitish or straw-coloured spine-like bristles, generally mixed with smaller bristles, dehiscing by 4–7 (–9) valves for a short distance just from the apex, the persistent style 6–10 mm long, stout, about half the length of the capsule, broadening towards the base, sometimes markedly so, often dark purplish brown or purplish black. 2n = 56 (Ratter 1968). (Figs 19, 20A, B).

DISTRIBUTION. Nepal (throughout the Nepalese Himalaya, although not verified from Mugu, Nuwakot, Sindhupalchok), NE India (Arunachal Pradesh, Sikkim), N Bhutan, N Myanmar, C, NE, S & SE Tibet (southeast as far as Kongbo and Tsari), NW China (Qinghai); 3700–5790 m (Map 16).

HABITAT. Rocky places, boulder fields, screes, stabilised moraines, rock ledges and crevices, rocky low scrub, occasionally close to streams.

FLOWERING. Late June–August (–September).

SPECIMENS SEEN. *AGSES* 616 (K); *F. M. Bailey* s.n. 3 April 1935 (BM), s.n. (BM, E); *Beer, Lancaster & Morris* 8333 (BM); *Cave* 55 (K), 6632 (BM), s.n. 1 Sept. 1919 (E); *R. E. Cooper* 1628 (K); *Curzon* 218 (K); *Cutting & Vernay* 68 (K), 76 (K); *Dhwoj* 41 (BM), 235 (BM) 369 (BM); *Dungboo* s.n. Oct. 1880 (K); *Einarsson et al.* 2958 (BM), 3012 (BM); *Farrer* 691 (BM); *B. J. Gould* 2218 (K), 2305 (K), 2343 (K); *Grey-Wilson et al.* 4277 (K); *Grey-*

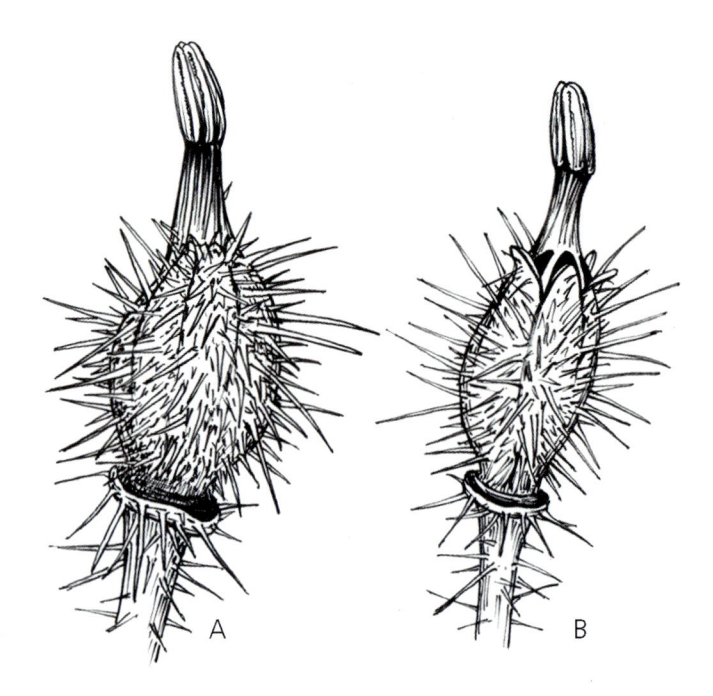

FIG. 19. *Meconopsis horridula* subsp. *horridula* fruit capsules: **A**, undehisced; **B**, dehisced. All × 2. DRAWN BY CHRISTOPHER GREY-WILSON

Wilson & Phillips 434 (K), 551 (BM, K), 650 (BM, K), 724 (K); *de Haas* 2227 (BM); *Halliwell* 175 (K); *S. Hedin* s.n. 1899 (K); *Hingston* s.n. 1924 Everest Exped. (K); *Ho et al.* 460 (BM), 754 (BM), 1652 (BM), 2074 (BM); *Hooker* s.n. Sikkim 14-17,000 ft (BM, K); *A. Horsfall* 51 (BM); *Kanai, Hara & Ohba* 72-1908 (BM); *KEKE* 511 (K); *Kingdon Ward* 5628 (K), 6006 (K), 6096 (K), 6126 (K), 6170 (K), 7091 (K); *Licent* 4671 (BM, K); *Litterdale* s.n. Aug. 1896 (K); *Lowdnes* 1094 (BM), 1134 (BM), 1242 (BM), 1309 (BM), *Ludlow* 45 (BM), 148 (BM); *Ludlow & Sherriff* 686 (BM), 796 (BM), 826 (BM), 850 (BM), 1980 (BM), 2442 (BM), 3028 (E), 3992 (BM); *Ludlow, Sherriff & Elliott* 15602 (BM), 15693 (BM); *Ludlow, Sherriff & Taylor* 6375 (BM, E), 6897 (BM); *S. B. Malla* 14225 (BM); *D. McCost* 374 (BM); *G. Miehe* 376 (BM); *Miyamoto et al.* 94-00050 (BM), 94-20210 (BM), 95-84139 (BM); *Mt Everest 1922 Exped.* 143 (K), 386 (K); *Polunin* 835 (BM), 1111 (BM), 1326 (BM), 1464 (BM), 1735 (BM), 1755 (BM), 1833 (BM); *Polunin, Sykes & Williams* 198 (BM), 2572 (BM), 2637 (BM), 3508 (BM), 4523 (BM), 5338 (BM); *Pradhan* 14225 (BM); *Przewalski* s.n. Tibet borealis (K); *Rock* 14339 (BM); *Ribu & Rohmoo* 5169 (K); *W. W. Rockhill* s.n. 1893 (K); *Rohmoo* 1032 (BM); *Schilling* 1008 (K); *K. N. Sharma* 11/94 (BM), E281 (BM), E283 (BM), E383 (BM), E384 (BM), E454 (BM); *Shipton* 113 (K), 146 (K); *T. B. Shrestha* 5289 (BM); *Stainton* 998 (BM); *Stainton, Sykes & Williams* 198 (BM), 1453 (BM), 1842 (BM), 2277 (BM), 3508 (BM), 3603 (BM), 4519 (BM), 4523 (BM), 4634 (BM), 5473 (BM), 6334 (BM), 8061 (BM), 9348 (BM); *W. G. Thorold* 134 (K); *J. B. Tyson* (BM); *Waddel* 23 (K); *L. R. Wager* 184 (K), 264 (K); *H. J. Walton* s.n. July 1904 (K); *Wollaston* 26 (BM, K); *Wraber* 280 (BM).

Meconopsis horridula is particularly common to the north of the main Himalayan watershed and in rain shadow areas in the prime monsoon belt and widespread and locally numerous over much of the central, southern and eastern Tibet, extending into the similar terrain of southern and central Qinghai. It is often found in fractured rocky, exposed habitats, sometimes sheltering amongst the larger rocks, at high altitudes: in fact *M. horridula* can without question claim the highest altitude achieved by any species of *Meconopsis*.

On the Qinghai Plateau (the north-eastern most extension of the Tibetan Plateau) particularly dwarf forms are to be found growing in desperately exposed sites between about 4000 and 4700 m altitude. These bear densely bristly narrow, oblanceolate to elliptical, leaves 5.8–11 × 0.6–1.7 (–2) cm and bunched scapes no more than 17 cm tall and often just 6–10 cm. The flowers

Meconopsis horridula subsp. *horridula*; Khung Khola, Dolpo District, W Nepal, 4600 m.
PHOTO: CHRISTOPHER GREY-WILSON

are rather small, just 4.5–6 cm across, generally of a rather pale blue or mauve-blue and unusual in facing upright or at least ascending (see photo p. 244). Normally the flowers of *Meconopsis horridula* are half-nodding or lateral-facing. In addition, the fruit capsules are on the small size and subglobose, 11–13 × 9–11 mm, with a relatively short 3–5 mm long persistent style. In addition, these plants bear concolourous bristles without darkened bases, unlike many of those found in the Himalaya. The status of this plant is uncertain and requires further investigation but these plants appear to be ecotypical variants evolving in a much drier and more severe terrain and more or less confined on the northern margin of the species distribution. It would not be surprising, for instance, to come across similar plants in remote parts of the plateau

42. MECONOPSIS HORRIDULA

within Tibet. Interestingly, *M. horridula* becomes increasingly scarce northwards across the Tibetan Plateau, the species not reaching the Kunlungshan, the mountain range that punctuates the northern margin of the Plateau along the boundary of Xinjiang (Sinkiang) Province, China's north-westernmost region that is dominated by a high, dry and desperately cold desert, the Taklimakan where few plants grow. There is a marked tendency for plants of *M. horridula* to be dwarfer, more densely spiny and with narrower, smaller leaves in the drier, most exposed habitats, while many parts of the Plateau are too dry and severe to support it at all, indeed this applies to much of the western, central and northern plateau. Towards the east and north-east of the plateau the mountains, although dry, receive at least a little benefit from the summer monsoon and this together with moisture in the soil from snowmelt is apparently enough to sustain the species.

Another variant is found in south-eastern Tibet, existing in photographs but without herbarium specimens as a back up. In this variant, photographed on the Potrang (Potrung) La, the leaves are deep, rather bright green, the lamina narrow, elliptic-lanceolate, with a slightly scalloped margin and adorned with sparse stout, pale bristles of more or less the same size. The bristles are whitish and arise from a raised, wart-like, bright green base. The flowers are pale blue, 4–5-petalled and mostly borne on basal scapes, but one of the accompanying photographs reveals a central stem bearing several bracted flowers (see photos pp. 244–245). Without further information and adequate herbarium specimens I am unable to advance this interesting plant further. In particular its relationship with subsp. *drukyulensis* requires particular attention.

Leaf indumentum varies from population to population, sometimes within a particular population. In some individuals plants are adorned with stiff pungent bristles, often densely so, especially on the leaves and pedicels (scapes). In others there is a mixture of stout bristles mixed with a few to many lesser, weaker and shorter bristles; such forms are more prevalent in the east of the distribution than the west. The stouter bristles generally arise from a raised wart-like base which may be concolorous or stained with brownish or purplish black, which gives the leaves in particular a dotted appearance.

Meconopsis horridula has proved to be extremely difficult in cultivation, despite the fact that seed has been introduced on numerous occasions from the Himalaya over the years and even more recent accessions have faired no better. Like many high Himalayan alpines it is tricky to master its cultivation even within the controlled environment of an alpine house. Seed germinates readily but many succumb at the pricking-

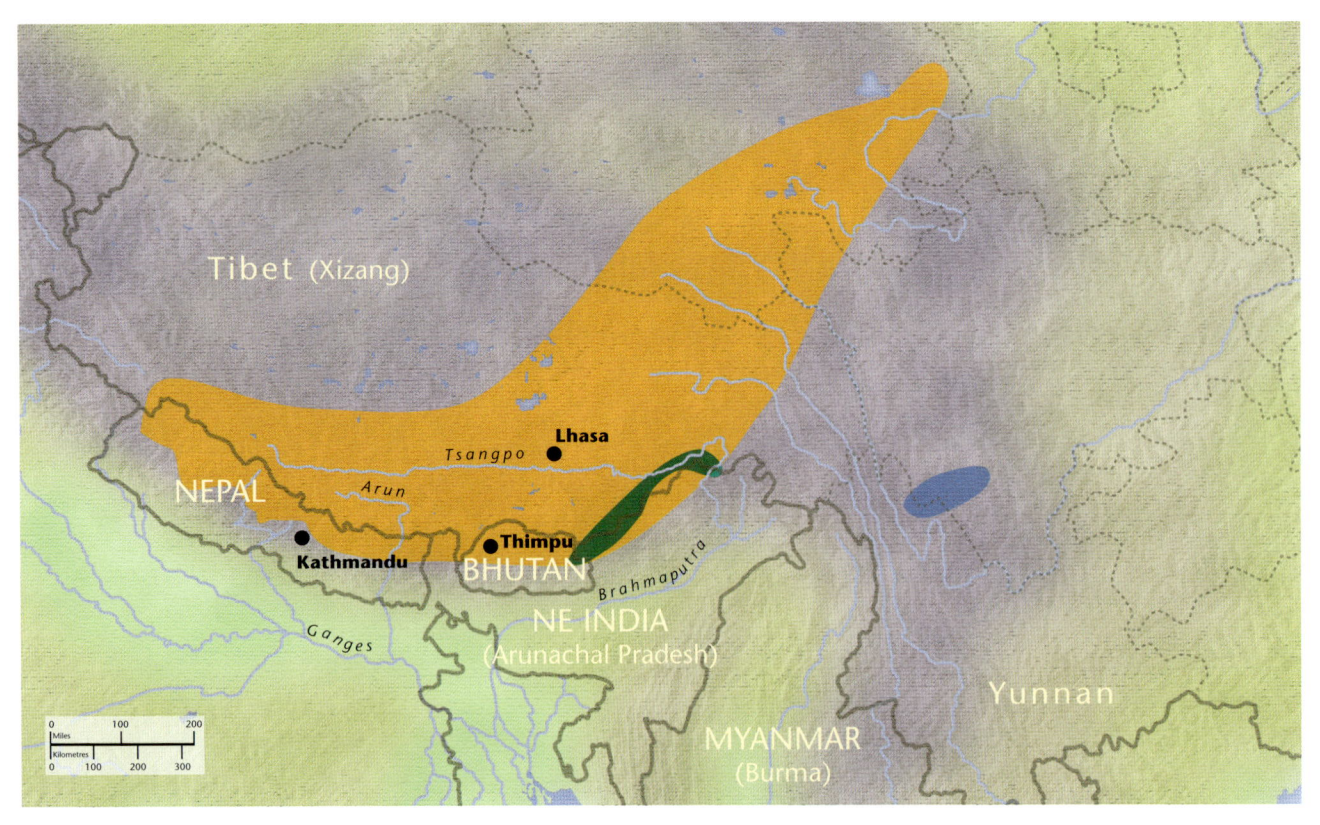

MAP 16. Species distribution: ■ *Meconopsis horridula* s.l., ■ *M. zhongdianensis*, ■ *M. prainiana*.

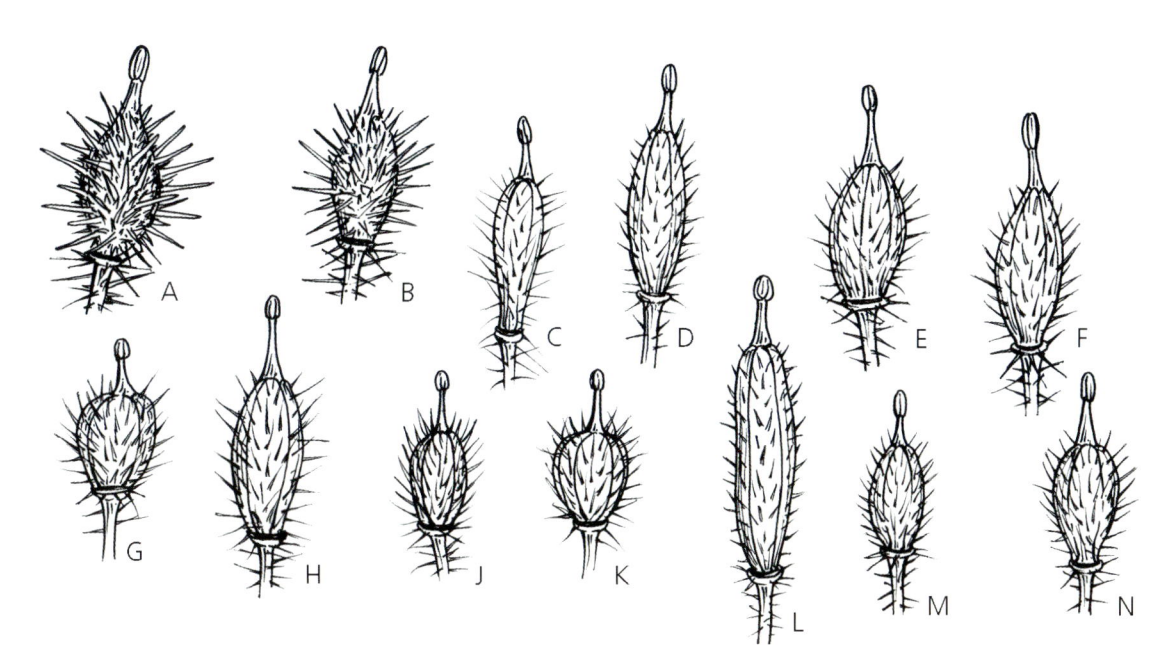

FIG. 20. Fruit capsules. **A–B**, *Meconopsis horridula* s.s.; **C**, *M. lhasaensis*; **D**, *M. prainiana*; **E–F**, *M. racemosa*; **G–H**, *M. zhongdianensis*; **J–K**, *M. rudis*; **L**, *M. bijiangensis*; **M–N**, *M. prattii*. All × 1. DRAWN BY CHRISTOPHER GREY-WILSON

out stage and very few have been brought to flowering maturity. However, there have been some successes, although short-lived: there is, for instance, a fine halftone photograph taken by William Purdom of a potful of genuine *M. horridula* in both flower and fruit in Reginald Farrer's *The English Rock Garden* published in 1928, the seed having been collected in Qinghai. The potful gives every indication that the plants have been sown and grown on without ever being pricked out, so perhaps this is an indication of how to succeed with this intractable species. If such a cultivation technique is pursued then very thin sowing or judicious thinning will be essential.

Many plants cited in literature, in plant catalogues and elsewhere, as *Meconopsis horridula* should be referred to *M. prattii* and *M. racemosa*, or perhaps *M. zhongdianensis*.

Above left: *Meconopsis horridula* forma; central Qinghai. PHOTO: ROSIE STEELE

Above right: *Meconopsis horridula* subsp. *horridula*; Langma La, S Tibet, 5320 m. PHOTO: JOHN & HILARY BIRKS

42. MECONOPSIS HORRIDULA

Above left: *Meconopsis horridula* subsp. *horridula*; Mi La, east of Lhasa, S Tibet, 4850 m. PHOTO: HARRY JANS

Above right: *Meconopsis horridula* subsp. *horridula*; Kangnai to Shun Tscho, SE Tibet, 4950 m; note insects sheltering within the flowers. PHOTO: HARRY JANS

According to the Royal Horticultural Society's *Dictionary of Gardening* published in 1951, *M. horridula* first flowered in cultivation in 1904 and this is confirmed in George Taylor's 1934 monograph of the genus. The origin of this introduction is unclear and the plant was almost certainly *M. racemosa* and not the true *M. horridula*, and as such it was written up by A. K. Bulley in *Flora and Silva* in 1905. In 1914 the closely related *M. rudis* was featured in *Curtis's Botanical Magazine*, the plant having been introduced as seed from the Lijiang mountains (Yulongxueshan) from collections of George Forrest, probably from his second expedition, 1910–11. However, David Prain (1915) comments that "Since 1908 both *M. Prattii* and *M. rudis* have been in cultivation in European gardens .... Though nearly allied, the two differ considerably in the consistence of their leaves, those of *M. Prattii* being softer in texture and of a different shade of green; in their prickles, those of *M. Prattii* are weaker and are always pale in colour, those of *M. rudis* usually being purple, at least at the base; in the anthers which are whitish or buff-coloured in *M. Prattii*, yellow in *M. rudis* .....". Successive expeditions collected seed of members of the *M. horridula* aggregate: for instance Forrest certainly collected seed of *M. prattii*, *M. rudis* and *M. zhongdianensis* (all under the name *M. horridula*) and Kingdon Ward certainly collected seed of *M. prattii* and *M. rudis*, while E. H. Wilson collected seed of *M. prattii* and *M. racemosa*. In more modern times there have been countless introductions of seed of the true *M. horridula* (primarily from the Nepalese Himalaya), of *M. racemosa*, *M. rudis* and *M. zhongdianensis*. However, '*M. horridula*' of gardens stems from the earlier collections. It is extremely variable in leaf form, height and inflorescence details which varies from lax to dense flowered, while the flower colour can range from the palest blue to deep blue, purple and lilac. George Taylor (1934) comments that "The species, under cultivation, is subject to very considerable variation, particularly in the colour and disposition of the flowers. Indeed, even if seed is carefully saved from good forms, it is rarely possible to raise a uniform batch of plants, and for this reason *M. horridula* is apt to prove a disappointment. Occasionally most attractive plants are produced when the central flowering-stem is damaged, as this appears to stimulate the growth of the lower and basal pedicels, which lengthen and produce flowers more or less at the same level, thus avoiding the rather unattractive fruiting-stem of the more normal plant. This suggests that *M. horridula* might be a better garden plant if induced to assume this form by pinching out the central flowering-stem as it is about to elongate".

The variability apparent in most garden stocks of '*Meconopsis horridula*' suggests strongly that plants are of hybrid parentage. What is quite apparent is that the true *M. horridula* is certainly not involved: this high Himalayan–Tibetan species is fiendishly difficult to cultivate and has never succeeded in the open garden, however, its allies have. Although intricate cytological examination is necessary to prove the point beyond question, I suspect that *M. horridula* of horticulture is a complex of hybrid forms that have arisen in gardens and been distributed as seed to other establishments. While some cultivated forms approach *M. racemosa* in some details, others approach *M. prattii* or *M. zhongdianensis*, and neither can one rule out the participation of *M. rudis* in the mix. Whatever their origin, the finest forms with deep blue flowers, shot through with hints of purple or lilac, are lovely plants, the tallest reaching 1 m in height, while some of the dwarfer forms with pale or mid blue flowers are equally appealing. Others have suggested to me that *M. aculeata* might also be involved in the parentage of these garden hybrids but the lack of lobing or dissection of the foliage would probably mitigate against this. On the other hand, both species

Below left: *Meconopsis horridula* subsp. *horridula*; upper Barun Khola, E Nepal, 4730 m. PHOTO: CHRISTOPHER GREY-WILSON

Below right: *Meconopsis horridula* subsp. *horridula*; lateral moraine Kangchenjunga glacier, E Nepal, 4700 m. PHOTO: TOSHIO YOSHIDA

are quite easy to cultivate, especially in humus beds in open woodland glades where, in quite a few gardens in Britain, they are known to self-sow. Interestingly, both *M. aculeata* and *M. horridula* hort. often behave as biennials in gardens rather than monocarpic plants flowering after several years from seed. Occasionally they have been witnessed flowering in the first year from seed. Such an annual or biennial lifestyle is unwitnessed in the wild where there is every indication that plants are essentially monocarpic in habit.

Interestingly, a number of different species in the complex have been cultivated in field plots at the research station at Napahai on the Zhongdian Plateau in NW Yunnan, including *Meconopsis zhongdianensis* itself (which naturally grows close by), as well as *M. rudis* and *M. prattii* and possibly *M. racemosa*. Left to themselves these species are no longer distinguishable on the plots, having hybridised freely to produce a multitude of forms from short plants no more than 20 cm tall to plants in excess of 90 cm and bearing lax to dense inflorescences in a range of colours from pale blue and mauve to rich blues, lavenders and purple. Some plants have produced flowers with supernumerary petals and some of the individuals are very similar to some of the forms of '*M. horridula*' in cultivation in western gardens. Such hybrids are unlikely to occur in the wild as the species involved do not grow together.

In Tibetan herbal medicine *Meconopsis horridula* is used to treat a variety of conditions from colds and fevers, to sinusitis, itching, bile, lung and skin diseases.

NOTE. Despite the fact that the *Flora of Nepal* does not record *Meconopsis horridula* from the large western district of Dolpo I can confirm its presence there. On my first expedition to Nepal in 1971 (led by Dr George F. Smith) we came across it on a number of occasions in Dolpo, notably in the upper reaches of the Khung Khola where I photographed it for the very first time. Its absence from the other districts specified above seems inconceivable, as does its apparent absence from the Himalaya to the immediate west of Nepal in NW India (Uttar Pradesh).

var. *spinulifera* L. H. Zhou, *Acta Phytotax. Sin.* 17(4): 113 (1979). Type: Qinghai, Yushu, 4000 m, *B. Z. Gua & W. Y. Wang Weiyi* 8354 (holtoype, PE). Unfortunately, I have not been able to examine the type specimen of this interesting variety; however, the original description and accompanying drawing clearly show a plant about 30 cm tall with a well-defined multi-flowered racemose inflorescence. These features place the plant in question not in *M. horridula* according to this account but in *M. racemosa* which is widely distributed in north-western

Meconopsis horridula forma; Amne Machin, SE Qinghai, NW China. PHOTO: RON MCBEATH

Meconopsis horridula forma; W of Potrang La, SE Tibet, 4720 m. PHOTO: PAM EVELEIGH

Sichuan, southern Gansu and much of Qinghai except the north-west. The distinctive, and indeed only feature that separates this plant is the presence of bristles on the outside of the petals towards the base. The presence of petal bristles is very rare in *Meconopsis* but is present in the unrelated *M. torquata* which is endemic to the hinterland of Lhasa in Tibet. For this reason var. *spinulifera* is probably best transferred to *M. racemosa* but this requires further investigation and I have not at present formalised this new combination.

Meconopsis horridula subsp. drukyulensis Grey-Wilson, **subsp. nov.** Similar to the typical plant, subsp. *horridula*, in habit but leaves often incisely dentate or shallowly lobed, adorned with somewhat raised wart-like, yellowish green bristle bases, the flowers relatively larger and nodding, scapose or pseudoracemose. Type: C Bhutan, Dungshinggang (Black Mountain), 15,000 ft, *Ludlow & Sherriff* 3308 (holotype, BM; isotype E).

DESCRIPTION. *Plant* 27–41 cm tall in flower. *Leaves* rather bright green, sometimes flushed with purple, mostly basal, elliptic to ovate-elliptic or oblanceolate, 7–14.5 × 1.8–3.3 cm, the margin unevenly incise-dentate to sublobed, sometimes slightly revolute, adorned with stout sharp bristles, underlain with some shorter ones, sometimes densely so, the larger bristles generally with a slightly raised, wart-like, yellowish green base which is particularly pronounced in young unflowered plants; cauline leaves few, borne towards the base of the plant, similar to the basal leaves but sessile or subsessile; petiole 3.2–5 cm long. *Inflorescence* racemose-agglutinated with up to 14 flowers, generally accompanied by basal or sub-basal scapose flowers, the flowers saucer- to bowl-shaped, 7–10 cm across. *Pedicels* (*scapes*) to 18 cm long, moderately to densely and unevenly bristly, especially dense just beneath the flowers, often purple-flushed. *Petals* 6–8, sky blue to turquoise, bright blue or lavender-blue, rarely purplish blue, rose-pink or rosy purple, rounded to ovate, 3.3–4.8 × 3.1–4.5 cm, with a slightly erose margin often. *Stamens* with filaments the same colour as the petals, although somewhat darker, the anthers yellow or golden. *Ovary* ovoid, with dense ascending or subappressed bristles, the style 3–6 mm

Right, from top:

Meconopsis horridula forma; W of Potrang La, SE Tibet, 4720 m.
PHOTO: PAM EVELEIGH

Meconopsis horridula forma; W of Potrang La, SE Tibet, 4720 m.
PHOTO: PAM EVELEIGH

Meconopsis horridula subsp. *drukyulensis*, young leaf-rosette; Om Tsho La, Tongsa district, C Bhutan, 4400 m. PHOTO: TIM LEVER

42. MECONOPSIS HORRIDULA

Above left: *Meconopsis horridula* subsp. *drukyulensis*; near Chukarpo, W Bhutan, 4595 mn. PHOTO: MARTIN WALSH

Above right: *Meconopsis horridula* subsp. *drukyulensis*; Tempe La, C Bhutan, 4350 m. PHOTO: MARTIN WALSH

Meconopsis horridula subsp. *drukyulensis*; Om Tsho La, Tongsa district, C Bhutan, 4400 m.
PHOTO: TIM LEVER

long, protruding well beyond the boss of stamens, greenish, the stigma 4–6 mm long, greenish or yellowish. *Fruit capsule* 14–20 × 8–9.5 mm, broadened towards the base, covered in dense stout bristles with an underlay of smaller bristles, ascending-appressed at first, later patent, the persistent style 5–9 mm long.

DISTRIBUTION. North-western, central and north-eastern Bhutan (Dungshinggang, Goktang La, Jui La, Me La, Namda La, Passu Sepu, Upper Pho Chu, Shingbe, Tang Chu), NE India (W Arunachal Pradesh); 4268–4905 m.

HABITAT. Rocky places, screes, cliff bases and ledges.

FLOWERING. June–August.

ETYMOLOLGY. Druk Yul is the alternative name for Bhutan.

SPECIMENS SEEN. *R. Bedi* 991 (K); *B. J. Gould* 549 (K), 1356 (K); *Ludlow & Sherriff* 374 (BM), 1034 (BM), 3308 (BM), 3376 (BM); *Ludlow, Sherriff & Hicks* 16637 (BM), 16707 (BM), 16911 (BM), 17229 (BM), 19806 (BM), 19439 (BM), 20761 (BM), 21198 (BM), 21330 (BM); *Sinclair & Long* 5273 (E, K).

NOTE. L & S 3308 (Dungshinggang (Black Mountain), 15,000 ft) & L, S & H 21330 (Narimthang, 14,000 ft) have exceptionally large flowers, scapose or with 2–3 flowers agglutinated!

THE GENUS MECONOPSIS | 247
42. MECONOPSIS HORRIDULA

Above: *Meconopsis horridula* subsp. *drukyulensis* herbarium sheet, *Ludlow & Sherriff* 3308 (BM), holotype.

Left, from top:

Meconopsis horridula subsp. *drukyulensis*; Tempe La, C Bhutan, 4350 m. PHOTO: MARTIN WALSH

Meconopsis horridula subsp. *drukyulensis*, pink-flowered form; near Chukarpo, C Bhutan, 4595 m. PHOTO: MARTIN WALSH

Subsp. *drukyulensis* is confined to Bhutan and the border area of Arunachal Pradesh, although it is not exclusive to that region because subsp. *horridula* can also be seen in parts of the north and north-east; however, it is perhaps the most beautiful taxon with exceptionally large flowers which, in the most striking forms are a delightful soft turquoise blue. Apart from its large flowers, subsp. *drukyulensis* is distinguished by its rather bright green leaves that usually sport several, rather uneven, somewhat incised teeth or short lobes along the margin. In addition, the large leaf bristles have a raised, wart-like, yellow-green base which is especially pronounced in young plants. The flowers are mostly borne in a lax but short raceme, generally accompanied by one or several basal or sub-basal scapes.

43. MECONOPSIS LHASAENSIS

Meconopsis lhasaensis Grey-Wilson **sp. nov.** Closely allied to *M. horridula* but a slighter plant with thin, less pungently bristly leaves and pedicels, the flowers borne on long, very slender, scapes that are sometimes agglutinated in the lower part. Type: S Tibet, Reting, 60 miles N of Lhasa, 15,000 ft, July 1942, *Ludlow & Sherriff* 8981 (holotype, E; isotypes, BM).

DESCRIPTION. *Small monocarpic, rather wispy herb* to 30 (–37) cm tall in flower; stem slender, 3–4 mm diameter towards the base, moderately to densely bristly, the bristles thin and rather weak. *Leaves* pale yellow-green or grey-green, spreading to somewhat ascending, mostly borne in a rather lax basal rosette, petiolate, the lamina elliptic to elliptic-oblanceolate, 3.8–8.5 × 0.7–1.7 cm, the base cuneate into the petiole, the apex acute to subacute or subacuminate, sparsely to moderately covered on both surfaces with pale rather weak bristles of varying lengths, the margin somewhat undulate, entire or occasionally shallowly and unevenly lobed; petiole 17–28 mm long, linear, slightly expanded at the base. *Inflorescence* racemose, accompanied by several basal scapose flowers, the flowers 5–19 relatively small, rather flat to saucer-shaped, 2.6–4.2 cm across, lateral facing or ascending; stem generally present between the lowermost leaves and the bracts but only the lowermost flowers bracteate. *Pedicels* ascending, green, often flushed with purple, 1.4–6.5 cm long (that of the terminal flower 4.5–6 cm long), rather sparsely patent bristly; basal scapes (pedicels) to 12 cm long. *Buds* pendent, pyriform, 8–10 mm long, green marked with purple, the sepals with sparse patent bristles. *Petals* 5–8, very pale to mid blue, occasionally mauve-blue to mauve-red, elliptic to elliptic-oblanceolate, 12–20 mm long, spreading widely apart, sometimes somewhat reflexed. *Stamens* with linear filaments the same colour as the petals although generally a little darker, the anthers yellow, c. 1 mm long. *Ovary* subglobose, densely appressed bristly, the bristle c. 2 mm long, the stigma cream. *Immature fruit capsule* narrow-ellipsoid, 18–19 × 5–6 mm, the styles linear, 3–4 mm long; stigma c. 1.5 mm. (Fig. 20C).

DISTRIBUTION. S Tibet, north and east of Lhasa (Nyenchengtan, Pempochakla, Reting, Serapu); 3965–4575 m.

HABITAT. Stony slopes, cliff ledges, boulder fields.

FLOWERING. June–July.

SPECIMENS SEEN. *Aufschnaiter* s.n. 1946–50, vicinity Lhasa (BM); *J. Guthrie* 20240 (K), s.n. 27 July 1946 (K); *Kennedy* 20 (K); *Ludlow & Sherriff* 8753 (BM, E), 8785 (BM), 8981[31] (BM, E), 9643 (BM), 9789 (BM), 9964 (BM, E), 9965 (BM, E); *Miehe* 1177 (BM); *Richardson* 229 (BM); *Spencer Chapman* 157 (BM), 296 (K), 394 (K); *Waddel* s.n. Sept. 1904 (K); *Walton* s.n. Aug. 1904 (BM, K).

Meconopsis lhasaensis is primarily found within an area stretching some 90 km into the mountains around Lhasa in southern Tibet. My attention was first drawn to it by the collections of Ludlow & Sherriff, but others have come across it in the years that followed their collecting. Most notable have been the travels of Toshio Yoshida who photographed it in a number of localities not far from Lhasa while preparing his monumental book *Himalayan Plants Illustrated* published in 2005.

Meconopsis lhasaensis; behind the Drepung Monastery, Lhasa, S Tibet, 4450 m. PHOTO: TOSHIO YOSHIDA

---

[31] The sheet at the British Museum (BM) is a mixed gathering, apparently collected near Reting, 60 miles N of Lhasa. The British Museum sheet contains three specimens: the central one conforms to *M. lhasaensis* and is therefore an isotype, however, the other two are dwarf scapose plants with deeply pinnately lobed leaves and are referrable to *M. speciosa* subsp. *cawdoriana*. This latter taxon is not found in the Lhasa region but well to the east in the wetter mountains bordering the Yalung Tsangpo. There is a strong suspicion that these have been fixed to the *L&S* 8981 sheet by mistake. Ludlow & Sherriff made almost 20 collections of *M. speciosa* subsp. *cawdoriana* in south-eastern Tibet.

Meconopsis lhasaensis can be likened to a rather refined version of *M. horridula* with thinner, less bristly leaves and long thin pedicels (scapes). The flowers are pale and rather flat or saucer-shaped, ascending to lateral-facing, the petals sometimes slightly reflexed. The fruit capsule is relatively narrow and with a linear persistent style. In contrast, *M. horridula* bears thicker more leathery leaves with rigid bristles and stouter more densely bristly pedicels (scapes). The flowers, that are generally larger (4.5–8.2 cm across rather than 2.6–4.2 cm) are nodding or half-nodding (except for forms found in the drier hinterland of the Qinghai Plateau). In addition the stout little fruits bear a cone-shaped persistent style whose base expands out over the top of the ovary.

44. MECONOPSIS RACEMOSA

Meconopsis racemosa Maxim., *Bull. Acad. Imp. Sci. Saint-Pétersbourg* 28: 310 (1877) and in *Fl. Tangut.*: 36 (1889). Type: NW China, Gansu (Kansu), near Ta-tung range, Chobsen, *Przewalski* (LE).

[syn. *Meconopsis horridula* var. *racemosa* (Maxim.) Prain, *J. Asiat. Soc. Bengal, Pt. 2, Nat. Hist.* 64 (2): 313 (1896)]

DESCRIPTION. *A tall rather leafy monocarpic plant* to 95 cm tall in flower, although generally 45–75 cm, and as little as 28 cm, with a stout dauciform taproot, 7–17 mm diameter at the top. *Stem* erect, to 10–12 mm diameter towards the base, beset with rather sparse to dense patent whitish or pale straw-coloured bristles, mostly 4–6 mm long but generally accompanied by fewer much shorter ones. *Leaves* mostly in a basal rosette, grey-green, the lamina elliptic-oblong to elliptic-oblanceolate, to linear-oblanceolate, occasionally ovate-elliptic, especially in young unflowered plants, 8–21.5 × 1.4–3.8 (–5.3) cm, attenuate into the petiole, the apex obtuse to subacute, sparsely covered by patent bristles similar to those on the stem, the margin entire, generally rather flat, the lower stem leaves similar to the basal but generally smaller, short-petioled or subsessile, the upper stem leaves merging with the bracts, increasingly smaller and sessile; petiole narrow 3–11.5 cm long. *Inflorescence* a slender 10–27-flowered raceme with only the lowermost flowers bracteate, often accompanied by one or several solitary flowered basal scapes; flowers lateral-facing, saucer-shaped to rather flat, 3.3–6 cm across. *Pedicels* 0.5–6.7 cm long (to 9 cm long in fruit), with sparse patent bristles, but densely so immediately beneath the flowers; basal scapes to 22 cm long. *Buds* obconical, pendent, patent bristly. *Petals* 6–8, bright azure blue to deep blue or purplish blue, rarely pinkish, ovate to obovate, 3–3.6 ×1.8–2.4 cm, the margin often rather undulate or somewhat fluted. *Stamens* with filaments the

Meconopsis lhasaensis; Guo La, N of Lhasa, S Tibet, 4200 m.
PHOTO: TOSHIO YOSHIDA

Meconopsis lhasaensis; Guo La, N of Lhasa, S Tibet, 4200 m.
PHOTO: TOSHIO YOSHIDA

same colour as the petals, although somewhat darker, 9–11 mm long, the anthers white or greyish, occasionally very pale yellow, 1–1.5 mm long. *Ovary* ovoid, beset with dense ascending to subappressed pungent bristles; style 5–7 mm long, broadening towards the base; stigma whitish or greyish, subclavate, 4–5 mm long, protruding beyond the boss of stamens. *Fruit capsule* obpyriform to obovoid, 9–17 × 7–9 mm, beset with patent subpungent bristles, the persistent style 6–12 mm long, linear but broadening slightly towards the base. (Fig. 20E–F).

DISTRIBUTION. NW China: SW Gansu, NW Sichuan, NE, E & SE Qinghai (Amne Machin and Dabanshan southwards); 3230–4700 m (Map 17).

HABITAT. Rocky and stony meadows, open scrub, often with shrubby *Potentilla*, rocky slopes, occasionally in rock pockets or in open *Picea* woodland.

FLOWERING. Mid-June–August.

SPECIMENS SEEN. *Bartholomew & Gilbert* 1258 (E); *Boufford et al.* 29486 (E), 29724 (E); *R. C. Ching* 655 (E); *W. P. Fang* 4097 (E); *Farrer* 136 (E), 691 E); *Ho, Bartholomew & Gilbert* 185 (BM, E), 916 (BM, E), 1258 (BM, E); *Ho, Bartholomew & Watson* 2478 (BM, E); *Long et al.* 284 (E), 444 (E), 515 (E); *Potanin s.n.* 1885 (K); *Purdom* 736 (K); *Rock* 192 (BM), 12621 (E, K), 13009 (BM), 13026 (E), 14434 (E); *Soulie* 2433 (K); *SBQE* 670 (E), 676 (E), 677 (E); *E. H. Wilson* 951 (K), 951a (K).

Meconopsis racemosa is a characteristic plant of the drier hills and moorland on the north-eastern fringes of the Tibetan Plateau, being especially prevalent in north-western Sichuan and the neighbouring regions in central and south-eastern Qinghai and southern Gansu, from whence it was first described. It is a handsome and bold species which, in the most striking forms, carries rich blue flowers. Plants bear quite dense racemes of flowers and can be found in the wild in small groups or large scattered colonies, often to the exclusion of other *Meconopsis* species, but sometimes in association with the small, sharply bristly north-eastern forms of *M. horridula*.

The species clearly finds its closest ally in *Meconopsis prattii* and the two have been greatly misunderstood and confused not only in literature, but also in horticulture. While the former was described from Gansu, the latter was described from a plant collected much farther south close to Tatsienlu (now Kanding) in western Sichuan. While both are slender, *M. racemosa* is a generally much taller and stouter plant carrying denser racemes of flowers than *M. prattii*, while the inflorescences of the latter are characteristically secund or semi-secund. Differences can be observed in the flowers themselves: those of *M. racemosa* generally have 6–8 petals that are broad and overlapping, those of *M. prattii* generally 5–6-petalled, the

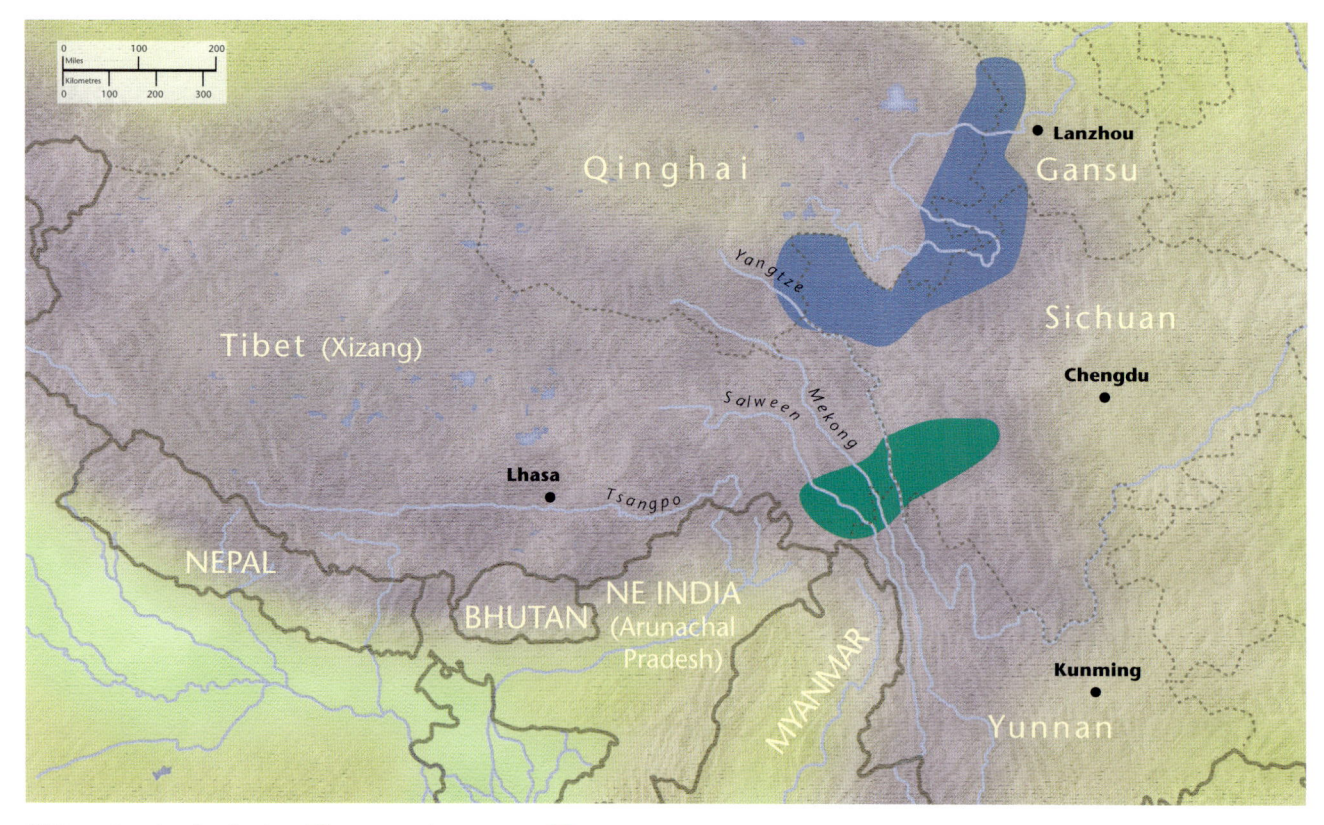

MAP 17. Species distribution: ■ *Meconopsis racemosa*, ■ *M. prattii*.

Meconopsis racemosa; between Aba and Longriba, NW Sichuan, 4100 m. PHOTO: CHRISTOPHER GREY-WILSON

petals often narrowed below and rather gappy, scarcely or not overlapping, although this is quite variable. The fruit capsules can also be distinguished fairly readily; those of *M. racemosa* are larger and relatively narrower with a longer persistent style that is linear for much of its length, while the style of *M. prattii* is a narrow cone-shape (i.e. expanding gradually towards the base).

Unfortunately, there is some confusion over the type specimen and type locality of *Meconopsis racemosa*. The species was based on specimens collected by the Ta-tung river near the temple at Chobsen, which is situated in western Gansu to the east of the great lake Kokonor (= Kuku-nor; today Qinghai Lake). Today the river the Datong Ho (Farrer's Da-Tung) drains the Dabanshan range (Farrer's Da-Tung Alps) which arcs north and east of the city of Xining. This locality cited as Kansu (= Gansu) by Farrer lies in the province of Qinghai, created since Farrer's day. The Dabanshan is a considerable and rather bleak range rising to just over 4500 m at its maximum. Although there exists in the herbarium at Leningrad a number of Przewalski specimens none appear to be annotated 'Chobsen or Chebson' and they do not exist in any other herbarium collection to my knowledge. David Prain (1915) was equally puzzled by this, remarking: "But the writer has not seen a specimen of the original *M. racemosa* from Chobsen, in Kansu, in any of the herbaria he has been able to examine, nor is there any indication that duplicates of this Chobsen gathering have ever been issued to other herbaria. It is not definitely stated that the figure which appears in the *Flora Tangutica was* prepared from a specimen of the original Chobsen gathering. Moreover, no one since Przewalski has collected anywhere in Kansu specimens of the Tibetan *Meconopsis* with dark blue flowers and yellow anthers, figured and distributed by Maximowicz as *M. racemosa*".

In *The English Rock Garden* (1928) Reginald Farrer comments that "The original specimen on which Maximowicz based *M. racemosa* is said to be labelled as coming from Chebson (Chobsen) Abbey on the western foothills of the Da-Tung Alps. But, in examining the specimens of *M. racemosa* in the Petrograd herbarium, I was not able to find any indication of this locality, all those given referring to the ranges south and west of Sining [Xining], and away to the Koko-nor. This exactly bears out my own experience: in the Da-Tung Alps I never saw a sign of *M. racemosa*, the prevailing Poppy of the high grass and shingles being exclusively *M. prattii*, while Chebson Abbey could in no case be the haunt of any prickly Poppy at all ....".

44. MECONOPSIS RACEMOSA

Above left: *Meconopsis racemosa*; near Longriba, NW Sichuan, 4150 m. PHOTO: CHRISTOPHER GREY-WILSON

Above right: *Meconopsis racemosa*; Huanglong, Songpan County, NW Sichuan, 3900 m. PHOTO: TOSHIO YOSHIDA

Not long after this was written, in 1914–15, Reginald Farrer explored the mountains of Gansu and this was later recorded in his book *The Rainbow Bridge* published in 1921. Farrer, who travelled in part with William Purdom, was equally puzzled by *Meconopsis racemosa*. "……. The original specimen of all on which the description of *Meconopsis racemosa* as a species was built, is one of Przewalski's collecting, and is said to bear 'Chebson' on its label. Now, around Chebson itself clearly there was no poppy of the Celestial group: therefore it seemed reasonable to suppose, on finding one of the group in the Alps behind, that this must be the missing *M. racemosa* — which I will henceforth be familiar with as the sapphire Poppy, seeing that, while *M. prattii* has blossoms of clear light blue with creamy drab anthers, *M. racemosa* tends to a deeper tone, with orange anthers gorgeously contrasting".

Farrer seems to have a preconceived idea on the characteristics of *Meconopsis prattii*; however, this species was described from a long way to the south, in fact some 700 km, from the mountains in the vicinity of Kangding (formerly Tatsienlu) in western Sichuan. Modern collections from the same mountains which are clearly the same taxon show that the anther colour of these plants is consistently yellow, occasionally yellowish grey (see under *M. prattii* for further comments). I strongly suspect that all Farrer's references to *M. prattii* in Gansu (including modern day Qinghai province) with cream anthers are in fact referable to *M. racemosa* and it is doubtful that *M. prattii* occurs in the region at all. I have travelled on several occasions in south-eastern Qinghai and north-western Sichuan where *M. racemosa* is a relatively common plant of the grassy moorlands and rocky outcrops and screes and, although like all members of the *M. horridula* complex it is a very variable plant, the anthers are consistently cream or greyish cream. Anther colour has to be judged at anthesis because anthers can often darken with age and may be masked by the pollen when that bursts forth.

This confusion is borne out by Reginald Farrer, who also wrote with great passion and insight, and had the following to say about *Meconopsis racemosa* (some of the information previously recorded in his book *The Rainbow Bridge* (see above)):

"*Meconopsis racemosa* may always be known from *M. prattii* by its golden, instead of creamy, anthers. It seems that one specific name ought to include *M. horridula* and *M. racemosa* as fluctuating forms of a single species. The original specimen on which Maximowicz based *M. racemosa* is said to be labeled as coming from Chanson (Chobsen) on the Western foothills of the Da-Tung Alps. But in examining the specimens of *M. racemosa* in the Petrograd Herbarium, I was not able to find any indication of this locality, all those given referring to the ranges south and west of Sining (Xining), and away to the Koko-nor. This exactly bears out my own experience; in the Da-Tung Alps I never saw a sign of *M. racemosa*, the prevailing Poppy of the high grass and shingles being exclusively *M. prattii*, while Chebson Abbey could in no case be the haunt of any Prickly Poppy at all, sheltering as it does in the green foothills, some 4 to 6 miles distant from the great Alps behind. On the other hand, the Poppy brought back by Purdom from the ranges of Kweite and the Koko-nor, was unmistakably and universally the genuine *M. racemosa*, darker blue in colouring, and golden anthered, in forms as often as not reverting wholly or partially to the trunkless, many-flowered development which is typically *M. horridula*."

[Reginald Farrer, *The English Rock Garden* 2: 479–80, 1928].

It is clear that the name Chebson was used by Przewalski as a general location and not meant to be specific and that Farrer's assertion was correct, for the species is found today in the Dabanshan immediately to the north along with *Meconopsis quintuplinervia*, in fact it is widely distributed in the mountains of eastern and south-eastern Qinghai and neighbouring Gansu, as well as north-western Sichuan. Farrer based much on anther colour aligning all those specimens with golden anthers in *M. racemosa* and all those with greyish or cream anthers in *M. prattii*.

In summary, the Gansu-Qinghai plants with well-developed inflorescences carrying flowers often of rich blue and bearing cream-coloured anthers all belong to *Meconopsis racemosa*. The closely allied *M. prattii*, which is found far to the south-west is a generally slighter plant, usually with yellow or golden anthers. It differs in others respects (see p. 255 and the key above). The other prickly poppies of the Qinghai Plateau area, which are generally rather dwarf and with scapose flowers, the scapes sometimes partly agglutinated, are referable to *M. horridula*.

Right (top and bottom): *Meconopsis racemosa*; Huanglong, Songpan County, NW Sichuan, 3980 m. PHOTOS: CHRISTOPHER GREY-WILSON

Both botanically and horticulturally the name *Meconopsis racemosa* has been widely used to cover virtually any plant in the *M. horridula* aggregate that happens to have a racemose inflorescence and this has been the prime source of misunderstanding over the years. Within the aggregate the following species bear a racemose inflorescence: *M. bijiangensis*, *M. georgei*, *M. prainiana*, *M. prattii*, *M. rudis* and *M. zhongdianensis*. These species, which all hail from western China and southeastern Tibet, occupy rather isolated positions from one another geographically, that is with the exception of *M. prattii* and *M. rudis* which overlap in distribution to some extent, although they have not been found growing together to my knowledge.

The true *M. racemosa* has certainly been in cultivation in the past but doubtfully so today, except at a few Botanic gardens (e.g. Edinburgh Botanic Garden). See also under the following species.

Meconopsis horridula Hook. f. & Thomson var. *spinulifera* L. H. Zhou (see *Acta Phytotax. Sin.* 17 (4): 112–113) described in 1979 almost certainly belong in *M. racemosa*, although I have not been able to examine the type material. This taxon is distinguished by the presence of a few sparse bristles towards the base of the petals on the exterior and apparently by its very slender filaments. Details of the type are as follows: Qinghai Province, Yushu, 4000 m altitude, *Gua Ben-zhau & Wang Wei-yi* 8354 (PE).

45. MECONOPSIS PRATTII

Meconopsis prattii Prain, *Curtis's Bot. Mag.* sub t. 8568 (1914), nomen; et t. 8619 (1915), descr.; *Bull. Misc. Inform., Kew* 1915: 149 (1915). Type: W China, Sichuan, Tachien-lu [Tatsienlu], *Pratt* 525 (K, holotype; BM isotype).
[syn. *M. sinuata* Prain var. *prattii* Prain, *J. Asiat. Soc. Bengal, Pt. 2, Nat. Hist.* 64 (2): 314 (1896) and *Curtis's Bot. Mag.* 140, sub tab. 8568 (1914), in obs.; *M. rudis* var. *prattii* mss?]

DESCRIPTION. *Monocarpic plant* to 40 cm tall in flower but as little as 15 cm, with a long slender, simple or somewhat divided taproot 8–12 mm diameter at the top. *Stem* erect, slender, 4–6 mm diameter at the base, sparsely to moderately covered in patent to somewhat deflexed, pungent straw-coloured stiff yet thin bristles, these 3–7 mm long. *Leaves* grey-green, often rather pale, mostly crowded in a basal rosette, the lamina narrow to broadly elliptic to elliptic-lanceolate or elliptic-oblanceolate, 5.5–16 × 0.9–3.5 cm, the base gradually attenuate into the petiole, the apex subobtuse to acute, sparsely to moderately adorned on both surfaces with patent bristles similar to the stem, somewhat undulate often, the margin generally somewhat sinuate to remotely and unevenly toothed, rarely more coarsely so; cauline leaves few, alternate, rarely more than 6, similar to the basal leaves but gradually decreasing in size up the stem, short-petiolate to subsessile, merging with the bracts, the pubescence the same as the basal leaves; petiole of basal leaves 1.5–5.2 cm long. *Inflorescence* racemose, more or less secund, bearing 7–17 (–20) flowers, generally borne from close to the base of the plant, with only the lowermost flowers bracteate, the raceme sometimes accompanied by one or several solitary flowers borne on basal scapes; flowers lateral-facing, saucer-shaped, 4–6 cm across. *Pedicels* patent to ascending, 8–26 mm long, the terminal one in the inflorescence to 5.8 cm long, patent-bristly, sometimes sparsely so but densest just below the flower; basal scapes, when present, to 11.5 cm long. *Buds* pendent or semi-pendent, obovoid to subpyriform, c. 10–15 × 7–10 mm, patent-bristly. *Petals* 5–6, pale to mid blue or purplish blue, occasionally claret-coloured, elliptic-oblanceolate, 2–3.2 × 1–1.8

Meconopsis racemosa; Huanglong, Songpan County, NW Sichuan, 3950 m. PHOTO: CHRISTOPHER GREY-WILSON

Meconopsis prattii; N of Ata Kang La, SE Tibet, 4500 m.
PHOTO: TOSHIO YOSHIDA

cm, somewhat ruffled. *Stamens* numerous, the filaments linear, 8–11 mm long, darker than the petals, the anthers cream or buff to pale yellow, occasionally greyish, 1.5–2 mm long. *Ovary* ovoid to subglobose, beset with subappressed stiff bristles; style linear or very slightly expanded at the base, 3–4.5 mm long; stigma whitish or greenish white, capitate c. 1 mm, just protruding from the boss of stamens. *Fruit capsule* ovoid, 10–12 × 9–11 mm, densely covered in patent to ascending, sharply pointed bristles, the persistent style 5-6 mm long with a short, narrow cone-like base. 2n = 56, as *M. horridula* sensu Ying *et al.* 2006. (Fig. 20M–N).

DISTRIBUTION. W & SW Sichuan, NW Yunnan (probably not found east of the Yangtse river) and SE Tibet; 3650–4700 m (Map 17).
HABITAT. Rocky slopes, meadows, cliff ledges.
FLOWERING. June–early August.
SPECIMENS SEEN. *R. Cunningham* 209 (E); *Forrest* 12834 (E, K), 13021 (E, K), *Hanbury-Tracy* 49 (BM); *Kingdon Ward* 762 (E), 891 (E), 4164 (E), 4171 (E), 4616 (E), 4701 (E), 5107 (E); *McLaren* 107 (E), s.n. 1938 (E); *A. E. Pratt* 525 (BM, K); *Rock* 17683 (K), 17776 (E), 23216 (E, K), 23416 (E), 23774 (E); *Soulié* 635 (K); *E. H. Wilson* 951 (E, K), 951a (K), 3030 (K), 3162 (K), 3163 (BM, K).

Meconopsis prattii has been the source of a great deal of confusion and misunderstanding over the years. The species was based upon a collection made by Pratt at the end of the nineteenth century near Tatsienlu (now Kangding) and described in 1896, surprisingly as a variety of the Himalayan *M. sinuata*. Realising his mistake, David Prain upgraded the plant to species level in 1914. From then on it has been variously equated with *M. horridula* and *M. racemosa*, sometimes as a variety, sometimes as synonymous. Reginald Farrer assumed that the plant which he came across in Gansu province was the same plant:

"*Meconopsis Prattii*. Seed was distributed as *M. rudis*, but this glorious blue Poppy is *M. prattii*. In Fedde's key to the race, *M. rudis* has stem-leaves up to the middle of the spire, while *M. racemosa* has neither bracts nor stem-leaves at all. Unfortunately, in the diagnoses of *M. racemosa*, a full description is given of the stem-leaves already declared to be non-existent! My quite different Kansu plant [F 136], sent out as *M. rudis*, is undoubtedly *M. Prattii*, and *M. Prattii* alone. The specimens and seedlings will, however, repay investigation, as these two Poppies are not as yet of any final and absolute distinctness. F 136, at least, takes two clearly-marked forms; so far as I can judge, from Thundercrown up into the foothills of the Min S'an, it is a dense and stocky plant forming a close 8- to 10-inch race of gorgeous dawn-blue blossoms, woven of silk and opals. In the highest craggy Alps above Ardjeri it takes a new character; the stems are taller, darker, barer, the pedicels are very much longer, so that the inflorescence is a loose and irregular broken flight of flowers, instead of a solid huddled mass.........in every variety this Poppy (or Poppies) it must be noted, stands apart from all its grass-loving kin, in being always and only found in the gaunt screes and stone-slopes and precipices of the highest limestone or shaly ridges from 12,000 to 14,000 feet. In other words, it is born and made for the moraine, and there should be sown again and again, that its biennial splendour may annually repeat the glory of light with which its dense spires of amassed azures illuminate the vast and lifeless stone-slopes on the highest crests of Tibet. Every part of the growth is virulently prickly, and the fierce hardened thorns of the fruiting stage make its sturdy pyramids of capsules an agony to collect, unless with a mailed fist and a pair of tongs. (The Celestial Poppy).

[Reginald Farrer, *The English Rock Garden* 2: 507, 1928]

45. MECONOPSIS PRATTII

As has been noted (p. 252) Farrer's plant is in fact the related *Meconopsis racemosa*, itself the subject of great confusion in literature. In his 1934 monograph of the genus, George Taylor makes little attempt to separate the members of the 'Horridula' complex, declaring that:

> "The polymorphic character of this species is reflected in the somewhat extensive synonymy, several of the forms having been given specific or varietal rank. It is in the disposition of the flowers that the most evident variation is displayed. These may be borne on simple basal scapes or on a central flowering axis which is bracteate towards the base and ebracteate towards the apex, but all transitions are found between these extremes. The western representatives are usually smaller in stature and show a high proportion of forms in which only basal scapes are produced, but these also occur in the extreme north-eastern limit of the species (in Kansu) and, it may be observed, all forms may be found growing in association in the field. Marked variation is also shown in the number and colour of the petals, in the texture of the leaves and colour of the spines, and also in the shape, pubescence, and dehiscence of the capsule".

In the light of more field data in recent times, but more especially by a plethora of excellent photographs taken in the wild, it is now possible to reassess this complex. At face value it seems ridiculous to place small high altitude plants no more than 15 cm tall and bearing scapose flowers in the same species as lower altitude plants in excess of 1 m bearing many-flowered racemes. It is the equivalent of placing all the blue *Gentiana* species in one solely because they have opposite leaves and blue flowers. The 'Horridula' complex has a very wide distribution in the wild, both from west to east and north to south and, judging by the evidence at hand, many of the variants seen in the wild have not yet properly evolved into distinct and easily definable species. Yet, at the same time, there are regional variations that are easy to define, so that within the complex distinct, albeit very closely related, species can be recognised. One of the problems apart from analysing a complex with sharp bristly stems and leaves and predominantly blue flowers, is that as already noted (p. 236), there has been a tendency in the literature to place any racemose plant in the complex under the name *Meconopsis racemosa* and this

Below, from left:

Meconopsis prattii; E of Zogong, SE Tibet, 5100 m.   PHOTO: TOSHIO YOSHIDA

Meconopsis prattii; N of Raog, SE Tibet, 4700 m.   PHOTO: TOSHIO YOSHIDA

Meconopsis prattii; Geniexueshan, W Sichuan, 3700 m.   PHOTO: TOSHIO YOSHIDA

Meconopsis prattii; Baimashan, NW Yunnan, 4100 m.
PHOTO: CHRISTOPHER GREY-WILSON

alone has created a great deal of ambiguity. *M. racemosa* was described in 1877 by Maximowicz from specimens collected by Przewalski in Gansu (then Kansu). The plant described is very different from the forms ascribed to *M. horridula* from the Himalaya and often named *M. horridula* var. *racemosa* but not that of David Prain.

This division is also maintained in cultivation: the higher altitude taxa such as *Meconopsis horridula* s.l., *M. balangensis* and *M. rudis* are extremely difficult to maintain in cultivation, indeed difficult often to get to flowering size, while those from lower altitudes and relatively drier habitats (*M. prattii*, *M. racemosa* and *M. zhongdianensis*, for instance) are far easier and have even been known to self-sow in favoured gardens.

Meconopsis prattii is fairly easily recognised. It is a rather slight plant with slender stems in which the majority of leaves are to be found at the base. The racemes of relatively small flowers are borne in a generally one-sided or semi-secund, often rather lax, raceme. The flowers are predominately 6-petalled as far as it is known and are usually rather gappy, the petals not overlapping or only very little, and narrowed at the base to leave a gap between the petals. The fruit capsule is small and barrel-shaped and beset with dense, rigid, sharply pointed, ascending bristles.

Unfortunately, some years ago it was I who applied the name *Meconopsis prattii* to plants growing on the Zhongdian Plateau, especially at the northern end and around Napahai. It is now clear that they differ in a number of respects, no less in the denser rosettes of rather undulate-margined leaves, but more especially in the rather dense, many flowered racemes in which the flowers are placed all around the stem, while the stem itself is far more leafy and generally devoid of flowers in the lower third. The fruit capsule of this plant (described in this work: see *M. zhongdianensis*) is very different in shape, being larger and relatively narrower and topped by a narrow cone-shaped persistent style.

Meconopsis prattii is found in an arc from the Kangding area of western Sichuan, southwards through the mountains around Litang and Batang, into north-western Yunnan, and extending westwards into south-eastern Tibet: it was, for instance, collected by Kingdon Ward north of the Ata Kang La in SE Tibet and, more recently, photographed by Toshio Yoshida N of Raog, also in SE Tibet.

A note to me from Toshio Yoshida adds additional information:

"Many plants of *Meconopsis* sp. A [= *prattii*] can be observed around 3700 m in altitude in valleys of the Geniexueshan, where it grows almost definitely on shady, steep slopes exposed only to morning sunshine, mostly among tall herbaceous plants or dwarf shrubs, occasionally on shady bare cliffs...... The leaves are thin and coarsely toothed. The spines are rather weak but not easy to break."

NOTE. The position of Kingdon Ward's *Meconopsis calciphila* needs to be considered. Based on a single gathering from the extreme north of Myanmar, this plant has a close affinity with *M. prattii* and may be an extreme form of that species; however there is insufficient material and information on which to make a positive judgement. It was collected in the upper Seinghku valley in furthermost Myanmar close to the border of both Tibet and Arunachal Pradesh (NE India) not far from the type localities of *M. violacea* and *M. impedita* subsp. *rubra* (syn. *M. rubra*).

Of it Kingdon Ward wrote in 1930:

"Another poppy of the Seinghku valley is *M. calciphila*, found above 12,000 feet, and almost confined to limestone screes. It is a prim and prickly creature with flowers of taciturn blue, not to be compared with the glorious sky-blue prickly poppy, *M. latifolia*, but no doubt quite as hard to grow! *M. calciphila* made so slight an impression on the alpine flora, that I shall say no more about it here."

Kingdon Ward, *Plant Hunting on the edge of the World*.

Meconopsis prattii; Hongshan, NW Yunnan. PHOTO: MARGARET NORTH

The field notes record "In rocky situations amongst limestone boulders and cliff crevices. In sheltered situations it reaches a height of 2–3 feet, but in open is quite dwarf. The entire leaves are characteristic. Petals dark blue, about 7 in number. Anthers bright yellow. Ovary and style and stigma pale green."

Meconopsis calciphila Kingdon Ward, *Gard. Chron.* 1926, Ser. 3, 82: 506 in obs. (1927) and *Ann. Bot.* 42: 856, 862 (1928). Type: N Myanmar (Upper Burma), valley of Di Chu, Burma-Tibet border, 12,000–13,000 feet, 27 July 1926, *Kingdon Ward* 7200 (holotype BM, isotype, K).

The basic details of the plants are as follows: *Leaves* all in a basal rosette c. 7–9, the lamina elliptic-oblanceolate, 4–7 × 1.4–1.8 cm, the base gradually attenuate into the petiole, the apex subacute, moderately patent bristly; petiole 2–6.5 cm long. *Inflorescence* a lax few-flowered raceme, ebracteate, with a moderately patent-bristly stem. *Pedicels* 4.5–6.5 cm long, bristly as for the stem but densely so immediately beneath the flowers. *Flowers* saucer-shaped to flattish, 3.5–5 cm across. *Petals* blue, 6–7, obovate to oblanceolate, 20–23 × 14–17 mm.

46. MECONOPSIS ZHONGDIANENSIS

Meconopsis zhongdianensis Grey-Wilson **sp. nov.** closely related to *M. prainiana* Kingdon Ward but differing in it is more numerous, more spreading, undulate leaves, in the more crowded inflorescences with flowers occupying at least two-thirds the plant's height, and in the greater number of petals; from *M. racemosa* it differs in its more undulate leaves, in the more crowded inflorescences and relatively shorter pedicels, in the subobtuse rather than obtuse apex to the petals and in the larger fruit capsules. Type: NW Yunnan, Napahai, N of Zhongdian, 27°54'N, 99°38'E, 3279 m, 2 July 1994, *ACE* (Alpine Garden Society China Exped.) 883 (E, holotype; K, isotype).

DESCRIPTION. *Monocarpic plant* to 1.2 m tall in flower, generally 40–78 cm. *Stem* stiffly erect, green, often flushed purple in the inflorescence or purplish-brown overall, 9–22 mm diameter near the base, densely patent-bristly, the bristles mostly 4–6 mm long, pale whitish or straw-coloured, but often intermixed with a few shorter bristles. *Leaves* matt greyish green, paler beneath, the basal leaves generally 9 or more, spreading to ascending, the lamina elliptic to elliptic-oblanceolate or oblanceolate, (8–)10.5–22 × (1.4–)2–4.2 cm, borne in a compact rosette, the base cuneate into the petiole, the apex acute to subacute, entire to sinuate and undulate on the margin, often markedly so, occasionally somewhat toothed, densely covered in straw-coloured bristles (setae), these semi-pungent, mostly 3–6 mm long; petiole rather narrow, to 6.5 cm long; cauline leaves sessile or subsessile, spirally arranged in the lower third of the plant, similar to the basal but decreasing in size up the stem, the uppermost merging with the bracts. *Inflorescence* a rather dense raceme with (11–)14–40 lateral-facing or slightly nodding flowers, each 3.8–7 cm across, saucer-shaped to rather flat, borne in the upper half to a third of the plant, only the lowermost flowers in the inflorescence bracteate. *Pedicels* slender, 0.6–5.2 cm long (to 7.2 cm long in fruit), that of the terminal flower to 7.2 cm long (to 9 cm long in fruit), arched towards the top but becoming erect in fruit. *Buds* nodding, ovoid, 12–15 mm long, densely patent subpungently bristly, the bristles often purplish at the base. *Petals* normally (5–)6–8(–9), purple or claret to indigo-mauve, or blue, sometimes with a purplish flush, rarely pale lavender or whitish with a bluish mauve flush, or pure white, ovate to elliptic-obovate 18–34 × 15–28 mm. *Stamens* numerous, with filaments the same colour as the petals although generally darker, the anthers white or cream, greyish on ageing, 1–1.5 mm long. *Ovary* ovoid, densely appressed-bristly, the style 3–5 mm long, stigma pale green or greenish yellow 1.5–3.5 mm long, subcapitate to subclavate, 3 mm across, not protruding beyond the boss of stamens. *Fruit capsule* ovoid-ellipsoidal, 15–25 × 8–11 mm, 5–7-valved, covered with rather sparse patent to ascending bristles, terminating in a narrow cone-shaped persistent style, 5–9 mm long, about half as long as the capsule, broadening towards the base. 2n = 56, as *M. racemosa* sensu Ying *et al.* 2006. (Fig. 20G, H).

DISTRIBUTION. NW Yunnan (centred on Zhongdian), SW Sichuan (centred on Muli and Yungning); 3000–3910 m (Map 16).

HABITAT. Rock slopes, cliffs, roadside screes and embankments, old quarry workings, rocky streamsides, open rock scrub, stony pastures, dry limestone slopes.

FLOWERING. June–July.

SPECIMENS SEEN. *ACE* 76 (E), 215 (E), 883 (E, K); *Delavay* 987 (K), s.n. 1883–1885 (K, P); *Forrest* 1931, 12664 (E, K), 14479 (E), 16603 (BM, E, K), 16615 (BM, E, K), 16618 (E, K), 16621 (BM, E), 16899 (BM, E, K), 16903 (E), 17275 (E, K), 20404 (E), 22123 (E), 28823 (BM, E), 28839 (BM, E, K), 30105 (E), 30109 (BM, E), 30112 (BM); *KEG* (Kunming Edinburgh Gothenburg Exped.) 615 (E), 1328 (E), 1331 (E), 1380 (E); *Kingdon Ward* 4171 (E), 4701 (E), 4616 (E), 4701 (E); *Kunming-Edinburgh Exped. Sichuan* 311 (E), 326 (E); *RBG Edinburgh Dêqên Exped.* 392 (E); *Rock* 192 (E), 2460 (E), 24608 (K), 25271 (E), 28823 (BM, E), 28839 (E, K); *Soulié* 2433 (E); *T. T. Yü* 10909 (BM), 10978 (BM), 10980 (BM, E), 10990 (BM, E), 13661 (BM), 13779 (BM), 13968 (BM, E), 14274 (BM, E).

Above: *Meconopsis zhongdianensis* habitat; Napahai, N of Zhongdian, NW Yunnan, 3350 m.
PHOTO: HARRY JANS

Below, from left:

Meconopsis zhongdianensis; by Napahai, Zhongdian, Xianggelila County, NW Yunnan, 3300 m.
PHOTO: TOSHIO YOSHIDA

Meconopsis zhongdianensis; Napahai, N of Zhongdian, NW Yunnan, 3350 m.
PHOTO: JOHN & HILARY BIRKS

Meconopsis zhongdianensis; Napahai, N of Zhongdian, NW Yunnan, 3350 m.
PHOTO: JOHN & HILARY BIRKS

46. MECONOPSIS ZHONGDIANENSIS

Above, from left:

Meconopsis zhongdianensis, young flowering plant showing basal leaf-rosette; Napahai, N of Zhongdian, NW Yunnan, 3350 m.
PHOTO: JOHN & HILARY BIRKS

Meconopsis zhongdianensis; Napahai, N of Zhongdian, NW Yunnan, 3250 m. PHOTO: HARRY JANS

Meconopsis zhongdianensis; near Bigu Tianchi, Xianggelila County, NW Yunnan, 3300 m. PHOTO: TOSHIO YOSHIDA

Meconopsis zhongdianensis is a bold and handsome species most frequent in the area centred upon the northern Zhongdian Plateau (Napahai and Tianchi in particular), but extending into south-western Sichuan as far as the Muli area. In the literature it has been frequently referred to *M. horridula* and, in recent years, to both *M. racemosa* and *M. prattii* (see p. 257). *M. zhongdianensis* undoubtedly finds its closest ally in *M. prainiana*, both being tall and rather slender plants. In *M. zhongdianensis* the basal leaves are grey-green, more spreading and broader with a markedly undulate margin normally, the inflorescences with denser more numerous flowers that generally have 6–8 petals, and white or cream anthers. *M. prainiana*, on the other hand, bears brighter green leaves that are characteristically fewer and rather upright in the basal rosette and flat-margined, the inflorescence laxer and fewer-flowered, the flowers nearly always 4-petalled and the anthers orange yellow or golden. Differences can also be noted in the fruit capsule, which in *M. prainiana* are pronouncedly broader. From *M. prattii*, with which it is sometimes confused with in the field, it differs in the generally stouter-stemmed plant, dense more numerous flowered racemes, the flowers generally with more petals, while the fruits are larger and ellipsoid rather than ovoid. In the former the style is slender, in the latter more conical. The inflorescence is also different in the disposition of the flowers: in *M. zhongdianensis* they are borne all around the stem, in *M. prattii* they tend to be borne more or less to one side. It is the dense basal leaf-rosette, the thick, stouter stem-inflorescence axis, the relatively short pedicels and many-flowered racemes that most characterise *M. zhongdianensis*.

It should also be noted that the closely related *Meconopsis prainiana*, confined to south-eastern Tibet and northern Arunachal Pradesh, the neighbouring part of eastern Bhutan and the extreme north of north-western Myanmar, varies considerably in flower colour, with blue, purple, claret, yellow and white forms being found in the wild.

NOTE. In recent years a botanic garden/ field station has been developed at the northern end of the Zhongdian Plateau very close to the classic site for *Meconopsis zhongdianensis*. Among the plants being grown there in field trials are members of *M.* 'horridula' complex, including *M. rudis*, *M. prattii* and *M. zhongdianensis*. These have more or less been allowed to hybridise resulting in a very mixed offspring, some tall, some short, some with dense inflorescences, others with extra petals. While these plant are presumably being grown for their medicinal value they pose a real problem to the wild population, especially as bees are actively seen pollinating around the plants in the field station. It would not take much for the adjacent wild population of *M. zhongdianensis* to become contaminated, especially in its *locus classicus* which lies within a very short distance of the field station. Incidentally, a similar problem has happened in cultivation in Britain and presumably elsewhere where various taxa (notably *M. prattii*, *M. racemosa* and *M. zhongdianensis*) have in effect become a hybrid swarm in which it is now difficult to recognise individual species, that is with the exception of plants grown from recently introduced wild seed. These in effect represent what is generally referred to as *M. horridula* of horticulture (see also p. 243).

47. MECONOPSIS PRAINIANA

Meconopsis prainiana Kingdon Ward, *Gard. Chron.* 1926, Ser. 3, 79:308, fig. 232 (1926) and *Ann. Bot.* 1926, 40: 540, tab. 16 fig.1 (1926). Type: SE Tibet, Temo la, 15–16,000 ft, *Kingdon Ward* 5909 (BM, holotype, E, isotype).

[syn. *M. horridula* subsp. *tibetica* K. B. Jørk mss!].

DESCRIPTION. *Monocarpic plant* to 87 cm tall in flower (to 1.2 m in fruit) with a stout taproot to c. 50 cm long, 1.3–2.8 cm diameter at the top. *Stem* erect greenish, rarely slightly purple flushed, especially towards the top, 12–22 mm diameter near the base, striate, moderately to densely covered in patent, rather slender, pale straw-coloured or pale brownish bristles, the bristles generally 5–7 mm long. *Leaves* deep green, somewhat paler beneath, the basal leaves 7–11 normally, borne in a lax, rather erect rosette, the lamina elliptic to elliptic-oblanceolate, 9–19.5 × 1–3.8 cm, narrowed below in the petiole, the apex obtuse to subobtuse, margin entire to very slightly toothed, often undulate, occasionally the margin slightly revolute in places, generally covered overall with pale, straw-coloured bristles, densest along the midrib beneath and along the margin; cauline leaves scattered on the lower half of the plant, similar to the basal leaves but gradually decreasing in size, strongly ascending, the lowermost short-petioled, the uppermost sessile; petiole to 5.5 cm long, often very short to obsolete, narrowly winged at the lamina base. *Inflorescence* a lax raceme, generally with 9–19 pendent or semi-pendent, cupped to flattish, flowers, 5–7.5 cm across, only the lowermost bracteate, modestly fragrant. *Pedicels* 2.5–9 cm long (8–21 cm long in fruit), pendent in the upper part at first but becoming erect in fruit. *Buds* nodding, ovoid, 18–21 × 14–18 mm, green, sometimes purple-flushed, sparsely to moderately short-setose, the bristles often extending from a purplish-brown base. *Petals* usually 4, the uppermost flower in the inflorescence sometimes 5–6 petalled, washy blue to pale blue to azure blue, mauve-blue, violet-blue or purplish blue, more rarely plum-purple or vinous red, white or pale sulphur yellow, suborbicular to somewhat obovate, 2.6–4.4 × 2.3–3.8 cm, slightly undulate at the margin. *Stamens* numerous, the filaments filiform, the same colour as the petals, violet-purple to blue, white or pale yellow, the anthers orange yellow to golden, greying upon ageing. *Ovary* green, oval, with appressed purplish bristles, the style pale green or yellowish green, 4–9 mm long, including the stigma, just protruding beyond the boss of stamens; stigma clavate, pale greenish yellow to deep green, clavate, obscurely 4–6-lobed. *Fruit capsule* narrow-ovoid, 18–34 × 8–17 mm, covered in patent-ascending blackish to violet-black bristles, 4–6-valved, the persistent style 8–14 mm long, to 2.5 mm diameter at the base. (Fig. 20D).

DISTRIBUTION. NE Bhutan (Orka La, Sakden); NW India (N & NW Arunachal Pradesh); SE Tibet (Xizang): Bimbi La, Milakatory La, Mira La, Nam La (Kongbo), Nyima La, Potrang La, Shagam La, Shoga Dzong (Kongbo), Temo La, Tsari; N Myanmar (Seinghku valley); 3048–4878 m (Map 16).

HABITAT. Grassy slopes, block boulder screes, open shrubberies, woodland clearings, cliff ledges.

FLOWERING. June–mid August.

This elegant species is restricted to a relatively small area in south-eastern Tibet and the neighbouring part of northern Burma (Myanmar) and NE India (Arunachal Pradesh) but has been collected over the years by a number of plant hunters including, most notably, Ludlow & Sherriff and their companions. In recent years it has been seen and photographed by various people travelling through the region, these including Toshio Yoshida, Ann Chambers, Harry Jans and John Mitchell. Interestingly, nearly all those photographed until very recently depict pale azure blue plants, occasionally those with plum-purple flowers; however, the yellow-flowered form has not been seen again until 2012 to my

THE GENUS MECONOPSIS
47. MECONOPSIS PRAINIANA

knowledge. This latter is the plant described by George Taylor as *M. horridula* var. *lutea* discovered by Ludlow & Sherriff on the Shagma La, Tsari, in south-eastern Tibet to the north of Arunachal Pradesh in NE India in 1936. Harold Fletcher recounts (*A Quest of Flowers*, 1975) that:

"….Fortunately, on 14 August, though feeling well below par, he [George Taylor] had summoned up the energy to explore the Mira La (15,800 feet) with Sherriff and to find some of his beloved meconopsis. He had found the little yellow poppy (6064) which Sherriff had gathered at Go Nyi Re on 21 July and which he was to regard as a previously unrecorded yellow form of *Meconopsis argemonantha*, giving it the varietal name of *lutea*. He had found another of Sherrif's discoveries, the yellow-flowered form of *M. horridula* which Sherriff had seen in very small quantity on the Shagma La in Tsari, about 80 miles to the south. But here, on the Mira La, *M. horridula* var. *lutea* (6062) was in abundance between 15,000 and 16,000 feet in block boulder-scree on a very steep grassy hillside. Plants were up to 3½ feet high, with pale sulphur-yellow petals, and were growing in association with the short-styled form of *M. integrifolia* [*M. pseudointegrifolia*] which was in immature fruit and commonly had but one flower. Not far away, amongst dwarf rhododendrons, there was *M. simplicifolia* (6083) whilst on grassy cliff ledges and on loose granitic scree *M. impedita* (6052, 6084) still carried a few blue-violet flowers".

Writing in *Gardener's Chronicle* in 1926, Frank Kingdon Ward comments that it [*Meconopsis prainiana*]:

"is a fine plant, as much as three feet high, with pale watery blue flowers, scentless, having the soft texture of Japanese silk. It has four, rarely five petals, and is thus related to the Himalayan set; we must look at *M. aculeata* and *M. latifolia* and *M. sinuata* for its affinity, but as a matter of fact that it is as near as we can get; we cannot match it. Though it has the green stigma of *M. aculeata*, instead of latifolia's purple prong, its entire narrow, linear leaves separate it from both. The very long style, too, easily distinguishes it from *M. latifolia*. From *M. sinuata* it differs not only in its foliage, but still more in its capsule, which is obovoid instead of obconic".

In fact various colour forms are to be found in the wild and it would be invidious to give all of them formal recognition, especially as some of the colour forms overlap in their distributions. However, some recognition seems appropriate including George Taylor's *Meconopsis horridula* var. *lutea* referred to above. This taxon, which has very pale yellow flowers is without question *M. prainiana* rather than *M. horridula* and is accordingly transferred here. As this plant grows in geographical isolation from the more usually blue, purplish or wine-purple variants (which often grow in mixed colonies) I have maintained it as a variety. Since then more evidence has come to light from eastern Bhutan (two collections, one made

Below, from left:

Meconopsis prainiana var. *prainiana*; Serkhyem La, SE Tibet, 4400 m. PHOTO: HARRY JANS

Meconopsis prainiana var. *prainiana*; basal leaf-rosette; Serkhyem La, SE Tibet, 4400 m. PHOTO: HARRY JANS

Meconopsis prainiana var. *prainiana*, detail of inflorescence; Serkhyem La, SE Tibet, 4400 m. PHOTO: HARRY JANS

Above left: *Meconopsis prainiana* var. *lutea*; SW of Yangser, W Arunachal Pradesh, NE India. PHOTO: DAVID & MARGARET THORNE

Above right: *Meconopsis prainiana* var. *lutea*; Pange La, W Arunachal Pradesh, NE India. PHOTO: TIM LEVER

by Kingdon Ward, the other by Ludlow & Sherriff (see footnote 32 below) clearly also belong to *M. prainiana* but are white-flowered. This has been substantiated in 2012 by a group that included Martin Walsh, Tim Lever and David and Margaret Thorne who ventured into western Arunachal Pradesh close to the eastern border of Bhutan. There they came upon both white and very pale yellow, sometimes almost cream-coloured, colonies of *M. prainiana*. Although they were unable to collect herbarium specimens they did manage to collect ample photographic evidence of these as well as red-flowered forms of *M. paniculata* and a new species (described here as *M. ludlowii*). I have amended the description below to include both yellow- and white-flowered forms.

var. prainiana

DESCRIPTION. Flowers washy blue to pale blue to azure blue, mauve-blue, violet-blue or purplish blue, more rarely plum-purple or vinous red.

DISTRIBUTION. SE Tibet (Xizang): Bimbi La, Nam La (Kongbo), Nyima La, Potrang La, Shoga Dzong (Kongbo), Temo La, Tsari; N Myanmar (Seingkhu valley).

SPECIMENS SEEN. *Kingdon Ward* 5717 (E, K), 5909 (BM, E, K), 7057 (K), 11648 (BM), 13727 (BM), 13727a (BM), 13727c (BM); *Ludlow & Sherriff* 623 (BM), 639 (BM), 659 (BM), 1095 (BM); 1767 (BM), 1789 (BM), 2139 (BM), 2159 (BM), 2207 (BM), 2286 (BM), 2522 (BM), 2564 (BM), 5612 (BM), 5550 (BM), 6236 (BM), 5612 (BM); *Ludlow, Sherriff & Elliot* 13140 (BM, E), 13257(BM), 13727 (BM), 14173 (BM), 14481 (BM, E); *Ludlow, Sherriff & Hicks* 14173 (E); *Ludlow, Sherriff & Taylor* 5061 (BM), 5061a (BM, E), 5612 (BM), 6056 (BM), 6062 (BM), 6236 (BM, E).

var. lutea (G. Taylor) Grey-Wilson, comb. nov., descr. emend.

[syn. *M. horridula* var. *lutea* G. Taylor, *New Fl. & Silva* 9: 158 (1937). Type: SE Tibet, Tsari, Shagam La, 16,000 ft, 20 June 1936, *Ludlow & Sherriff* 2188 (BM, holotype; E, isotype)].

DESCRIPTION. *Flowers* pale sulphur-yellow, cream or white; anthers orange-yellow. *Style* green with a pale yellow stigma.

DISTRIBUTION. NE Bhutan (Milakatong La, Orka La, Sakden); NW India (NW Arunachal Pradesh); SE Tibet (Mira La, Shagam La).

SPECIMENS SEEN. *Kingdon Ward* 13727a (BM), 11648[32] (BM); *Ludlow & Sherriff* 659[32] (BM), 2188 (BM); *Ludlow, Sherriff & Taylor* 6056 (BM, E), 6062 (BM, E).

---

[32] The eastern Bhutanese material from Mago and the Milakatong La, along with plants recently photographed in neighbouring Arunachal Pradesh are white-flowered populations which correspond in all other details to var. *lutea* and so I have included them under that variety rather than create another taxon for yet another colour variant.

47. MECONOPSIS PRAINIANA

The field notes accompanying the type collection read: "Corolla pale golden yellow. Filaments pale yellow: anthers golden yellow. Style green. Stigma dark green. Fragrant. On cliff faces, south face. Very few seen".

This interesting variety is similar in most respects to the type plant (var. *prainiana*) except that it bears pale sulphur-yellow to white flowers. It has been recorded from just two localities in Tibet in the absence of the more common 'blue-flowered' variant, on the Shagma La in Tsari and on the Mira La in Kongbo, localities about 128 km apart. The inclusion of material from eastern Bhutan and the neighbouring area of Arunachal Pradesh extends its distribution well to the south of the Tsari locality.

The presence of both blue- and yellow- or red- and yellow-flowered forms in *Meconopsis* species is not unique to *M. prainiana* for it is also found in *M. georgei*, while recently pink- and red-flowered forms of the normally yellow-flowered *M. paniculata* have been found in Bhutan and Arunachal Pradesh. At the same time there is some dispute whether or not the Nepalese *M. regia* also has red-flowered forms in the wild. Conversely, *M. discigera* which is often recorded in literature as being either yellow- or blue-flowered has been divided into two closely allied species based on sound morphological evidence, the blue-flowered version now correctly identified as *M. bhutanica*.

The white-flowered form was noted by Ludlow & Sherriff in north-eastern Bhutan: "On the Milakatong La [north of Sakden], where they marked a very beautiful white form of *Meconopsis horridula* [*M. prainiana* s.l. sic] for future seed collecting". (Fletcher, *A Quest for Flowers*: 54, 1975). It was later recorded that they had failed to collect any seed of the 'white poppy'.

Photographs from various sources show that *Meconopsis prainiana* is a very elegant plant with attractively cupped flowers borne stiffly aloft on a slender leafy stem. It would without question be an exciting addition to the garden. Seed has been introduced in the past from Ludlow & Sherriff collections, both of the clear blue as well as the yellow versions. In his original description, George Taylor allied this plant to *M. horridula* which he saw as a widely distributed and extremely variable species.

Below, from left:

Meconopsis prainiana var. *lutea*; Tsena to Sela, W Arunachal Pradesh, NE India, 4300 m. PHOTO: MARTIN WALSH

Meconopsis prainiana var. *lutea*; Tse La, W Arunachal Pradesh, NE India. PHOTO: TIM LEVER

Meconopsis prainiana var. *lutea*; Tse La, W Arunachal Pradesh, NE India. PHOTO: TIM LEVER

48. MECONOPSIS RUDIS

Meconopsis rudis (Prain) Prain, *Ann. Bot.* 20: 347 (1906). [syn. *M. horridula* var. *rudis* Prain, *J. Asiat. Soc. Bengal, Pt. 2, Nat. Hist.* 64 (2): 314 (1896). Type: China, NW Yunnan, 1884, *Delavay* 39 (P, holotype)]

DESCRIPTION. *Monocarpic plant* (14–)20–57 cm tall in flower, somewhat taller in fruit with a stout taproot to 12 mm diameter at the top. *Stem* stiffly erect, striate, glaucous, often purple flushed towards the top, 8–15 mm diameter near the base, sparsely covered with patent, fawn or straw-coloured stiffish bristles, each 4–5 mm long. *Leaves* in a lax basal rosette at first but in flowering plants spirally scattered on the lower half of the stem, all but the uppermost petiolate, the lamina of basal and lower leaves ovate-elliptic to elliptic-oblanceolate, 5–15 × 1.5–5.5 cm, often rather abruptly narrowed at the base into the petiole, the apex obtuse to subobtuse, margin entire to somewhat lobed or toothed, sometimes slightly undulate, markedly glaucous, occasionally deep green, overall, but paler beneath, sometimes with a purplish or pinkish flush, rather sparsely covered in patent bristles like the stem but, above at least, each bristle arising from a slightly raised black or purple-black base; cauline leaves few, decreasing markedly up the stem, the uppermost small and bract-like; petiole linear, strap-like, (0.5–)2.2–10.5 cm long, slightly winged, 3–11 mm wide. *Inflorescence* a 7–19-flowered raceme covering the whole height of the plant, only the lowermost bracteate; flowers cupped to saucer-shaped, half-nodding to lateral-facing, 4.5–8.4 cm diameter. *Pedicels* ascending but arched towards the top in flower, becoming fully erect in fruit, 1–8.5 cm long (to 15 cm in fruit), generally purple-flushed, bristly like the stem, rather densely so at the top. *Buds* pendent, obovoid to ellipsoid-obovoid, dark green or purplish with dense dark purplish black ascending bristles. *Petals* 6–8, most often 6, pale blue to azure, steely blue, purple- or violet-blue, rounded to obovate, 23–41 × 19–35 mm, slightly erose towards the apex. *Stamens* numerous, the filaments the same colour as the petals although often darker, the anthers 0.75–1.25 mm, whitish or cream, greying on ageing. *Ovary* narrow-ovoid, deep green or purplish-black, beset with dense more or less appressed deep purple bristles, the style linear, 1–5 mm long; stigma whitish, subcapitate, 1.5–2 mm long, protruding beyond the boss of stamens. *Fruit capsule* depressed-globose, 10–15 × 9–14 mm, beset with dense ascending spine-like bristles, narrowed abruptly above into the 3–8 mm long persistent style. 2n = 56, as *M. racemosa* sensu Ying *et al.* 2006. (Figs 3B, 20J–K).

Meconopsis rudis with *Paraquilegia microphylla*; Baimashan, NW Yunnan, 4350 m. PHOTO: JOE ATKIN

Meconopsis rudis; Baimashan, Deqin County, NW Yunnan, 4550 m. PHOTO: TOSHIO YOSHIDA

48. MECONOPSIS RUDIS

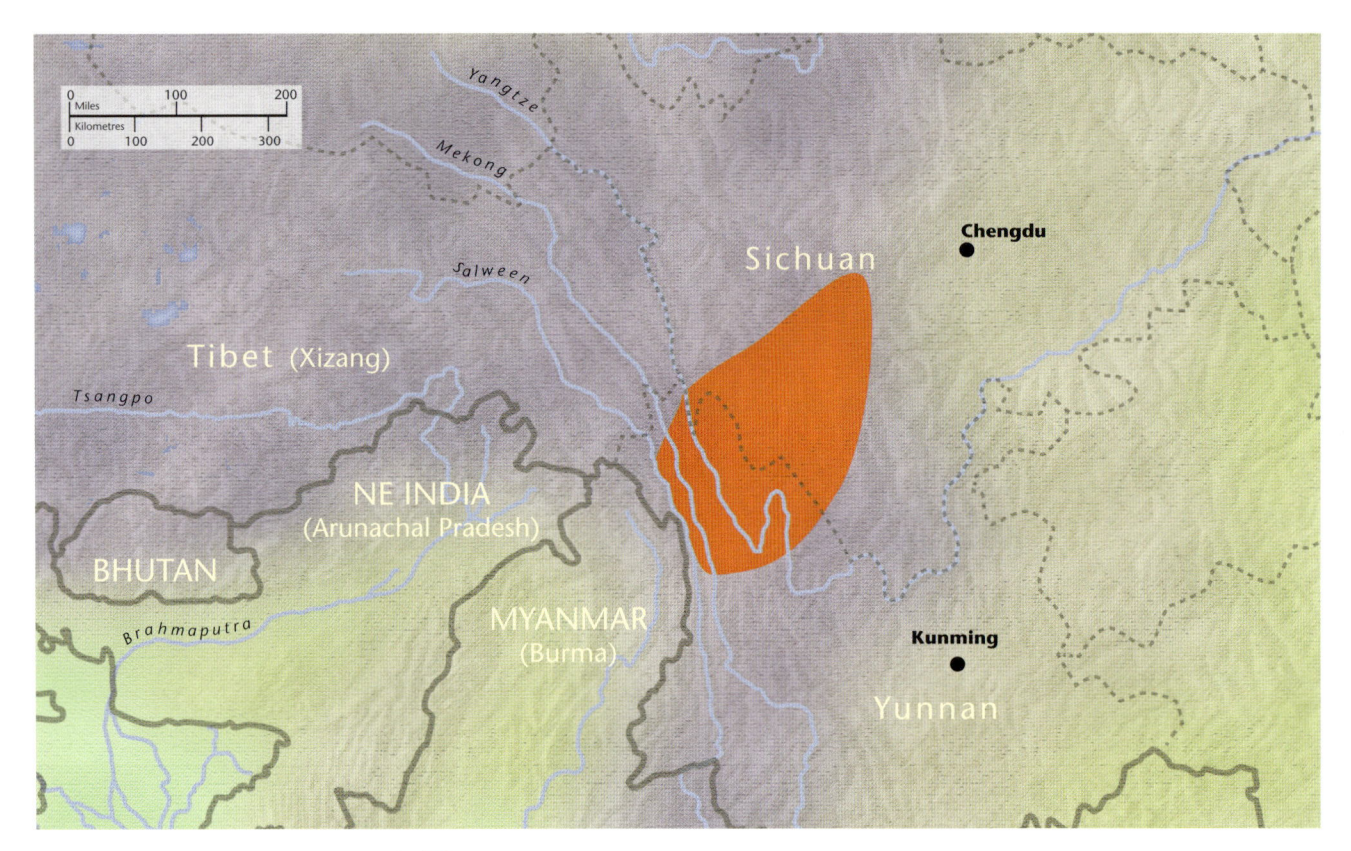

MAP 18. Distribution of *Meconopsis rudis* ■.

DISTRIBUTION. SW Sichuan (south of Kangding southeast to Muli region) and NW Yunnan (especially Baimashan, Habashan, Hongshan and Yulongxueshan, but also on other mountains on the Yangtse-Mekong and Mekong-Salween divides), possibly also in extreme SE Tibet; 3400-4500 m (Map 18).
HABITAT. Rocky alpine meadows, stabilised screes and moraines.
FLOWERING. Late-May–July.
SPECIMENS SEEN. *ACE* 1345 (E), 1542 (E), 1567 (E), 1707 (E), 1708 (E), 1709 (E), 1723 (E), 1773 (E), 2147 (E); *Delavay* 39 (BM, P), s.n. 13 July 1886 (K, P) s.n. Aug. 1886 (K); *Forrest* 2623 (BM, E), 6586 (BM, E), 10469 (BM, E), 13233 (BM, E, K), 16902 (E), 19649 (BM, E), 20070 (BM, E), 20170 (BM, E), 20200 (BM, E), 22505 (E, K), 22634 (E), 28227 (E), 28225 (BM, E), 28524 (BM, E), 28660 (BM, E), 29137 (BM, E), 29198 (BM), 30077 (BM), 30079 (BM, E), 30089 (BM, E), 30092 (BM, E), 30093 (BM, E), 30104 (BM, E), 30105 (BM, E), 30112 (BM, E), 30603 (BM, E); *Kingdon Ward* 172 (E), 3994 (E), 4164 (E), 5245 (E), 5365 (E), 6170 (E); *McLaren* 107 (BM, E, K), 133D (BM), 134D (BM), 144 (BM, E), 241 (E); *Rock* 4535 (E), 4692 (E), 5555 (E), 6104 (E), 6797 (E), 9315 (E, K), 9636 (E), 9840 (E, K), 9992 (E), 16550 (K), 16858 (BM), 17776 (E), 18267 (E), 22250 (E), 22679 (E), 22881 (BM, E, K), 23096

Meconopsis rudis, leaf-rosette; upper Gangheba, Yulongxueshan, NW Yunnan, 4550 m.
PHOTO: CHRISTOPHER GREY-WILSON

(BM, E, K), 23399 (E), 23409 (E), 23416 (BM, E, K), 23774 (E), 24345 (E), 24608 (BM, K), 24754 (BM, E, K), 24756 (BM, E, K), 24942 (E), 25134 (E), 25248 (E), 25287 (E), 25289 (E); *Schneider* 1808 (K), 3360 (K); *Sino-British Exped. Lijiang* 659 (E), 577 (E); *H. Smith* 4411 (BM), 11461 (BM), 11463 (BM); *T. T. Yü* 10784 (E), 13661 (E), 13779 (E), 15359(BM, E), 22643 (E).

Meconopsis rudis, for long hidden under the *M. horridula* umbrella, is a very distinct and handsome species first recognised as long ago as 1906 by David Prain. In its typical form *M. rudis* produces rather spreading leaf-rosettes that are characteristically glaucous overall, the leaf-laminas relatively broad, often somewhat lobed or toothed at the margin and with a rather sparse scattering of pungent bristles, each of which arises from a slightly raised black or purple-black base. The racemose inflorescence is quite dense and occupies much of the height of the plant. Differences can also be noted in the anther and stigma colours, those of *M. rudis* being cream or white, that of rather similar looking *M. balangensis* yellow or orange-yellow. In the former the fruit capsule is smaller and depressed globose, in the latter more ellipsoidal. The leaf and inflorescence characteristics serve to distinguish *M. rudis* from all the other racemose species in the 'Horridula' aggregate with the exception of *M. balangensis* with which it was for long confused; however, a fine piece of field botany by Toshio Yoshida was clearly able to separate the two, this culminating in his 2011 paper written in conjunction with H. Sun and D. E. Boufford. Both species bear significantly glaucous foliage which is generally adorned with dark-based bristles. The prime and over-riding difference in the two species can be observed in the centre of the flower: in *M. rudis* the filaments are spreading and linear, while in *M. balangensis* the inner filaments are inflated like tiny sausage balloons and curved inwards to enclose the ovary.

In the wild *Meconopsis rudis* has its prime focus in the mountains of north-western Yunnan, especially around Lijiang and Zhongdian, with outliers in the mountains further north around Muli in Sichuan, and as far north as the Kangding area. I have been lucky enough to see it on a number of occasions. The first time was at the upper end of the Gang-he-ba north of Lijiang in the Yulongxueshan where it was growing beneath towering limestone cliffs in a much-fractured boulder scree along with other alpine delights such as bright yellow *Parrya forrestii* and white-flowered *Saxifraga calcicola*. *M. rudis* was scattered in ones and twos across the scree, some in flower but many more were in the early rosette stage and would probably flower in two or three years' time. It is also scattered on the high screes and moraines of the

Meconopsis rudis; Shikashan, Zhongdian Plateau, NW Yunnan, 4410 m.
PHOTO: HARRY JANS

Meconopsis rudis; Yulongxueshan, NW Yunnan, 4700 m.
PHOTO: TOSHIO YOSHIDA

48. MECONOPSIS RUDIS

Above left: *Meconopsis rudis*; upper Gangheba, Yulongxueshan, NW Yunnan, 4400 m. PHOTO: CHRISTOPHER GREY-WILSON
Above right: *Meconopsis rudis* forma; W Haba Shan, NW Yunnan, 4120 m. PHOTO: ALAN DUNKLEY

Baimashan, sometimes occupying cliff crevices with one of Yunnan's other mountain gems, *Paraquilegia microphylla*.

Although I have not visited the Hongshan, close to Zhongdian, I have seen ample photographs of *Meconopsis rudis* from that mountain. Although the Hongshan plants resemble those from elsewhere in most respects they tend to have rather narrower and greener leaves. Such plants require further detailed investigation.

Meconopsis rudis has been flowered successfully in Britain from various seed collections in recent years, perhaps most noticeably from ACE (Alpine Garden Society China Expedition 1994). Although it has been grown in the open garden, the greatest success has been when grown within the confines of a part-shaded alpine house where the plants can be kept cool and moist during the summer months and over-wintered relatively dry, although not so dry that they become desiccated. Plants flower in their second or, more normally, third, year from seed. It is without doubt a very pretty species which, in its best forms, has deep azure blue or purplish blue flowers centred by a boss of numerous white or cream anthers.

49. MECONOPSIS BIJIANGENSIS

Meconopsis bijiangensis H. Ohba, Tosh. Yoshida & H. Sun, *J. Jap. Bot.* 84: 294–302 (2009). Type: China, NW Yunnan, Bijiang, upper Pi-he Valley, Biluoxueshan, 3700–4000 m, 8 July 2008, *Yoshida* K1 (KUN, holotype & isotype; E, TI, isotype)

DESCRIPTION. *Monocarpic plant* 15–54 cm tall in flower, covered overall with pale brown to amber bristles. *Taproot* dauciform or fusiform, 15 cm long or more, 5–10 mm diam. near the top. *Stem* short, 2.8–5.8 cm long, 8–11 mm diam. at the base, sparsely to rather densely covered in stiff, patent bristles 2–3 mm long, these usually with an elliptic black or purple-black raised, wart-like base. *Leaves* thick and rather leathery, borne in a basal rosette or crowded towards the base of the stem, the lamina narrow-ovate to lanceolate, linear-elliptic to linear-oblong, or more or less strap-shaped, 2.5–14.5 × 0.7–1.4 cm, deep green to yellowish green and somewhat shiny above, paler beneath, sometimes flushed purple along the margin, with sparse bristles on both surfaces like those on the stem but with a rounded, raised, wart-like, black or purple-black base, the margin entire to undulate or sinuate, sometimes with a few coarse teeth or shallow lobes, the apex obtuse to subacute; petiole, broad-linear, 3–10.7 cm long. *Inflorescence* essentially racemose, leafy in the lower quarter, otherwise ebracteate, with 7–12, rarely up to 16, nodding flowers, the basal flowers borne at the uppermost leaf-axils, free or partly agglutinated to the inflorescence rachis. *Pedicels* 2–8.5 cm long (to 15 cm long in fruit), the terminal one 5–9.5 cm long, moderately to densely bristly, expanded towards the top, often flushed with purple, nodding near the top but becoming erect as the fruit develops, decurrent down the rachis giving it a ribbed appearance. *Flowers* deeply cupped, 4.2–6.9 cm across, nodding, generally several open at a time, the buds subglobose to obovoid or ellipsoid. *Sepals* covered with dark-based patent spines. *Petals* 4, more rarely 5–6 on the terminal flower, celestial blue, pale mauve or blue-mauve, broad-ovate or rounded to broad-elliptic, 20–35 × 21–31 mm, glabrous, slightly undulate at the margin. *Stamens* numerous, the filaments filiform, 7–10 mm long, darker than the petals, the anthers orange-yellow, 1.75–2 mm long. *Ovary* ellipsoid, densely bristly with stiff ascending spines, the style 3–5 mm long with a capitate stigma, 1–1.5 mm diameter. *Fruit capsule* ellipsoid-cylindrical to cylindrical, 25–38 mm long, 7–10 mm diam., brownish purple, covered in rather sparse patent bristles 2–3 mm long, with dark purple or blackish raised bases, the persistent style 5–9 mm long. (Fig. 20L).

DISTRIBUTION. NW Yunnan (Biluoxueshan, Gaoligongshan) and NE Myanmar (Chimili); 3700–4000 m (Map 19).

HABITAT. Stony mountain slopes, rocky places, dwarf *Rhododendron* scrub, in humusy soils or pockets.

FLOWERING. Late June–August.

Despite the fact that the original description states that *Meconopsis bijiangensis* has 'Inflorescence a simple raceme occupying more than one thirds of stem' and indeed the type specimen shows this quite clearly, one of the accompanying photographs reveals a plant with multiple basal scapes (14 in all!). While the specimen in question undoubtedly belongs to this species it does reveal once again the tendency for some species to produce racemose and scapose forms. In some the racemose form is accompanied by one or several basal scapes on

Above left: *Meconopsis bijiangensis* subsp. *bijiangensis* habitat; Bijiang, Biluoxueshan, NW Yunnan, 3700 m. PHOTO: TOSHIO YOSHIDA
Above right: *Meconopsis bijiangensis* subsp. *bijiangensis* leaf-rosette; Bijiang, Biluoxueshan, NW Yunnan, 4000 m. PHOTO: TOSHIO YOSHIDA

49. MECONOPSIS BIJIANGENSIS

Above left: *Meconopsis bijiangensis* subsp. *bijiangensis*; Bijiang, Biluoxueshan, NW Yunnan, 3900 m. PHOTO: TOSHIO YOSHIDA

Above right: *Meconopsis bijiangensis* subsp. *bijiangensis*; Bijiang, Biluoxueshan, NW Yunnan, 3900 m. PHOTO: TOSHIO YOSHIDA

the same plant. In this instance the specimen may be an aberration, perhaps induced in the field by grazing or physical damage that has destroyed the original inflorescence as it emerged.

Toshio Yoshida reports that:

"In Biluoxueshan, the plants grow, often gregariously, on the west-facing (Nujiang-side) gentle stony-slopes near the ridge thickly covered with dried peaty soils made of half-decomposed mosses and other plants, sometimes among dwarf *Rhododendron* bushes, just above steeps at the head of the valleys where ascending foggy winds of south-west summer monsoon gather and swiftly flow over the slopes with watering the plants growing there."

The habitat is an exposed one close to the mountain ridge and subjected to swirling mists, especially from mid-morning onwards. Toshio Yoshida reports that when he visited the site in April 2005 the slopes inhabited by *Meconopsis bijiangensis* were still deep in snow, this presumably melting away during May and early June but affording the plants protection during the winter and the pre-monsoon months. Yoshida also records seeing an occasional bumble bee visiting the flowers of this species which grows in close proximity to *Rhododendron rupicola* which flowers at the same time and which is very attractive to the bees both for nectar and pollen: the meconopsis can only supply pollen!

Meconopsis bijiangensis was described from close to the Myanmar border in north-western Yunnan, in the narrow mountain divide separating the Salween and Mekong rivers. Bijiang County lies about midway along the Nushan, south of the Nujiang Nature Reserve, which itself lies west of the town of Weixi, a town familiar to a number of plant collectors of the past including George Forrest and Joseph Rock. It inhabits the same general locality as *M. georgei*, although the two species are not found growing together as far as I am aware. The two species are clearly closely related but the narrow, dark-spotted leaves and generally fewer petals of *M. bijiangensis* are very distinctive, as are the blue or mauve-blue flowers. Plants from further west inside eastern Myanmar are distinguished as follows:

subsp. bijiangensis

Plant 15–40 cm tall in flower, the stem c. 8 mm diameter at the base. *Leaves* basal or a few scattered on the lower half of the stem, adorned like the leaves and pedicels with slender bristles 2–3 mm long, the lamina narrow-ovate to lanceolate to linear-oblong, 2.5–10 × 0.7–1.4 cm, cuneate at the base into a short petiole, the margin entire to undulate, sometimes with a few coarse teeth or shallow lobes; petiole 3–5 cm long. *Pedicels* 2–5 cm long, the terminal one 7–9.5 cm (to 15 cm long in fruit). *Flowers* 7–12 (–16), the petals rounded to ovate, 25–35 × 21–23 mm. *Fruit capsule* ellipsoid-cylindric to ellipsoid-obovoid, 25–38 mm long, 7–10 mm diameter, the persistent style 7–8 mm long.

DISTRIBUTION. NW Yunnan: the old Bijiang (Zhiziluo) County, Biluoxueshan (26°37'N, 99°00'E to 26°31'N, 99°00'E) and Gaoligongshan; 3700–4000 m (Map 19a).

SPECIMENS SEEN. *Yoshida* K1 (E, KUN, TI); *Nujiang Team* 0848, 1135 (KUN); *Wu Su-gong* 8832 (KUN)

NOTE. *Nujiang Team* 0848 was collected in Biluoxueshan on 16 June 1978 before flowering with flower-buds held near the base. *Nujiang Team* 1135 was collected in Gaoligongshan on 14 July 1978 with flowers. *Wu Su-gong* 8832 was collected in Biluoxueshan on 12 Sept. 1964 with mature fruits, to 3.5 cm long. The unique characters in the leaf-shapes and the 'spines' of the new species can be observed in these specimens.

subsp. chimiliensis Grey-Wilson, subsp. nov.

Differs from the typical plant, subsp. *bijiangensis*, in being a taller plant, in the larger, longer-petioled leaves and in the relatively narrower fruit capsules. Type: NE Myanmar (Burma), Chimili Valley, Hpawshi, 11–12,000 ft, *Farrer* 1159 (E, holotype, isotype).

[Syn. *Meconopsis cyanochlora* Farrer in mss!]

DESCRIPTION. *Plant* to 54 cm tall in flower, the stem c.11 mm diameter at the base. *Leaves* basal or a few scattered on the lower half of the stem, adorned like the leaves and pedicels with slender bristles 2–3 mm long, the lamina linear-elliptic to almost strap-shaped, to 14.5 × 1.1 cm, cuneate at the base into a long petiole, the margin subentire to sinuate or slightly lobed; petiole 10–10.7 cm long. *Pedicels* 2.2–8.5 cm long, the terminal one 5–8.5 cm (to 12.6 cm long in fruit). *Flowers* 7–10, the petals rounded to ovate, 20–25 × 21–23 mm. *Fruit capsule* ellipsoid-cylindric to ellipsoid-obovoid, 27–34 mm long, 7–8.5 mm diameter, the persistent style 5–9 mm long.

DISTRIBUTION. NE Myanmar (Chimili) (Map 19b).

SPECIMENS SEEN. *Farrer* 1159 (E); *Forrest* 26831 (BM, K), 29934 (BM, E), 30099 (BM, E), 30117 (BM).

Farrer 1159. Notes on herbarium sheet:

"Leaves bright green, pale glaucous beneath, often with dark warts from which the bristles spring. Scape, pedicels and capsules dark glaucous blue: bristles brown: stamens [filaments sic] blue, anthers not orange or gilvous, but yellow: petals normally 4 but not infrequently 5, 6 or many, pale and soft celestial blue. I have not certainly seen any independent scapes around the base of the main one, as all such seem really to belong to it below ground. Alpine meadows, moors and (rarely) rocks".

Farrer clearly thought that this plant represented a distinct taxon for he attached the name *Meconopsis cyanochlora* to it; however, this has never been validly published and remains only in manuscript.

The Chimili valley lies in the extreme north-east of Myanmar (formerly Upper Burma) close to the Yunnan border and about 100 km to the south west of the Weixi and the Bilouxueshan, the home of subsp. *bijiangensis*. In the future, when the Myanmar-Yunnan border area, that to the west of the Salween (Nu Jiang) river is more thoroughly explored, intermediates between these two subspecies may be found, especially in the southern Gaoligongshan. In that instance it may then prove difficult to uphold this new subspecies.

Above left: *Meconopsis bijiangensis* subsp. *bijiangensis*; Bijiang, Biluoxueshan, NW Yunnan, 3900 m. PHOTO: TOSHIO YOSHIDA

Above right: *Meconopsis bijiangensis* subsp. *bijiangensis* fruit detail; Bijiang, Biluoxueshan, NW Yunnan, 3700 m. PHOTO: TOSHIO YOSHIDA

50. MECONOPSIS GEORGEI

Meconopsis georgei G. Taylor, *The Genus Meconopsis*: 86 (1934). Type: NW Yunnan, Fuchuanshan, Mekong-Yangtse Divide, 27°N, 99°30'E, 12–13,000 ft, *Forrest* 30100 (BM, holotype, isotype; E, isotype).
[syn. *M. lancifolia* var. *georgei* (G. Taylor) G. Taylor mss!]

DESCRIPTION. *A monocarpic herb* 20–40 cm tall in flower, to 50 cm, occasionally more, in fruit, with an elongated, narrow-dauciform to fusiform taproot 5–25 cm long, 7–12 mm diameter at the widest, the stem often rather poorly developed, 6–30 cm long, to 10 mm diameter near the base, leafy in the lower third to half, moderately to densely hirsute with rather sharp patent, pale brown or reddish brown hispid hairs, these mostly 2–4 (–5) mm long and scarcely expanded at the base. *Leaves* mostly arranged in a lax basal rosette, with a few carried spirally on the stem, the lowermost generally withered by flowering time, deep green above, glaucous beneath, the lamina ovate-elliptic to lanceolate or somewhat oblanceolate, 7–11.5 × 1.2–2.2 cm, the base attenuate into the petiole, the apex acute to subobtuse, margin entire to slightly repand, often sinuate, occasionally slightly to rather coarsely, lobed, glabrous or moderately covered above and beneath with stiff, rather sharp hispid hairs similar in colour to those of the stem, the upper surface generally with a denser indumentum than the lower; petiole broad-linear, mostly 2.5–7 cm long, although the uppermost leaves are increasingly smaller with broader, shorter petioles, the uppermost subsessile or sessile, somewhat expanded at the base. *Flowers* 5–18 borne in a part-bracteate racemose inflorescence, lateral-facing to half-nodding, saucer-shaped to cupped, 3.6–7 cm diameter, the lowermost bracts (leaves) sometimes with a vestigial flowerbud present. *Pedicels* ascending to erect, rather slender, 0.7–3.2 cm long, the terminal pedicel up to 7 cm long (to 15.5 cm long in fruit) patent-hispid, sometimes very densely so, especially just beneath the flowers. *Buds* ovoid to oval, 15–18 × 14–17 mm, the sepals adorned with patent bristles, sometimes densely so. *Petals* 5–8 (–9), primrose-yellow, bluish purple, ruddy purple, deep wine-purple, deep purple-crimson to dark red, sometimes with a brownish flush, rounded, ovate to broad-elliptic, 25–38 × 20–30 mm, subacute to subobtuse, the margin often minutely denticulate, otherwise entire, sometimes recurved at the upper margin. *Stamens* numerous, with filiform whitish, yellowish, purplish, or reddish filaments and orange-yellow anthers which grey or blacken with age. *Ovary* oblong to ellipsoid, densely covered with stiff, sharp, straw-coloured, ascending to subappressed bristles; style 3.5–6 mm long, terminated by a capitate to subclavate

Meconopsis georgei forma *castanea* habitat; S of Lawushan, Biluoxueshan, Fugong County, NW Yunnan, 3750 m. PHOTO: TOSHIO YOSHIDA

stigma, 1.5–2.5 mm across. *Fruit capsule* ellipsoid to ellipsoid-obovoid to cylindric-ellipsoid, (15–)20–34 × 6.5–12 mm, 4–5 mm diameter at the base, covered in dense patent to ascending, sharply pointed bristles, 3-5-valved, splitting for a short distance from the apex at maturity, the persistent, relatively stout, linear style 4–8 mm long. *Seeds* falcate-ellipsoid, irregularly rugose, minutely pitted.

DISTRIBUTION. NW Yunnan (Fuchuanshan, Weixi, Mekong-Yangtse Divide); 3658–4420 m (Map 19).

HABITAT. Stony and rocky alpine meadows and slopes, screes, often rather wet and mossy.

FLOWERING. June–July (–early August).

This species was named in honour of George Forrest who discovered it growing in rocky alpine meadows in the Fuchuanshan on the Mekong-Yangtse Divide (now the Lancan Jiang-Jinsha Jiang divide), just west of Weixi. It is an attractive little species which, for many years was thought to exist only as a yellow-flowered plant. However, George Forrest did not connect this interesting species to other specimens collected in the same vicinity with red or purple-red flowers rather than yellow. However, the fact is that he collected consecutive numbers (30100 — yellow and 30101 — purple-crimson) with the same locality and co-ordinates of latitude and longitude. This is reinforced by a similar consecutive numbering at the hands of Joseph Rock (22696 — yellow and 22697 purplish red). Rock's collections are backed by another, 22699 with dark purple flowers: all were collected in the Fuchuan range, part of the Hengduan Mountains that stretch down from the north-west, forming the upper watersheds of the Mekong, Salween and Yangtse rivers. In addition, a further Rock collection, number 23288 (previously misidentified as *Meconopsis speciosa*), also from the Fuchuan range, shows a fine specimen with well-developed fruits. Toshio Yoshida's collection of the plant placed under *M. castanea* (red flowers) was gathered slightly further south and west on the west side of the Mekong River (on the Mekong-Salween divide in Fugong, Biluoxueshan). Another Forrest collection (21986) stretches the known distribution of the species a little further south in the region of the Chienchuan-Mekong Divide (Jian Chuan-Lancang Jiang Divide).

H. Ohba, Toshio Yoshida and Hang Sung were also unaware of the yellow forms when in 2009 they described *Meconopsis castanea*, based upon a Biluoxueshan collection.

This species agrees in all details with Taylor's *Meconopsis georgei* except in the colour of the flowers, so that it is not possible to maintain it except at forma level.

Meconopsis georgei forma *castanea*; S of Lawushan, Biluoxueshan, Fugong County, NW Yunnan, 3750 m. PHOTO: TOSHIO YOSHIDA

I have at the same time extended the original description to include those plants with bluish purple, ruddy purple, deep wine-purple or deep purple-crimson flowers, as well as originally described with dark red ones.

Perhaps surprisingly George Taylor assigned the red- and purple-flowered collections to *Meconopsis lancifolia*: he remarks under *M. lancifolia* (1934: 88) that "In one of the extreme forms, of which most of the specimens from Upper Burma [Myanmar] are represented, there is a strong development of spreading spines on the leaves, stems, pedicels and fruits, but this spiny character is so intimately intergraded through all degrees of spinosity that it is quite unreliable as a basis for any subdivision of the species". However, it is quite clear to me that these collections belong to *M. georgei* forma *castanea* and once placed there *M. lancifolia* (subsp. *eximia*) becomes a much more homogeneous taxon. Such colour variants are fairly well recorded in the genus. The Nepalese *M. regia* is stated to exist primarily in the wild as a yellow-flowered species but red-flowered ones have been recorded and the same is true of *M. paniculata*. *M. prainiana* from SE Tibet exists primarily in blue or purple variants, although yellow-flowered (assigned to *M. horridula* var. *lutea* by Taylor!). and white ones have been recorded.

Meconopsis georgei probably comes closest to *M. bijiangensis* and *M. rudis*, differing from the former in having leaves without the tell-tale dark-based bristles and by having a greater number of petals on average; from *M. rudis* it differs primarily in its narrow, strap-like leaves, in flower colour (*M. rudis* is primarily blue) and in the larger and relatively longer fruit capsules.

forma georgei

DESCRIPTION. *Flowers* primrose yellow.
DISTRIBUTION. Restricted to the Fuchuanshan, the Mekong-Yangtse divide.
SPECIMENS SEEN. *Forrest* 30100 (BM, E), 30595 (BM, E); *McLaren* s.n. 'D' coll. 1932 (E), 133 (E, K), 134 (E), 318 (E); *Rock* 22696 (BM, E, K), 23287 (E, K), 23288 (E, K).

forma castanea (H. Ohba, Tosh. Yoshida & H. Sun) Grey-Wilson **comb. & stat. nov.**

[syn. *Meconopsis castanea* H. Ohba, Tosh. Yoshida & H. Sun, *J. Jap. Bot.* 84: 294–302 (2009). Type. NW Yunnan, Fugong County, S of Lawushan (Laowoshan), southern Biluoxueshan, 3650–3850 m, 17 July 2008, *T. Yoshida* K3 (KUN, holotype; E, TI, isotypes)].

DESCRIPTION. *Flowers* bluish purple, ruddy purple, purple-rose, deep wine-purple, deep purple-crimson to dark red.
DISTRIBUTION. Forma *castanea* has a wider distribution than forma *georgei*, being found in the Fuchuanshan on the Mekong-Yangtse (Lancan Jiang-Jinsha Jiang) divide, in the vicinity of Weixi (Weishi) on the Mekong-Salween divide. Two Kunming sheets (0201326 & 0201327) are marked 'Lijiang County' and appear to extend the distribution well to the east of Weixi.
SPECIMENS SEEN. *Forrest* 1159 (E), 9848 (BM), 19790 (BM, E), 24886 (K), 26902 (BM, K), 27277 (BM, K), 27284 (BM, K), 30101 (BM, E), 38099 (E); *McLaren* 134 (E); *Rock* 17029 (E), 18361 (K), 22696 (E), 22697 (E), 22699 (BM, E, K), 23253 (BM, E, K), 23286 (E); *Yoshida* K3 (E, KUN). In addition the following Kunming Botanic Garden numbers: 0201326 (KUN), 0201327 (KUN), 0201676 (KUN), 0201677 (KUN), 0201689 (KUN), 0201693 (KUN).

Forrest 19790 ("Mekong-Salween Divide" with "Flowers pale to dark purple-rose") is a remarkable specimen and would have make a tremendous plant in cultivation for it bears five shoots, each bearing a raceme of 4–6 flowers. Filed in the Edinburgh herbarium (E), the sheet bears a *Meconopsis lancifolia* determinavit slip signed by George Taylor. The plant is clearly an aberration and without additional non-flowering (vegetative) shoots would appear to be monocarpic. The stems are in fact leafy towards the base and this and the flower colour point clearly to *M. georgei* forma *castanea*, and the locality and co-ordinates also match those for this particular taxon. Other aberrations have been observed in herbarium specimens and one can only conjecture that these have arisen by some physical damage to the plant prior to flowering: for instance, if the primary inflorescence is badly damaged (by grazing or trampling, for instance) then the plant may compensate by producing secondary inflorescences, especially if the lowermost nodes remain intact.

According to Toshio Yoshida this taxon inhabits granitic boulder slopes enveloped with mosses and lichens, where it grows in the humus-filled gaps and on the ledges of both western and south-eastern aspect. Like *Meconopsis bijiangensis*, also recorded from the Biluoxueshan, the habitat of forma *castanea* is snow-covered during the winter months and shrouded by mist a great deal in an average summer monsoon.

Meconopsis georgei forma *castanea*; S of Lawushan, Biluoxueshan, Fugong County, NW Yunnan, 3750 m. PHOTO: TOSHIO YOSHIDA

series Heterandrae

This series is very closely related to the previous series and indeed its two species could be easily mistaken for members of it; however, they differ in one fundamental respect and that is in the filaments to the stamens. In both species, while the outer filaments are linear and spreading, the inner one or two whorls are inflated like minute sausage balloons and incurved at the same time so as to closely envelop the ovary. The purpose of this phenomenon is unclear but it is very distinctive. The two species in the series are closely related and superficially similar in leaf and flower, but whereas *Meconopsis balangensis* produces a well-developed racemose inflorescence, the flowers of *M. heterandra* are all borne on basal scapes. Both are confined to the mountains of western Sichuan but do not overlap in distribution.

KEY TO SPECIES OF SERIES HETERANDRAE

1. Flowers all scapose, the petals rounded to broadly oval, widely overlapping . . . . 52. **Meconopsis heterandra** (W Sichuan — Mianning)
+ Flowers borne in a well-developed raceme, the petals elliptic-obovate, little or not overlapping. . . . . 51. **Meconopsis balangensis** (W Sichuan — Balangshan, Siguniangshan, Xaiojinshan)

51. MECONOPSIS BALANGENSIS

Meconopsis balangensis Tosh. Yoshida, H. Sun & Boufford, *Acta Bot. Yunnan.* (Plant Diversity and Resources) 33 (4): 409–413 (2011). Type W China: Sichuan, Balangshan, Wengchuan Xian (between Rilong in Xiaojin Xian and the Wolong valley), 4250–4300 m, 19 July 2010, *T. Yoshida* K39 (holotype & isotype KUN).

DESCRIPTION. *Monocarpic herb* to 40 cm tall in flower, often half that height (to 50 cm tall in fruit), with an elongated subcylindrical, firm taproot to 25 cm long, 6–10 mm diameter at the top; plants forming a spreading leaf-rosette in the years before flowering, most parts covered in straw-coloured or reddish brown thin pungent, narrowly pointed bristles of uneven length, the longest up to 9 mm, the base often thickened and blackish, wart-like. *Leaves* in mature plants crowded towards the base, markedly glaucous on both surfaces, although paler beneath, with a paler midrib, petiolate, the lamina elliptic to oblong or oblanceolate, sometimes rather narrowly so, 4–15 × 1.3–5.5 cm, the base cuneate or attenuate into the petiole, the apex subobtuse to acute, the margin entire to somewhat repand, occasionally shallowly and unevenly sinuate, often slightly undulate, covered on both surfaces with rather uneven, well-spaced, patent bristles, more sparsely so beneath; stem leaves few, spirally arranged, decreasing in size up the stem, the lowermost very similar to the basal, the upper merging with the bracts that are no more than 5 cm long and 3–10 mm wide; petiole linear, strap-like, 3–8 cm long, to 3 mm wide, somewhat expanded towards the base. *Inflorescence* a 6–20-flowered raceme with only the lower flowers bracteate, often supported by one or several flowers arising from the basal leaf-rosette; flowers bowl- or saucer-shaped, 3.2–8 cm across, mostly lateral-facing sometimes slightly nodding. *Pedicels* ascending, 1–8 cm long (extending up to 14 cm long in fruit), arched towards the top, patent bristly like the rachis, expanded somewhat immediately below the flowers. *Buds* nodding, ellipsoid to ovoid, the sepals oval, 15–18 mm long, patent-bristly. *Petals* commonly 5–8, occasionally up to 11, blue, bluish purple, bluish lavender to dark purple, occasionally maroon or pale to medium pink, rounded to obovate, occasionally oval,

Meconopsis balangensis, leaf-rosette; Balangshan, Xiaojin County, W Sichuan, 4450 m.
PHOTO: HARRY JANS

276 | THE GENUS MECONOPSIS
51. MECONOPSIS BALANGENSIS

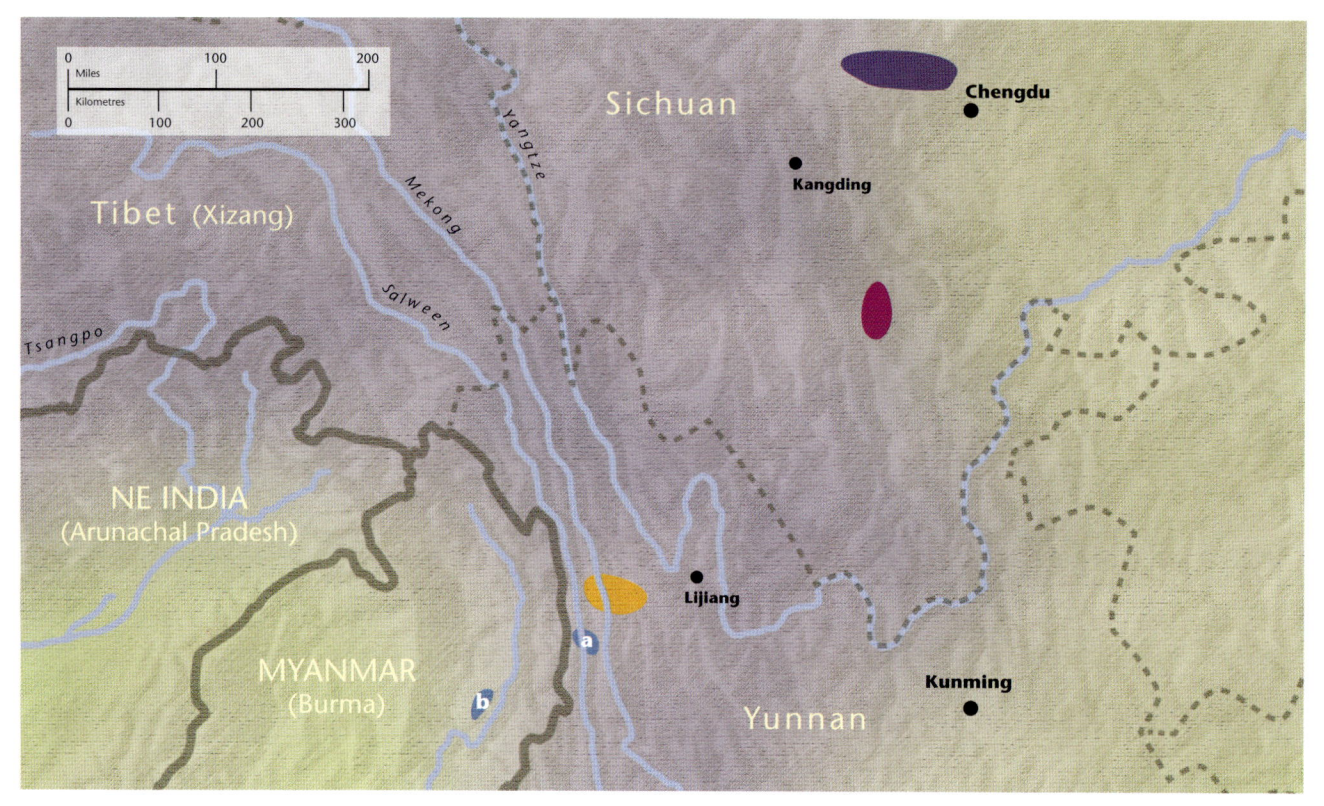

MAP 19. Species distribution: ▪ *Meconopsis bijiangensis* **a** subsp. *bijiangensis*, **b** subsp. *chimiliensis*, ▪ *M. georgei* (incl. forma *castanea*), ▪ *M. heterandra*, ▪ *M. balangensis* s.l.

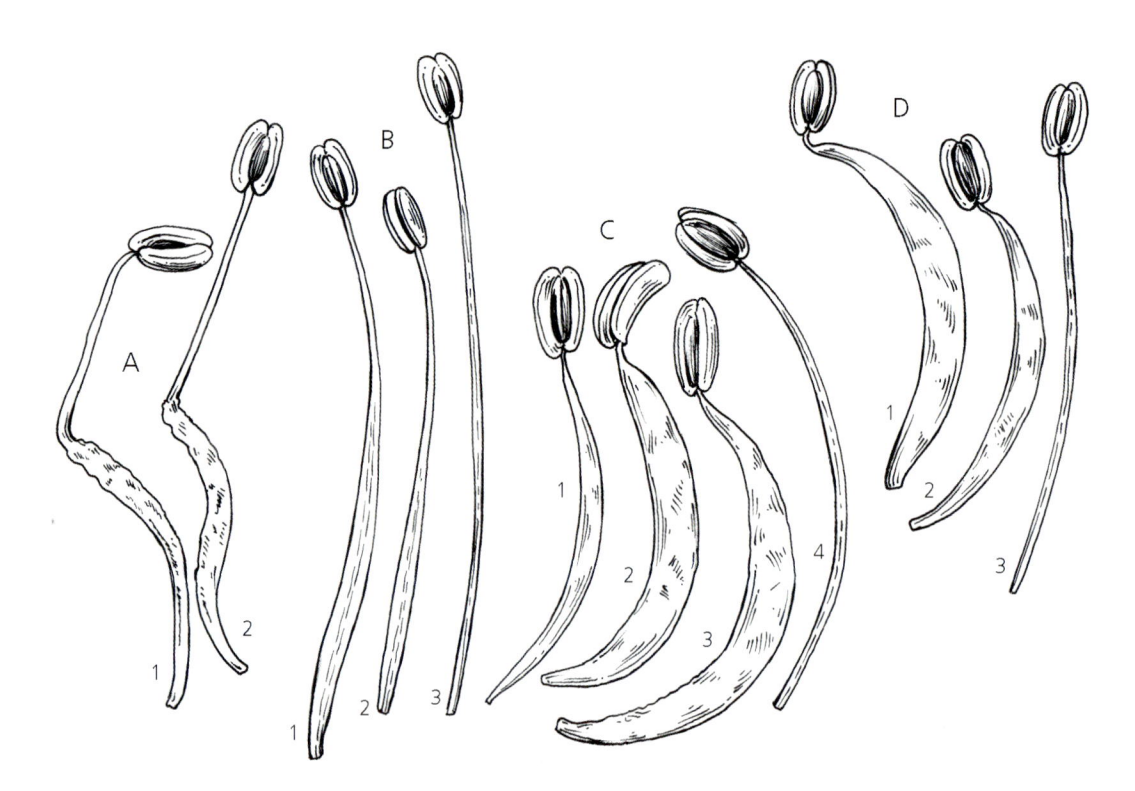

FIG. 21. *Meconopsis* stamens: **A1–2**, *M. henrici*; **B1–2**, *M. sinomaculata*, inner stamens; **B3**, *M. s.*, outer stamen; **C1–3**, *M. heterandra*, inner stamens; **C4**, *M. h.*, outer stamen; **D1–2**, *M. balangensis*, inner stamens; **D3**, *M. b.*, outer stamen. All × 5. DRAWN BY CHRISTOPHER GREY-WILSON

1.5–4 × 1–2.8 cm. *Stamens* numerous, the filaments similarly coloured to the petals, often slightly darker, dimorphic, 5–12 mm long, the outer linear, straight and spreading, the inner incurved, narrowly inflated, to 1.5 mm wide, and forming a 'jacket' around the ovary; anthers oblong, 1.5–2 mm long, purplish with orange-yellow pollen. *Ovary* ovoid, 4–7 mm long, adorned with dense, ascending pungent bristles, the style 2–5 mm long, the stigma protruding beyond the boss of stamens, linear-clavate, 2–3 mm long, yellow or greenish yellow. *Fruit capsule* ellipsoidal, 15–20 mm long, with a rounded base and subobtuse top, covered in patent to ascending pungent bristles, borne on erect pedicels. (Fig. 21D1–3).

DISTRIBUTION. W China, Sichuan, Balangshan, Baoxing Xian, Jiajinshan, Kangding Xian, Lianghekou environs, Siguniangshan, Wengchuan Xian; 3650–4300 m (Map 19).

HABITAT. Rocky slopes and rock fissures, amongst low herb communities, sometimes close to rivulets or seepage zones in the company of coarser vegetation amongst the rocks, in moist dark soils, generally on slopes facing other than north.

Meconopsis balangensis; Balangshan, Xiaojin County, W Sichuan, 4050 m. PHOTO: TOSHIO YOSHIDA

Above: *Meconopsis balangensis*; Balangshan, Xiaojin County, W Sichuan, 4300 m.
PHOTOS: CHRISTOPHER GREY-WILSON

FLOWERING. June to mid-August.

This interesting species has only recently been described; however, the plant has been known and well photographed over the past 30 years, especially on the Balangshan to the west of Chengdu, its prime locus and from whence it was described. It was long included under *Meconopsis rudis* and were it not for the stamen details would almost certainly still be included there. It was at one time thought to be a hybrid between the recently described *M. heterandra*, described from well to the south of the Balangshan area, and *M. rudis*, but this theory has been discarded, especially in the light of the fact that neither of these species are found in these particular mountains centred upon the Balangshan. Close examination by Toshio Yoshida and others have revealed significant differences between these species which leaves *M. rudis* distributed now well to the south in Sichuan province and in north-western Yunnan. Both species bear rosettes of broad glaucous leaves with a scattering of stiff bristles that are darkened and somewhat raised at the base, and both have similar inflorescences and petal numbers. However, the significant difference in the filaments is reinforced by orange yellow anthers and a linear yellow or greenish yellow stigma in *M. balangensis*, and white or cream anthers and a subcapitate whitsh stigma in *M. rudis*. In addition, while both have pungently bristly fruit capsules, those of *M. balangensis* are larger and ellipsoid, those of *M. rudis* squatter and subglobose.

I have been fortunate enough to see and photograph *Meconopsis balangensis* on the Balangshan on a number of visits over the years. The Balangshan is a fine mountain that arises above Wolong to in excess of 5000 m. Above the tree line there are extremely rich meadows where slipper orchids (*Cypripedium tibeticum* forma), *Ajuga lupulina*, *Cardamine macrophylla*, *Corydalis calycosa*, and *Geranium pogonanthum* provide a tapestry amongst many other meadow species. *M. balangensis* favours the rockier meadows, inhabiting gullies, often close to running water, and cliff ledges or crevices, sometimes growing in the company of *M. integrifolia* subsp. *souliei*. The plants grow in ones and twos, occasionally in small colonies but never in large numbers but, at the same time, there are always plenty of unflowered leaf-rosettes to be seen. While the flower colour is fairly consistently blue or bluish purple, occasional pink-flowered forms can be seen.

Like *Meconopsis rudis*, *M. balangensis* is a highly desirable species horticulturally. It has been grown and flowered in cultivation but under the name *M. rudis*; however it does not appear to be in cultivation at the present time.

Toshio Yoshida (2011) has noted that: "Many flies, as well as bumblebees and other insects, visit the laterally-facing flowers during fine weather".

var. balangensis

DESCRIPTION. *Plants* adorned with straw-coloured, thin, pungent bristles that usually have a blackish raised base, most noticeable on the upper surface of the leaves, the bristles tapering to the tip. *Flowers* with blue to bluish-purple or bluish-lavender, rarely pink, petals.

DISTRIBUTION. W Sichuan: Balangshan, Siguniangshan on the boundaries of the Kangding Xian, Wenchuan Xian and Xiaojin Xian; 3650–4300 m.

SPECIMENS SEEN. *K. Y. Lang, L. Q. Li & Y. Fei* 1134 (KUN); *E. H. Wilson* 951 (BM); *T. Yoshida* K39 (KUN), K45(K).

Below, from left:

Meconopsis balangensis; Balangshan, Xiaojin County, W Sichuan, 4350 m. PHOTO: CHRISTOPHER GREY-WILSON

Meconopsis balangensis, flowering plant along with several young plants in leaf-rosette; Balangshan, Xiaojin County, W Sichuan, 4200 m. PHOTO: CHRISTOPHER GREY-WILSON

Meconopsis balangensis, variant with nodding flowers and narrow petals; Balangshan, Xiaojin County, W Sichuan, 4400 m. PHOTO: CHRISTOPHER GREY-WILSON

52. MECONOPSIS HETERANDRA

Meconopsis heterandra Tosh. Yoshida, H. Sun & Boufford, *Acta Bot. Yunnan*. 32 (6): 505 (2010). Type: SW China, S Sichuan, N of Lamagetou, Yele Xiang, Mianning Xian, 29°00'13"N, 102°11'19"E, 4200–4450 m, 4 August 2009, *T. Yoshida* K22 (holotype & isotype KUN).

DESCRIPTION. *Monocarpic more or less acaulous plant* to 20 cm tall at the most at flowering time (to 30 cm in fruit), generally shorter, with a cylindrical taproot, sometimes of several strands, to 28 cm long, 4–8 mm diameter at the top, tapering to the tip, most parts of the plant moderately to rather sparsely covered with thin weakish, purplish to brownish patent bristles to 7 mm long, these slightly raised at the base, the base sometimes blackish. *Leaves* all borne in a spreading basal rosette, glaucous overall, somewhat fleshy, sometimes flushed with purple above, paler beneath, the lamina oval to almost oblong, elliptic-oblanceolate to rhombic-elliptic, 3.8–9.5 × 1–4.3 cm, the base abruptly narrowed, cuneate or somewhat attenuate into the petiole, the apex rounded, obtuse to subacute, the margin with several coarse uneven teeth, usually 2–4 on each side, occasionally sinuate, generally sparsely covered on both surfaces with thin bristles which are generally slightly, but not prominently, darkened at the base; petiole narrow, linear, 2–7 cm long, not more than 4 mm wide, somewhat expanded at the base. *Flowers* 7–12, scapose or several from the upper part of the leaf-rosette agglutinated forming a pseudo-raceme, saucer-shaped, lateral-facing to half-nodding, 4–7 cm across. *Pedicels* (scapes) slender, erect to ascending, 5–12 cm long (to 20 cm long in fruit), expanded below the flower, moderately to sparsely patent-bristly (setose), the bristles mostly 4–6 mm long. *Buds* nodding ovoid, the sepals oval, patent-bristly. *Petals* 5–8, semi-translucent blue or purple-blue, oval to suborbicular or obovate, 20–35 mm long, broadly overlapping, somewhat fluted at the margin, finely erose at the top. *Stamens* numerous, dimorphic, with 7–10 mm long filaments of similar colour to the petals or slightly pinkish, the outer stamens with linear, straight to slightly curved, filaments, the inner stamens somewhat shorter, to 1 mm across, the filaments markedly incurved and inflated, forming a cocoon around the ovary; anthers 1.75–2 mm long, purple with orange pollen. *Ovary* ovoid to subglobose, 5–7 mm long, densely covered with ascending to subappressed purplish-brown pungent bristles; style 2–3 mm long with a whitish 4–7-lobed subcapitate oblong stigma. *Fruit capsule* ovoid, c. 10–15 mm long, 7–11 mm, diameter, densely adorned with ascending pungent bristles. (Fig. 21C1–4).

Meconopsis balangensis var. *atrata*; Jiajinshan, W Sichuan, 4000 m.
PHOTO: TOSHIO YOSHIDA

var. atrata Tosh. Yoshida, H. Sun & Boufford, *Acta Bot. Yunnan*. (Plant Diversity and Resources) 33 (4): 413 (2011). Type: China: W Sichuan, Xaiojinshan, Jiajinshan, 30°52'18"N, 102°41'02"E, 3950 m, 20 July 2010, *T. Yoshida* K41 (holotype & isotype KUN).

DESCRIPTION. *Plants* very similar in most respects to var. *balangensis*, but the pungent bristles on the stems and leaves are unicoloured or with a reddish brown, slightly raised base, the bristles not tapering towards the tip. *Flowers* with dark purple or dark maroon petals.

DISTRIBUTION. W Sichuan: Xiaojin Xian and Baoxing Xian boundary, primarily on the slopes of the Jiajinshan; 3950–4000 m.

SPECIMENS SEEN. *Tzupu Soong* 39109 (KUN); *T. Yoshida* K41 (KUN).

52. MECONOPSIS HETERANDRA

Above left: *Meconopsis heterandra*, leaf-rosettes; Yele Xiang, Mianning County, S Sichuan, 4350 m. PHOTO: TOSHIO YOSHIDA

Above right: *Meconopsis heterandra*, flower details, showing dilated inner filaments; Yele Xiang, Mianning County, S Sichuan, 4350 m.
PHOTO: TOSHIO YOSHIDA

DISTRIBUTION. S Sichuan: Mianning Xian, to date being only known from around the type locality in the vicinity of the northern area of the Yele (Lamagetou) Nature Reserve; 4200–4450 m (Map 19).

HABITAT. Amongst rocks and in deep crevices of steep south-west facing mountain slopes; 4350–4400 m.

Meconopsis heterandra; Yele Xiang, Mianning County, S Sichuan, 4350 m. PHOTO: TOSHIO YOSHIDA

FLOWERING. July–August.

SPECIMENS SEEN. *Yoshida* K22 (KUN).

This interesting and apparently very localised species was discovered by Toshio Yoshida recently, in 2009, and was at once allied to *Meconopsis balangensis* (described the following year), or at least to plants known from the Balangshan further north, because of the presence of dilated inner filaments to the flowers. The two species differ most obviously in the way the flowers are borne: in *M. heterandra* on basal scapes, in *M. balangensis* in well-defined racemes. Were this the only difference then it might be wiser to consider them as subspecies but other differences can be observed. In *M. heterandra* the leaves are generally smaller and often with a few coarse teeth and the fruit capsules are also smaller with a subcapitate whitish (at least at first) stigma; in *M. balangensis* the stigma is linear, yellow or greenish yellow.

At present *Meconopsis heterandra* is only known from the type locality on Mianning Xian in southern Sichuan. Toshio Yoshida notes that it is found on "steep, south-west facing slopes exposed to moist currents in summer, growing on a metaliferous mountain in the vicinity of several pits where iron is mined". Like many of the scapose species of *Meconopsis* the flowers are borne from a very condensed stem so that the upper flowers that arise are clearly borne from the axils of the uppermost (innermost) leaves, which thus act as bracts. In this respect it can be said that the flowers are in reality borne in a much-reduced raceme in which the inflorescence rachis is very condensed.

section IMPEDITAE

This important section contains many of the smaller species and is confined to the higher mountains in south-eastern Tibet, northern Myanmar and south-western China. They are typified by primarily monocarpic, rarely polycarpic, species with small leaf-rosettes that die back in the autumn to an over-wintering bud or buds at or just below the soil surface. The species bear a sparse to moderate indumentum of simple weak hairs, or are occasionally glabrous. Flowers are borne on basal scapes, one or more per leaf-rosette, but on occasion several of the central scapes in a rosette are partly agglutinated to form a pseudo-inflorescence. The stamen filaments are linear or dilated in the lower half. In the section the species bear flowers in shades of blue, purple, reddish purple or occasionally blackish purple.

KEY TO SERIES OF SECTION IMPEDITAE

1. Polycarpic plants with several leaf-rosettes . . . . series **Delavayanae**
+ Monocarpic plants with a single leaf-rosette 2

2. Filaments of all stamens dilated in the middle or bottom half . . . . series **Henricanae**
+ Filaments of all stamens linear throughout . . . . series **Impeditae**

series Impeditae

This series contains all the monocarpic species in section *Impeditae* with linear undilated filaments. With six species, it is one of the largest series in the genus: they are primarily centred upon SW Sichuan and NW Yunnan, with one species, *Meconopsis impedita*, being found further west in south-eastern Tibet and northern Myanmar. There is a marked tendency in most of the species to have transitional leaves on the same plant, the lowermost entire, those above, often larger and progressively more lobed or divided. On occasions plants with all the leaves entire, or conversely all lobed or dissected, can be observed, these often occurring in mixed colonies of leaf forms. While the species primarily bear scapose flowers there is also a tendency, most marked in *M. pseudovenusta* and *M. muscicola*, for the central scapes to become partly agglutinated.

KEY TO SPECIES OF SERIES IMPEDITAE

1. Leaves glabrous or subglabrous 2
+ Leaves moderately to fairly densely pubescent 4

2. Mature fruit capsules 3.5–9.3 cm long, at least 5 × longer than wide; petals 4 . . . .
 . . . . 58. **M. venusta** (NW Yunnan)
+ Mature fruit capsule generally less than 2.8 cm long, less than 4 × long than wide; petals 4–8 (–10) . . . 3

3. Flowering and fruiting scapes not more than 7 per plant; fruit capsules not more than 5 mm diameter, the persistent style 1.5–5 mm long . . . . 55. **M. concinna** (SW Sichuan, SE Tibet & NW Yunnan)
+ Flowering and fruiting scapes at least 8, and up to 17, per plant; fruit capsules 7–9 mm diameter, the persistent style 4–6 mm long . . . . 59. **M. pseudovenusta** (NW Yunnan)

4. Petals normally 5–10, very rarely 4; pedicels (scapes) decumbent at the base or not 5
+ Petals normally 4 (rarely 5–6 in the first flower to open); pedicels (scapes) not decumbent at the base . . 6

5. Plant generally more than 15 cm tall in flower, bearing 7–20 flowers; pedicels (scapes) generally decumbent towards the base; ovary and style usually blackish purple, the style 4–7 mm long in flower . . . .
 . . . . 53. **M. impedita** (N Myanmar, SW Sichuan, SE Tibet & NW Yunnan)
+ Plant not exceeding 10 cm tall in flower, bearing up to 9 flowers; pedicels not decumbent at the base; ovary and style cream, the style 2–3 mm long . . . 54. **M. xiangchengensis** (SW Sichuan, NW Yunnan)

6. Flowers borne in an agglutinated raceme, the lowermost borne in the axils of leaf-like bracts; leaf-lamina acutely pinnately lobed, occasionally crenate at the margin. . . . . 56. **M. muscicola** (NW Yunnan (Laojunshan) and SW Sichuan (E of Yunning))

+ Flowers solitary, borne on basal scapes; leaf-lamina entire to pinnately lobed, if the latter then lobes more or less at right angles to the leaf-rachis. 7

7. Fruit capsule 4–5-valved, 3–5.5 mm diameter; petals spreading widely apart; leaf-lamina very varied from entire to pinnately or bipinnately lobed. . . . . 55. **M. concinna** (SW Sichuan, SE Tibet & NW Yunnan)

+ Fruit capsule 3–4-valved, 5–8 mm diameter; petals ascending; leaf lamina always entire. . . . . 57. **M. pulchella** (S Sichuan; Mianning Xian)

53. MECONOPSIS IMPEDITA

Meconopsis impedita Prain, *Bull. Misc. Inform.*, Kew 1915: 162 (1915). Types, NW Yunnan: *Maire*, China 1913 (E, K); *Monbeig*, China 1894 (K, P); *Forrest* 13314 (BM, E, K); *Forrest* 25522 (BM, E, G); *Kingdon Ward* 792 (E, K). Lectotype: *Forrest* 13314 chosen here; Mekong-Salween Divide, 28°10'N, 13,000 ft (E, lectotype; BM, K, isolectotypes).

[syn. *M. impedita* var. *morsheadii* Prain, *Bull. Misc. Inform.*, Kew 1915: 163 (1915). Type: SE Tibet, E Kongbo, Varma-la [Nyima La], 15,000 ft, *Bailey* 9 (K)]; *M. morsheadii* (Prain) Kingdon Ward, *Ann. Bot.* 62: 859 in obs. (1928), in obs.]

DESCRIPTION. *Monocarpic plant* to 36 cm tall in flower, generally far less, sometimes as little as 10 cm tall in flower, stemless, with a stout dauciform taproot up to 30 cm long, often less, 6–16 mm diam, covered at the top in the remnants of the persistent leaf-bases. *Leaves* all in a lax basal rosette or tuft, deep green and sometimes streaked with red above but glaucous beneath, the lamina very variable in shape, linear-elliptic, linear-lanceolate to spathulate or oblanceolate, 4.0–11.5 × 0.8–3.0 cm, entire, sinuate to pinnatisect with up to 4 forward-pointing, incised, lobes on each side, the lobes with an obtuse to subacute apex, the margin often markedly undulate, sparsely to densely beset with weakish bristles on both surfaces or glabrous beneath, these often pale reddish or yellowish brown with a somewhat raised brownish or purplish base; petiole linear, 3–11 cm long, gradually expanding into the lamina. *Flowers* (4–)7–20, solitary on basal scapes, nodding to half-nodding, 36–67 mm diameter. *Pedicels* (scapes) slender, 8–38 cm long (to 45 cm long in fruit), erect to ascending but generally decumbent towards the base, moderately to densely covered in stiff hairs, especially towards the top, occasionally subglabrous. *Buds* spherical to ovoid, 11–17 × 10–13 mm, nodding at first, the sepals ovate-oblong, 8–14 mm long, moderately to densely patent bristly. *Petals* (4–)5–10 (commonly 6–8), deep blue to violet-blue, dark mauve, purple, rosy-purple, violet-purple, reddish purple, purple-maroon, wine-purple, ruby-red, reddish black or violet-black, suborbicular to obovate, occasionally elliptic-oblanceolate, 17–32 × (8–)12–24 mm. *Stamens* shorter than or equalling the gynoecium, with filaments similarly coloured to the petals, occasionally whitish or whitish at the base, and with cream, yellow, buff-yellow or orange-yellow anthers. *Ovary* ellipsoid-oblong, deep green to almost black, dark violet or deep purplish-red, often flushed greenish blue, sparsely to densely white-bristly, occasionally subglabrous; style 4–7 mm long, blackish purple normally, occasionally greenish, with a greenish, creamy yellow or whitish, rarely brownish, capitate stigma. *Fruit capsule* green with purple-black ribs usually, narrow-obovoid, to oblong-ellipsoid, 18–36 × (5–)6–8 mm, 3–4-valved normally, often narrowed at the base, beset with fairly dense pale or whitish, fine bristles that are at right angles to the capsule, splitting for only a short distance from the apex when mature, the persistent style 4–11 mm long. (Fig. 3A).

DISTRIBUTION. N Bhutan, SE Tibet, SW Sichuan, NW Yunnan and the neighbouring parts of N Myanmar (Upper Burma); commonest along the Mekong-Salween Divide and in the Kongbo and Tsarung regions of Tibet; 3355–4878 m (Map 20).

HABITAT. Open stony alpine or thinly grassy, generally moist, meadows, rocky and gravelly places, moraines, margins of low scrub or banks, occasionally on grassy cliff ledges, mostly over limestone, occasionally over granitic rocks.

FLOWERING. June–August.

Meconopsis impedita was described by David Prain in 1915 based on various collections from NW Yunnan by Forrest and Kingdon Ward as well as two French collectors Maire and Monbeig. It is a very variable species in the wild especially as regards its leaves. These can range from entire to sinuate or incisely lobed and are often undulate as well, these varying conditions being present on distinct individuals, but on occasions the leaves can range from entire to incisely lobed on the same individual. While the leaves are quite variable, the wide bowl-shaped, pendulous flowers seem more uniform, except that the colour ranges widely from deep blue or violet-blue through to ruby red, although dusky and rather sombre colours seem to predominate.

Like the majority of the scapose species of *Meconopsis* in which all the leaves are clustered into a basal rosette, plants do bear a stem, but a very reduced one which is generally wholly or partially below ground, and along which can be found the remains of petioles from previous seasons.

Meconopsis impedita finds its closest allies in *M. venusta* and *M. pseudovenusta*. All three species produce multiple scapes per plant. *M. venusta* differs in generally bearing 4-petalled flowers and in its exceptionally long cylindrical fruit capsules. In addition, the leaves are glabrous or almost so and generally more deeply divided as they are in *M. pseudovenusta*. *M. pseudovenusta* differs primarily in the deeper and finer dissections of its leaves that are glabrous or more or less so, in the occasional agglutinisation of the scapes; in *M. impedita* the scapes are more wiry and generally rather decumbent towards the base.

As already mentioned *Meconopsis impedita* displays a remarkable degree of variation in the shape and degree of dissection of its leaves, often within the same colony. It can be readily mistaken for *M. concinna* in its entire-leaved forms; however, the two species are very different in most other respects. The flowers of the former are generally more deeply and sombrely coloured and are strictly pendent, while those of the latter face sideways or are only half-nodding. In addition, *M. impedita* has usually more petals (often 6–8 as compared to 4). The dark protruding style of *M. impedita* is also very distinctive, while the fruit capsules are relatively broader and more densely bristly and are often, but not always, deep green with purple-black ribs before dehiscence.

While the differences between these species may seem more quantitative rather than qualitative there are clear differences that are especially apparent in the photographs of them taken in the wild. While *Meconopsis venusta* and *M. pseudovenusta* are localised endemics confined to relatively few known sites in NW Yunnan, both *M. concinna* and *M. impedita* have a wider range that stretches from Yunnan into Sichuan, south-eastern Tibet and, in the latter case, northern Myanmar.

Meconopsis impedita has been in and out of cultivation on a number of occasions from seed collected by Forrest,

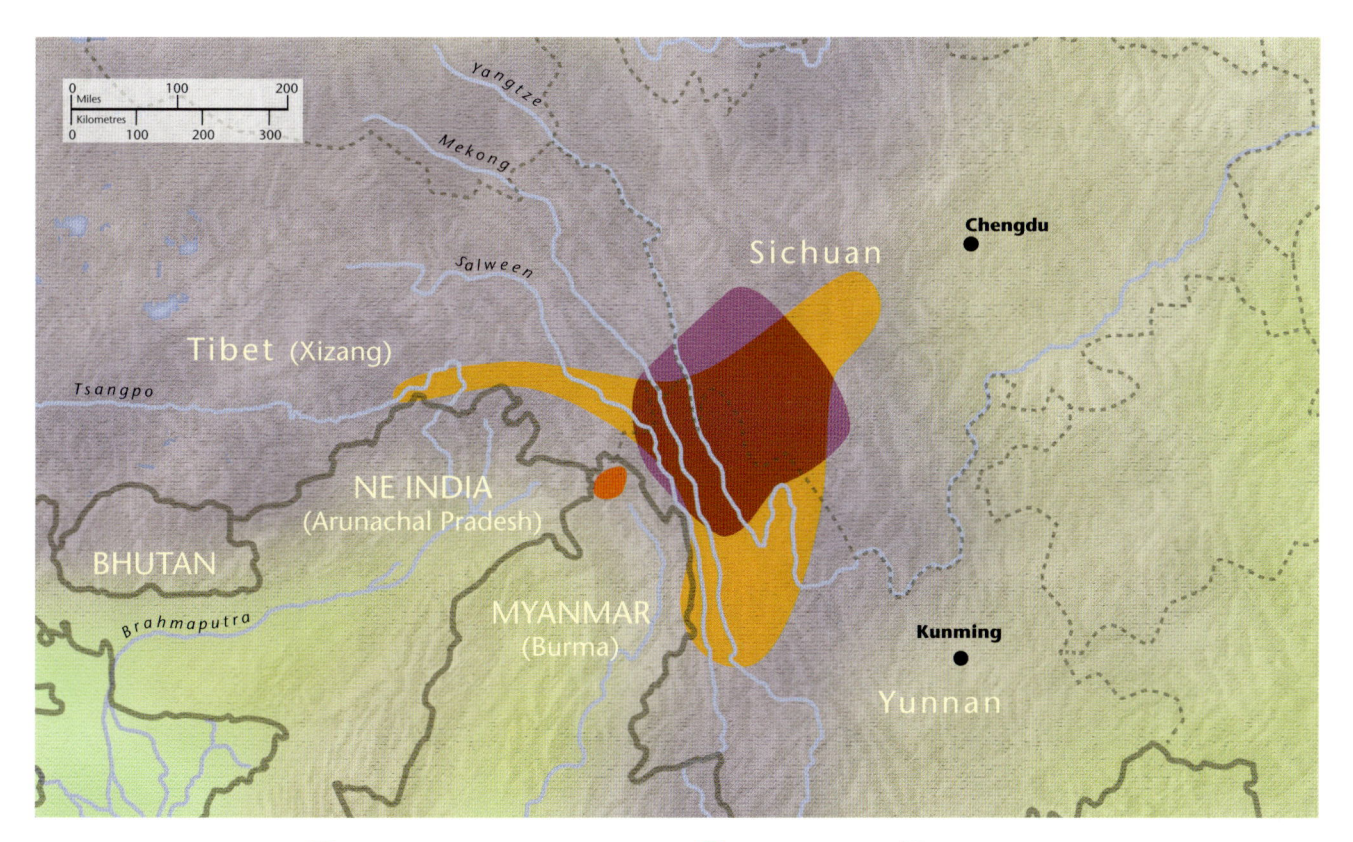

MAP 20. Species distribution: ■ *Meconopsis impedita* subsp. *impedita*, ■ *M. i.* subsp. *rubra*, ■ *M. concinna*.

Rock and others but has never persisted for long. Being a small species it is clearly a subject for a trough or raised bed in the open garden given ample moisture in the summer months, but probably best protected from excessive winter wet. It would probably repay pot culture in a well-ventilated alpine house or cold frame. These small monocarpic species have proved particularly difficult to maintain in cultivation and even when they achieve flowering size (generally in the third or fourth year from seed) it is often difficult to get them to set viable seed. However, with its nodding, wide bell-shaped flowers and neat, often crimped, leaf-rosettes, *M. impedita* would make a fine addition to any collection of rare and unusual alpines.

KEY TO SUBSPECIES OF MECONOPSIS IMPEDITA

1. Leaf-lamina entire to irregularly and rather obliquely pinnatisect, with up to 4 pairs of forward-pointing lobes with acute sinuses; petals deep blue to violet-blue, dark mauve, purple, rosy-purple, violet, dark purple to reddish purple or blackish purple; anthers cream to bright yellow, occasionally buff; style in fruit 4–9 mm long. . . . . subsp. **impedita** (W & SW Sichuan, SE Tibet, NW Yunnan)
+ Leaf-lamina usually regularly pinnatisect, occasionally subentire, with up to 3 pairs of lateral lobes with obtuse to subobtuse sinuses; petals ruby red; anthers orange-yellow; style in fruit 9–11 mm long . . . . subsp. **rubra** (N Myanmar)

subsp. impedita

DESCRIPTION. *Leaf-lamina* entire to irregularly and rather obliquely pinnatisect, with up to 4 pairs of forward-pointing lobes; sinuses acute. *Petals* deep blue to violet-blue or deep violet, dark mauve, purple, rosy-purple, dark purple to reddish purple or blackish purple, sometimes with a greyish bloom on the outside. *Anthers* cream to bright yellow, occasionally buff, the androecium equalling or longer than the gynoecium. *Style* dark blackish purple or blackish violet, 4–9 mm long in fruit. (Figs 22C1–3, 23).

DISTRIBUTION. SW China (W & SW Sichuan, SE Tibet, NW Yunnan), E Myanmar; 3400–4878 m (Map 20).

HABITAT. As above.

SPECIMENS SEEN. *ACE* 814 (E); *Bailey* 9 (K); *Boufford et al.* 28160 (E); *Forrest* 459 (E, K), 9875E (E), 13314 (BM, E, K), 14234 (BM, E, K), 14414 (BM, E, K), 14901 (E, K), 17383 (BM, E, K), 18967 (BM, E, K), 18969 (BM, E, K), 19705 (BM, E), 19771 (BM, E), 20012 (BM, E), 20281 (E), 20408 (BM, E), 21592A (E), 22298 (E), 23421 (BM, E), 25214 (E), 25522 (BM, E, G), 25700 (BM, E, K), 25890 (E, K), 25898 (E, K), 30084 (BM, E), 30085 (BM, E), 30094 (E), 30095 (BM), 30097 (BM); *Handel-Mazzetti* 9621 (E, K); *Kingdon Ward* 792 (E, K), 4008 (E); 4229 (E), 4640 (E), 5750 (BM, E, K), 5808 (BM, E, K), 7571 (BM, W), 9730 (BM), 9852 (BM), 6993 (K); *Ludlow, Sherriff & Elliot* 12259(BM), 13259 (BM), 13878 (BM), 14175 (BM); 15338 (BM, E), 15340 (BM), 15375 (BM, E), 15459 (BM, E), 15773 (BM); *Ludlow, Sherriff & Hicks* 15773 (BM); *Ludlow, Sherriff & Taylor* 5024 (BM), 5024A (BM), 5940 (BM), 6052 (BM, E), 6084 (BM, E), 6234 (BM); *Maire* s.n. 1913 (E, K); *McLaren* 6 (BM), 142 (BM, E, K), s.n. 1938 (E), s.n. 'Lichiang' (BM); *Monbeig* s.n. (K, P); *Rock* 9480 (BM, E, K), 9971 (E), 16313 (K), 16315 (E), 17164 (E, K), 22164 (K), 22376 (BM, E, K), 22377 (BM, E, K), 22379 (BM), 22410 (BM, E, K), 22411 (BM, K), 22411b (E), 22698 (BM, E, K), 22880 (BM, E, K), 23055 (E), 23133 (BM,

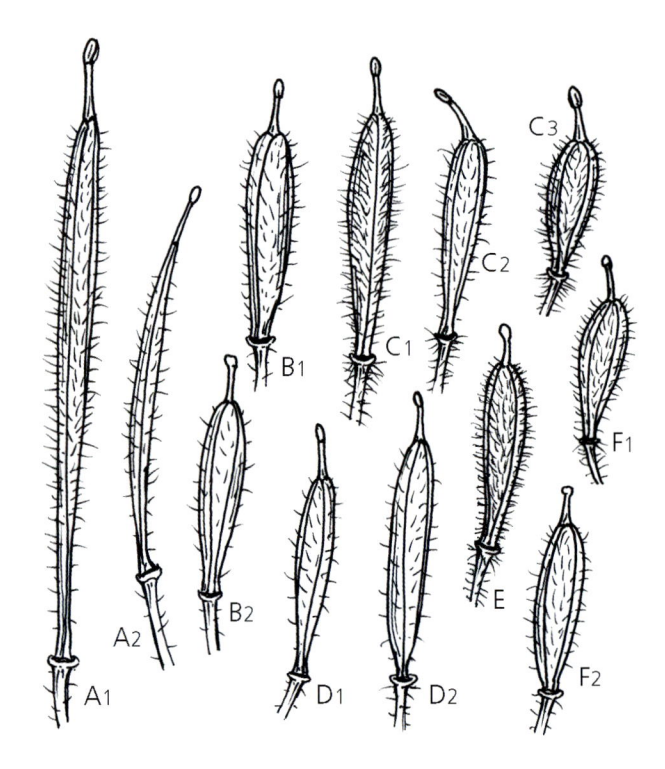

FIG. 22. Fruit capsules: **A1–2**, *Meconopsis venusta*; **B1–2**, *M. pseudovenusta*; **C1–3**, *M. impedita* subsp. *impedita*; **D1–2**, *M. concinna*; **E**, *M. pulchella*; **F1–2**, *M. muscicola*. All × 1. DRAWN BY CHRISTOPHER GREY-WILSON

Above, from left:

Meconopsis impedita; Serkhyem La, SE Tibet, 4500 m. PHOTO: JOHN MITCHELL

Meconopsis impedita; Serkhyem La, SE Tibet, 4500 m. PHOTO: HARRY JANS

Meconopsis impedita; Serkhyem La, SE Tibet, 4500 m. PHOTO: HARRY JANS

E, K), 23138 (E, K), 23214 (BM, E, K), 23252 (BM, E, K), 23410 (BM, E, K), 23440 (BM, E, K), 23463 (BM, E, K), 23605 (BM, E, K), 23606 (BM, E, K), 24305 (E), 25017 (BM, E, K), 25105 (BM, E, K), 25411 (E); *E. H. Wilson* 1371 (K); *Yoshida* K4 (E); *T. T. Yü* 19773 (E), 22165 (E), 22561 (E), 23209 (E).

Subspecies *impedita* is by far the most widespread and common of the two subspecies being found in a narrow arc that encompasses north-western Yunnan and the extreme south-eastern mountains of Tibet. It is one of the most commonly collected *Meconopsis* in the wild with numerous collections made by Joseph Rock and George Forrest in particular. It is most frequent on the mountains of the great divides in NW Yunnan, the Yangtse-Mekong and Mekong-Salween (especially the Hengduanshan and Gaoligongshan): Toshio Yoshida especially reports its common appearance in the northern Gaoligonshan, spanning the Tibet-Yunnan frontier. In this general region the majority of herbarium specimens were gathered in the past, particularly by Forrest and Kingdon Ward. In north-western Yunnan it is more local further east, for instance in the Cangshan by Dali and the Yulongxueshan close to Lijiang. In south-eastern Tibet it is found locally on many passes including the Kucha La, Mira La, Nambu La, Nyima La, Rupina La, Serkhyem La, Temo La in the altitudinal range 4115–4878 m.

It is very variable in flower colour, but typically has rather dark and sombre hues in deep mauve, dark purple, reddish and blackish purple range. Although it is often cited in literature as being present in south-western Sichuan there are surprisingly few authenticated records of it from that province; however, there are a number of specimens collected in the mountains close to Kanding and on the Gonggashan (Minyakonka snow range).

Subsp. *impedita* shows a wide range of leaf form from entire to markedly incisely lobed, the lamina ranging from narrow-elliptic, sometimes almost linear-elliptic, to broadly elliptic or oblanceolate. The petiole can be short or long and the leaf surface can be moderately to rather densely soft-bristly (with rather weak bristles), or

FIG. 23. *Meconopsis impedita* subsp. *impedita* leaf-laminas, lower petioles and pubescence not shown. All × 1¹/₃. DRAWN BY CHRISTOPHER GREY-WILSON

subglabrous. In size, plants can be very dwarf (as little as 10 cm tall in flower) or more usually they are larger, sometimes 30 cm or more in flower, even taller at the fruiting stage.

Meconopsis morsheadii was upgraded from Prain's variety to a species in 1928. The material that Prain based his original diagnosis upon had been collected by Capt. F. M. Bailey on the Nyima La in SE Tibet. Bailey's type material is poor and imperfect; however, Kingdon Ward visited the type locality, the Nyima La at 14–15,000 ft, and was able to secure better material under his number 5808: the field notes read: "Plant with a long taproot. Flowers dark violet. Style black-violet". The western extension of the distribution of *M. impedita* is an interesting one and Kingdon Ward makes much of the specific credentials of the western form in a paper published in the *Annals of Botany* (1928); although seldom collected, it appears to be quite widespread on the mountains of SE Tibet close to the Tsangpo valley, especially in Kongbo and Tsarung regions. It is sometimes incorrectly spelled *morsheadiana*. However, Kingdon Ward's account is far from persuasive and the details and figures he presents allow for a considerable overlap in characters. Having examined all the available material and photographs taken in recent years I see no reason to uphold *M. morsheadii*, even at varietal level. On the other hand, the evidence for Kingdon Ward's *M. rubra* detailed in the same paper (apparently confined to northern Myanmar) is far more persuasive and the plant is accordingly presented here at subspecific level.

According to the field notes accompanying the collections of Ludlow, Sherriff & Taylor from the type locality, *Meconopsis impedita* [subsp. *morsheadii*] is "Abundant on the Nyima La" in SE Tibet.

subsp. rubra (Kingdon Ward) Grey-Wilson, **stat. nov.** [syn. *M. impedita* var. *rubra* Kingdon Ward, *Gard. Chron.*, Ser. 3, 82: 151 (1927), in obs.; *M. rubra* (Kingdon Ward) Kingdon Ward, *Gard. Chron.* 1927, Ser. 3, 82: 506, in obs.; and *Ann. Bot.* 42: 857 (1928). Type: N Myanmar (Upper Burma), Seinghku Wang, 28°08'N, 97°24'E, 12–13,000 ft, *Kingdon Ward 6974* (K, holotype)]

DESCRIPTION. *Leaf-lamina* elliptic-lanceolate, 2.5–5.4 × 1.8–2.6 cm, regularly pinnatisect with up to three pairs of narrow-oblong lobes which decrease in size towards the leaf-tip, the lobes and sinuses inbetween obtuse or subobtuse. *Pedicels* 10–18 cm long in flowering specimens, patent bristly, very densely so immediately beneath the flowers. *Sepals* densely subpungent, patent pubescent. *Petals* 6–8, ruby red, suborbicular, 23–29 × 18–26 mm. *Stamens* shorter than the gynoecium, the anthers orange-yellow. *Style* deep green, 9–11 mm long in fruit. (Fig. 24).

53. MECONOPSIS IMPEDITA

DISTRIBUTION. N Myanmar = Upper Burma (Adung & Seingkhu Valleys): 3355–3965 m (Map 20).
HABITAT. Stony river banks, screes, steep earthy banks.
FLOWERING. June–July.
SPECIMENS SEEN. *Kingdon Ward* 6932 (BM, K), 6974 (K), 6993 (K).

I have examined Kingdon Ward's Myanmar (Upper Burma) material very closely and while the general characters of the plant, especially its habit, correspond with *Meconopsis impedita* sensu stricto, there are clear differences in leaf characters and flower colour. Kingdon Ward in his various writings on the subject (some quotes from which are cited below) stresses these differences. In his 1928 paper published in the *Annals of Botany* he discusses this in full, including two tables of comparisons with *M. impedita* and *M. morsheadii*.

The field notes to KW 6974 read: "Petals 6, rich reddish purple, almost the colour of madeira. Anthers golden; style and stigma green, ovary likewise; filaments purple. Prickles straw-coloured. In crevices and on grassy ledges of granitic cliffs".

In addition, there is a half-tone photograph of subsp. *rubra* Kingdon Ward's book (first edition) *Plant Hunting on the Edge of the World*: 107 (1930). Kingdon Ward was very much attracted by this little plant and wrote about in on a number of occasions. He initially treated it as a variety of *Meconopsis impedita* but later raised it to specific status, however, the overlap of some characters with *M. impedita* sensu stricto, along with its geographic isolation warrants, in my view at least, subspecific recognition.

Meconopsis impedita subsp. *impedita*, leaf-rosette; Serkhyem La, SE Tibet, 4500 m.
PHOTO: JOHN MITCHELL

FIG. 24. *Meconopsis impedita* subsp. *rubra* leaf-laminas, lower petioles and pubescence not shown. All × 1¹/₃. DRAWN BY CHRISTOPHER GREY-WILSON

53. MECONOPSIS IMPEDITA

Meconopsis impedita subsp. *impedita*; Zachung La, northern Gaoligongshan, W Yunnan. PHOTO: TOSHIO YOSHIDA

Meconopsis impedita subsp. *impedita*; Meilixueshan, NW Yunnan, c. 4250 m. PHOTO: MARGARET NORTH

"By this time I was clear of the larger scrub, and glancing up the screes, I was just in time to see a flower twinkle, as a bayonet of sunlight stabbed the clouds. Approaching, I stood in silent wonder before the Ruby Poppy. In Sino-Himalaya, the mountain poppies are generally blue, sometimes yellow, very rarely red. Therefore a red poppy here is as exciting as a blue poppy in England. Also it was exquisite. A sheaf of finely drawn olive green jets shot up in a fountain from amongst the stones, curled over, and where they reached the ground again, splashed into rubies. Thus one might visualise *Meconopsis rubra*, if one imagines a fountain arrested in mid-career, and frozen. But it was not till later, when on a stormy day I saw whole hillsides dotted with these plants, that I really believed in *M. rubra*. As the wind churned up the clouds, a burning brand was lit, and touching off the flowers one by one, up they went in red flame!......"

"Another wonderful little alpine poppy is *M. rubra*, which grows scattered up and down the long screes, being very abundant between 11,000 and 12,000 feet. *M. rubra* is like *M. morsheadii*, a plant common on the high mountains of Kongbo, in Tibet, except for its deep ruby-red petals — they are violet in *M. morsheadii*. It is still more like *M. impedita* from the mountains of Northwest Yunnan, but that too has violet petals. The first time I saw *M. rubra* peeping from a crack in the bare gneiss cliff, with a solitary red flower glaring balefully at me, I thought it was a freak: for the colour of these alpine Poppies is not always fixed. But it was not a freak, for no flower of the hundreds seen ever opened any other colour; in the Seinghku valley [N. Myanmar] *M. rubra* has definitely replaced *M. morsheadii* and *M. impedita*."

Kingdon Ward *Plant Hunting on the Edge of the World* pp. 107, 301, 1930.

"The first flowering specimens encountered were leaning from crevices in the gneiss cliff at 14,000 feet altitude, a never-to-be-forgotten sight. The foliage was more like that of the dainty, blue-flowered *M. bella* from Sikkim, but the habit was perfect impedita. To the colour of the flower, however, I at first attached little importance, thinking that I grazed upon a freak: but when ten days later every scree and rubble-chute three up fountains of blood-red Poppy flowers, no longer could I conceal from myself that I had stumbled upon a prize .... *M. rubra* like *M. horridula*, sometimes tries to build up a central stem of agglutinosed scapes. The usual form, however, is a number, up to twenty even, of independent one-flowered, basal scapes, not above six or eight inches high. To say that the flowers are red is tame enough, and yet startling. There was no questions of freaks; I saw hundreds of plants, and the colour never varied a semitone. When, as often happened on a steep face, a sun shaft, loosed from the cloud rack, drove clean through a flower, it flared up in a jet of scarlet flame. Even on the dankest day, when the recking valley smoked and dripped forlornly, and the cliffs were blotted out in the driving drizzle, the colour was that of a glowing port-red-wine, which nothing could dim. Towards the middle of July, every stone-chute and gravel-bank between 11,000 and 12,000 feet incandesced with scores of twinkling red lights as a flash of sunshine touched off the flowers."

Kingdon Ward, *Burmese species of Meconopsis* in *Gardener's Chronicle* Ser. III, 82: 151 (1927).

54. MECONOPSIS XIANGCHENGENSIS

Meconopsis xiangchengensis R. Li & Z. L. Dao, *Novon* 22: 180–82 (2012). Type: SW Sichuan, Xiangcheng Co., Shagong township, Wuming Snow Mtns Pass, alpine meadow on slope, 29°08'15.5"N, 100°02'38.6"E, 4684 m, 19 July 2007, *R. Li & Z. L. Dao 036* (holotype, KUN)]

DESCRIPTION. *Small monocarpic scapose herb* to 10 cm tall in flower, with a branched, cylindrical, fleshy taproot to 20 cm long and 1 cm diam., the top covered in the persistent remains of petioles from previous seasons. *Leaves* all borne in a solitary basal rosette, mid green above, paler and somewhat glaucous beneath, the lamina elliptic to elliptic oblanceolate or oblanceolate or linear-oblanceolate, the base gradually attenuate into the petiole, the apex obtuse to subobtuse, the margin entire, or remotely sinuate-toothed, often rather undulate, with sparse yellow-brown, weak bristles on both surfaces, densest along the margin, occasionally subglabrous; petiole to 2.5 cm long, whitish green, somewhat expanded towards the base. *Flowers* solitary, scapose, 4–9 per plant, saucer-shaped, lateral-facing to somewhat ascending, 2.5–5 cm across. *Scapes* (pedicels) erect, pale whitish green, 6–10 cm long, glabrous or with a few scattered bristles. *Flower buds* nodding, patent-bristly, the sepals ovate-oblong, 9–12 mm long. *Petals* 6–10, deep purple-blue, oval to obovate, 12–25 × 10–18 mm, the margin entire. *Stamens* numerous, 7–8 mm long, about one third the length of the petals, the filaments linear, the same colour as the petals or darker, the anthers cream. *Ovary* cream, ellipsoid-oblong, adorned with sparse yellow-brown bristles to subglabrous, the style 2–3 mm long, terminating in a capitate cream stigma which just protrudes beyond the boss of stamens. *Fruit capsule* unknown.

DISTRIBUTION. SW Sichuan (Wuming Snow Mts Pass, Xiangcheng); NW Yunnan (Zhongdian, near Landdu pass); 4385–4685 m.

HABITAT. sloping alpine meadows.

FLOWERING. July.

SPECIMENS SEEN. *R. Li & Z. Dao 009C* (MO, paratype), *039* (KUN).

Little is known of this interesting species published in 2012 from two fairly close localities on either side of the Sichuan-Yunnan border. The authors ally it to *Meconopsis impedita* which is widely distributed, especially in north-western Yunnan. *M. xiangchengensis* differs primarily in its small stature, in its short petioles and style and in the glabrous or subglabrous pale scapes. The greatest difference can be seen in the photographs of the two species: *M. impedita* has fully nodding, sombrely coloured flowers, those of *M. xiangchengensis* are lateral-facing and brightly coloured. There is no indication that the two species grow together in the wild nor even in close proximity.

Although the leaves are said to be entire in the original description, photographs show that they can be somewhat sinuately toothed in much the same manner as those found in some forms of *Meconopsis impedita*. The exact relationship between these two species requires further investigation.

The superficial resemblance of *Meconopsis xiangchengensis* to *M. psilonomma* (syn. *M. barbiseta*) should also be noted. The latter is found in the wild well to the north in Sichuan, and the neighbouring parts of Gansu and

Meconopsis xiangchengensis; SW Sichuan, Xiangcheng Co., Wuming Snow Mtns Pass, 4684 m.
PHOTO: RONG LI

Qinghai and differs significantly in its more robust nature and much larger flowers that are generally solitary, one per plant. In addition, the latter has a considerably longer style, while details of the petals and ovary are at variance with those of *M. xiangchengensis*.

55. MECONOPSIS CONCINNA

Meconopsis concinna Prain, *Bull. Misc. Inform., Kew* 1915: 163 (1915). Type: China, NW Yunnan: mountains NE Yangtse Bend, 27°45'N, *Forrest* 10404, China 1913 (BM, K, P, syntypes); mountains NE of Yangtse Bend, 27°45'N, *Forrest* 10979 = 10404 in fruit, (BM, P, syntypes); Mountains W of Fengkou, 27°40'N, *Forrest* 12670, China 1914 (BM, K, syntypes); *Forrest* 12706, China 1914 (BM, K, syntypes). Lectotype *Forrest* 12670 (BM; isolectotype K) chosen here.
[Syn. *M. lancifolia* Franch. ex Prain var. *concinna* (Prain) G. Taylor, *The Genus Meconopsis*: 90 (1934)]

Meconopsis concinna; Habaxueshan, NW Yunnan, 3700 m.
PHOTO: TOSHIO YOSHIDA

DESCRIPTION. *Monocarpic, rarely polycarpic, plant 4.5–13.5 cm tall in flower (to 30 cm in fruit), stemless or with a very short trunk[33], with a small, swollen, carrot-like, napiform to fusiform rootstock, 1.8–6 cm (occasionally up to 10 cm) in length, 0.4–1.2 cm diameter, simple or sparingly branched, very rarely plants with a somewhat extended 'hypocotyl trunk' to 25 mm long. Leaves* pale to mid green, glaucous beneath, sometimes with a reddish or purplish flush above, all basal, forming a lax rosette, the lamina linear to oblong, ovate or oblanceolate, rarely spathulate, 12–52 × 4–16 mm, usually pinnately- or bipinnately-lobed (with 2–4 pairs of pinnae), sometimes but by no means always present on the same plant, entire or shallowly sinuately lobed (plants in mixed populations can have transitional leaves or they may on occasion all be entire or all divided), the base wedge-shaped or attenuate into the petiole, the lobes oblong to rounded, entire or occasionally the lowermost asymmetrically bi- or tri-lobed, the apex obtuse to subacute, glabrous to sparsely or moderately hairy on both surfaces; petiole linear, longer than the lamina, to 65 mm long. *Flowers* (1–)3–7 per plant, 24–58 (–64) mm across, borne on erect to ascending basal scapes, very rarely one or two scapes partly agglutinated, half-nodding. *Pedicels* (scapes) 3.2–17.5 cm long (extending to 22 cm, rarely as much as 33 cm, long in fruit), the pedicels rarely partly agglutinated (as in *McLaren* 77, BM!), glabrous to moderately hairy. *Buds* oblong-obovoid, 8–12 × 7–8 mm, glabrous or with a few scattered hairs. *Petals* 4, rarely 5–6, violet to blue-violet to purple-blue, deep purple, ruddy-purple or purplish-black, sometimes very deeply coloured, ovate to suborbicular, (11–)15–31 × (7–)12–29 mm, margin erose. *Stamens* with linear filaments, 8–10 mm long, the same colour as the petals or darker, and cream, creamy yellow or greyish yellow, occasionally orange, c. 1 mm long anthers. *Ovary* narrow-ovoid or ellipsoid, glabrous or sparsely bristly; style green or purplish black, 2–4 mm long, with a cream or greenish stigma. *Fruit capsule* narrow obovoid to ellipsoid or subcylindrical, 17–40 × 3–5.5(–7) mm, 3–5-valved normally, moderately to sparsely adorned with a few scattered patent bristles[34], the persistent style 2.5–6 mm long. (Figs 22D1–2, 25).

---

[33] In older plants a short stem-like trunk develops as the lower leaves die away, especially in plants that take several years to reach flowering maturity. Both *Forrest* 12670 and *Rock* 24769 exhibit this feature well.

[34] The fruit capsules of *Rock* 24305 ("Muti Konka, E of Yalung, 14,500 ft" (BM)) are unusually long and narrow (c. 40 × 3.5–4 mm) and moderately patent, somewhat reflexed-bristly. This collection requires further investigation.

DISTRIBUTION. SW Sichuan (Litang, Muli, Yalung (Multikonka)), NW Yunnan and neighbouring areas of Tibet; particularly in the Yulongxueshan, Habashan and the mountains of the Zhongdian Plateau and those mountains close to Yungning; 3658–4573 m (Map 20).

HABITAT. Stony alpine meadows, screes, moraines, gravelly places, cliff ledges, banks, rhododendron moorland, margins of conifer forest treeline, open forest on mossy rocks.

FLOWERING. Late May–July.

SPECIMENS SEEN. *ACE* (Alpine Garden Society China Exped.) 327 (E); *Forrest* 2730 (BM), 2745 (BM), 10404 (BM, E, K), 10777 (E), 10979 (BM, E), 10979 (BM), 12670 (BM, E, K), 12706 (BM, E, K), 12706 (BM, E, K), 12796 (BM, E, K),15369 (BM, E, K), 20527 (E), 28491 (BM, E), 29234 (BM, E), 30086 (BM), 30088 (BM, E), 30106 (BM, E), 30113 (BM, E); *Kingdon Ward* 5191 (E), 5239 (E), 5286 (E); *McLaren* 77 (BM, E, K); *Rock* 9377 (E), 9685 (E), 9686 (E), 16553 (K), 17640 (E), 17941 (E), 17943 (E), 24305 (BM), 24745 (BM, E, K), 24749 (BM, E, K), 24769 (BM, E, K), 25294 (BM, E, K), 25295 (E), 25296 (K), 25309 (E), 25296 (BM, E); *Schneider* 15369 (K).

Meconopsis concinna was included under *M. lancifolia* as a subspecies in George Taylor's 1934 monograph; however, the dissection of the leaves, the clearly scapose nature of the flowers, the longer fruit capsules and the generally fewer petals per flower, clearly distinguish this taxon and I have no hesitation in returning it to specific status.

Interestingly, although there is every indication that *Meconopsis concinna* is a monocarpic plant, one specimen, *ACE* 327, collected on the Habashan, NW Yunnan, reveals an individual with three shoots, only one of which is flowering and is clearly more perennial in nature.

The species is fairly widespread in NW Yunnan, particularly in the mountains in close proximity to the Zhongdian Plateau, Habashan and the Baimashan. It is less common in Sichuan, although collections have been made on several occasions around Yungning and Muli and as far north as Litang and the Minya Konka. It does not appear to have been recorded very often from Tibet.

The leaves of this rather localised little species show a great range of size and shape, especially in the amount of dissection of the lamina. Some specimens show a range of leaf shapes from entire (outer leaves) to neatly pinnatisect (upper most in the rosette). One good example of this is seen in *McLaren* 77 (from Yungkingshan), but also *Forrest* 12706, 15369 & 20527. Such variation is seen in other species, particularly *Meconopsis bella*, although in this latter instance individuals can be found in which all the leaves of an individual are entire. Interestingly,

FIG. 25. *Meconopsis concinna* leaf-laminas, lower petioles and pubescence not shown. All × 1⅓. DRAWN BY CHRISTOPHER GREY-WILSON

Meconopsis concinna; Habaxueshan, NW Yunnan, 4000 m. PHOTO: JOE ATKIN

55. MECONOPSIS CONCINNA

Above left: *Meconopsis concinna*, entire-leaved form; Black Lake, Habashan, NW Yunnan, 4000 m, flowering in late July. PHOTO: ALAN DUNKLEY

Above right: *Meconopsis* aff. *concinna*; Habaxueshan, NW Yunnan, 3650 m. PHOTO: CHRISTOPHER GREY-WILSON

McLaren 77 (gathered in the Lijiang mountains, the Yulongxueshan) has several plants on the one sheet all with entire leaves, but the plants in question undoubtedly belong to *M. concinna*. Flower and fruit size can vary considerably between individuals even in the same population but also, more importantly, on the same plant: the first flowers to open being the largest, as are the resultant fruit capsules. On the Habashan close to the Zhongdian Plateau and adjacent to the Yangtse river, entire, dissected and intermediate forms can be found within easy walking distance from one another on the mountain, at or just above the tree-line. Unlike *M. bella* in which the leaf-form kinds have a geographical distribution I can find no such separation in *M. concinna*.

I can find no evidence that this little species has ever been in cultivation, although it may have been introduced as seed at some stage under the name *Meconopsis lancifolia* with which it was at one time closely associated. Being a small species it is a very difficult plant to find in fruit in the wild and no authenticated seed has been introduced in recent years to my certain knowledge.

The record of *Meconopsis concinna* in the *Flora of Bhutan* is erroneous. It is based on a single collection, *Ludlow & Sherriff* 642 (BM), collected on the Orka La in eastern Bhutan close to the border of Arunachal Pradesh. This collection has proved to be a new species (see *M. ludlowii*, p. 350) that has subsequently (2012) been found and photographed in Arunachal Pradesh.

56. MECONOPSIS MUSCICOLA

Meconopsis muscicola Tosh. Yoshida, H. Sun & Boufford, *Acta Bot. Yunnan.* (Plant Diversity & Resources) 34 (2): 145 (2012). Type: SW China, Yunnan, Lijiang Xian, Laojunshan, 26°38'04"N, 99°42'54"E, 3800 m, 10 July 2009, *T. Yoshida* K11 (holotype KUN, isotypes E, KUN).

DESCRIPTION. *A monocarpic herb* to 35 cm tall in flower, but as little as 15 cm, covered in most parts by straw-coloured or reddish, soft bristles to 5 mm long; taproot more or less fusiform, 10–20 cm long, to 10 mm diameter close to the top, extended above into a short stem 2–5 cm long, 2.5–3.5 mm diameter. *Leaves* mostly borne in a basal rosette, long-petiolate, ascending to arching outwards, the lamina papery, mid-green above, paler beneath, lanceolate to oblanceolate, rarely subovate, 3–9 cm long, 1–2.5 cm wide, coarsely crenate to pinnatifid with up to 5 pairs of lobes or occasionally subentire (especially the smaller lowermost ones), the lobes oblique, somewhat forward directed, obtuse, with a rounded sinus, the terminal lobe deltoid-ovate, the lamina moderately covered in patent bristles, sparsely so beneath; petiole linear, 2–8 cm long, to 3 mm wide. *Inflorescence* agglutinated-racemose with up to 15 spirally arranged flowers, patent bristly, the lowermost 2–4 flowers borne in the axils of leaf-like bracts, the flowers half-nodding or lateral-facing, cupped to almost flat, 3–4.5 cm across. *Pedicels* slender, 1–10 cm long (to 15 cm long in fruit), pubescent as for the inflorescence rachis. *Buds* nodding, the sepals oval, 8–14 mm long, patent-bristly. *Petals* 4, rarely 5, deep violet-blue to wine-purple, ovate to obovate or more or less rounded or somewhat rhombic, 15–22 × 13–20 mm, somewhat fluted and undulate, the apex obtuse to rounded, finely erose. *Stamens* numerous, much shorter than the petals (about a quarter the length), the filaments similar in colour to the petals or more bluish or purplish, 4–6 mm long, the anthers dark purple with orange yellow pollen, c. 1.5 mm long. *Ovary* ellipsoid, to 5 mm long, covered with short whitish, patent to ascending bristles, with 3–4 longitudinal ribs; style 1–2 mm long with an ovoid stigma of about the same length. *Fruit capsule* obovoid-oblong, 15–27 mm long, 4–8 mm diameter, moderately to densely patent-bristly, 3–4-valved, borne on stiffly erect pedicels, the persistent style 4–5 mm long, including the stigma. (Fig. 22F1–2).

DISTRIBUTION. SW China: NW Yunnan (Laojunshan, Chienchuan-Mekong divide) and SW Sichuan (E of Yunning); 3300–4270 m. (Map 21).

HABITAT. *Abies* forest underlain with rhododendrons and other shrubs, growing in mossy places on humusy banks amongst rotting tree trunks and granitic boulders, often along streamsides, also moorland, generally on slopes of north-west aspect.

FLOWERING. Late June–August.

SPECIMENS SEEN. *Forrest* 21592 (E), 21986 (E), 31592 (K); *Rock* 25095 (BM, E, GH, K); *T. Yoshida* K11 (E, KUN).

This interesting and rather unusual, recently described, species was first collected as far as it is known by George Forrest in the mountains east of Yungning (SW Sichuan) in July 1922, and also in August of the same year in NW Yunnan on the Chienchuan-Mekong divide. These two widely separated Forrest collections, which have long-harboured in the Edinburgh Herbarium under *Meconopsis lancifolia*, clearly belong to *M. muscicola*; both are recorded from "open alpine meadows", rather than in the upper woodland zone as is Yoshida's type collection. *Forrest* 21592 collected in NW Yunnan, Chienchuan[Jianchuan]-Mekong divide, is described as a "Plant of 7–12 inches. Flowers ruddy purple, anthers orange"; *Forrest* 21986 collected in SW Sichuan, east of Yunning [Yongning], as a "Plant of 9–16 inches. Flowers deep wine-purple.....Aug. 1922". Joseph Rock's collection (25095) was collected at "Lao-chun shan", Toshio Yoshida's type locality (Laojunshan) for *M. muscicola*. It remained harbouring under *M. lancifolia* since determined by George Taylor; the determinavit slips on the relevant specimens show that they were

Meconopsis muscicola, young leaf-rosettes; Laojunshan, NW Yunnan, 3800 m.
PHOTO: TOSHIO YOSHIDA

56. MECONOPSIS MUSCICOLA

identified as such in 1933 and included within *M. lancifolia* in George Taylor's monograph published the following year. Since then a great deal more material, backed by field observations and modern photography has made it possible for these specimens to be reassessed.

Meconopsis muscicola is a fairly distinctive monocapic species with characteristically thin, rather pale leaves with at least the inner rosette leaves with slightly forward-directed, shallow lobes and with a rather modest covering of weak bristles. The small flowers are rather flat or saucer-shaped and borne on long slender stalks.

It has been noted that the species is known from a few widely scattered localities, primarily in north-western Yunnan; however, it is very likely that it will be found in other localities in the region, especially in the upper forest zone.

In leaf this interesting and restricted species could be mistaken for *Meconopsis impedita*, but the leaves are thinner and flatter and more papery, paler and with rather more substantial bristles, although they are far from the pungent bristles found in species such as *M. horridula* or *M. rudis*. In fact the leaf lobes, when present, in *M. impedita* point obliquely forward, while in *M. muscicola* they are more or less at right angles to the midrib or slightly forward-directed. Besides this, *M. muscicola* is decidedly a forest or meadow dweller, while *M. impedita* is generally a species of greater elevations in more exposed, often rocky, habitats outside the forest. The two species also differ in their inflorescence and floral characters: while *M. muscicola* normally has 4-petalled flowers that generally face sideways or are somewhat inclined, those of *M. impedita* are strictly nodding and generally with 5 or more (up to 10) petals. Furthermore, while the flowers of *M. impedita* are borne strictly on basal scapes those of *M. muscicola* are basal or borne in agglutinosed racemes. The stem like extension of the plant (probably the hypocotyl) is concealed in the mossy litter in which the plant grows.

The inflorescence of *Meconopsis muscicola* requires closer examination. In some individuals the flowers are clearly borne on basal scapes or, more often, several of the central flowers in the leaf-rosette have partly agglutinated pedicels or scapes. In others the flowers appear to be in a more or less well-organised ebracteate raceme but the manner in which the pedicels arise from the inflorescence rachis is more suggestive of the

Above: *Meconopsis muscicola*; Laojunshan, NW Yunnan, 3800 m. PHOTOS: TOSHIO YOSHIDA

THE GENUS MECONOPSIS
56. MECONOPSIS MUSCICOLA

Meconopsis muscicola; Laojunshan, NW Yunnan, 3800 m.
PHOTO: TOSHIO YOSHIDA

agglutinated state than a true raceme. In the true raceme or panicle common to many species of *Meconopsis*, the pedicels or peduncles arise at an obtuse angle or at more or less right angles to the inflorescence rachis, while in the agglutinated state they arise at a very acute angle, almost vertical in some instances. A similar situation can be seen in some forms of *M. pseudovenusta*.

In the original description Yoshida *et al.* ally their new species to *Meconopsis lancifolia* and to *M. impedita*, stressing that the prime difference from *M. lancifolia* is in the pinnately lobed rather than entire leaves; however, the two are very different in most respects including the thicker, less papery leaves, in the few-flowered ebracteate inflorescence which is sometimes reduced to a single flower, and in the shorter and weaker hairs on leaves and pedicels. In the characters of the fruit capsule *M. muscicola* comes closest to *M. impedita*.

Meconopsis muscicola shares a habitat preference in common with *M. concinna*, although they have not been found growing together in the wild; typically both are found in damp places within the forest zone, although often close to the tree line, occasionally on grassy banks out of the forest; however, all the other members of series *Impeditae*, to which these two species belong, are plants of more exposed, alpine habitats above the tree line.

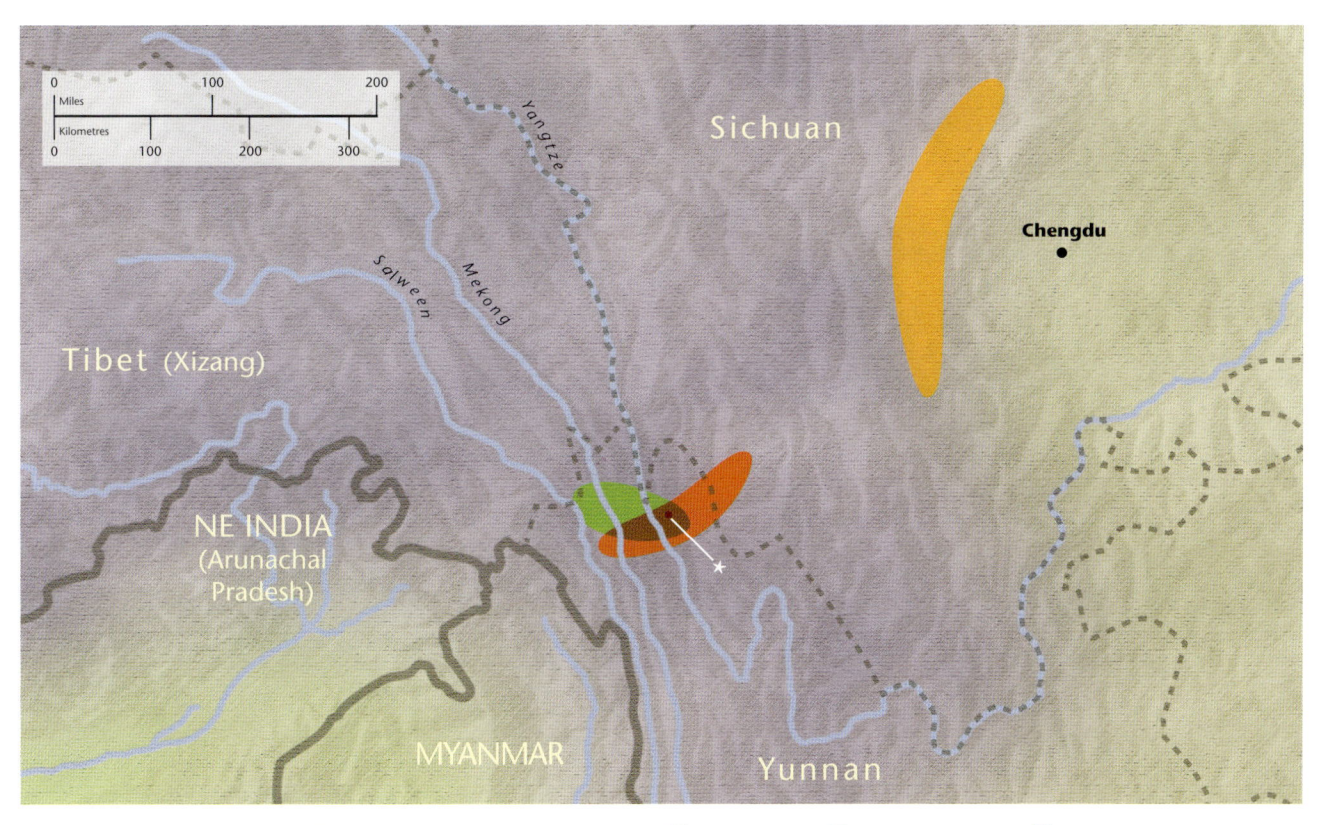

MAP 21. Species distribution: ■ *Meconopsis pulchella* (speculative), ■ *M. muscicola*, ■ *M. pseudovenusta*, ■ * *M. venusta*.

57. MECONOPSIS PULCHELLA

Meconopsis pulchella Tosh. Yoshida, H. Sun & Boufford, *Acta Bot. Yunnan.* 32 (6): 503–507 (2010). Type: SW China, S Sichuan, N of Lamagetou village, Yele Xiang, Mianning Xian, 29°00'05"N, 102°11'03"E, 4150–4300 m, 4 August 2009, *T. Yoshida* K21 (KUN, holotype & isotype).

DESCRIPTION. *Monocarpic herb* to 18 cm tall in flower (to 28 cm in fruit), with a fusiform rootstock 4–10 cm long, to 6 mm diameter at the top, most parts of the plant covered with straw-coloured, softish bristles which are sometimes darker or blackish at the base, 2–6 mm long. *Leaves* all borne in a lax basal rosette, rarely more than 12, petiolate, deep green, sometimes purple-flushed above, paler beneath, the lowermost in flowering plants smaller than the upper, the lamina lanceolate-elliptic to lanceolate-oblong or subspathulate, 1.5–6 × 0.7–1.5 cm, the base somewhat abruptly and slightly rounded to attenuate into the petiole, the apex subobtuse to subacute, the margin entire, sometimes slightly repand, soft-bristly above and beneath, the bristles above sometimes with a dark blackish base; petiole linear 2–5 cm long, not more than 1.5 mm wide, but expanded somewhat at the base. *Flowers* scapose, 3–9 per plant borne from the axils of the rosette leaves, cupped to bowl-shaped, half-nodding, 2.5–5 cm across. *Pedicels* (scapes) slender, erect, straight but slightly curved at the base, more so immediately below the flowers, 4–18 cm long (to 28 cm in fruit), to 1.5 mm diameter (2 mm in fruit), patent to patent-retrorse bristly, denser towards the slightly expanded top. *Buds* nodding, narrow-ellipsoidal, the sepals 8–12 mm long, patent-bristly. *Petals* normally 4, occasionally 5–6, magenta or purplish blue or reddish purple, obovate to almost rounded, 17–30 × 14–25 mm, the apex obtuse to rounded. *Stamens* numerous, the filaments filiform, 5–7 mm long, similar to the petals in colour, although generally darker, the anthers rounded, c. 1 mm, dark purplish red or magenta with greyish or pale rusty brown pollen. *Ovary* ellipsoidal, 5–7 mm long, with dense ascending bristles, the style 2–3 mm long with a subcapitate 1–2 mm stigma, somewhat exceeding the boss of stamens. *Fruit capsule* narrow obovoid, 20–25 mm long, 5–8 mm diameter, densely patent bristly, 3–4-valved. (Fig. 22E).

Right, from top:

Meconopsis pulchella; Balangshan, W of Chengdu, W Sichuan, 4250 m.
PHOTO: CHRISTOPHER GREY-WILSON

Meconopsis pulchella; Yele Xiang, Mianning County, SW Sichuan, 4300 m. PHOTO: TOSHIO YOSHIDA

57. MECONOPSIS PULCHELLA

Meconopsis aff. *pulchella*; Mt Siguniang, Xiaojin County, W Sichuan, 4300 m. PHOTO: TOSHIO YOSHIDA

DISTRIBUTION. W China: S & W Sichuan (Balangshan, Jiajinshan, Mianning Xian); 4150–4300 m. (Map 21).
HABITAT. Humusy pockets amongst rocks on west and south-west facing slopes.
FLOWERING. July–August.
ETYMOLOGY. *Pulchella* means "small and beautiful", but Toshio Yoshida also says that the name alludes to that of *Primula pulchella*, also found in Yunnan, that has similarly coloured flowers.

This is another recently described and rather distinctive species characterised primarily by its small, simple, entire leaves and scapose 4-petalled flowers. Toshio Yoshida, who discovered it, records that on Mianning Xian the plant grows "on a south-west facing stony slope of a metaliferous mountain where iron ore is mined from several pits around the habitat of *Meconopsis pulchella*".

In the original description Toshio Yoshida and colleagues ally *Meconopsis pulchella* to *M. impedita* commenting that it "..... differs from the latter in the half nodding flower posture (nodding downwards in the latter), in petal numbers (4, usually 5–9 in the latter), colour (magenta or reddish purple, not bluish purple as in the latter), and base of petals ascending (not spreading as in the latter), and also in the entire leaves (often lobed in the latter)". While these two species are clearly allied, *M. pulchella* probably finds its closest ally in *M. concinna*. The latter species, which can have entire leaves, more usually has leaves that are variously dissected, sometimes with a range of transitional forms on the same plant. Both species have primarily 4-petalled flowers borne on basal scapes, more cupped and nodding in the former species. A marked difference can be seen in the fruit capsules which, in *M. pulchella*, are relatively broader and densely beset with patent bristles; in *M. concinna* the capsule can be subglabrous or adorned with just a few bristles. While the petal number is most often 4, and recorded as such in the original description, flowers with 5 or 6 petals have been observed and photographed. Indeed Toshio Yoshida's photographs reveal several flowers with 5 petals. The presence or absence of dark-based bristles to leaves and stems is clearly variable and both forms have been observed growing side by side in the wild.

Although there are apparently no herbarium specimens to substantiate the occurrence of *Meconopsis pulchella* away from the type locality, it is almost certain that this interesting and little known species is also found in other mountains around Kangding and on the Balangshan west of Chengdu. In the latter locality I have photographed it on several occasions, as have others, at about 4250 m altitude, just below the upper limit of *M. balangensis*. On the Balangshan *M. pulchella* grows in much the same situation as it does further south at Mianning, that is in earthy pockets amongst rocks or at the base of cliffs of a westerly aspect. This plant, earmarked as a new species in the 1990s, was not described at the time because of a lack of herbarium specimens (collecting was and still is heavily restricted in China!). Fortunately, Toshio Yoshida managed to secure specimens legally from the Mianning Xian and so was able to compose a formal description. Although *M. pulchella* appears to be very localised and little collected, there is the suspicion that it is more common in the mountains to the north-east and south of Kanding than is generally supposed. To my certain knowledge some plants recorded and photographed in the field as *M. lancifolia* or *M. impedita* have proved to be *M. pulchella*. The paucity of authenticated herbarium specimens does little to support this argument and more fieldwork in the area is required. Nonetheless, this is an attractive and relatively distinct little species.

58. MECONOPSIS VENUSTA

Meconopsis venusta Prain, *Bull. Misc. Inform., Kew* 1915: 164 (1915) emend. in *Hooker's Icon. Pl.* 31: tab. 3036 (1915) pro parte. Type: NW Yunnan, mountains NE of Yangtse Bend, 27°45'N, *Forrest* 10408 (E, syntype; K, P, isosyntypes); NW Yunnan, mountains of the Chungtien Plateau, 27°30'N, *Forrest* 12993, China 1914 (E, K); *Forrest* 11008, China 1913 (E, P); NW Yunnan, mountains of the Chungtien Plateau, 27°N, *Forrest* 12685, China 1914 (E, K); *Forrest* 16658, China 1918 (K) — referable to *M. pseudovenusta*. Lectotype chosen here: *Forrest* 12685 "Mountains of the Chungtien Plateau ... 13,000ft" (E, lectotype; E, K, isolectotypes). [Syn. *Meconopsis leonticifolia* Hand.-Mazz., *Anz. Akad. Wiss. Wien, Math.-Naturwiss. Kl.* 57: 49 (1920). Type: China, 1914, *Handel-Mazzetti* 4700 (P, holotype)].

DESCRIPTION. *Monocarpic or polycarpic* plant 5–25 cm tall in flower, with a stout elongated or long-fusiform taproot, 14–23 cm long, 0.8–2 cm diameter, bearing a single or up to 10 shoots in a congested clump. At the most, covered at the top by the remnants of old leaf bases where the root merges into a short thick stem which is not more than 10 cm long at the most and often more or less undistinguished. *Leaves* in a lax basal rosette or tuft, somewhat fleshy, green above, glaucous beneath, up to 13.5 cm in length overall, the lamina ovate to elliptic or spathulate, 15–37 × 10–24 mm, occasionally entire but more often trilobed or pinnately-lobed (pinnatifid to pinnatisect) with 2–3 pairs of rounded, obtuse lobes, the terminal lobes larger than the others, rounded to ovate to broad elliptic, 10–19 × 9–10 mm, the base abruptly narrowed, often cordate, or more gradually attenuate into the petiole, the apex obtuse to rounded, glabrous or with a few stiff hairs beneath; petiole slender, linear, (10–)25–80 mm long, somewhat expanded at the base. *Pedicels* (scapes) slender, erect to ascending, 6–19.5 cm long (10.5–28.5 cm long in fruit), glabrous or with a few scattered stiff hairs. *Flowers* generally (4–)7–16, 32–74 mm diameter. *Buds* ovoid-oval, 8–12 mm long, the sepals glabrous to sparsely bristly. *Petals* 4, pale blue to pale lilac, deep wine-purple, suborbicular to obovate, 15–40 × 11–36 mm, the apex rounded to subacute. *Stamens* about one-third the length of the petals, with linear filaments the same colour as the petals, the anthers orange to orange-yellow. *Ovary* ellipsoid to oblong-ellipsoid, sparsely to moderately densely covered with spreading bristles; style 4–5 mm long with a subclavate to subcapitate, 3–4-lobed stigma. *Fruit capsule* subclavate to subcylindric, (27–)36–93 × 4–7 mm, at least 5 × longer than wide, (2–)3–4-valved, with a 6–10 mm long persistent style, splitting for up to a third of its length from the apex when mature, with scattered, sometimes dense, spreading bristles, or glabrescent. (Figs 22A1–2, 26).

Meconopsis venusta; Xianggelila County, NW Yunnan, 4100 m. PHOTO: TOSHIO YOSHIDA

DISTRIBUTION. NW Yunnan; confined to the mountains north and north-east of the Yangtse bend, particularly in the proximity of the Zhongdian (Chungtien) Plateau and Laboshan (Lo-bo Shan), generally on limestone; 3300–4700 m. (Map 21).

HABITAT. Stony alpine meadows, humusy pockets in boulder fields, moraines, screes and cliff ledges, on limestone.

FLOWERING. Late June–early August.

SPECIMENS SEEN. *Forrest* 10408 (E, K), 11008 (E), 12685 (E, K), 12686 (E, K), 12993 (E, K), 15089 (E, K), 20531 (E), (E), 30078 (E), 30114 (E); *Handel-Mazzetti* 4700, China 1914 (P); *Schneider* 3128 (K).

FIG. 26. *Meconopsis venusta* leaf-laminas, lower petioles and pubescence not shown. All × 1 ¹/₃.
DRAWN BY CHRISTOPHER GREY-WILSON

George Taylor rightly stresses in his 1934 monograph of the genus that *Meconopsis venusta* probably behaves as a monocarpic species, suggested by the long cord-like taproot and the leaf remains present around the neck of the plant, yet it can also behave as a true perennial. He cites one specimen in particular [*Forrest* 30078, from 'Lobo Shan' *sic*] which clearly shows a clump-forming plant with at least six stout branches, several of which are bearing young buds, while another specimen on the same sheet sports fruit capsules from the previous year. However, the same sheet also includes two single-rosetted specimens with an equally stout taproot: all the other herbarium specimens conform to this latter type yet the stout taproot, which sports at its top the petiole

Below left: *Meconopsis venusta*; Xianggelila County, NW Yunnan, 4000 m. PHOTO: TOSHIO YOSHIDA
Below right: *Meconopsis venusta* in young fruit; Yulongxueshan, Lijiang County, NW Yunnan, 3950 m. PHOTO: TOSHIO YOSHIDA

58. MECONOPSIS VENUSTA

Meconopsis venusta in fruit; Xianggelila County, NW Yunnan, 3850 m.
PHOTO: TOSHIO YOSHIDA

remains from several years, probably indicates that the species is often perennial rather than monocarpic. While this is indicative it is by no means proven: in cultivation some species (most notably in the *M. horridula* complex) that are truly monocarpic can produce a few late flowers in addition to those borne in the early summer and thus mature fruits and flowers can appear together while, at the same time, the occasional plant may carry on after flowering to produce one or two flowers the following year before dying. In addition, damage to the emerging inflorescence of monocarpic species can result in aberrant flowering, the development of subsidiary inflorescences or scapes, or even propel a plant into prolonging life for another season.

Interestingly, *Meconopsis venusta* is found in a rather limited area centred on the mountains of the Zhongdian (Chungtien) Plateau (primarily between 27°30' and 28°12'N) and, despite the fact that the region has been quite well explored in recent years, little has been seen of it, although Toshio Yoshida has seen and photographed it on both Xianggelila and in the Yulongxueshan, both in north-western Yunnan. Nearly all the material in British herbaria consist of collections made by George Forrest; of these only *Forrest* 20531 strays east of 100°, the sheet marked 'Gaodu Shan'. In addition, there is a single gathering collected by Handel-Mazetti in the Paris herbarium.

At one time all the material now referred to *Meconopsis pseudovenusta* was included in *M. venusta*; however, George Taylor in his 1934 monograph separates them out as distinct yet closely allied species. The prime areas of distinction are to be found in the character of the flowers and fruits. In *M. venusta* the flowers are always borne on basal scapes and are consistently 4-petalled, while the long fruit capsule is more or less cylindrical and at least five times longer than broad. In *M. pseudovenusta*, on the other hand, although the flowers can be borne on basal scapes, some of these may at times be partly agglutinated into a pseudoraceme. In addition, the flowers generally have more than 4 (up to 10) petals, while the fruit capsule is distinctly shorter and more obovoid, no more than four times longer than broad. Recent photographic evidence from the wild would support this view although there is no doubting the fact that these two species are closely related.

Although described first, *Meconopsis venusta* seems to be far scarcer in the wild than *M. pseudovenusta* and has rarely been seen in recent years, although as already stated, Toshio Yoshida, who has explored the mountains of north-western Yunnan extensively, has seen and photographed it. From this it is possible to see the differences between the two close allies, especially the pale 4-petalled flowers and the slender fruit capsules of the former, which are only partly developed as seen in the photographs.

Meconopsis venusta and *M. pseudovenusta* are often allied to the widespread *M. horridula* but the true allegiance would seem to be with *M. impedita* and the other members of section *Impeditae*: *M. horridula* falls in section *Horridulae* along with a number of other species from south-eastern Tibet and western China, in particular *M. prattii, M. prainiana* and *M. rudis*. *M. impedita* is readily distinguished by its rather undulate leaves which can be more or less entire or with several forward-directed short lobes, and by the more sombrely coloured flowers that range from red-purple to dark violet or even blackish-violet. The foliage of both *M. venusta* and *M. pseudovenusta* is often glabrous or almost so.

59. MECONOPSIS PSEUDOVENUSTA

Meconopsis pseudovenusta G. Taylor, *The Genus Meconopsis*: 85 (1934). Type: NW Yunnan, NE Chungtien, 28°N, 13–14,000 ft, 1918, *Forrest* 16658, (E, holotype; BM, E, K, isotypes).

[Syn. *M. venusta* Prain, *Bull. Misc. Inform.*, Kew 1915: 164 (1915) *pro parte*].

DESCRIPTION. *Monocarpic* plant to 8–25 cm tall in flower, with a stout, elongated, tapered rootstock, occasionally branched, to 13 cm long, 1–2.5 cm diameter, with the withered fibrous remains of leaf bases at the top where the root merges into the base of a short, generally undistinguished, stem. *Leaves* in a lax basal rosette, 10.5–15.5 cm in length overall, occasionally one or two borne shortly above on a reduced stem; lamina ovate to elliptic or oblanceolate, 2.6–9.0 × 1.4–4.8 cm, entire to pinnately or bipinnately-lobed, with usually 3–4 pairs of pinnae, the lobes rounded to subacute at the apex, the terminal lobe rounded to subovate, 3.5–4.0 × 1.7–1.9 cm, the base of the lamina shortly attenuate into the petiole, glabrous or with a few scattered stiff hairs above or on the midrib beneath; petiole linear, 4.5–12.0 cm long, slightly expanded towards the base. *Pedicels* (scapes) erect to ascending, slender, 6–20 cm long (8.5–35 cm long in fruit), separate or rarely partly agglutinated (as in *Forrest* 12686 (BM!)), glabrous or sparsely to moderately

Meconopsis pseudovenusta, plant just coming into flower; Shikashan, Zhongdian, NW Yunnan, 4410 m. PHOTO: HARRY JANS

Meconopsis pseudovenusta; Shikashan, Zhongdian, NW Yunnan, 4290 m. PHOTO: HARRY JANS

59. MECONOPSIS PSEUDOVENUSTA

Above left: *Meconopsis pseudovenusta*; Shikashan, Zhongdian, NW Yunnan, 4290 m. PHOTO: HARRY JANS

Above right: *Meconopsis pseudovenusta*; Shikashan, Xianggelila County, NW Yunnan, 4400 m. PHOTO: TOSHIO YOSHIDA

covered with stiff, slender hairs. *Flowers* solitary, 8–17 per plant, 41–64 mm diameter, facing laterally or slightly nodding. *Buds* subglobose to ovoid, glaucous, nodding at first, the sepals sparsely bristly on the outside, or glabrous. *Petals* 4–8 (–10), deep blue to purple, purple-pink, lilac-purple, wine-purple or ruddy-purple, elliptic to obovate to suborbicular, with a somewhat cuneate base, 20–32 × 12–28 mm. *Stamens* with linear filaments the same colour as the petals, the anthers orange, orange-yellow, or brownish-orange, greying with age. *Ovary* ellipsoid to ellipsoid-oblong, with sparse to dense spreading to appressed bristles, occasionally subglabrous; style 3–4 mm long with a capitate or subclavate stigma. *Fruit capsule* ellipsoid to narrow-obovoid, 14–27 × 7–9 mm, not more than 4 × longer than wide, 3–4-valved, splitting for a very short distance from the apex, covered with sparse to rather dense spreading bristles, the persistent style 4–6 mm long. 2n = 56. (Figs 22B1–2, 27).

DISTRIBUTION. SW Sichuan and NW Yunnan; locally common particularly in the area around the Zhongdian (Chungtien) Plateau and the Baimashan (Yangtse-Mekong Divide) as well as the Chienchuan-Mekong divide, west as far as the Doker La on the Mekong-Salween (Lancangjiang-Nujiang) divide; 3658–4268 m. (Map 21).

HABITAT. Stony and rocky alpine meadows, screes and rock ledges.

FLOWERING. June–July (–early August).

SPECIMENS SEEN. *Forrest* 14901 (E), 16658 (BM, E, K), 16893 (E, K), 19668 (E), 23423 (E), 30111 (E); *T. T. Yü* 13653 (E).

Meconopsis pseudovenusta is an altogether delightful little species that finds its closest ally in *M. venusta*. As with that species it is a plant of very limited distribution in the wild, with a few scattered localities in the region of the Zhongdian Plateau, the Baimashan and in the mountains around Weixi (Mekong-Yangtse Divide). Even so, it is more widely distributed than *M. venusta* and has been seen and photographed a number of times in recent years, particularly on Shikashan near Zhongdian where it is a plant of rather coarse screes, growing in very open and exposed situations.

Although the *Flora of China* states that it is found in south-eastern Tibet (Bomi) I have not seen any authenticated material to substantiate this claim. In fact *Meconopsis impedita* is recorded from a number of

FIG. 27. *Meconopsis pseudovenusta* leaf-laminas, lower petioles and pubescence not shown. All × 1¹/₃.
DRAWN BY CHRISTOPHER GREY-WILSON

localities in SE Tibet (especially on the passes to the north and west of Namcha Barwa) and may have been mistaken for the closely related *M. pseudovenusta*.

Meconopsis pseudovenusta clearly finds its closest ally in *M. venusta* under which it harboured until recognised by George Taylor (1934). In his monograph Taylor remarks that "The collections now referred to *M. pseudovenusta* were previously included in *M. venusta*, from which they differ most markedly in the character of the fruit, and as their removal from that species leaves it a remarkably homogeneous unit, the course adopted appears to be desirable". The two species are undoubtedly closely related and it is remarkable that they both grow in the same general vicinity in north-western Yunnan; however, the characters that separate the two taxa appear to be constant and no intermediates have been found, neither have the two been found growing side by side in the wild to my knowledge. The prime differences are to be seen in consistently 4-petalled flowers of *M. venusta*, while those of *M. pseudovenusta* generally bear 4–8 petals, those of the former generally paler. In addition, the fruit capsules are notably different: those of the former are long, slender and cylindrical, those of the latter more ellipsoidal, much shorter and relatively broader. Differences can also be noted in the leaves: despite the fact that both can bear more or less entire leaves, the lamina generally has fewer lobes in *M. venusta*.

At the same time, these two species have an obvious alliance with other members of the *Impeditae*, notably *Meconopsis concinna* and *M. impedita* and the differences are outlined in the key to these species found on p. 281. In general *M. concinna* is a smaller and more delicate species. *M. impedita* has a wider range from NW Yunnan and W Sichuan westwards through northern Myanmar

Meconopsis pseudovenusta in flower and immature fruit; Shikashan, Xianggelila County, NW Yunnan, 4400 m. PHOTO: TOSHIO YOSHIDA

and into south-eastern Tibet, west as far as the Serkhyem La. In *M. impedita* the leaves are generally more substantial, greener and less glaucous, and sparsely to moderately bristly, while the base is gradually attenuated into the petiole (generally more or less glabrous in *M. venusta* and *M. pseudovenusta* and more or less abruptly narrowed into the petiole) and (in the pinnately lobed variants) with the lobes incised and forward projecting. In addition, the flowers usually bear 6–10 dark, dusky petals, while the stout style projects well beyond the boss of stamens.

It is perhaps unfortunate that the type specimen of *Meconopsis pseudovenusta* (*Forrest* 16658) cited by George Taylor is also a syntype of Prain's *M. venusta*. Both species are based upon specimens from the Chungtien (Zhongdian) Plateau region. The field notes for *Forrest* 16658 record that: "Plant 3–12 inches. Flowers wine-purple, anthers orange. On open stony pasture. 13–14,000 ft."

Interestingly, both *Forrest* 12686 (Chungtien) and *Forrest* 23423 (Chienchian-Mekong divide) bear specimens in which several of the scapes are partly agglutinated, but in most collections the flowers are borne on separate scapes directly from the basal leaf-rosette.

Both *Meconopsis pseudovenusta* and *M. venusta* are delightful high alpine gems of south-western China and would be greatly cherished by growers of alpine plants. As George Taylor comments, re the former "has the appearance of a desirable garden plant". Neither is in cultivation at the present time to my knowledge. Both species are very localised in the wild and probably rare, never being found in large numbers. The fact that they grow in remote mountain regions probably saves them from the hands of collectors and extinction. There is also the possibility that further colonies of both species will be found on some of the higher less botanised mountains of north-western Yunnan, especially in the mountains surrounding the Zhongdian Plateau.

Meconopsis pseudovenusta; Shikashan, Zhongdian, NW Yunnan, c. 4200 m. PHOTO: JOHN & HILARY BIRKS

series Henricanae

This series, which is closely related to the previous, is at once distinguished on the character of the stamens in which the filaments, especially of the outer stamens, are dilated in the lower half, sometimes pronouncedly so. The three species that make up the series all have simple undivided leaves and scapose flowers that are relatively large in contrast to the small, relatively few-leaved basal rosettes. The series is distributed from western and south-western Sichuan, northwards to southern Gansu and the neighbouring regions of south-eastern Qinghai.

KEY TO SPECIES OF SERIES HENRICANAE

1. Flowers deeply cupped, the petals with a dark basal blotch; anthers blackish purple; ovary and young fruit finely pubescent. .62. **M. sinomaculata** (W China; NW Sichuan, SE Qinghai)
  + Flowers saucer-shaped, the petals unmarked at base; anthers buff or orange-yellow; ovary and young fruit with scattered bristles. 2

2. Scapes slender, 2–4 mm diameter towards the base; petals 1.4–2.6 cm wide; anthers bright yellow; mature fruit capsules 7–11 mm diameter . . . . .60. **M. henrici** (W China; W, SW Sichuan)
  + Scapes relatively stout, 4–7 mm diameter towards the base; petals 2.3–3.5 cm wide; anthers buff; mature fruit capsules 11–18 mm diameter. . . . . 61. **M. psilonomma** (S Gansu, NW Sichuan, SE Qinghai)

60. MECONOPSIS HENRICI

Meconopsis henrici Bureau & Franch., *J. Bot. (Morot)* 5: 19 (1891). Type: W Sichuan, Tatsienlu [Kangding], *Bonvalot & Henri d'Orleans* (P, holotype).

[syn. *M. principis* Bulley, *Fl. & Silva* 3: 84 (1905); *M. wardii* Kingdon Ward, *Gard. Chron.*, Ser. 3, 72: 268 (1922), nomen nudum; *M. henrici* var. α *genuina* G. Taylor, *The Genus Meconopsis*: 80 (1934)]

DESCRIPTION. *A monocarpic herb* to 35 cm tall in flower (55 cm in fruit) with a swollen, napiform, often branched, taproot, 4–7.5 × 0.8–1.2 cm. *Leaves* all basal, withering in the autumn leaving little trace, the plant over-wintering as a small congested bud just below ground level, the lamina bluish green above, paler and somewhat glaucous beneath, oblong-oblanceolate to oblanceolate or linear-oblanceolate, 3–14.5 × 0.5–2.3 cm, the base attenuate into the petiole, the apex subobtuse to rounded, margin entire, or rarely slightly serrate, often somewhat undulate, moderately to densely covered above with yellowish brown, somewhat curved, soft bristles, especially in the marginal area, sparsely bristly to subglabrous beneath; petiole linear, 2–6 cm long, slightly expanded at the base. *Flowers* cupped to saucer-shaped, lateral-facing, 4.8–8.5 cm across, (1–)3–11 normally, borne on simple basal scapes, occasionally in an ebracteate inflorescence due to partial agglutination of several pedicels; central scape the stoutest, free or partly agglutinated to some or all of the secondary scapes present. *Pedicels* (scapes) 6–35 cm long (to 45 cm long in fruit), 2–4 mm diameter, erect, moderately to densely covered with yellowish brown patent, straight to curved soft bristles, densest just beneath the flowers, generally somewhat deflexed. *Buds* ovoid, yellowish brown bristly, often rather sparsely so. *Sepals* oval-ovate, 17–20 mm long, 10–12 mm wide, with dense, curved bristles on the outside, the margin thin, membranous. *Petals* 5–10 (often 6 or 8), violet-blue, mid to dark purple or indigo, ovate to obovate, 2.9–4.2 × 1.4–2.6 cm, the apex subobtuse to rounded, often rather fluted, especially at the margin but not erose. *Stamens* with filaments the same colour as the petals, 10–13 mm long, the filaments contorted, filiform in the upper third but markedly dilated, shiny and semi-translucent below; anthers yellow to amethyst-blue, 1–1.25 mm long. *Ovary* ovoid or subglobose, with dense pale yellowish brown subappressed bristles, occasionally subglabrous; style well developed, 4.5–8 mm long, the stigma cream or yellow, clavate or sublobed, the lobes partly or wholly united. *Fruit capsule* narrow-obovoid or obovoid-oblong, 18–36 × 7–11 mm, sparsely to densely bristly, with ascending bristles, 4–6-valvate, the persistent style 6–11 mm long. (Fig. 21A1–2).

Meconopsis henrici habitat; Zheduo Pass, Kanding, W Sichuan, 4298 m. PHOTO: HARRY JANS

60. MECONOPSIS HENRICI

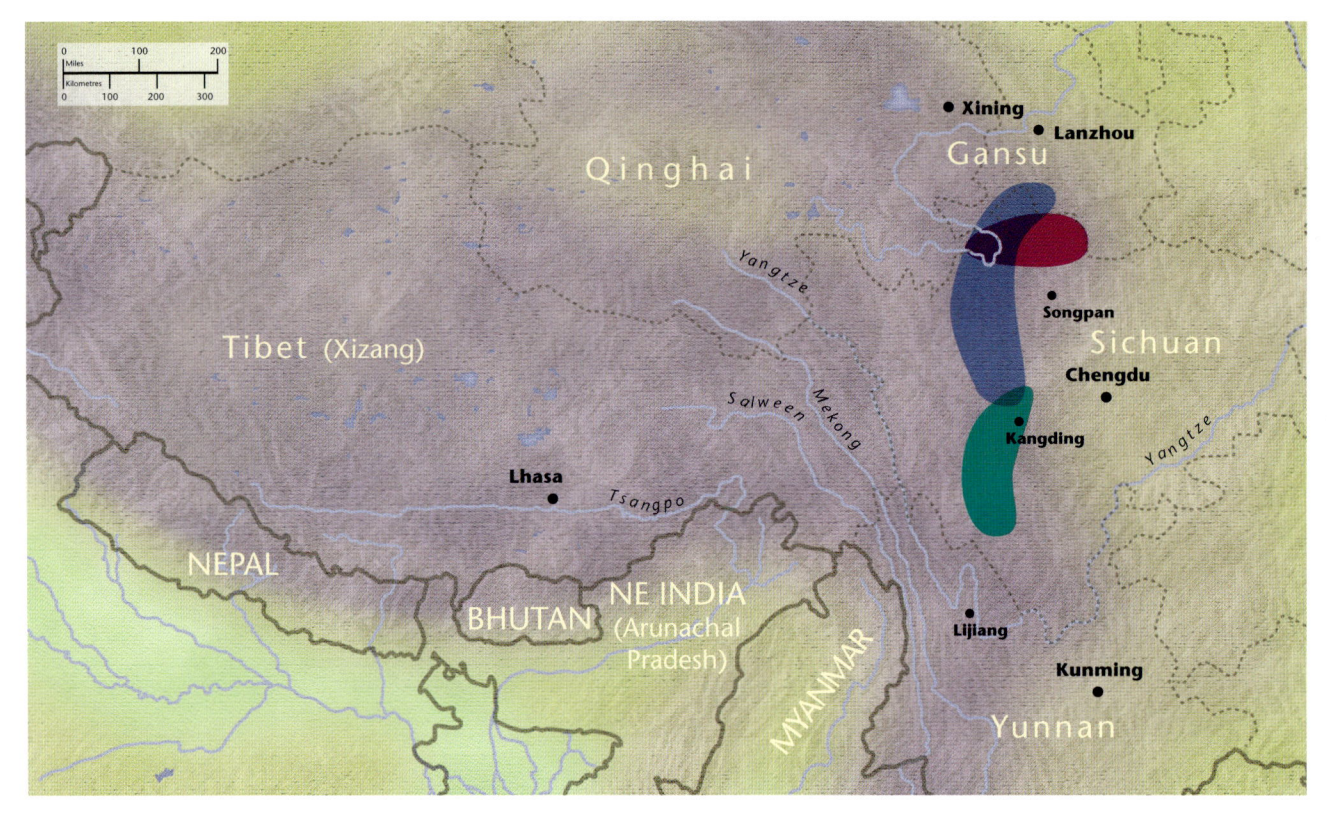

MAP 22. Species distribution: ■ *Meconopsis henrici*, ■ *M. psilonomma*, ■ *M. sinomaculata*.

DISTRIBUTION. W & SW Sichuan (especially Daxueshan-Shaluishan, Kangding, Muli), possibly just extending into NW Yunnan; 3200–4660 m (Map 22).

HABITAT. Alpine meadows, grassy and rocky slopes, rock ledges, open low shrubberies, moorland.

FLOWERING. June–August.

SPECIMENS SEEN. *R. Cunningham* 185 (E), 360 (E); *Kingdon Ward* 4421 (E); *McLaren* 6 (E); *Pratt* 25 (K), 600 (BM, K); *Rock* 17829[35] (E, K), 23765[35] (BM, E, K); *H. Smith* 10676 (BM), 10688 (BM), 11509 (BM), 11316 (BM), 11478 (BM), 11509 (BM), 12496 (BM); *Soulié* s.n Tibet orientale (K); 523 (K); *E. H. Wilson* s.n. (K), 957 (K), 3028 (BM, E, K), 3166 (BM, K), 4353 (K), 4535 (BM, K).

Meconopsis henrici is a very pretty little species that often forms scattered colonies in the alpine haunts that it frequents. The species was discovered in the late nineteenth century close to Tatsienlu (now Kangding) by Bonvalot in the company of Prince Henri d'Orleans after whom the species is named. The brightly coloured flowers were at once distinguishable by their staminal filaments that are clearly dilated in their lower half and for a long time this was the only species in the genus known to possess this feature. However, since then others have been added, notably the closely related yet distinct *M. sinomaculata* and even more recently by the publication of *M. balangensis* and *M. heterandra*. The latter two species are very different in a number of features and the filament details are quite distinct: in both the latter species only the inner filaments are dilated and they are air-filled like minute sausage balloons and curved inwards to form a protective jacket around the ovary; the outer filaments are linear and spreading. In both *M. henrici* and *M. sinomaculata* and to some extent *M. psilonomma* also, the most prominently dilated filaments are those of the outer stamens and they are not so obviously air-filled but simply expanded in the lower half, and neither do the stamens form a protective jacket around the ovary.

Meconopsis henrici is rather a uniform species in the wild, forming small, spreading neat leaf-rosettes from which arise a number of solitary-flowered scapes. As with many of the small monocarpic scapose species the first flower(s) to open are rather larger and borne on more substantial scapes than those that follow. The

---

[35] Both *Rock* 17829 and 23765 gathered close to Muli in southern Sichuan closely match *Meconopsis henrici* in most details except that the flowers are generally larger and apparently more cupped, although the latter may be due to the way in which the specimens were pressed. I have not seen any more recent collections of *M. henrici* from the Muli region but its occurrence there forms a southerly outlier from its prime haunts in the Kangding, Minya Konka region of western Sichuan.

Above: *Meconopsis henrici*; Zheduo Pass, Kangding, W Sichuan, 4340 m.　PHOTOS: HARRY JANS

tongue-like leaves are generally somewhat channelled with a rather undulate margin and differ mainly in the amount of pubescence from plant to plant, some having rather dense pubescence on the upper surface, others being moderately covered or even subglabrous. The flowers are generally of an intense deep violet-blue, although those with indigo or purple can be seen in the wild. The species typically forms scattered colonies where it is found and probably, in the wild at least, flowers in its third or fourth year from seed.

Ernest Henry (Chinese) Wilson sent seed back of *Meconopsis henrici* back to Britain from western Sichuan in 1904 and David Prain recorded (*Ann. Bot.* 220: 329 (1906)) that it flowered in cultivation two years later; however, the plant does not appear to have persisted from this introduction. In recent times it has been introduced again but once more its continuation in cultivation has failed. However, one such introduction flowered at the Royal Botanic Gardens Kew, being cultivated under alpine house conditions; however, it

Below, from left:
Meconopsis henrici; Zheduo Pass, Kangding, W Sichuan, 4340 m.　PHOTO: HARRY JANS
Meconopsis henrici; Gaoersi Pass, Kangding to Yajian, W Sichuan, 4100 m.　PHOTO: HARRY JANS
Meconopsis henrici; Huaihaizi (Grey Lake), W Sichuan, 4070 m.　PHOTO: HARRY JANS

failed to set seed and was soon lost from the collections. It is a desirable little species with its deep violet-blue flowers set off by a prominent boss of golden stamens. *M. henrici* would make a great addition to the alpine garden but it is not easy and perhaps can only be successfully grown within the careful confines of an alpine house or similar, or perhaps in a trough outdoors if summer heat and winter wet can be controlled. If grown outdoors, some overhead protection is generally advisable during the winter months (e.g. a cloche or pane of glass secured overhead) and in the spring as the new growth appears plants should be guarded against the depredation of both slugs and snails.

In his 1934 monograph of the genus George Taylor recognised two varieties var. *genuina* (i.e. var. *henrici*) and var. *psilonomma*, described from a Reginald Farrer collection. The two are readily distinguished and are geographically isolated from one another and I see no reason to continue including them within a single species: see *Meconopsis psilonomma* below.

61. MECONOPSIS PSILONOMMA

Meconopsis psilonomma Farrer, *Gard. Chron.* 1915, Ser. 3, 62: 110; Prain, *Bull. Misc. Inform.*, Kew 1915: 160 (1915). Type: SW Gansu, Minshan, above Ardjeri, 1914, *Farrer* 255 (BM, E, K).
[syn. *Meconopsis henrici* Bureau & Franch. var. *psilonomma* (Farrer) G. Taylor, *The Genus Meconopsis*: 81 (1934); *M. barbiseta* C. Y. Wu & H. Chuang ex L. H. Zhou, *Acta Phytotax. Sin.* 17 (4): 113–114 (1979). Type: Qinghai, Juizhi, 4400 m, *Exped. Bot. Guoluo* 438 (PE, holotype)]

DESCRIPTION. *A monocarpic* herb to 25 cm tall in flower, with a napiform rootstock 18–30 × 10–13 mm, abruptly contracted distally into a slender taproot. *Leaves* all borne in a small basal rosette, rather few, often 6–9 at flowering stage, fresh green, the lamina elliptic-oblanceolate to oblanceolate, 3.2–7 × 0.8–1.4 cm, the base long attenuate into the petiole or more or less cuneate, the apex rounded to subobtuse, subglabrous, although often with some hairs along the raised midrib beneath, the margin entire; petiole slender, 2–5.8 cm long, broadening towards the base. *Flowers* solitary, rarely 2–3, scapose, saucer-shaped, lateral-facing, 8.5–11.5 cm across. *Scapes* erect 23–28 cm long (to 38 cm long in fruit), 4–7 mm diameter, green, often flushed with purple in the upper part, patent bristly, the bristles often somewhat deflexed. *Buds* nodding to half-nodding, narrow-ovoid, the sepals pale green, white patent-hairy. *Petals* 6–8, deep purple or purple-blue, broad-elliptic to elliptic-ovate, 4.2–5 × 2.3–3.5 cm, overlapping, the margin more or less entire to somewhat erose. *Stamens* numerous held together in a quite tight group around the ovary, the filaments 12–14 mm long, generally dilated in the lower half, deep purple-blue, the anthers 1–1.25 mm long, buff. *Ovary* narrow-ovoid, fairly densely subappressed bristly, the style green, 6–9 mm long; stigma cream. *Fruit capsule* broad-ovoid, 20–27 × 11–18 mm, 4–6-valved, thin patent-bristly to subglabrous, the persistent style 9–12 mm long of which half represents the stigma.

Meconopsis psilonomma; Huanglong, Xueshanliang, NW Sichuan, 3950 m. PHOTO: HARRY JANS

DISTRIBUTION. S Gansu, extreme SE Qinghai and NW & W Sichuan[36]; 3500–4115 m (Map 22).
HABITAT. Alpine meadows and moorland, low shrubberies, often on very exposed slopes.
FLOWERING. Late June–early August.
SPECIMENS SEEN. *Farrer* 209 (E), 255 (BM, E, K); *SBQE* 594 (E); *H. Smith* 3832 (BM); *E. H. Wilson* 3028 (BM).

---

[36] In recent times most often seen and photographed in the mountains above Huanglong, Deng Zhangwa.

Meconopsis henrici, *M. psilonomma* and *M. sinomaculata* have been much confused in the field and in the literature. To add to this confusion a fourth species, *M. barbiseta*, was described by two Chinese botanists, C. Y. Wu & H. Chuang in 1979. In addition, many photographs in print and on various websites purporting to be *M. henrici*, particularly from north-western Sichuan, are in fact referable to *M. psilonomma*. At first, in order to clear up this muddle, it is necessary to look at the origins of the first two species. The first, *M. henrici*, was described by Bureau & Franchet in 1891 based upon a collection made by Bonvalot and Prince Henri d'Orleans in the region of Tatsienlu (now Kangding) in western Sichuan and it was named in honour of the latter. Later collections by Kingdon Ward, Joseph Rock and others extended the known distribution into south-western Sichuan. *M. henrici* is an attractive small species bearing a small tuft of leaves from which arise up to 11 solitary scapose flowers. These latter are lateral-facing, rather flat or saucer-shaped and usually of a generous deep violet-purple colour. The single most distinguishing feature of *M. henrici* are the filaments which are filiform in the upper half but noticeably dilated in the lower half.

Meconopsis psilonomma, described in 1915, is very close in general appearance and stature to *M. henrici* and for this reason the two have been much confused. The species was based upon some rather poor, yet adequate, and incomplete specimens collected in the Minshan, SW Gansu, by Reginald Farrer in 1914. According to George Taylor (1934) "Farrer's original set of specimens were stolen, but those available show such a close resemblance to *M. henrici* that it is not considered desirable to treat *M. psilonomma* as a separate species. The collector noted that *M. psilonomma* invariably has single basal pedicels, and this character, in conjunction with the geographical segregation of the plant, appears sufficient to sustain varietal rank……". Thus in his 1934 monograph Taylor transferred *M. psilonomma* to *M. henrici* as a variety.

Despite the loss of the original set of specimens there are just a few Farrer sheets extant, primarily at the Royal Botanic Garden Edinburgh but also at RBG Kew and the British Museum (Natural History). To the type specimens, *Farrer* 255, is attached the following note:

"A very indifferent specimen (the set having been stolen) of a superb species seen only once, on one portion of a great grass slope in the Thibetan Alps of Ardjeri, beginning at the topmost limit of *M. punicea* (11,500 feet), and ascending to the highest ridges at 12,500 where *M. quintuplinervia* seemed pale and poor: scattered freely in the grass, with *Allium kansuense*, *Gentiana* 217, *Primula* 13 and *P. tangutica* etc. July 30: seed mid-September. Note that it is invariably single-scaped and single-flowered: this specimen quite understates its size and stature."

Above left: *Meconopsis psilonomma*; Huanglong, Xueshanliang, NW Sichuan, 4100 m. PHOTO: HARRY JANS

Above right: *Meconopsis psilonomma*; Huanglong, NW Sichuan, 4100 m. PHOTO: CHRISTOPHER GREY-WILSON

61. MECONOPSIS PSILONOMMA

It is worth examining the type sheet at Kew which is accompanied by a half-tone photograph taken by Farrer in the field and by a watercolour painting drawn by Matilda Smith from Farrer's material, dated February 27th 1915. Several facts become clear: first the plant in question is very similar in general appearance to *Meconopsis henrici* but produces solitary flowers (i.e. one per plant) and, while the flowers are open and rather flat, like *M. henrici*, the stamens have undilated or part-dilated filaments. Farrer comments on the anthers being distinctly pale buff-coloured and this feature is borne out by recent photographs from a number of different localities in north-western Sichuan. At this point it is worth quoting Farrer's own musings on the plant in question:

> "A magnificent one-flowered biennial in the group of *M. delavayi* from the high Alpine turf on one Thibetan mountain only at 11–12,000 feet .....only on one hillside above Ardjeri, in the grass beginning at the upper limit of *M. punicea* at 12,000 feet and sharing the topmost crest at 13–14,000 feet with *M. quintuplinervia* .....Primulinae group akin to *M. delavayi* but larger. The best specimens were stolen and these are not good enough..... The anthers in this species are a very pale buff, unlike those of any other member of the Primulinae group except *M. lepida*."
>
> Selected from the field notes of the Farrer-Purdom Expedition to Kansu (Gansu), 1914–1915.

Farrer had a completely different conception of the *Primulinae* than that interpreted today!

Writing in the *Gardener's Chronicle* in 1915 Farrer writes admiringly about this species:

> "And then, among these suddenly, rain-laden and heavy, like a gigantic specimen of the purple *Anemone coronaria*. It was another new *Meconopsis* (now, at least, to Fedde, for I do not know the diagnosis of *M. wardii*). Like the last, it belongs to the biennial Primulina group; but, leaving to the last its monopoly of small and dainty grace, this one stands pre-eminent in grandeur and opulence. Few and small are the leaves, all at the base, glaucescent, narrow, almost hairless; the naked scape is invariably solitary, and invariably carries only one very large flower with six, seven or eight very broadly ovate rhomboidal, crimply petals of rich lavender purple. The solitary imperial eye of colour at the top of each stout scape of 6 to 10 inches reminded me of the 'solitary eye' which blind old Oedipus said they had torn from him when they took away Antogone; accordingly, until it is put in its place or more authoritatively and illustriously named, I think of this treasure as *Meconopsis Psilonomma*. It seems a rare plant: never again on any of the Thibetan downs did I see it, and on this huge ridge of several miles only on this one particular slope. It begins about the upper limit of *M. punicea* at some 11,000 feet and ascends in the fine turf to the gaunt topmost ridge at 12,000, where it stands gorgeous from the broken tussocky ledges, in and out amid the browning cushions of Rhododendron, making *M. quintuplinervia* here look but a poor starved anaemic cinderella. That *M. Psilonomma* may burst ere long upon the world I have my hopes; also I have *in situ* photographs of its glory that I keep as yet for a riper occasion and a tantalised judgement, for who will dare send photographs, no matter how beautiful, to an Editor who is apt to declare that they leave much to be desired."

The dilation of the filaments, an obvious feature in the closely allied *Meconopsis henrici*, is less well-marked in *M. psilonomma* and Farrer does not comment on it, neither is it shown in the drawing by Matilda Smith referred to above. Having examined all the available herbarium sheets closely it is clear that the filaments are more weakly dilated than those of *M. henrici* and it is not always easy to observe, particularly as it often only the outer series of stamens that are affected, the inner generally bearing linear filaments; however, under close scrutiny all the

Meconopsis psilonomma; Huanglong, Xueshanliang, NW Sichuan, 4100 m. PHOTO: HARRY JANS

specimens examined were shown to have at least some dilated filaments.

Re-examination of the material from the region (primarily NW Sichuan and SW Qinghai) has shown that plants that I had formerly attributed to *Meconopsis henrici* are in fact all referable to *M. psilonomma*.

Moving further south into western Sichuan the confusion was compounded by the fact that both *Meconopsis henrici* and *M. psilonomma* can be found there, although they do not appear to grow in close proximity to one another but again, because of their rather similar appearance plants have nearly always been assigned to the former. To sum up, the overriding difference between the two species is to be seen in close examination of the stamens: in *M. henrici* all the filaments are dilated in the low half forming a distinctive muff around the base of the ovary, while in *M. psilonomma* the filaments less markedly dilated and this affects only the outer stamens. Other differences can be observed: in *M. henrici* the anthers and stigmas are yellow, in *M. psilonomma* they are creamy buff, while in the former the petals are relatively narrower and have entire margins, in the latter they are finely erose towards the top. As I have not come across any authenticated herbarium specimens of *M. psilonomma* from western Sichuan I am unable to confirm that the outer filaments are consistently dilated or not.

Meconopsis barbiseta was described from a Chinese collection (Expedition Botanic Guoluo) from Juizhi (Jigshi) in south-eastern Qinghai province and there is a good drawing accompanying the type description. This plant closely matches the Farrer material in nearly all details. Furthermore, the same plant has been photographed in a number of localities in neighbouring north-western Sichuan (north of Songpan, Longriba and close to Juizhaigou for instance) and I have no doubt that all these represent a single taxon. There is also little doubt that this taxon, for which the earliest name is *M. psilonomma*, is distinct from *M. henrici*. The latter is restricted to western and south-western Sichuan, while *M. psilonomma* is found in one or two localities in western Sichuan, while its prime distribution is in the high rugged mountains in the Gansu-Qinghai-Sichuan borderlands. Farrer's photograph, although a half-tone plate, closely matches more recent colour photographs taken in the field. In fact I have seen *M. psilonomma* on a number of occasions in the wild, not least in the vicinity of Jiuzhi, the type locality of *M. barbiseta*, where is is a not infrequent member of the alpine moorland community, often growing in association with *M. punicea*. It is, as Farrer describes, a charming and colourful little plant with well-rounded flowers.

Meconopsis psilonomma; Huanglong, NW Sichuan, 4100 m.
PHOTO: CHRISTOPHER GREY-WILSON

The last in this quartet of confusion is *Meconopsis sinomaculata* which I described in 2002 and whose history is described under that species (see p. 312). The fact that it grows in the same general area as *M. psilonomma* has led to further confusion and doubts expressed by some on the authenticity of this species. Yet, *M. sinomaculata* is the most distinct species in this lax alliance. It is particularly distinctive with its large deeply cupped flowers in which the petals have a characteristic dark basal zone. These features alone separate it from the others. It shares in common with *M. henrici* distinctive dilated filaments to the stamens and with *M. psilonomma* the solitary scapose flowers. Again I have observed this species on a number of occasions in the wild, both in north-western Sichuan and south-eastern Qinghai where it is a plant of alpine meadows and open shrubberies often growing in association with *M. punicea* and *M. integrifolia* subsp. *integrifolia*, and sometimes also with *M. quintuplinervia*. I have never seen it growing in association with *M. psilonomma* and I have not seen any reports of it doing so in the wild.

Meconopsis psilonomma; Huanglong, Xueshanliang, NW Sichuan, 4100 m. PHOTO: HARRY JANS

It is unfortunate that there is little herbarium material of *Meconopsis psilonomma* and *M. sinomaculata* to study, yet there are ample photographs taken in recent years. The prime reason for this is the heavy restriction placed by Chinese authorities on collecting samples, even, or perhaps especially, for scientific purposes.

Meconopsis henrici is endemic to western and south-western Sichuan, while *M. psilonomma* is endemic to western and north-western Sichuan and the neighbouring part of Qinghai province as well as south-western Gansu.

Meconopsis psilonomma has been introduced as seed collections in recent years but I have no record to date of plants having reached maturity in cultivation.

62. MECONOPSIS SINOMACULATA

Meconopsis sinomaculata Grey-Wilson, *Plantsman* N.S. 1 (4): 221–227 (2002). Type: China, NW Sichuan, top of Gonggan Len Pass between Juizhaigou and Songpan, 33°00'46"N, 103°42'53"E, 3400–3600 m, 29 June 2000, *Sino-British Qinghai Alpine Garden Society Exped*. SBQE 500 (E, holotype; E, GB, HNWP, WSY isotypes).

[syn. *Meconopsis psilonomma* var. *sinomaculata* (Grey-Wilson) Ohba, *J. Jap. Bot.* 81 (5): 296 (2006).

DESCRIPTION. *Monocarpic plant* to 70 cm tall in fruit, generally less. *Rootstock* fleshy, napiform, sometimes two-lobed, 17–45 mm long, 7–15 mm diameter, giving rise to thin fibrous roots. *Leaves* rather few, often 5–9, pale green, all borne in a lax basal rosette, the lamina ovate to ovate-lanceolate, ovate-oblanceolate or narrow oblanceolate, 3.5–14.5 × 0.7–2.4 cm, gradually narrowed at the base into the petiole, the margin entire, sparsely to densely bristly above and beneath or subglabrous, the bristles non-barbellate, the lamina weakly 3-veined from the base; petiole 1.5–5.2 cm long, expanded and part-ensheathing at the base. *Flowers* solitary (generally only one per plant), scapose, horizontal to half-nodding, deeply cupped, generally 7–12 cm across when flattened, with the petals erect to ascending rather than spreading widely apart. *Scape* erect, 20–55 cm long in flower, elongating somewhat in fruit to 70 cm, (3–)4–9 mm diameter near the base but narrowed gradually and purple-flushed towards the generally arched top, clothed with numerous horizontal to somewhat downturned whitish to fawn-coloured, rather sharp, bristles, especially dense just beneath the flower. *Buds* nodding or half-nodding, the sepals 2, boat-shaped, 25–34 mm long, with patent rather pale bristles. *Petals* often 6 (but 4–7), azure to deep purple or deep lavender-purple, with a prominent maroon-black or purple-black blotch at the base, very rarely pure white, suborbicular to obovate or elliptic-obovate, (3.6–)4.5–8 × 2.5–5.6 cm, the margin undulate to minutely and unevenly toothed. *Stamens* numerous with filaments the same colour as the petals or blackish purple, 11–18 mm long, at least the outer somewhat dilated in the lower half or third, the anthers blackish-purple or black, 1.2–2.5 mm long; pollen cream or white. *Ovary* ovoid, ascending-bristly, 3–5-valved, with a 10–13 mm long style that terminates in a creamy-white, 3–5-lobed stigma that protrudes just beyond the stamens, the lobes of the stigma eventually parting and partly recoiling, equal in length to the style. *Young fruit capsule* narrow-ovoid, 8–14 mm long, probably 3–5-valved, bristly, the hairs blackish and rather shiny; mature fruit capsule unknown. (Fig. 21B1–3).

DISTRIBUTION. NW Sichuan, SE Qinghai and SW Gansu, centred on the southern Minshan and adjacent areas south as far as Juizhaigou in northern Sichuan; 3350–4420 m (Map 22).
HABITAT. Low montane scrub and grassy slopes in alpine tundra, and drier parts of peat bogs, generally as scattered colonies, often in association with other *Meconopsis* species.
FLOWERING. Mid-June–early August.
SPECIMENS SEEN. *Rock* 12613 (BM, E, K), 13632 (E); *SBQE* 500 (E), 501 (E), 513 (E), 569 (E).

Meconopsis sinomaculata is a very beautiful plant with large cupped flowers held aloft on long scapes well above the rather insubstantial basal leaf-rosettes (rather more tufts than rosettes in effect). The species is rather restricted in the wild, being confined, so far as it is known, to the rugged mountains of the Gansu-Qinghai-Sichuan, borderlands. In the wild it can be found in quite large colonies at times, often growing in association with *M. punicea* and *M. quintuplinervia*, and occasionally *M. integrifolia* subsp. *integrifolia*. The large purple blobs of its blooms bob around enticingly in the slightest breeze. During inclement weather insects, notably flies, harbour within its blooms seeking shelter until the sun reappears. Seed was collected by Toshio Yoshida but I have not heard of the plant yet blooming in cultivation, although it would make a fine sight should it do so.

This species finds it closest ally in *Meconopsis henrici* which is found further to the south but with its large deeply cupped, dark-centred flowers it is quite unmistakable.

The type specimen *SQAE* 500, bears a fairly full field description: "Biennial up to 60 cm. Rootstock swollen. Leaves narrowly oblanceolate, sparingly set with long hairs, pale green. Stem [scape] covered in white or pale brown hairs. Flowers pendent, narrowly campanulate, deep purple, petals broadly obovate. Anthers black; filaments blackish purple; stigmas white."

SQAE 569 (E), collected at the top of the Huangshan Guan, east of Chuanzhusi, bears two specimens one normal, the other a depauperate plant just 15 cm tall and smaller in all its parts from the norm. Such depauperate specimens have been found in a number of species of *Meconopsis*, notably *M. grandis*, *M. horridula* sensu stricto, *M. prainiana* and *M. sulphurea* and doubtless others. Such plants seem to reflect poor habitat conditions. It is also possible that they are young plants coming into flower too soon before having had time to build up a decent sized basal leaf-rosette.

I described this species in 2002 based on material gathered by the *Sino-British Qinghai Alpine Expedition (SQAE)* the year previously. Two years before that,

Meconopsis sinomaculata in habitat; Chuanzhusi, Gonggan Len, NW Sichuan, 3500 m.
PHOTO: CHRISTOPHER GREY-WILSON

Meconopsis sinomaculata in habitat; Juizhaigou to Gonggan Len, NW Sichuan, 3400 m, late June. PHOTO: HARRY JANS

62. MECONOPSIS SINOMACULATA

Clockwise from top left:

Meconopsis sinomaculata just after sepal drop; Chuanzhusi, Gonggan Len, NW Sichuan, 3500 m.
PHOTO: CHRISTOPHER GREY-WILSON

Meconopsis sinomaculata in full flower; Chuanzhusi, Gonggan Len, NW Sichuan, 3500 m.
PHOTO: CHRISTOPHER GREY-WILSON

Meconopsis sinomaculata flower detail; Chuanzhusi, Gonggan Len, NW Sichuan, 3500 m.
PHOTO: CHRISTOPHER GREY-WILSON

Meconopsis sinomaculata, flower detail, reverse; Chuanzhusi, Gonggan Len, NW Sichuan, 3500 m.
PHOTO: CHRISTOPHER GREY-WILSON

Meconopsis sinomaculata, white form; Juizhaigou to Gonggan Len, NW Sichuan, 3400 m, late June.
PHOTO: HARRY JANS

whilst leading a group for the Alpine Garden Society we had come upon the same plant in the mountains above Juizhaigou in north-western Sichuan and were then able to photograph but not to collect it. It was at first mistaken by me and others to be the same as Reginald Farrer's *Meconopsis psilonomma* collected further north in the Minshan of southern Gansu; however, once herbarium specimens were gathered several years later by the *Sino-British Qinghai Expedition (SBQE)* it became quite clear upon closer inspection that there were two distinct taxa involved. To add to this, further trawling through herbarium material at Edinburgh Botanic Garden revealed two Joseph Rock collections of the 1920s which closely matched that of the SBQE. Joseph Rock's material, like Farrer's, was also from SW Gansu (Kansu) and was at that time residing under *M. henrici* at both Kew and the British Museum (Natural History) and did not at that time come into the equation: there are excellent herbarium specimens under the number 12613 at Edinburgh (E) and Kew (K) consisting of three individuals in flower and identified as *M. henrici* var. *psilonomma*, SW Gansu, Mt Lissedzadza, 12,500 ft ('Plant to 1.5 ft, deep lavender-purple, especially at corolla base'). *Rock* 13632 (E) also belongs here, a fruiting specimen, although the fruit appears to have been lost, which is a pity, as mature fruits of this species have not been recorded.

series Delavayanae

This series contains a single species. In most respects it resembles the previous series (*Impeditae*) but differs significantly in its polycarpic habit and in the long cylindrical, glabrous, fruit capsules.

63. MECONOPSIS DELAVAYI

Meconopsis delavayi (Franch.) Franch. ex Prain, *J. Asiat. Soc. Bengal, Pt. 2, Nat. Hist.* 64: 311 (1896).
[syn. *Cathcartia delavayi* Franch., *Bull. Soc. Bot. Fr.* 33: 390 (1886). Type: *Delavay* s.n., China, 1883–5 "Les ravines .... du glacier Likiang. 3800 m. Juillet 1884" (P, holotype; BM, K, isotypes)]

DESCRIPTION. Small *polycarpic perennial* to 27 cm tall in flower, but as little as 10 cm, with a taproot to 25 cm long, laxly branched above and somewhat woody, to 4 mm diameter, generally supporting several leaf-rosettes subtended by petiole base remains of the previous years. *Leaves* all basal, deep bluish green above, paler and glaucous beneath, the lamina lanceolate-elliptic, ovate-elliptic, to rhombic-ovate or spade-shaped, 2.1–7 × 1–2.8 cm, gradually tapering at the base into the petiole or rather abruptly decurrent, the apex subacute to obtuse, the margin entire, glabrous or sparsely weak-bristly above and beneath; petiole 1.5–12 cm long, broadening somewhat at the base. *Flowers* nodding or semi-nodding, scapose, one per leaf-rosette (up to 8 per plant), bowl-shaped, 3.6–6 cm across. *Pedicels* (scapes) ascending to erect, slender, 10–30 cm long (to 50 (–60) cm in fruit) to 3.5 mm diameter, glabrous to sparsely patent-bristly. *Buds* nodding at first, elliptic-obovoid, 13–14 mm long, subglabrous, occasionally with a few scattered bristles. *Petals* 4 (–8), deep indigo-purple to violet-purple, satiny rose-purple or purple-blue, ovate to obovate, 25–33 × 18–28 mm, the apex acute to subobtuse. *Stamens* numerous, the filaments linear, the same colour as the petals although often darker, 7–10 mm long, the anthers orange, oblong, 1–1.25 mm long. *Ovary* narrow-ellipsoidal, greenish, glabrous, the style 1.5–5.5 mm long, rarely almost obsolete, with a capitate to subclavate stigma. *Fruit capsule* oblong-cylindrical, 49–73 × 3–7 mm, glabrous, 3– (4–5–)valved, splitting for a short distance from the top at maturity, the persistent style to 10 mm long. *Seeds* falcate, the testa longitudinally striate, sometimes rather obscurely so. $2n = 56$. (Fig. 3C).

DISTRIBUTION. NW Yunnan (Yulongxueshan, Lijiang; Chienchuan-Mekong Divide[37]); 3353–4270 m.

HABITAT. Stabilised moraines and screes, rocky alpine meadows over limestone, open scrub.

FLOWERING. May–July.

SPECIMENS SEEN. *CLD* 90 (E, K), 1171 (K), *Delavay* s.n. "Likiang", July 1884, 3800 m (BM, K, P); *Forrest* 2272 (BM, E, K), 5618 (BM, E), 10128 (BM, E, K), 15157 (BM, E, K), 15374 (E), 20691 (BM, E), 21933 (E, K), 23191 (E, K), 23422 (BM, E), 28224 (BM)[38], 30075 (BM, E), 30080 (BM, E); *Kunming-Edinburgh Yulong Exped.* 4 (E), 193 (E); *McLaren* 69 (BM, E, K); *Rock* 3795 (E, K), 3860 (E), 4265 (E), 4368 (BM, E, K), 4470 (E), 4540 (E), 8689 (E), 10856 (E), 24840 (BM, E, K), 24957 (BM, E, K); *C. Schneider* 1788 (K), 3577 (K), 3634 (K).

Meconopsis delavayi; Yulongxueshan, Lijiang County, NW Yunnan, 4000 m. PHOTO: TOSHIO YOSHIDA

---

[37] Nearly all the known gathering of *Meconopsis delavayi* are from the Yulongxueshan (north to 27°40'N), with the exception of two Forrest collections, numbers 23191 & 23422, both gathered along the Chienchuan-Mekong Divide (26°30'N, 99°04'–99°20'E) well to the south-west of the prime location, and one collection, *Forrest* 15157 annotated Chungtien (now Zhongdian) without further qualification.

[38] *Forrest* 28224 is a mixed gathering with *Meconopsis delavayi* and *M. lancifolia* included on the same sheet (BM) and marked 'Tali Range'. While the latter is recorded from this range (today the Cangshan), *M. delavayi* is not and must have crept onto the sheet in error.

63. MECONOPSIS DELAVAYI

This pretty little species was discovered by the French missionary and botanist Père Delavay (1834–1895) who made extensive gatherings in Yunnan towards the end of the nineteenth century and whose prime collections are deposited at the Paris herbarium. It is fitting that this, one of the most delightful and distinctive of the smaller *Meconopsis* species, should commemorate one of the most intrepid and prolific collectors to have explored south-western China. His exploits predate most of the great plant explorers that followed him, notably George Forrest, Frank Kingdon Ward and Joseph Rock. Delavay found it growing on the eastern flanks of the Lijiang Range (Yulongxueshan) in 1884 and most subsequent sightings and collections have been made in the same general vicinity. Further collections were made by Forrest, Handel-Mazzetti and Joseph Rock and, in more recent times, by several expeditions to the Yulongxueshan, most especially the 1984 Alpine Garden Society Expedition (ACE). It was thought to be endemic to the Yulongxueshan at one time but George Forrest (number 23422) also found it growing some distance to the south-west on the Chienchuan-Mekong Divide. Even so, this is a very restricted and localised species which must be considered to be vulnerable in the wild especially in its lower altitude haunts such as the Gangheba, close to Lijiang.

Few of the smaller *Meconopsis* species are truly perennial (*M. bella* being another exception), so that gardeners should find them more accommodating and reliable plants. Alas however, such is not the case and, while the *M. delavayi* is at the least scarce in cultivation, *M. bella* appears to be no longer so and that despite the fact that seed has been introduced from the wild on many occasions over the years. Seed of *M. delavayi* was initially introduced into Britain by George Forrest and first flowered at the Royal Botanic Garden Edinburgh in 1913 but, unfortunately, the plants were lost soon after this. R. D. Trotter of Brin House in Inverness was extremely successful in growing *M. delavayi* in the 1930s, discovering, apart from getting plants into flower, that the species could be propagated from portions of thicker roots inserted top upright in a suitably sandy medium. Subsequently, Jim Jermyn, formerly the proprietor of Edrom Nursery in Berwickshire grew the species

Above, from left:

Meconopsis delavayi; Gangheba, Yulongxueshan, Lijiang County, NW Yunnan, 3150 m. PHOTO: CHRISTOPHER GREY-WILSON

Meconopsis delavayi; Gangheba, Yulongxueshan, Lijiang County, NW Yunnan, 3100 m. PHOTO: HARRY JANS

Meconopsis delavayi; Gangheba, Yulongxueshan, Lijiang, NW Yunnan, 3100 m. PHOTO: CHRISTOPHER GREY-WILSON

successfully in a stone trough and this was repeated with great success by Ian and Margaret Young in Aberdeen, also in a stone trough, from seed imported from China in the 1980s, primarily resulting from the ACE Expedition (Alpine Garden Society China Expedition). It is not without significance that in Britain at least it has only been successfully grown and propagated (from seed as well as root cuttings) in the cooler north of the country, all three growers cited above having gardens in Scotland. This is perhaps not surprising as *M. delavayi*, like *M. bella*, requires cool moist summers and cold winters with snow cover being an added and perhaps necessary bonus.

Henry and Margaret Taylor commenting in the *Bulletin of the Alpine Garden Society* in 1995 that "The small *Meconopsis delavayi* reintroduced by Ron McBeath and others from China can be grown at the front of the peat bed or in a trough. It is perennial and again, to get viable seed, you have to cross-pollinate separate plants. Our first three plants all flowered quickly from seed but all at different times. We obtained no seed from the first to flower but kept some of the anthers in a little bottle in our fridge and used these to pollinate the next plants to flower, with good results. It is worth a little bit of effort to ensure that special plants stay in cultivation".

> "Gardens in northern England and Scotland have acted as a repository for several species otherwise lost to cultivation, but the exquisite *Meconopsis delavayi* had vanished from these refuges too, though it has now been restored from a succession of seed collections made in NW Yunnan. Significantly, a few growers have been able to persuade their plants to set seed reliably, and are growing a mixture of clones to ensure the continuance of this much-improved state of affairs. In particular, the Aberdeen garden of Ian and Margaret Young has at least one trough planted exclusively with this normally recalcitrant species."
>
> Robert Rolfe, *The Alpine Gardener* (Bull. Alpine Gard. Soc. Gr. Brit.) 70 (3): 354 (2002).

Nonetheless, *Meconopsis delavayi*, is a charming species that is well worth persevering with and, unlike many of the smaller species, it has been successfully cultivated and maintained over a number of years. Apart from keeping plants growing, it is important that, as indicated, every attempt should be made to get plants to set seed. This can be accomplished in part by growing several plants in close proximity to one another and also by hand pollinating, transferring pollen from one plant to another. Fruit set and seed development is a bit of a gamble especially as a sudden hot dry spell at the wrong moment can cause the fruits to abort.

Meconopsis delavayi; Wutodi, Yulongxueshan, Lijiang, NW Yunnan, 4100 m.
PHOTO: CHRISTOPHER GREY-WILSON

The affinity of *Meconopsis delavayi* has been disputed over the years. With its perennial, tufted habit, entire leaves and nodding, essentially violet, flowers it certainly has a distinctive look. Amongst the small species the only other one that is truly perennial (polycarpic) is *M. bella* and that also presents some interesting features that give it a rather isolated position within the genus. Phylogenetic studies appear to place *M. delavayi* fairly close to *M. lancifolia* and *M. impedita* but the overall characters of these three taxa are very distinct and, while a good case can be made for including all three within section *Impeditae* (see p. 281), it seems, at the same time, that a reasonable argument can be made for placing each within a separate series, *M. delavayi* being the only representative of series *Delavayanae*. Both series *Impeditae* and series *Lancifoliae* contain only monocarpic species. In the former the flowers are borne on basal scapes, in the latter in ebracteate racemes which are generally rather few-flowered. *M. bella* appears to be even more isolated within the genus and phylogenetic studies thus far appear to reveal that its affinities lie closer with the large monocarpic species such as *M. discigera* and *M. napaulensis* (Kadereit *et al*. 2011; see also under *M. bella* for further discussion). It should be added that pollen data (Henderson 1965) show that both *M. bella* and *M. delavayi* are classified in the 'Horridula-type' (pollen grains 3-colpate, spheroidal (15–28 μ diameter) and with the tectum adorned with minute spinules). Within this type are included *M. aculeata*, *M. argemonantha*, *M. bella* (subsp. *subintegrifolia*), *M. delavayi*, *M. forrestii*, *M. georgei*, *M. henrici*, *M. horridula* (probably *M. prattii* from a cultivated source, Edinburgh Botanic Garden), *M. lancifolia*, *M. pseudovenusta*, *M. quintuplinervia*, *M. speciosa* and *M. venusta*.

Apart from the polycarpic habit, *Meconopsis delavayi* possesses very distinctive slender, cylindrical, glabrous fruit capsules that cannot be confused with any other species in the genus.

section FORRESTIANAE

This is an interesting section comprising just three species confined to western Sichuan, north-west and north-east Yunnan and the neighbouring part of north-western Myanmar, with an outlier in south-western Gansu. The section is characterised by a monocapic habit, the plants with rather small, generally few-leaved basal rosettes and leafless stems except in *Meconopsis yaoshanensis* which has one or two leaves low down on the stem. The flowers are borne in ebracteate racemes and possess 4 or more petals. The fruit-capsules are generally long and slender, held on stiffly erect pedicels.

KEY TO SPECIES OF SECTION FORRESTIANAE

1. Flowers borne only towards the top of the inflorescence: fruit capsule narrow-clavate to subcylindric, not more than 6 mm diameter, with a very short of obsolete style, not more than 1.5 mm long. . . . . 66. **M. forrestii** (SW Sichuan, NW Yunnan)
+ Flowers scattered along much of the length of the inflorescence, often accompanied by one or several basal 1-flowered scapes; fruit-capsule ellipsoid to obovoid or oblong-cylindric, 5–11 mm diameter, with a 2–7 mm long persistent style. 2

2. Inflorescence with 8–16 flowers, the lowermost bracteate; fruit-capsule 35 mm long or more . . . . 65. **M. yaoshanensis** (NE Yunnan)
+ Inflorescence with up to 8 flowers, often 3–5, all ebracteate; fruit-capsule not more than 29 mm long . . . . 64. **M. lancifolia** (SW Gansu, W & SW Sichuan, NW Yunnan)

64. MECONOPSIS LANCIFOLIA

Meconopsis lancifolia (Franch.) Franch. ex Prain, *J. Asiat. Soc. Bengal, Pt. 2, Nat. Hist.* 64 (2): 311 (1896), in adnot.
[Syn. *Cathcartia lancifolia* Franch., *Bull. Soc. Bot. Fr.* 33: 391 (1886). Type: NW Yunnan, Yangtsehay (Yangtse Hay), supra Langkong [Mosoying], 3200 m, June 1886, *Delavay* 2080 (P, holotype; E, K, isotypes)]

DESCRIPTION. *Monocarpic plant* to 35 cm tall in flower (to 42 cm tall in fruit), although often only 8–20 cm, with a dauciform taproot 2.5–10 × 10–16 mm. *Stem* erect, 3.7–25 cm long, 5–13 mm diameter near the base, with ascending to patent-reflexed, tawny-coloured, soft bristles, sometimes densely so, but occasionally more or less glabrous. *Leaves* all borne in a basal rosette, green or grey-green above, paler, often glaucous beneath; lamina elliptic-lanceolate to elliptic-oblanceolate or narrow oblanceolate, more rarely linear-lanceolate, (2–) 4–16 × 0.5–2.2 cm, tapering gradually at the base, or cuneate, into the petiole, the margin entire to slightly sinuate, subglabrous or sparsely to moderately soft-bristly on both surfaces; petiole (1.5–)2.7–7.2(–9) cm long, linear, somewhat expanded and sheathing at the base. *Inflorescence* an ebracteate raceme with up to 8 flowers, the stem thickening towards the base, or sometimes with, in addition, one or a few (up to 6) basal scapes; buds nodding. *Pedicels* 3–9.6 cm long (to 14.7

cm long in fruit), sparsely to moderately soft-bristly, often more densely so immediately beneath the flower, the upper pedicels partly agglutinated. *Buds* nodding at first, oblong to obovoid or subglobose, 15–22 × 12–20 mm, the sepals sparsely to densely bristly, occasionally subglabrous. *Flowers* 2.6–8.2 cm across, laterally-directed to ascending. *Petals* (4–)6–11, satiny, deep blue to violet-blue, purple-blue, rose-lavender, purple or ruddy-purple, indigo-purple or pinkish-purple, more rarely pinkish or pale rose-lavender, ovate to suborbicular or obovate, 13–32 × 10–30 mm, the margin entire or finely denticulate. *Stamens* with linear filaments the same colour as the petals normally and with greyish, greyish-white, yellowish-grey or orange anthers, darkening with age. *Ovary* ellipsoid-oblong to ovoid, more or less glabrous to moderately or densely appressed bristly; style 1–4 mm long, with a capitate to subclavate stigma, 3–6-lobed. *Fruit* capsule green with purple ribs at first, later brown, oblong-ellipsoid to subcylindric or obovoid, 15–29 × 5–11 mm, 3–6-valved, glabrous or glabrescent to moderately bristly, the bristles spreading to ascending, splitting for a quarter to a third of its length from the top when ripe, the persistent style 2–7 mm long.

DISTRIBUTION. W China (SW Gansu, NW Yunnan, NW, W & SW Sichuan, E & SE Tibet; locally common in the Cangshan, Yulongxueshan and the mountains of the Yangtse-Mekong and Mekong-Salween divides), N Myanmar (N Upper Burma); 3350–4600 m (Map 23).

HABITAT. Alpine moorland, stony alpine meadows, rocky slopes, moraines, stony and rocky pastures, open low moorland scrub.

FLOWERING. June–early August.

Meconopsis lancifolia is a distinctive and attractive species which is very variable in its many forms seen in the wild: it can be a small plant no more than 10 cm tall bearing just two or three flowers, or more substantial to 35 cm and with a raceme of up to eight flowers. In addition to the main flowering axis, plants may also produce subsidiary flowers borne on basal scapes from the upper axils in the leaf-rosette; these flowers are generally smaller than those borne in the raceme and are usually present on the more vigorous plants but not invariably so. On occasions the inflorescence axis is so condensed that the flowers appear all to be borne on basal scapes, while at the same time depauperate plants bearing a solitary flowers can be found in the wild but this is not the norm.

There has been a great deal of confusion between *Meconopsis lancifolia* and *M. georgei*, although the two are not particularly closely related. This primarily stems from the fact that purple-flowered forms of *M. georgei*

Meconopsis lancifolia habitat; Shikashan, Zhongdian, NW Yunnan, 4100 m. PHOTO: HARRY JANS

64. MECONOPSIS LANCIFOLIA

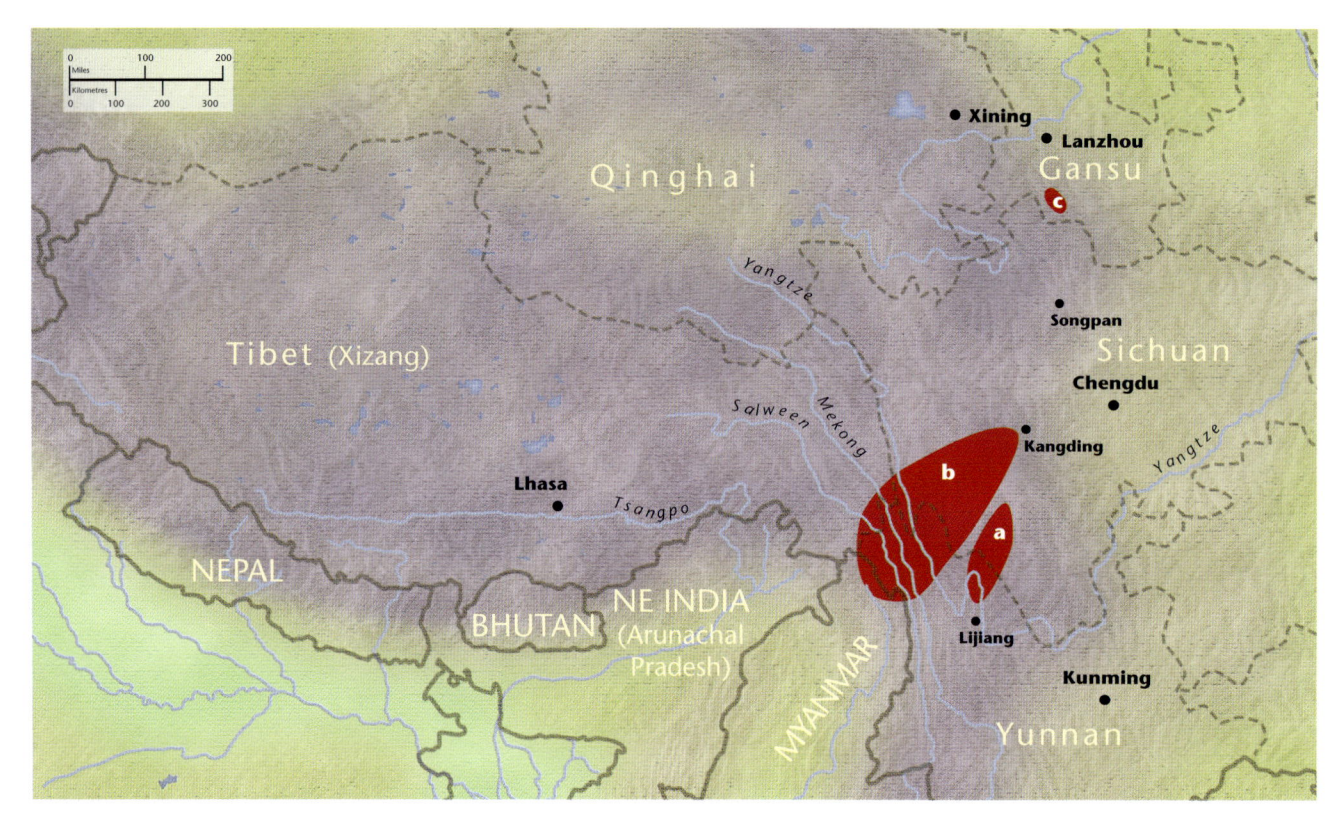

MAP 23. Distribution of *Meconopsis lancifolia*: ■ *M. lancifolia* **a** subsp. *lancifolia*, **b** subsp. *eximia*, **c** subsp. *lepida*.

were included in the description of *M. lancifolia* in George Taylor's 1934 monograph of the genus. This is substantiated by the fact that the sheets in question have been annotated by George Taylor as *M. lancifolia* (see p. 273).

I can find no evidence that this little species has ever been in cultivation, although it may have been introduced as unidentified seed from the wild. However, George Taylor (1934) pointed out: "So far as I am aware there is no authenticated record of this species in cultivation, as the plant figured in 1923 [*The Garden* 137: 293, fig.] as *M. lancifolia* does not appear to be accurately identified, and probably represents *M. delavayi*. There is no doubt that the form which has been described as *M. eximia*, with large deep purple flowers, would be a very desirable addition to gardens, but no success attended Forrest's efforts to introduce it. In several gardens plants have been grown under the name *M. eximia*, but all that I have had the opportunity of examining proved to be forms of *M. horridula*".

Meconopsis lancifolia in its various guises has a fairly wide distribution in the wild and it must be possible to get a form that will respond well to cultivation, even if it is a monocarpic species. Of the subspecies presented below the boldest is certainly subsp. *eximia* which, in the brightest forms, has sumptuous rich royal purple blooms. It also happens to be the most widespread of the subspecies and the one that most of those exploring the mountains of south-western Sichuan and north-western Yunnan are likely to come upon.

KEY TO SUBSPECIES OF MECONOPSIS LANCIFOLIA

1. Inflorescence devoid of flowers in the lower half; petals (6–)8–11 . . . . subsp. **lepida**
+ Inflorescence with flowers for all its length, the lowermost often borne in the axils of basal leaves on separate scapes; petals 4–8 2

2. Petals 6–8, 18–39 mm long, deep purple to purple-crimson or indigo-purple; buds and fruit capsules patent-bristly . . . . subsp. **eximia**
+ Petals 4–6, 15–22 mm long, pale to mid-blue, pink or pinkish-purple; buds and fruit capsules subglabrous . . . . subsp. **lancifolia**

subsp. lancifolia

[syn. *Meconopsis lancifolia* var. *solitariiflora* Fedde, *Repert. Spec. Nov. Regni Veg.* 17: 197 (1921). Type: E Tibet, 1914, *Limpricht* s.n.; *M. lancifolia* var. *limprichtii* Fedde ex Limpr. f., *Repert. Spec. Nov. Regni Veg. Beih.* 12: 383 (1922), *laps pro* var. *solitariiflora*]

DESCRIPTION. Generally rather small plants, seldom exceeding 24 cm in height in flower, but often as little as 8–15 cm. *Inflorescence* a 2–5-flowered raceme, with one or several flowers borne from the lowermost leaf-axils, sometimes accompanied by one or several basal scapes, rarely all flowers scapose or solitary; flowers 2.9–4.8 cm diameter. *Petals* 4–7, 15–22 mm long, pale to mid-blue, pink or pinkish purple, to intense violet. *Anthers* whitish to greyish yellow. *Fruit capsule* subglabrous or with sparse scattered bristles, the persistent style 2–3 mm long.

DISTRIBUTION. NW Yunnan (Yulongxueshan, Yongning), S & SW Sichuan (Fenchingshan, Muli, Mts W of the Yalong river); 4270–4575 m (Map 23a).

SPECIMENS SEEN. *Delavay* s.n. 1887 (E, K), 2080 (E, K, P); *Forrest* 16657 (BM), 21230 (E); 28388 (BM, E), 30086 (BM, E), 30090 (BM, E); *Kunming Edinburgh Exped. Sichuan* 383 (E); *Rock* 5501 (E, K), 5557 (E), 17833 (E), 18113 (K), 23711 (BM, E, K), 23946 (BM, E, K), 27711 (BM).

NOTE. Due to confusion with subsp. *eximia*, especially in the literature, the precise distribution is unclear, but appears to stretch northwards from Yulongxueshan to the mountains around Muli in southern Sichuan.

Subsp. *lancifolia* is a slighter plant than subsp. *eximia* and has a more limited distribution in the wild. Although both are found in Yunnan and Sichuan they have discreet distributions and there is no indication to my certain knowledge of them growing in the same locality. Subsp. *lancifolia* has a far more restricted distribution and is, on the whole, a more uniform taxon.

Above left: *Meconopsis lancifolia* subsp. *lancifolia*: herbarium sheet *Rock* 23946 (K), SW Sichuan, Muli Territory

Above right: *Meconopsis lancifolia* subsp. *lancifolia*: herbarium sheet *Rock* 23711 (K), SW Sichuan, Muli Territory

subsp. eximia (Prain) Grey-Wilson, **comb. & stat. nov.**

[syn. *Meconopsis eximia* Prain, *Bull. Misc. Inform.*, *Kew* 1915: 159 (1915). Type: *Rock* 5501, China 1922 (syntypes, E, K); NW Yunnan, *Forrest* 12691, mountains of the Chungtien Plateau, 1914 (syntypes, E, K); *Forrest* 13020, Mekong-Yangtse Divide, 1914 (syntypes, E, K); *Forrest* 13238, Mekong-Salween Divide, 1914 (syntypes, E, K); *Forrest* 13352, China 1914 (syntype, K). Lectotype *Forrest* 13020 (E, lectotype; K, isolectotype) chosen here]

DESCRIPTION. A more robust plant than the other two subspecies, to 35 cm tall in flower with a thick, bristly stem (peduncle), 7–13 mm diameter close to the base. *Inflorescence* a stout, (2–)4–10-flowered raceme, the stem often markedly thickening towards the base, usually with several flowers (1–5) borne on separate scapes in the axils of the upper leaves of the basal rosette; flowers 3.8–8.2 cm diameter. *Buds* patent-bristle, sometimes densely so. *Petals* (4–)6–8, 18–39 mm long, deep purple to purple-crimson or indigo-purple, purple-blue or satiny deep blue. *Anthers* cream, greyish or greyish yellow. *Fruit capsule* 15–29 × 6–11 mm, moderately to densely patent-bristly, occasionally subglabrous, the persistent style 4–7 mm long.

DISTRIBUTION. SW China: W Sichuan (Kangding area southwards), NW Yunnan (Baimashan, Cangshan, Daxueshan, Zhongdian as well as various mountains on the Mekong-Salween and Mekong-Yangtse divides); N Myanmar (Upper Burma; Maikha); 3810–4600 (5180) m (Map 23b).

Right: *Meconopsis lancifolia* subsp. *eximia* leaf rosette; Shikashan, Zongdian, NW Yunnan, c. 4000 m. PHOTO: HARRY JANS

Below, from left:

Meconopsis lancifolia subsp. *eximia*; Shikashan, Zongdian, NW Yunnan, c. 4000 m. PHOTO: JOHN & HILARY BIRKS

Meconopsis lancifolia subsp. *eximia*; Baimashan, Deqin, NW Yunnan, 4050 m. PHOTO: JOHN & HILARY BIRKS

Meconopsis lancifolia subsp. *eximia*; Hongshan, Zhongdian, 4535 m. PHOTO: HARRY JANS

SPECIMENS SEEN. *ACE* (Alpine Garden Society China Exped.) 568 (E, K), 744 (E), 1305 (E); *Forrest* 1950 (BM, E, K), 1999 (BM, E, K), 12691 (E, K), 13020 (BM, E, K), 13238 (E, K), 13714 (E), 14087 (E, K), 14088 (E), 14473 (BM, E, K), 14625 (BM, E, K), 16657 (E, K), 16659 (E, K), 18242 (BM, E, K), 19683 (BM, E), 19694 (BM, E), 20032 (BM, E), 20399 (BM, E), 20406 (BM, E), 21230 (K), 21543 (E), 21576 (E, K), 21579 (E), 22642 (E), 23425 (BM, E), 25522 (K), 30110 (BM, E), 30114 (E), 30117 (E), 30946 (BM, E); *Kingdon Ward* 5334 (E); *McLaren* 134 (K), 183b (BM, E, K); *Rock* 6324 (E), 9316 (BM, E), 9994 (E, K), 10339 (E, K), 22768 (E), 23370 (E), 25071 (E), 25192 (BM, E, K), 25214 (BM, K); *E. H. Wilson* 957 (K), 3027 (K), 3028 (K); *T. T. Yü* 13790 (E); *Zhang-ting et al.* 10CSZ195 (k).

In its largest and most brightly coloured forms subsp. *eximia* is a very striking plant. In the most robust the flowers are generally borne on the central rachis, however, in others, often those with a weaker axis, the raceme is often accompanied by several basal scapes. In depauperate specimens, or those just coming into flower, the central axis can appear to be absent but it can usually

Below, from left:

Meconopsis lancifolia subsp. *eximia*; Baimashan, Deqin, NW Yunnan, 4220 m. PHOTO: JOHN & HILARY BIRKS

Meconopsis lancifolia subsp. *eximia*; Balangshan, Wolong, W Sichuan, 4100 m. PHOTO: HARRY JANS

Meconopsis lancifolia subsp. *eximia*; Balangshan, Wolong, W Sichuan, 4100 m. PHOTO: HARRY JANS

Above left: *Meconopsis lancifolia* subsp. *eximia*; Balangshan, Wolong, W Sichuan, 4100 m. PHOTO: HARRY JANS

Above right: *Meconopsis lancifolia* subsp. *eximia*; Shikashan, Zhongdian, NW Yunnan, 4100 m. PHOTO: HARRY JANS

Above, from left:

Meconopsis lancifolia subsp. *eximia*; Daxueshan, NW Yunnan, 4150 m. PHOTO: CHRISTOPHER GREY-WILSON

Meconopsis lancifolia subsp. *eximia*; Daxueshan, NW Yunnan, 4250 m. PHOTO: CHRISTOPHER GREY-WILSON

Meconopsis lancifolia subsp. *eximia*, pink form; Juinzikou Pass, Yajiang to Li. PHOTO: HARRY JANS

be detected on close inspection. In these instances, as the terminal flower opens the central axis expands pushing the raceme clear of the basal leaf-rosette. This can also happen when the central axis is damaged as it develops, as for instance by grazing, trampling or other effect which has prevented its normal development; this has been observed in other species such as *Meconopsis georgei*, *M. racemosa* and *M. rudis*.

The accompanying basal scapes, when present, develop last and these scapose flowers may not open until the raceme has almost past the flowering phase.

Of the three subspecies of *Meconopsis lancifolia* currently recognised, subsp. *eximia*, holds the greatest potential as a horticulturally desirable plant, although like so many of the small monocarpic species maintaining plants in cultivation, particularly ensuring regular seed set, is probably the prime hurdle to success or failure.

subsp. lepida (Prain) Grey-Wilson, **comb. & stat. nov.** [syn. *Meconopsis lepida* Prain, *Bull. Misc. Inform.*, *Kew* 1915: 158 (1915). Type: SW Gansu, Mountains of Thundercrown (Lei-go-shan), Siku Alps, 12–13,000 ft, June 1914, *R. Farrer* 123 (E, holotype, isotypes; BM, K isotypes); *M. eucharis* Farrer ex Irving, *Quart. Bull. Alpine Gard. Soc. Gr. Brit.* 1: 192 (1932), in obs., nom nudum.]

DESCRIPTION. A slighter plant than the other two subspecies, to 17 cm tall in flower; stem slender, with a few scattered hairs. *Leaves* subglabrous, ovate to spathulate, 1.6–4.5 × 0.6–1.5 cm. *Inflorescence* bearing 2–6 flowers well clear of the basal leaf-rosette on a slender stem bearing a few scattered hairs, not accompanied by basal scapes. *Pedicels* 1.3–2.8 cm long, but that of the terminal flower up to 6.8 cm long. *Buds* subglabrous. *Flowers* relatively small, 2.6–6.3 cm across. *Petals* (6–)

8–11, 26–30 mm long, lavender-purple. *Fruit capsule* subglabrous, the persistent style 4–6 mm long.

DISTRIBUTION. SW Gansu: 'Thundercrown', Siku Alps, Leigoshan; c. 3660–3965 m (Map 23c).

HABITAT. Rock ledges on cooler slopes.

FLOWERING. June–July.

Subsp. *lepida* is isolated, well to the north of the other two subspecies and was discovered there by Reginald Farrer in the first year of the First World War when he accompanied William Purdom to the mountains of south-western Gansu. Farrer's collection is in fact the only one available. At Kew it is accompanied by a fine watercolour drawing by Matilda Smith based upon Farrer's collections and notes. Subsp. *lepida* is altogether a more wispy and slighter plant than either subsp. *eximia* or subsp. *lancifolia*. At present it is best regarded as a subspecies of *Meconopsis lancifolia*, that is until such time that further collections can add to our knowledge: then a proper reassessment can be made.

> "A most lovely little biennial akin to *M. delavayi*. Rock ledges and edges and banks of limestone cliffs at 12,000 ft. All the seed that Summer and hail has left. Distribute but with intense economy. It confines itself to the cooler aspects …….. dainty texture of delicate silken lavender purple."   R. Farrer

Reginald Farrer, who came across this plant in SW Kansu [Gansu] and collected under his number 123, writes about it at length in the Appendix to his monumental work *The English Rock Garden* published in 1928, referring to it as "The Dainty Poppy":

> "Meconopsis sp. (F123) inhabits the upper alpine banks and ledges on Thundercrown, markedly preferring the cooler westerly aspects. It is not found in the open turf, but often occurs at its fringes round the base and up the gullies of little limestone outcrops in the huge grassy flanks of the mountain at 12,500 feet, not steadily abounding, but appearing in sporadic outbursts. It is a most lovely little biennial of some 4 to 8 inches, with all the narrow, rather glaucous foliage at the base, and the naked stem carrying 1 to 6 large flowers, made up of some 6–11 rhomboidal petals of lavender purple silk, arranged in a whirling catherine-wheel round the creamy crowded boss of stamens. These flaunt their frail and filmy loveliness in June ……nothing more daintily beautiful exists in the race, as you see its great whirling heads poised delicately amid the fine grasses, the golden Gages and Fritillaries, the innumerable purple Irids that enamel the grassy rocky ribs of Thundercrown".

David Prain in a letter attached to *Farrer* 123 (E) states that Reginald Farrer called the plant *"Euchonis"* which I have translated to *lepida"*, dated 22 March 1915.

SPECIMENS SEEN. *Farrer* s.n. (E), 123 (BM, E, K; also listed under the same number in the herbarium of William Purdom).

NOTE. Some specimens from the Cangshan west of Dali in NW Yunnan differ from subsp. *eximia* in their broader obovate, entire, leaf-laminas, in the generally longer pedicels, and in the consistently 4-petalled flowers. *Forrest* 15502 labelled, W flank Tali Range (Cangshan), 25°40'N, 12,000 ft, "Plant of 6–12 inches. Flowers deep satiny-purple. Anthers grey. Open stony pasture. July 1917"; and *Forrest* 13517 from the same vicinity "Open stony pastures. 25°40'N. Alt. 12,000 feet. Plant of 6–14 inches. Flowers deep blue-purple. Anthers grey. Aug. 1914." These require further observation. *Meconopsis lancifolia* subsp. *eximia* is also recorded from the Cangshan. Three specimens can be referred to: *Forrest* 13517 (BM, E, K), 15502 (BM, E, K), 28224 (BM, E).

65. MECONOPSIS YAOSHANENSIS

Meconopsis yaoshanensis Tosh. Yoshida, H. Sun & Boufford, *Acta Bot. Yunnan.* (Plant Diversity & Resources) 34 (2): 148 (2012), fig. 2–4. Type: SW China, NE Yunnan, Qiaojia Xian, Yao Shan, 27°12'40"N, 103°04'31"E, 3750 m, 7 July 2011, *T. Yoshida* K55 (holotype KUN; isotypes E, KUN).

DESCRIPTION. *Monocarpic herb* to 50 cm tall in flower, as little as 20 cm (to 60 cm in fruit), with an elongated, cylindrical, gradually tapering taproot to 20 cm long, 6–13 mm diameter at the top, sometimes with side branches close to the top, the plant adorned in most parts by patent golden- or reddish brown, softish, non-pungent, bristles, mostly 5–6 mm long. *Stem* short, 2–8 cm long, to 10 mm diameter (mostly underground in the living plant). *Leaves* spreading, petiolate, mostly in a basal rosette, matt yellow-green above, paler and somewhat glaucous beneath, the lamina subcoriaceous, elliptic-oblong to narrow oblanceolate, 4–17 × 1–3 cm, the base somewhat cuneate or attenuate into the petiole, the apex acute, or occasionally slightly acuminate, margin entire to somewhat repand, undulate, patent hairy on both surfaces although less densely so beneath; petiole strap-like, 2–8 cm long, to 3 mm wide. *Inflorescence* an 8–16-flowered erect raceme, ebracteate except for the lowermost 2–3 flowers which are borne in the axils of the uppermost leaves (bracts) very close to the basal rosette, the rachis with sparsely to moderately patent, softish bristles; flowers cup to saucer-shaped, nodding to lateral-facing, 2.5–4.2 cm across. *Bracts* similar to the leaves but gradually smaller up the inflorescence, short-petioled to sessile. *Pedicels* ascending (stiffly erect at the fruiting stage), 2–7 cm long (to 12 cm long in fruit), downcurved

65. MECONOPSIS YAOSHANENSIS

towards the top in the flowering phase, patent-bristly as the rachis, somewhat swollen immediately beneath the flower, decurrent onto the rachis. *Buds* nodding, yellow-green, the sepals oval 10–15 mm long, patent-bristly. *Petals* 4–5 (–6), pale blue to violet, elliptic to ovate or obovate, 17–24 × 8–15 mm, slightly fluted, the apex rounded to obtuse, finely to rather coarsely erose. *Stamens* numerous, the filaments filiform, 7–11 mm long, similar in colour to the petals, the anthers 1–1.5 mm long, ellipsoid, slightly curved, dark purple with yellow or orange pollen. *Ovary* ellipsoid, 5–7 mm long, densely ascending bristly, the style short, 1–2 mm long with a similarly sized capitate stigma. *Fruit capsule* narrow oblong-cylindrical, 35–52 mm long, 6–10 mm diameter, somewhat curved, slightly narrowed at the base, more so at the apex, retrorse-bristly overall, 3–5-valved, borne on stiffly erect pedicels that often twist slightly around the infructescence rachis; persistent style 3–7 mm long. *Seeds* blackish, 2.3–2.9 mm long.

DISTRIBUTION. SW China: NE Yunnan (Qiaojia Xian, Yaoshan), 3650–3800 m.

HABITAT. Basaltic boulder slopes with scattered herbs and grasses and other vegetation, primarily of southerly, south-easterly and easterly aspect.

FLOWERING. June–July.

SPECIMENS SEEN. T. *Yoshida* K46 (KUN), 55 (E, KUN).

This recent species, described in 2012, is currently only known from a limited area centred around the type locality. In its stiffly erect infructescences and narrow subcylindrical fruits it most closely resembles *Meconopsis forrestii*. It is an interesting species in the rosette stage for, at a glance, the leaves are very reminiscent of a fern, notably those of the hart's-tongue, *Asplenium scolopendrium*. Although the original description states that the flowers are primarily 4-petalled, photographic evidence supports the view that 4 or 5 petals may be the norm, while the terminal flower may have 6 petals.

Although the original description places *Meconopsis yaoshanensis* close to *M. castanea* (*M. georgei* sensu lato in this monograph) and *M. bijiangensis*, the general features of it seem more closely akin to those of *M. forrestii* and *M. lancifolia*, especially in the softer, more papery and less densely bristly leaves. In the characters of the infructescence *M. yaoshanensis* perhaps most closely resembles *M. forrestii* which, however, has fewer capsules crowded towards the stem top rather than scattered along most of its length. It also resembles *M. forrestii* in its predominately 4-petalled flowers. While the capsules of *M. yaoshanensis* bear an obvious style, those of *M. forrestii* are very short or more or less obsolete.

Below, from left:
Meconopsis yaoshanensis; Yaoshan, Qiaojin County, NE Yunnan, 3750 m. PHOTO: TOSHIO YOSHIDA
Meconopsis yaoshanensis; Yaoshan, Qiaojin County, NE Yunnan, 3750 m, flowering early July. PHOTO: TOSHIO YOSHIDA
Meconopsis yaoshanensis in young fruit and flower; Yaoshan, Qiaojin County, NE Yunnan, 3750 m. 3750 m, early July. PHOTO: TOSHIO YOSHIDA

Meconopsis yaoshanensis; Yaoshan, Qiaojin County, NE Yunnan, 3750 m, flowering in early July. PHOTO: TOSHIO YOSHIDA

Meconopsis yaoshanensis is a little known taxon and one of the most easterly species in the genus, an outlier from the main concentration of species to the west and north-west. Photographs show it to be a pretty plant and certainly a desirable one to have in cultivation if seed can ever be introduced. It is the only species of *Meconopsis* so far described that inhabits basaltic rocks, in fact basalt is rare in south-western China. In this locality, to the east of Kunming, the basalt (which is believed to have formed in the bottom of the Tethys Sea geologically) forms a gently inclined slope. The region is known to have a very specialised and unique flora to which *M. yaoshanensis* can be added.

66. MECONOPSIS FORRESTII

Meconopsis forrestii Prain, *Bull. Misc. Inform.*, *Kew* 1907: 316 (1907). Type: China, NW Yunnan, Lijiang Range [Yulongxueshan], June 1906, 10–11,000 ft, *Forrest* 2314 (E holotype, isotypes; BM, K, P, isotypes).

DESCRIPTION. *Monocarpic plant* to 50 cm, generally 18–45 cm tall in flower (to 60 cm tall in fruit), often only half that height, with a dauciform or napiform taproot 2–10 cm long, but not more than 1.5 cm diameter, adorned with slender fibrous roots. *Stem* erect, 2–6 mm diameter, with spreading to somewhat deflexed bristles, often dense. Leaves mostly or all basal, in a lax, generally rather few, rarely more than 9, borne in a spreading rosette, to 15 cm in length overall, withering in the autumn to leave a small resting bud at ground level or just below; lamina elliptic to oblanceolate, elliptic-oblong, linear-oblong or linear-lanceolate, 5.6–17 × 0.8–2.8 cm, the base attenuate into the petiole, the apex acute or subacute, the midrib prominent, especially beneath, sparsely adorned on both surfaces with stiff, subappressed golden-brown hairs, the margin entire or remotely sinuate; petiole linear, thinly winged, with a sheathing part-membranous base, 2.5–7 cm long. *Inflorescence* a 3–7-flowered raceme, ebracteate, erect, tapering from a rather stout base, devoid of flowers in the lower half, not accompanied by basal scapes, the rachis sparsely to moderately covered with stiff, spreading or somewhat deflexed, slender bristles. *Pedicels* very slender in flower, 1.2–4.6 cm long (to 10.5 cm long and thickening somewhat in fruit), pubescent like the stem, nodding at first, becoming spreading to ascending. *Flowers* nodding to half-nodding, saucer-shaped, 2.8–4.8 cm diameter, nodding in bud. *Buds* nodding, ovoid to subobovoidal, 9–12 mm long, the *sepals* oval-ovate, softly brown-bristly. *Petals* 4, occasionally 5, pale blue or whitish-blue, purplish, purplish-blue, ruddy purple, pale purplish-rose or reddish, ovate to obovate, 14–25 × 11–20 mm. *Stamens* with linear filaments 5–10 mm long, the same colour as the petals although generally darker, the anthers c. 1 mm long, orange or orange-yellow. *Ovary* narrow-ellipsoid, glabrous or sparsely bristly, with a very short or obsolete style, not more than 1.5 mm long; stigma subcapitate, 1–1.5 mm across, 2–4-lobed. *Fruit capsule* borne on stiff, erect pedicels, narrow-clavate or subcylindric, 32–65 × 3–6 mm, narrowed at the base, 2–4-valved, splitting for about a third of its length from the apex, glabrous to sparsely patent-bristly.

DISTRIBUTION. NW Yunnan and neighbouring parts of SW Sichuan; very local, especially noted in the Yulongxueshan (eastern slopes), Yangbi and the margins of the Zhongdian Plateau, also in the vicinity of Muli in southern Sichuan; 3048–4420 m.

HABITAT. Alpine meadows, grassy slopes, stony pastures, moraine and scree margins, edges of woodland and thickets, rocky places.

FLOWERING. Late-May–July.

SPECIMENS SEEN. *Chamberlain, Ming & Yuan et al.* 625 (E); *Delavay* s.n. July 1888/89 (BM), s.n. August 1888 (BM); *Forrest* 2314 (BM, E, K), 2748 (E), 5689 (BM, E), 8993 (E), 10799 (BM), 12507 (BM, E, K), 12672 (BM, E, K), 16309 (BM, E, K), 16799 (E, K), 21251 (E, K), 24637 (K), 30107 (BM, E); *Kunming-Edinburgh Yulongshan Exped.* 4 (E); *McLaren* 53 (BM, E, K); *Rock* 4378 (E), 8681 (BM, E, K), 9635 (E), 16483 (E), 24637 (BM, E), 24943 (E); *C. Schneider* 1514 (E, K).

66. MECONOPSIS FORRESTII

Above, from left:

Meconopsis forrestii; Bigu Tianchi, Xianggelila County, 3750 m, flowering in mid June. PHOTO: TOSHIO YOSHIDA

Meconopsis forrestii; NE Tianchi Lake, Zhongdian, NE Yunnan, 3820 m, flowering in early July. PHOTO: ALAN DUNKLEY

Meconopsis forrestii; Tianchi, Zhongdian, NW Yunnan, 3650 m. PHOTO: PHILLIP CRIBB

This elegant little species was based on specimens collected by George Forrest on the eastern flanks of the 'Likiang Range' (Yulongxueshan) in 1906 and seed was collected and distributed in 1914 by George Forrest. It is recorded that, although the seed germinated well, the plants failed to reach maturity. However, its discovery must be attributed to Abbé Delavay who collected it in the Lijiang area some 18 years earlier and included in his collections for 1888 and 1889. Delavay had collected vigorously in NW Yunnan and it is not surprising that he came across this species: in fact it grows close to *Meconopsis delavayi* (discovered by Delavay in 1884!) in some of its locations in the Yulongxueshan. Since Forrest's introduction, subsequent collections were made by Heinrich Freiherr Handel-Mazzetti, Joseph Rock and Camillo Karl Schneider. In more recent times, seed has been collected on a number of occasions but the species has still failed to become established in cultivation. Unfortunately, a number of the most recent collections purporting to be *M. forrestii* have proved to be other species, particularly the far more widespread *M. lancifolia*, to which it bears a passing resemblance in some of its forms. The similarity of these two species is no better exemplified than in the two sheets of Abbé Delavay in the British Museum (Natural History) denoted "Aout 1888, Yentzehay, 3500 m": in the flowering stage the two species look very similar, however, the 4-petalled flowers and near obsolete style are reliable diagnostic characters of *M. forrestii*. In the fruiting stage *M. forrestii* is very distinctive and cannot be mistaken for any other species, the long narrow fruit capsules being held in a near-fastigiate infructescence.

Meconopsis forrestii is a very distinct and localised species unlikely to be confused with any other at maturity. It is an elegant wisp of a plant with a sparse inflorescence of dainty, relatively small flowers which, in the darker forms, are very attractive, but with scarcely the impact of its larger-flowered brethren. The sparse-flowered inflorescence, devoid of flowers in its lower half and the long narrow fruit capsules, with scarcely a hint of a style, clearly distinguish it from its closest allies, *M. lancifolia* in particular. The flowers are almost always 4-petalled. Being a very slender plant it is very easily overlooked at the fruiting stage when all the basal leaves will have withered away.

It is perhaps a pity that so slight a species should bear the name of one of the greatest plant hunters who explored south-western China and who has done so much to enhance our knowledge of the genus. Yet *Meconopsis forrestii* has an ethereal grace and charm which few of its cousins possesses and of course George Forrest's name is also commemorated in the relatively little known and misunderstood *M. georgei*, also found in north-western Yunnan, although closer to the Myanmar border region. *M. forrestii* has a distinctive look in flower, tall and elegant with a long bare stem surmounted by several little lampshade flowers often of rather pale appearance in shades of purple, lavender or reddish. The leaf rosette is small and rather unsubstantial.

James Cobb (1989) reports that seed recently brought back from China germinated well and "The leaves have a few golden hairs and are a glaucous green and the plants have produced a typical taproot after the first year of growth. They become winter dormant much later than related species with leaves present into November".

section CUMMINSIA

Section *Cumminsia* contains about 10 species including *Meconopsis primulina* Prain, *M. argemonantha* Prain and *M. florindae* Kingdon-Ward and an aggregate of 7 other closely related species which are here referred to as the *Meconopsis lyrata* aggregate (series *Lyratae*). While there is little doubt of the close relationship of the members within the aggregate, its relationship to *M. primulina* is more questionable and the series may require redefining when it has been more fully analysed. In the meantime the details of the section as presently defined are presented here.

Section *Cumminsia* is characterised by small, generally slender, monocarpic plants, rarely more than 50 cm tall (often 10–30 cm), usually with a rather short, napiform or subnapiform taproot. Plants are moderately to sparsely hairy or subglabrous, the hairs weak and easily rubbed off. The leaves are generally rather few in number, membranous, generally lobed, crenate or sinuate at the margin, but sometimes entire. The 1–5 flowers are borne on a leafy stem both terminal and in the axil of leaf-like bracts, but they sometimes appear to be solitary and scapose because the leaves are congested close to the base of the plant (prominent in plants which appear to have all their leaves in a basal rosette). The fruit capsule is generally slender, subcylindric to ellipsoid-oblong or narrow-obovoid, glabrous to moderately but weakly bristly, often 4-, but 3–6-valved.

series Primulinae

The three species of this series bear a slender, tapering, dauciform taproot. The lateral-facing to half-nodding, saucer-shaped flowers have (4–)5–9 yellow, white, blue or purple-blue petals. The fruit-capsule is narrow-obovoid to oblong-ellipsoid, often hairy or bristly. The species are restricted to the region that encompasses Bhutan, north-western Arunachal Pradesh (NE India) and adjacent parts of Tibet.

KEY TO SPECIES OF SERIES PRIMULINAE

1. Flowers pale blue, lavender-blue, mauve-blue or pink; leaves always entire. . . . . 69. **M. primulina** (Bhutan and neighbouring parts of Tibet)
+ Flowers white, cream, or yellow; leaves often pinnately lobed, occasionally more or less entire 2
2. Fruit capsule densely bristly, 16–25 mm long; petals white, cream or lemon-yellow . . . . 67. **M. argemonantha** (SE Tibet)
+ Fruit capsule glabrous to sparsely bristly, c. 27 mm long; petals lemon-yellow. . . . . 68. **M. florindae** (SE Tibet)

67. MECONOPSIS ARGEMONANTHA

Meconopsis argemonantha Prain, *Bull. Misc. Inform.*, *Kew* 1915: 161 (1915). Type: SE Tibet, Mipak, Tawang District, 13,000 ft, 17 Sept. 1913, *Capt. Bailey 6* (K, holotype).

DESCRIPTION. *A slight, monocarpic plant* to 25 cm tall in flower, with a napiform to dauciform, rather swollen taproot, 18–31 × 4.5–9 mm. *Stem* to 13 cm long, sometimes very short, glabrous or sparsely bristly, the bristles non-barbellate. *Leaves* rather few, congested towards the base or at the base of the stem, alternate, bluish-green, paler or glaucous beneath, very sparsely bristly overall, sometimes glabrous; basal leaves ovate to oblong, 1.5–3.2 × 0..8–17 cm, entire to sinuate or somewhat pinnately lobed to pinnatisect, with 3–7 pairs of elliptic-oblong lobes, the lowermost sometimes asymmetrically bi- or tri-lobed, the base attenuate into a petiole up to 5.5 cm long, the apex obtuse; middle and upper leaves somewhat larger, to 2.7–6.5 (–11.5) × 1.0–2.0 cm, pinnately lobed towards the base, sinuately lobed above, sometimes sub-bipinnately lobed, with 3–5 pairs of lobes overall, the lobes subrounded to linear-elliptic, the terminal lobe larger than the adjacent ones, the base attenuate into a short, winged petiole to 2.5 cm long. *Flowers* 3–5 (–7), rarely solitary, borne from the leaf-axils, the upper two or three sometimes with partly agglutinated pedicels, saucer-shaped, 1.7–5.8 cm diameter, lateral-facing or ascending. *Pedicels* 4–15.5 cm long (9–25 cm long in fruit), sparsely patent-bristly, denser towards the top. *Buds* suborbicular to obovoid, 7–12 × 6–10 mm, the sepals bright green with dark red margins, often streaked brown at the base, adorned with scattered pale brownish hairs. *Petals* 4–7, white, lemon or pale yellow, ovate to elliptic-obovate, 17–28 × 10–22 mm, markedly fluted, the apex sometimes slightly apiculate, the margin fimbriate-dentate (erose). *Stamens* with yellow or golden anthers, the filiform filaments about one-third the length of the petals, white or pale yellow. *Ovary* pale green, appressed-bristly with reddish brown hairs, the style very slender, 2.5–6 mm long, greenish; stigma capitate. *Fruit capsule* dull purple-green at first, browning with age, ellipsoidal, (14–)16–25 × (4–)5–7 mm, broadest above the middle, somewhat narrowed near the base, 4-valved, adorned with spreading to somewhat deflexed reddish brown bristles, those at the bottom of the fruit (at least when young) often whitish, the persistent style 4–8 mm long. (Fig. 28).

DISTRIBUTION. SE Tibet; Bimbi La, Mira La, Pang La (Migyitum), Tsari valley, Tum La, Nyang Chu, Phu Chu, Paka Phu Chu; 3658–4575 m (Map 24).

HABITAT. Growing amongst rocks and in rock crevices, cliff ledges amongst grasses and mosses or on steep open, grassy slopes, juniper scrub.

FLOWERING. Late June–early August.

NOTE. Leaves are often transitional from entire at the base of the plant to pinnately-lobed or pinnatisect, the lowermost leaves often withering by flowering time. *Ludlow, Sherriff & Taylor* 5790 (the type of var. *lutea*) and 6064 includes such transitional forms.

This interesting and relatively little known species was first collected by Colonel F. M. Bailey in 1913, incidentally on the same expedition to SE Tibet when he discovered the fabled blue poppy that bears his name, *Meconopsis*

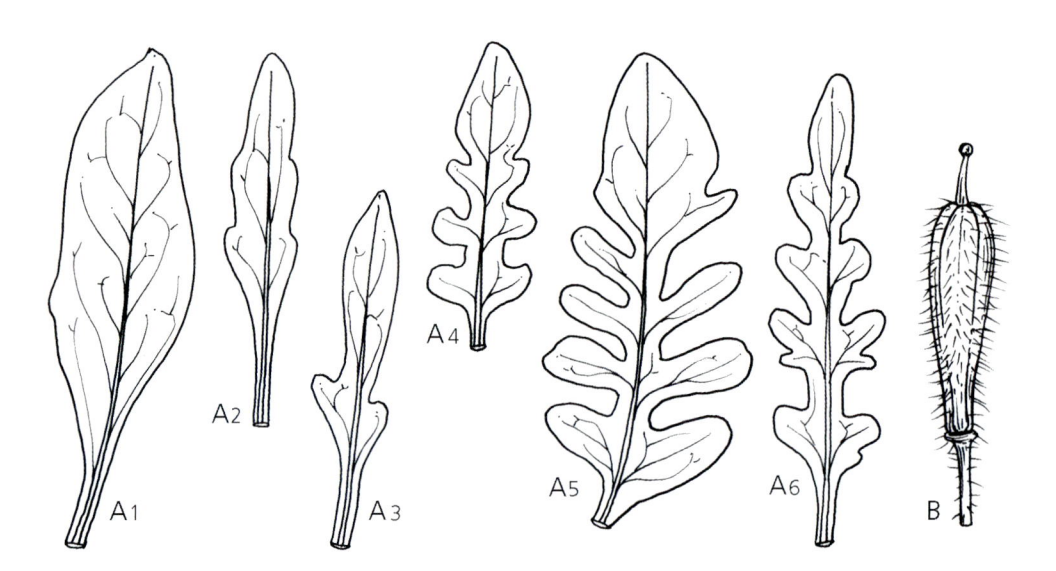

FIG. 28. *Meconopsis argemonantha*: **A1–3**, lower cauline leaves; **A4–6**, middle and upper cauline leaves; **B**, fruit capsule. Pubescence not shown, except for the fruit capsule. All × 1¹/₃. DRAWN BY CHRISTOPHER GREY-WILSON

baileyi. The original, rather brief, description was based on a fragmentary specimen consisting of two leaves and two flowers and the author, George Taylor, was unable to provide a full description of the plant. Furthermore, and perhaps more importantly because of this, he was unable to judge the affinities of *M. argemonantha* within the genus: in his monograph of 1934: George Taylor places it towards the end of his revision under the heading 'Insufficiently Known Species'.

Since its initial collection further, more complete, gatherings have been made, notably those of Kingdon Ward who made some excellent gatherings in the Tsari Valley in July 1935, followed by George Sherriff (under Ludlow & Sherriff numbers) a year later in the same general vicinity. Ludlow, Sherriff & Taylor made further collections on the Mira La in Kongo Province, SE Tibet in 1938. As a result Taylor was able to elaborate his original description and to assign the species to Prain's series *Primulinae*, stressing its allegiance in particular to *M. primulina*.

Meconopsis argemonantha is a very pretty and very distinct little species which, judging from remarks made by those who have observed and collected it in the wild, is both a local and rather rare plant. Seldom seen in recent years it was however photographed in the wild in south-eastern Tibet by Anne Chambers during an expedition to the Tsari valley and the adjacent Bimbi La. In *The Alpine Gardener* (75 (3): 377, 2007) in an article entitled 'Alpines in south-east Tibet' she comments that "While porterage was arranged for the Bimbi La ascent, we explored the surrounding valleys. Three of the *Meconopsis* species seen in the area, *M. simplicifolia*, *M. paniculata* and *M. betonicifolia* [now *M. baileyi*], are common enough in Tibet but the fourth, *M. argemonantha*, was new to us. A species apparently confined to south-east Tibet, a few plants of it grew on cliffs at the eastern end of the Tsari valley……… Rare perhaps, but certainly not showy, its white flowers are about 5 cm in diameter on a stem less than 30 cm tall".

Kingdon Ward records (*KW*11870) the "Flowers nodding, snowy white with golden stamens" and that the plant inhabited grassy cliff ledges.

Meconopsis argemonantha is an attractive little species, though it can in no way claim the visual impact of some of its larger cousins. However, it has a charm and distinctively individual look and, with the exception of *M. florindae*, cannot be confused with any other species known to date. Plants tend to open one flower at a time, rarely up to three open on a single plant at the same time, and these appear to be fairly short-lived. This fact means that out of flower the species is rather inconspicuous and readily overlooked and this may partly account for

Meconopsis argemonantha; Tsari valley, eastern end, SE Tibet.
PHOTO: ANNE CHAMBERS

the fact that relatively few collections have been made over the years. There is little doubt that *M. argemonantha* belongs to series *Primulinae* particularly in its short stem, in the usual transition of leaves from entire or subentire to pinnatisect on the same plant and, furthermore, by the fact that the flowers are borne from the axils of the uppermost leaves (bracts). Within *Primulinae* the white or yellow flowers are unique, but in the number of petals it closely matches *M. primulina* itself, but that species bears mauve-blue or purplish flowers and has leaves that are entire to somewhat sinuate.

These details probably make it seem that *Meconopsis argemonantha* is a fairly uniform species but plants can be quite variable, especially in height and, while a stem is often present and generally short, some specimens are apparently acaulous and this feature can also be readily observed in allied species, *M. primulina* and *M. wumungensis* in particular. Perhaps the most obvious variation is seen in the leaves. As already stated leaves can vary from entire (the lowermost) to strongly pinnately lobed on the same plant, but in some individuals all the leaves may be entire or subentire: *Ludlow, Sherriff & Taylor* 6064 is one such collection in which several plants on the sheet have leaves that are entire or scarcely lobed at all while, at the same time, there may or may not be a stem present. These details bring *M. argemonantha* very close to *M. florindae*, also described from south-eastern Tibet, but some ten years later. The prime distinguishing features between the two species is in the fruit capsules, somewhat larger in the latter and 4–6-valved, glabrous or sparsely bristly, 4-valved and densely bristly in the former. Both *M. florindae* and *M. argemonantha* var. *lutea* have pale yellow flowers.

var. argemonantha

[syn. *M. argemonantha* var. *genuina* G. Taylor, *J. Roy. Hort. Soc.* 82: 167 (1947)]

DESCRIPTION. As species description above; flowers white with white filaments.
DISTRIBUTION. SE Tibet; Bimbi La, Mipak (Tawang District), Pang La (Migyitum), Tama La, Tsari valley, Nyang Chu, Phu Chu (Map 24a).
SPECIMENS SEEN. *Capt. Bailey* 6 (K); *Bowes-Lyon* 11101 (E); *Kingdon Ward* 11870 (BM); *Ludlow & Sherriff* 2190 (BM), 2531 (BM, E), 2792 (BM); *Ludlow, Sherriff & Taylor* 6343 (BM).

var. lutea

G. Taylor, *J. Roy. Hort. Soc.* 82: 167 (1947). Type: SE Tibet, Tum La, Nayu Chu, 12,500 ft, July 1938, *Ludlow, Sherriff & Taylor* 5790 (BM, holotype; E, isotype).

DESCRIPTION. Similar to the typical plant (var. *argemonantha*), but flowers lemon-or pale yellow with golden yellow anthers and pale yellow filaments.
DISTRIBUTION. Restricted to the Paka Phu Chu and above Tse (Tsangpo valley), principally on the Mira La, Tra La and Tum La; 3810–4575 m (Map 24b).
SPECIMENS SEEN. *Kingdon Ward* 6038 (BM); *Ludlow, Sherriff & Taylor* 4589 (BM), 4589a (BM, E), 4589b (BM, E), 5790 (BM, E), 5898 (BM, E), 6064 (BM, E).

There are some fine half-tone photographs accompanying the Ludlow & Sherriff collections at the British Museum (Natural History) and the Royal Botanic Garden Edinburgh. The field notes accompanying *Ludlow, Sherriff & Taylor* 5790 read as follows:

"Corolla pale lemon yellow, filaments rather paler, anthers golden yellow. Ovary and style green. 4–6-petalled. Sepals green, striated dark red-brown at the base. Basal pair of leaves entire, or nearly so. Commonly 3 scapes to a plant. Not common and growing singly on grass-covered cliff ledges, roots often between stones or in crevices."

Seed was introduced from the Ludlow, Sherriff & Taylor expedition in 1938 but it failed to establish in cultivation and it has been seen very little in the wild since that time.

68. MECONOPSIS FLORINDAE

Meconopsis florindae Kingdon-Ward, *Gard. Chron.* 1926, Ser. 3, 79: 307, fig. 232, in obs. (1926); *Ann. Bot.* 40: 537 (1926); *J. Roy. Hort. Soc.* 52: 23, 233 (1927). Type: SE Tibet, Tra La, 12,000 ft, 1924, *Kingdon Ward* 6038 (BM, E, K syntypes), Pa La (= Tra La), 13–13,000ft, *Kingdon Ward* 6206 (E, K, syntypes). Lectotype *Kingdon Ward* 6038 (BM lectotype; E, K, isolectotypes) chosen here.

DESCRIPTION. *A monocarpic herb* to 45 cm tall, with a napiform taproot to 3.7 cm long. *Stem* slender, rather tenuous, glabrous or with a few sparse hairs. *Basal leaves* few, soon withering, generally absent at flowering time, the lamina oblanceolate to elliptic-oblanceolate, 3.5–8.0

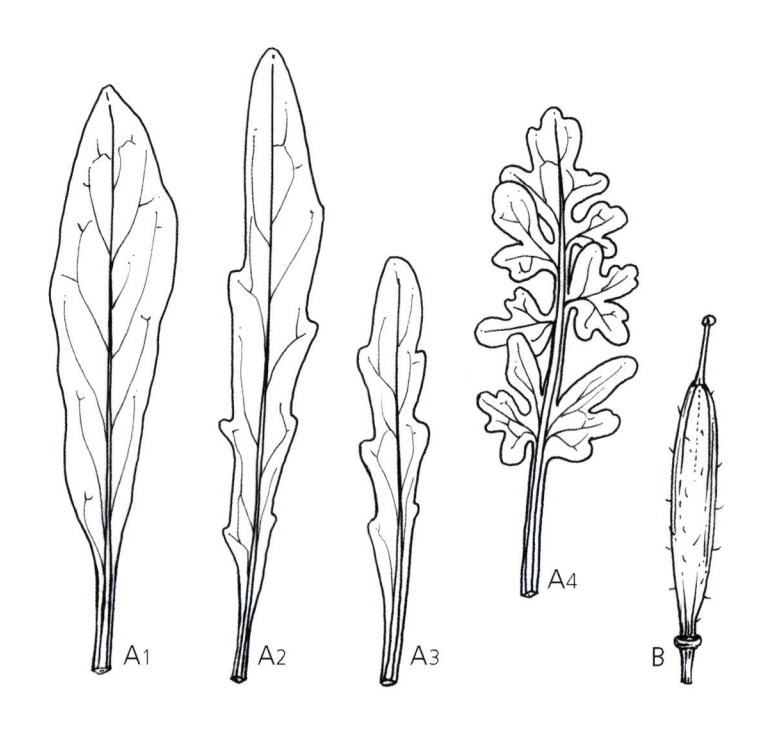

FIG. 29. *Meconopsis florindae*: **A1–4**, lower and middle cauline leaves; **B**, fruit capsule. Pubescence not shown except for the fruit capsule. All × 1¹⁄₃. DRAWN BY CHRISTOPHER GREY-WILSON

× 0.5–1.6 cm, green above, glaucous beneath, the base attenuate into the petiole, the apex acute to obtuse, margin entire to shallowly crenate to pinnately or bipinnately lobed or bipinnatisect, often rather undulate, glabrous, or rarely sparsely pubescent beneath; lower cauline leaves similar to the basal, the middle and upper cauline leaves gradually reduced, the uppermost often bract-like, sessile and with a semi-amplexicaul base, entire or serrate at the margin; petiole of lower leaves 1.3–3.5 cm, slender, ampliate towards the base. *Flowers* up to 6 (commonly 2–4), half-nodding, 32–50 mm across, borne in succession from the axils of the upper leaves, sometimes appearing to be ebracteate because of displacement of the upper leaf-bracts. *Pedicels* very slender, 1.5–9 cm long (to 15 cm long in fruit), pubescent-setose and somewhat expanded at the top, otherwise glabrous. *Buds* pendent, sparsely patent-bristly outside. *Petals* (4–) 5–7, lemon-yellow, ovate to obovate, 16–25 × 8–12 mm, with a subacute to rounded apex, the margin unevenly erose. *Stamens* with linear translucent, whitish filaments; anthers pale orange. *Ovary* ovoid to ovoid-oblong, glabrous to sparsely setose, the style slender, 3–6 mm long, the stigma yellow, subclavate, 3–6-lobed. *Fruit capsule* oblong-ellipsoid, c. 27 × 5 mm, 3–6-valved, glabrous to sparsely setose, splitting for about a third of its length from the apex at maturity. (Fig. 29).

DISTRIBUTION. SE Tibet (Rong Chu: Tra La (Pa La), Qomzo La); 3300–3900 m.

HABITAT. Woodland, boulder scrub.

FLOWERING. June–July.

SPECIMENS SEEN. *Kingdon Ward* 6038 (BM, E, K), 6206 (E, K).

NOTE. KW 6206 (middle specimen K sheet) has bipinnately-lobed leaves and may not be *Meconopsis florindae*.

Meconopsis florindae, named in honour of Frank Kingdon Ward's first wife, has its closest ally in the rather better known, and certainly better collected *M. argemonantha*. Both share in common pale yellow flowers (those of the latter can also be white) with usually 5–7 petals that are distinctly erose at the margin. Both species bear a raceme of up to 6 flowers, well-spaced on the upper part of the stem and each arising from a leaf-like bract. The prime difference is seen in the fruit capsules which are relatively longer and thinner in *M. florindae*, often glabrous, but sometimes with a few patent bristles; moderately to densely patent-bristly in *M. argemonantha*. *M. florindae* is a generally taller and more slender plant; however, once further exploration of the mountains of south-eastern Tibet takes places then these two taxa may well be linked by intermediate types and in such an instance *M. argemonantha* has priority. It appears from field records and other observations that both are very localised and scarce species in the wild.

Seamus O'Brien is one of very few people fortunate enough to have seen *Meconopsis florindae* in the wild since it was first collected more than 80 years ago. He came across it on the Qomzo La just of the Rong Chu, in south-eastern Tibet, a valley that was Lord Cawdor and Frank Kingdon Ward's base in 1924 and very close to the Tra La where Kingdon Ward had first come across it in that same year. Seamus comments that "I spotted it by the way on the 17th July 2001 in full bloom, a single plant at the edge of a thicket, beneath an entire boulder scree of *Rheum nobile* (in full flower). In the thicket, beneath the rheums, it grew with *Berberis temolaica*, *Primula cawdoriana*, *Saxifraga signata*, *Diapensia himalaica*, *Meconopsis speciosa*, *M. baileyi* (in shades of violet and blue) and *Cassiope wardii* ……. You will see from Kingdon Ward's writings that he praises it highly and maintains it was common in the area. It is not, it is extremely rare (I only ever saw one plant) and from a horticultural point it is not all that exciting. I reckon Kingdon Ward was trying to 'sell' it because it had been named for his first wife."

69. MECONOPSIS PRIMULINA

Meconopsis primulina Prain, *J. Asiat. Soc. Bengal, Pt. 2, Nat. Hist.* 64 (2): 319 (1896). Type: India 1879, *Dungboo* s.n. (P, holotype).

DESCRIPTION. *A slight monocarpic herb* 30 cm tall at the most, with a dauciform to napiform swollen taproot, 3–7.5 × 0.5–0.8 cm, beset at the top with fibrous leaf-base remains. *Stem* short, often rather obscure, to 5 cm long, glabrous to somewhat setose. *Leaves* yellow- or blue-green, paler beneath, the basal leaves spathulate, withered by flowering time, but cauline leaves crowded at stem base giving the appearance of basal leaves, oblanceolate to narrow elliptic-oblong, 1.5–6.4 × 0.8–1.4 cm, the margin entire to shallowly scalloped or sinuate, rarely with one or two marginal teeth, subglabrous to moderately setose on both surfaces; petiole generally 0.8–3.7 cm long, but the uppermost leaves sessile. *Flowers* 1–5, often 3, semi-pendulous, appearing scapose on short-stemmed plants, but borne from the uppermost leaf-axils, 3.6–5.3 cm diam. (when flattened). *Pedicels* erect, slender, 8–24.5 cm long (to 36 cm long in fruit), glabrous to reflexed setose. *Sepals* glabrous to sparsely setose. *Petals* (4–)5–8, blue, mauve, mauve-blue or purplish, narrow-ovate to obovate, 20–25 × 13–20 mm. *Stamens* numerous (c. 16 or more), the

69. MECONOPSIS PRIMULINA

filaments the same colour as the petals, anthers yellow or golden. *Ovary* green, narrow-obovoid, glabrous or with a few setae, the style 3–5 mm long, purple or purple-green, bearing a ± clavate, purple stigma, the lobes more or less decurrent. *Fruit capsule* narrow-obovoid to elliptic-cylindric, 22–39 × 3.5–6 mm, 4-valved, glabrous to slightly setose, the persistent style 5–8 mm long, broadened towards the base.

DISTRIBUTION. W & NW Bhutan (Lingshi, Lingshi La, Shinje La, Phile La, Yale La, Kangla Karchu La, Tremo La), SE Tibet (Chumbi; Phari); 3190–4600 m (Map 24).
HABITAT. Alpine slopes amongst low shrubs, especially dwarf rhododendrons, mossy or grassy banks, screes and moraines, rock crevices, cliff ledges.
FLOWERING. May–July.
SPECIMENS SEEN. *R. E. Cooper* 1606 (BM, E, K), 1713 (BM, E, K); *Dungboo* s.n. Chumbi, July 1879 (K); *Gould* 985 (K); *Ludlow, Sherriff & Hicks* 16310 (BM), 16405 (BM), 16569 (BM, E), 17338 (BM, E), 17370 (BM, E), 17426 (BM); *C. Sargent* 170 (E).
NOTE. L, S & H 16569 (BM sheet), collected on the Kachu La, Kangla, in Bhutan, is an excellent collection with nine specimens affixed. These show a range of variability: while some specimens are clearly stemmed, others have a very reduced stem, while both the upper leaf surface and pedicels can be glabrous or moderately hairy. In addition, while most specimens are single-flowered, several sport 2–3 flowers.

Meconopsis primulina is a pretty little species which, in the finest forms, bear flowers of a good rich blue. More often, however, the flowers are paler, purplish or bluish mauve. The affinity of *M. primulina* has been questioned in the past, primarily because plants rarely produce a significant stem and in photographs at least, this is generally hidden in the leafy and mossy litter in which the plant grows; however, close examination of all the available herbarium material shows that most specimens bear a stem, albeit a short one and that the flowers arise from the uppermost leaf-axils. While some would argue no doubt that the species probably finds its closest ally in *M. impedita* and *M. concinna*, it seems to me that the general features of leaves, flowers and fruits are more akin to *M. florindae* and *M. argemonantha*, indeed the latter species can also lack an obvious stem in some specimens. George Taylor (1934) assigns *M. florindae*, *M. lyrata* and *M. primulina* to series *Primulinae* (first distinguished as section *Primulinae* by David Prain's in 1896). Later Taylor added *M. argemonantha* to series *Primulinae*.

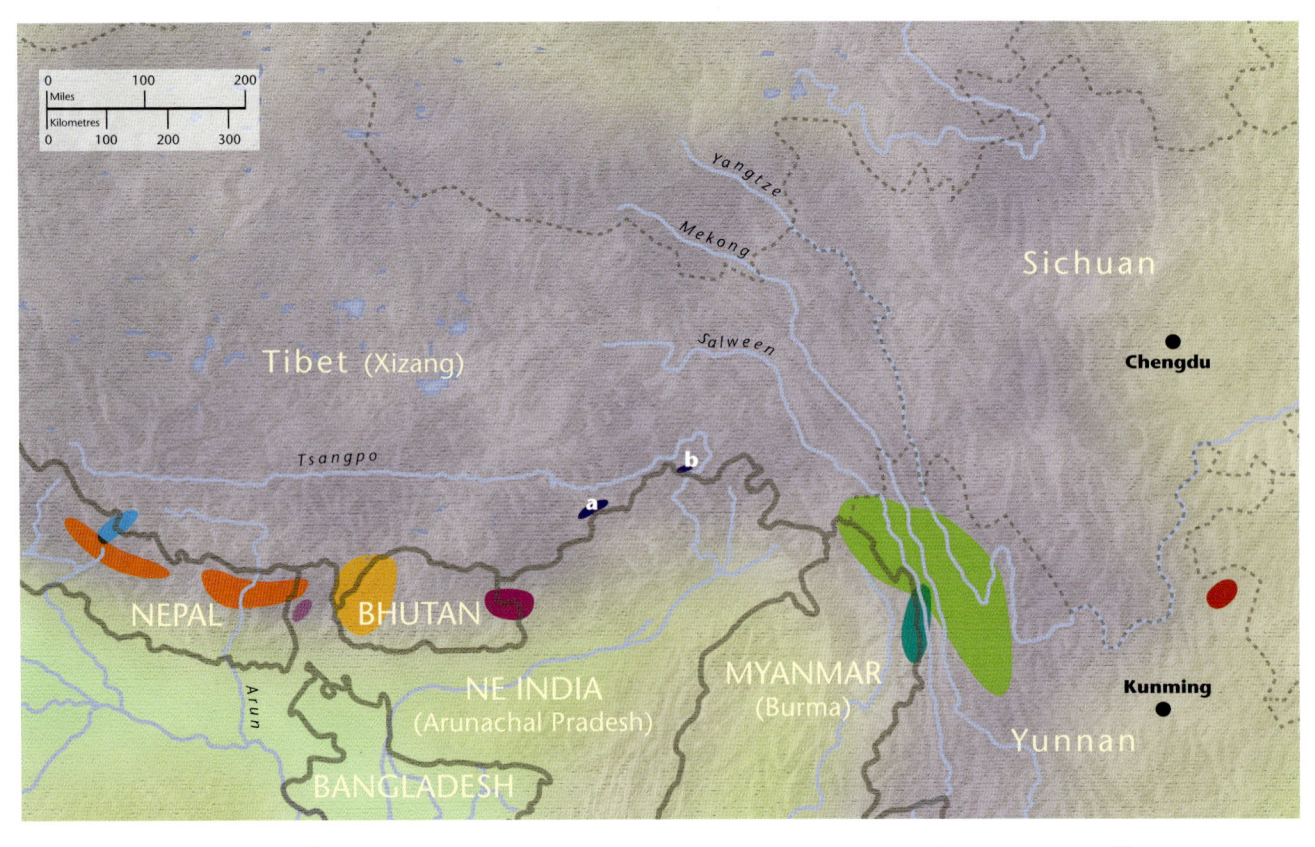

MAP 24. Species distribution: ■ *Meconopsis primulina*, ■ *M. argemonantha* a var. *argemonantha*, b *M. a.* var. *lutea*, ■ *M. lyrata*, ■ *M. bulbilifera*, ■ *M. exilis*, ■ *M. lamjungensis*, ■ *M. compta*, ■ *M. ludlowii*, ■ *M. wumungensis*.

THE GENUS MECONOPSIS
69. MECONOPSIS PRIMULINA

Above left: *Meconopsis primulina*, leaf-rosette; Shodu to Barshong, W Bhutan, 3900 m. PHOTO: DAVID & MARGARET THORNE
Above right: *Meconopsis primulina*, leaf-rosette; upper Mo Chu, above Linshi, NW Bhutan, 4250 m. PHOTO: DAVID & MARGARET THORNE

Modern research, especially that of Kadereit *et al.* (2011), has also shown the close allegiance of *M. primulina* to *M. lyrata*, *M. sinuata* and *M. wumungensis*, determined by molecular analysis, a phylogenetic reconstruction using 65 internal transcribed spacer sequences of some 62 taxa of Old World Papaveroideae and three outgroup taxa. The reader is referred to Kadereit *et al.* (2011) and Carolan *et al.* (2006) for the materials, methods and results of these analyses which do not need to be repeated here.

Meconopsis primulina was described from material gathered by a native collector in the latter part of the nineteenth century and was said to have come from India. However, Dungboo, the collector, was known to have gathered specimens in western Bhutan and Chumbi from where *M. primulina* has been recorded. There were no further records of the species or its distribution until R. E. Cooper came across and collected it in Bhutan in 1914. While there is relatively little herbarium material available for examination today, the species has been well photographed in recent years, particularly along the Chomolhari trail in western Bhutan and much has been gleaned from these photos. *M. primulina* is confined to the region of western and north-western Bhutan and the

Meconopsis primulina; Yaksa, Thimphu district, W Bhutan, 4200 m. PHOTO: TIM LEVER

adjacent region of Chumbi in southern Tibet where it appears to be locally common but never prolific.

George Taylor (1934) drew close comparisons between *Meconopsis primulina* and *M. lyrata* sensu lato (see *M. lyrata* aggregate below), commenting that "when more material becomes available for examination, it may prove to be a geographical form. At present it appears to differ from that species in having a dense tuft of persistent leaf bases, in the shape of the leaves, and in having the flowers borne on long, usually almost basal, pedicels". While these characters can be still upheld in the light of more modern collections and photographic evidence, other differences can be observed, not least the shape and colour of the flowers and in the greater number of petals in *M. primulina*. There is no doubt that *M. primulina* is a distinct species which is probably endemic to Bhutan, for the specimens recorded for Chumbi, including the type collection, are in all probability within today's Bhutan territory.

Meconopsis primulina is recorded in the *Gardener's Chronicle* in 1924 as having flowered in Britain. George

Meconopsis primulina, immature fruit capsule; north of Nyile La, NW Bhutan, 4400 m. PHOTO: DAVID & MARGARET THORNE

Taylor points out that the plant in question is almost certainly *M. lancifolia* var. *concinna* (now *M. concinna*), having been raised from seed collected in the Litang region of south-western Sichuan. As far as I am aware the delightful little *M. primulina* has never been successfully reared in cultivation. If growable, it would certainly make a welcome addition to the woodland garden.

Above left: *Meconopsis primulina*; Jagothang, Thimphu, W Bhutan, 4000 m. PHOTO: TIM LEVER
Above right: *Meconopsis primulina*, semi-mature fruit-capsule; Nyile La Shong, W Bhutan, 4650 m. PHOTO: DAVID & MARGARET THORNE

series Cumminsia

As currently understood this series contains 9 species with a distribution from central Nepal eastwards through the Himalaya, into adjacent Tibet, east as far as north-eastern Yunnan. The species are characterised by a short napiform taproot, a short to long, slender, laxly leafy stem bearing 1–3, rarely more, small campanulate, nodding flowers. The flowers generally have 4, occasionally 5–6 petals in the terminal one, that are generally rather pale, bluish lilac to pale rose, occasionally more richly coloured. The fruit capsule is narrow-oblong to subcylindrical, often glabrous, occasionally sparsely pubescent.

The members of series *Cumminsia* are slender and rather inconspicuous plants, especially when they are not in flower. On the whole they are the smallest and slightest members of the genus *Meconopsis*. For this reason they are often overlooked in the field and certainly under-collected, especially in the fruiting stage. Horticulturally they hold out little appeal except to *Meconopsis* aficionados; however, to the botanist they are an intriguing and rather perplexing cluster of species. Although the taxonomy of the series has recently been revised and updated by the present author along with Toshio Yoshida & Huang Sun (Yoshida *et al.* 2012) more close field work needs to be done to ascertain both the variability and true distribution of the various taxa. In addition, the allegiance of this series to series *Primulinae* and the other series in subgenus *Cumminsia* is still speculative and can probably only be properly resolved by detailed and well-sampled DNA analysis.

With a wide distribution from central Nepal to Sikkim, Bhutan, southern and south-eastern Tibet (Xizang), eastern Myanmar and north-western Yunnan, this small complex of taxa presents considerable problems of interpretation. In his monograph of the genus published in 1934, George Taylor highlights some of the variation found in the aggregate and the difficulty of interpretation; however, he concludes by treating all as a single species without subdivision. As there is considerable variation involved, some clearly geographically based, it is difficult to accept this conclusion, particularly in view of the material and photographs that have come to light in recent years. In analysing this, two courses have presented themselves, either to recognise a single species with a series of subspecies, or to recognise a series of closely related species. The latter course was decided upon, but no doubt there will be those who would draw a different conclusion.

The aggregate, which extends in distribution from central Nepal to SW China, clearly shares a common ancestor but geographical fragmentation and isolation and ongoing diversification has produced the situation as we see it today. This would be clearer if occasional intermediate types did not appear in the wild, especially in the Yunnan-Myanmar area. However, the vast majority of specimens examined (both in the field and in herbaria) clearly fit within the species concept as it is outlined here. It has been suggested that these intermediates types may represent plants of hybrid origin but there is absolutely no evidence to back this hypothesis. Perhaps more likely is the occasional reappearance of the ancestral form. While undoubtedly chromosome details and DNA studies are beginning to prove an important adjunct to studying the genus, much of this work is still in its infancy, allowing the taxonomist to base interpretation primarily on morphological studies and other data. While, at the same time, many

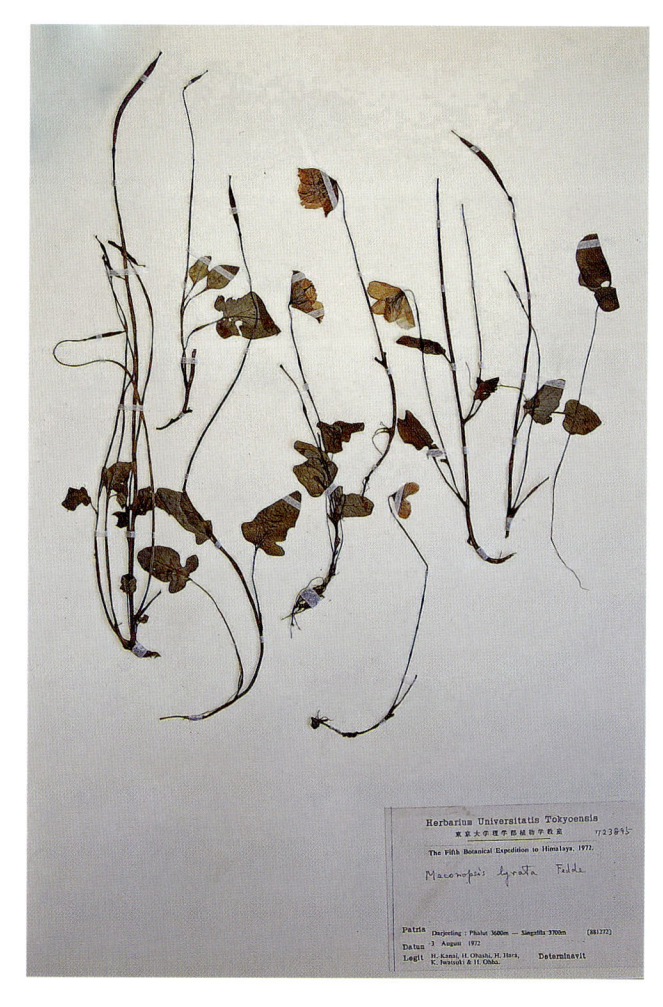

Meconopsis lyrata herbarium sheet, *H. Kanai et al.* 723895 (TI), collected at the locus classicus

THE GENUS MECONOPSIS
SERIES CUMMINSIA

recent studies have concentrated on the larger and more spectacular species (the bold evergreen monocarpic species in series *Polychaetia* and *Robustae*, or the famous 'blue poppies' of series *Grandes*), many of the smaller taxa have been virtually ignored. Among these latter, few have been neglected more than the demure species of section *Cumminsia*, yet they form a significant and interesting association; undoubtedly field observations by keen-eyed naturalists in recent years have provided a great deal more useful information. Although they would be of interest to some horticulturists and gardeners, primarily in the alpine garden, none of the species in the series are presently in cultivation.

As has been stressed, the members of the *Meconopsis lyrata* aggregate, that includes several recently described species, are particularly difficult to observe in the wild. They generally grow amongst rather coarse, lush herbage and are easily overlooked in flower and very difficult to locate at the fruiting stage; they are not nearly so conspicuous in the field as are other species of *Meconopsis* which are generally prominent at both the flowering and fruiting phases. Certainly more field observations and collections of these various taxa would be a valuable and an important addition to the ongoing studies of the genus as a whole.

The floral characters of the *Meconopsis lyrata* aggregate are remarkably similar: all have rather pale, nodding, basically 4-petalled (rarely 5–6) flowers, bearing relatively few stamens (generally 12–24 (–36)) compared to other species in the genus. In addition, the fruit characters are rather similar, being oblong- to linear-cylindrical and generally 3–4-valved, the top of the capsule splitting for a quarter to about one-third of its length at maturity. These factors have undoubtedly swayed interpretation in the past and yet the plants can look very different in other respects, particularly in the stem, inflorescence, and leaf characteristics as the following key reveals.

Meconopsis bulbilifera herbarium sheet, *F. Miyamoto et al.* 9400079 (TI), holotype.

KEY TO SPECIES OF SERIES CUMMINSIA

1. Leaves all petiolate, shallowly to deeply trilobed with a much larger terminal lobe, or trifoliate; stem decumbent towards the base, with adventitious buds or fibrous-roots often borne at the lower nodes. . 2
+ Leaves all petiolate or the upper sessile or subsessile, entire to pinnately divided, not trilobed; stem erect to ascending, rarely decumbent at the base, without adventitious buds or fibrous-roots at the lower nodes 3

2. Delicate plants with thread-like, somewhat zigzag stems, narrowed and extended at the base, often bearing tiny bulbils at the nodes; lower-surface of leaves finely stellately patterned; flowers solitary, generally smaller, the petals 12–17 mm long . . . . .71. **M. bulbilifera** (Nepal, Sikkim)
+ More robust plants with shorter and firmer stems more or less uniform throughout, without bulbils at the nodes; lower surface of leaves not stellately patterned; flowers 1–5, generally larger, the petals 13–22 mm long . . . . .70. **M. lyrata** (Darjeeling District of India)

3. Ovary and fruit capsule bristly; flowers 3–8 borne in an evenly spaced bracteate raceme (bracts often displaced so that the flowers may appear to be ebracteate) . . . . 78. **M. sinuata** (C & E Nepal, NE India (Sikkim), Bhutan, SE Tibet)
+ Ovary and fruit capsule glabrous or subglabrous; flowers solitary or several clustered at the top of the stem, bracteate 4

4. Plants 6–25 cm tall in flower; upper leaves petiolate or subsessile; lamina pinnately lobed, coarsely toothed or crenate, rarely entire 5
+ Plants 20–50 cm tall in flower; upper leaves sessile; lamina entire, sinuate or crenate, not lobed 7

5. Flowers 2–8, the petals 13–22 mm wide; stamens c. 36; style 2–4 mm long . . . .
. . . . 77. **M. ludlowii** (E Bhutan, NE India, Arunachal Pradesh)
+ Flowers 1–3, the petals 7–12 mm wide; stamens 10–24; style 1–3 mm long 6

6. Inflorescence often scapose with elongate pedicels and a reduced stem; leaf-lobes often rounded with contracted bases (especially in scapose plants); petals with an obtuse to acute apex . . . .
. . . . 75. **M. wumungensis** (SW China; N Yunnan)
+ Inflorescence not scapose, usually with a single terminal flower borne on a leafy stem; leaf-lobes broadly ovate to oblong; petals with a rounded to obtuse apex . . . .
. . . . 76. **M. compta** (SE Tibet, N Myanmar, SW China: NW Yunnan)

7. Flowers solitary, terminal; stem leaves (including bracts) 3–4, not subtending shoots; lamina of upper leaves with semi-amplexicaul base; petals narrow-obovate to narrow-elliptic, the apex acute to acuminate. . . .
. . . . 74. **M. polygonoides** (W Bhutan, S Tibet)
+ Flowers both terminal and axillary; stem leaves (including bracts) 3–8, often subtending a lateral shoot with a flower and 0–2 leaves (bracteoles); lamina of upper leaves with or without a semi-amplexicaul base; petals obovate or broad-obovate, the apex rounded to subacute 8

8. Uppermost 2–3 leaves subopposite or in a pseudowhorl normally; lamina of lower leaves with a rounded to attenuate base, those of the upper leaves with a semi-amplexicaul base . . . .
. . . . 72. **M. exilis** (SW China: NW Yunnan, NE Myanmar)
+ All leaves alternate; lamina of lower leaves with shallowly cordate to truncated base, those of the upper leaves without a semi-amplexicaul base . . . . 73. **M. lamjungensis** (C Nepal: Lamjung Himal)

70. MECONOPSIS LYRATA

Meconopsis lyrata (H. A. Cummins & Prain) Fedde ex Prain in Engl., *Pflanzenreich, Papav. Hypec. & Papav.*: 246 (1909). Types: *Dr King's Collector* s.n., Phulloot (Phalut), 12,000 ft, Aug 1887 (isosyntypes BM, K), s.n. 13 July 1884 (isosyntype BM).
[syn. *Cathcartia lyrata* H. A. Cummins & Prain ex Prain, *J. Asiat. Soc. Bengal, Pt. 2, Nat. Hist.* 64 (2): 325 (1896); *M. lyrata* sensu Taylor, *The Genus Meconopsis*: 73 (1934), pro parte].

DESCRIPTION. *Monocarpic herb*, 8–15 cm tall in flower, to 25 cm tall in fruit, with a napiform taproot often bearing adventitious fibrous-roots at the lower nodes; rootstock ovoid to more or less cylindrical, 6–20 mm long, 3–6 mm diameter, bearing adventitious fibrous-roots on the surface and partly covered with the remnants of old petioles, the entire plant scattered with 2 mm long yellow-brown, weak but somewhat crisp hairs. *Stem* rather firm, 1–2 mm diameter, obscurely 4-ridged and striate, erect or ascending except the decumbent base, subglabrous to pubescent, occasionally with adventitious buds and adventitious fibrous-roots born at the lower nodes, occasionally branched near the base above the ground. *Leaves* yellowish green, subglabrous or sparsely pubescent, generally with a few scattered hairs beneath, the basal leaves 3–5, often withered at flowering time, the petioles elongate to 13 cm long; cauline leaves 2–4 (often 3), the uppermost leaves occasionally subopposite, the lamina thin, ovate to oblong-ovate, or lyre-shaped, 1.5–5 × 0.7–3 cm, broad-cuneate, subcordate or cordate at the base, rounded to acute at the apex, entire to deeply 3-lobed or trifoliate with much larger terminal lobes (sometimes transitional on the same plant), the terminal lobes ovate with entire or coarsely and shallowly toothed margins, the petiole narrow-linear, 1–6 cm long, often narrowly winged towards the base. *Pedicels* 4–11 cm long in flower, to 17 cm long in fruit, subglabrous to sparsely pubescent, the hairs patent to subappressed. *Flowers* 1–5, semi-pendulous, the lower lateral flowers accompanied by 1–2 bracteoles. *Sepals* subglabrous. *Petals* 4, rarely 5, pale blue, lavender-blue, pale purple, mauve, obovate to elliptic, 13–22 × 7–15 mm, rounded or subacute at the apex, the upper margin minutely toothed. *Stamens* 16–24, about half the length of the petals, erect and surrounding the pistil, the filaments linear, 4–7 mm long, with gradually narrowed apex, similar to or darker than the petals in colour, the anthers c. 2 mm long, orange-yellow. *Ovary* narrow-oblong to cylindrically-

70. MECONOPSIS LYRATA

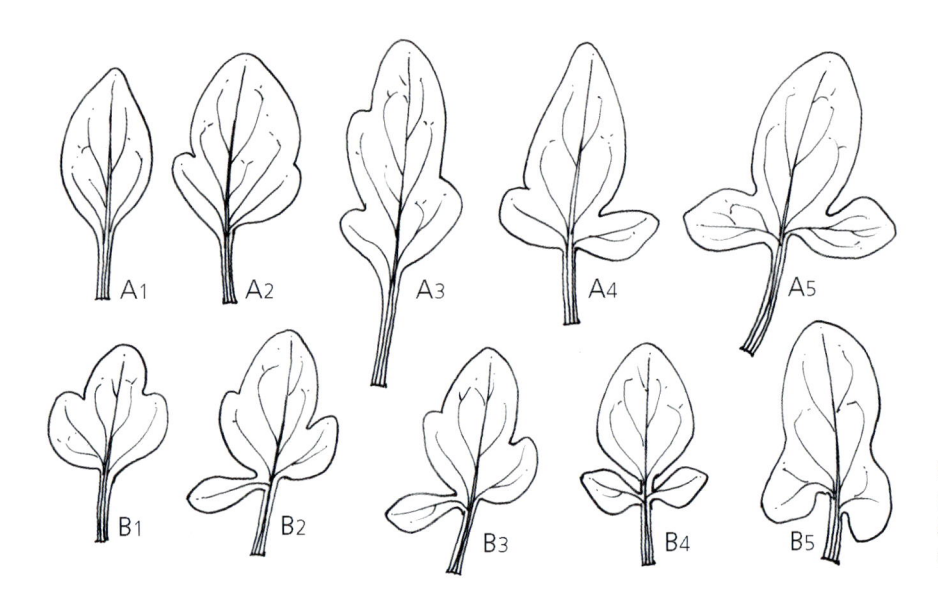

FIG. 30. Leaf-laminas, lower petioles and pubescence not shown: **A1–5**, *Meconopsis lyrata*; **B1–5**, *M. bulbilifera*. All × 1¹/₃.
DRAWN BY CHRISTOPHER GREY-WILSON

ellipsoid, 5–7 mm long, 1–2 mm diameter, glabrous. *Style* 2–3 mm long in flower, to 3 mm long in fruit. *Stigma* capitate, 0.7–1 mm diameter. *Fruit capsule* narrow oblong-cylindrical, 30–58 mm long, 3–5 mm diameter, 3–4-valvate, glabrous, splitting for a third of its length from the apex when mature. (Fig. 30A1–5).

DISTRIBUTION. Darjeeling District of India; only known from the limited area around the peak of Mt Phalut; c. 3596 m (Map 24).

HABITAT. Grassy and stony slopes, low open shrubberies, cliffs, banks.

FLOWERING. July–August.

SPECIMENS SEEN. *Cave* s.n. 24 July 1919(E), s.n. 25 July 1919 (E), s.n. 1 Sept. 1919 (E); *Kanai et al.* 723894 (TI), 723895 (KATH, TI).

Meconopsis lyrata is a little known species recorded only from the very limited area of Darjeeling district in India and little seen in recent years. It is without question very closely realted to *M. bulbilifera*: see under that species for further details and comparisons.

71. MECONOPSIS BULBILIFERA

Meconopsis bulbilifera Tosh. Yoshida, H. Sun & Grey-Wilson, *Curtis's Bot. Mag.* 29 (2): 7 (2012). Type: Central Nepal, Bagmati Zone, Rasuwa District, Ganesh Himal, Seto Kund to Chyauche Kharka, 3910 m, 11 August 1994, *F. Miyamoto et al.* 9400079 (TI, holotype; KATH, isotype).

[syn. *M. lyrata* sensu Taylor, *The Genus Meconopsis*: 73 (1934), pro parte]

DESCRIPTION. *Slender herb* to 35 cm tall in flower but as little as 6 cm, with a swollen taproot, the rootstock ovoid to more or less cylindrical, 6–20 mm long, 3–5 mm diameter, bearing adventitious fibrous-roots and covered with remnants of the old petioles, occasionally with a tail-like extension, being a slender stolon with short internodes, the entire plant scattered with yellow-brown hairs, weak but somewhat crisp, to 1.3 mm long. *Stem* slender and delicate, 0.3–1 mm diameter, rounded in section, erect or ascending (with the support of surrounding herbs), decumbent towards the base, often zigzagged above, subglabrous to pubescent, narrowed and extended underground to the root, usually bearing bulbils in the axils of the cauline leaves or at the bare nodes near the base; the bulbils ovoid-conical, 1.5–2.5 mm long, often with a remnant petiole at the apex, occasionally developing and enlarged at a node near the stem base, bearing fibrous adventitious roots. *Leaves* 3–7, alternate, yellowish green, subglabrous or sparsely pubescent, the basal 3–7, often withered at flowering time; cauline leaves 2–6, alternate, the lamina thin, ovate to oblong-ovate, triangular-ovate or lyre-shaped, rarely rounded, 1–2.5 × 0.5–2 cm, broad-cuneate, subcordate or cordate at the base, rounded to acute at the apex, the margin entire to deeply 3-lobed or trifoliate with an enlarged terminal lobe and often with a subopposite pair of oval-ovate lobes at the base (sometimes transitional on the same plant), the terminal lobes ovate to ovate-lanceolate with entire, crenate or with 1–2 marginal lobes, the lower-surface of lamina whitish in dried specimens, glabrous to rather sparsely hairy, drying to a minutely stellate-pattern; petiole narrow-linear or filiform, 0.5–6 cm long, often with an expanded base part-covering the bulbils, upper cauline leaves smaller and with shorter petioles. *Pedicels* 3–10 cm long, very slender (less than 0.8 mm diameter), subglabrous to sparsely pubescent, the hairs patent to subappressed. *Flowers* semi-pendulous, solitary, terminal. *Sepals* subglabrous. *Petals* 4, very

Table 6
COMPARISON OF CHARACTERISTICS SEPARATING *MECONOPSIS LYRATA* AND *M. BULBILIFERA*

	Meconopsis lyrata	*Meconopsis bulbilifera*
Plant	8–15 cm tall in flower	6–35 cm tall in flower
Stem	rather firm, 1–2 mm diam., obscurely 4-ridged and striate, occasionally branched near the base, more or less uniform throughout	slender and delicate, 0.3–1 mm diam., rounded in section, usually unbranched near the base, narrowed and extended underground to the root
Stolon	absent	occasionally present
Bulbils	absent	mostly present
Cauline leaves	2–4 (often 3), uppermost leaves occasionally subopposite	2–6, alternate
Terminal lobes of cauline leaves	ovate with entire or coarsely and shallowly toothed margins	ovate to ovate-lanceolate with entire, crenate or 1–2-lobed margins
Lower-surface of leaves	somewhat paler, scattered with yellow-brown hairs to 2 mm long, not stellately patterned	whitish in dried specimens, glabrous or with scattered yellow-brown hairs to 1.3 mm long, minutely stellately-patterned
Petioles of cauline leaves	narrow-linear, often narrowly winged towards the base	narrow-linear or filiform, often expanded at the base and part-covering the bulbils
Flowers	1–5 per individual	solitary
Petals	13–22 mm long, 7–15 mm wide, rounded to subacute at the apex	12–17 mm long, 6–11 mm wide, rounded to acute at the apex
Filaments	4–7 mm long, with gradually narrowed apex	3.5–4 mm long, more or less uniform throughout
Anthers	c. 2 mm long	1.5–2 mm long
Ovary	narrow-oblong to cylindrically-ellipsoidal, 5–7 mm long, 1–2 mm diam.	narrow-ovoid to cylindrically-ellipsoidal, 4–6 mm long, 1–1.5 mm diam.

rarely 5, pale blue, pale purple, mauve, rarely white, obovate, elliptic, or ovate-elliptic, 12–17 × 6–11 mm, rounded, obtuse or acute at the apex, the upper margin minutely erose. *Stamens* 16–24, half or slightly less than the length of the petals, erect and surrounding the pistil, the filaments filiform, 3.5–4 mm long, more or less uniform throughout, similar to or darker than the petals in colour, the anthers 1.5–2 mm long, orange-yellow. *Ovary* narrow-ovoid to cylindrically-ellipsoid, 4–6 mm long, 1–1.5 mm diameter, glabrous, the style c. 2 mm long, terminating in a capitate, c. 0.7 mm diameter, stigma. *Immature fruit capsule* narrow oblong-cylindrical, 3–4-valvate, glabrous (mature fruits unknown). (Fig. 30B1–5).

DISTRIBUTION. C & E Nepal (Dhading, Kaski, Lamjung, Nuwakot, Rasuwa, Sankhuwasabha, Solukhumbu & Taplejung districts; Annapurna Himal, Ganesh Himal, Khumbu Himal, Kanchenjunga Himal), NE India (Sikkim); 3200–4600 m (Map 24).

HABITAT. Grassy slopes, amongst densely growing coarse herbs on wet, stony slopes, banks, or in low open shrubberies.

FLOWERING. July–early September.

SPECIMENS EXAMINED. *L. W. Beer* 10027 (BM), *H. A. Cummins* s.n.[39] Sikkim 1893 (K), s.n. Tuko La, Sikkim (K); *Long et al.* 504 (E); *Lowndes* 734 (E); *F. Miyamoto et al.* 9400077 (BM, E, KATH, TI), 9400079 (KATH, TI),

---

[39] *H. A. Cummins* s.n. collected in 1893 in Sikkim was designated an isosyntype of *Meconopsis lyrata* by David Long in the *Flora of Bhutan* but is here transferred to *M. bulbilifera*.

9484087 (E), 9584087 (E, KATH, TI); *Ribu & Rohmoo* 5091 (K); *Rohmoo Lepcha* 1120 (E); *K. N. Sharma* 40/94 (BM), E402 (BM); *Stainton* 902 (BM, E); *Stainton, Sykes & Williams* 1954 (BM), 6599 (BM, E, KATH), 8583 (BM); *Wakabayashi et al.* 9715077 (KATH, TI), 9720304 (TI), 9730137 (TI).

Meconopsis bulbilifera is similar to *M. lyrata*, but differs in its fine thread-like stems from whose nodes tiny bulbils are borne, especially the lower nodes. Plants are not self-supporting but scramble through the surrounding herbage, often rooting at the lowermost nodes. The undersurface of the leaf-lamina is very pale, often whitish in *M. bulbilifera* and, in the dried state at least, is covered in a characteristic minute stellate-pattern. The prime differences between *M. lyrata* and *M. bulbilifera* are presented in Table 6.

This species is widespread in central and eastern Nepal and neighbouring Sikkim and although there are relatively few herbarium collections it is probably far more common than this might suggest for it is easily overlooked in the field. The production of bulbils, which presumably fall to the ground in the autumn when the thin stems wither away, must help in spreading this delicate species which is typical of wet grassy habitats and other coarse herbage.

72. MECONOPSIS EXILIS

Meconopsis exilis Tosh. Yoshida, H. Sun & Grey-Wilson, *Curtis's Bot. Mag.* 29 (2): 11 (2012). Type: China, Yunnan Province: on a western slope of Biluoxueshan in Bijiang region, Fugong Xian, 26°35'05"N, 99°00'29"E, 3700–3800 m elevation, 9 July 2008, *Yoshida* K2 (KUN, holotype; KUN, TI. isotypes).
[syn. *M. lyrata* sensu Taylor, *The Genus Meconopsis*: 73 (1934), pro parte]

DESCRIPTION. *Plant* monocarpic, 12–45 cm tall in flower, with a napiform taproot, 6–15 mm long, 4–6 mm diameter, with a caudal extension and fibrous roots, hairy or subglabrous, the hairs weak but somewhat crisp, to 1.5 mm long, easily rubbing off. *Stem* erect, 2–3 mm diameter near the base, obscurely 4-angled, scattered with retrorse hairs or glabrous, bearing a few vestiges of old petioles at the base. *Leaves* yellowish green above, somewhat paler beneath; basal leaves 1–2, often withering before the flowers appear, the petiole 2–5 cm long, partly buried under ground, the lamina membranous, rounded to ovate, 7–15 mm long, the base rounded to attenuate, the apex obtuse, margin entire or shallowly sinuate; cauline leaves 4–7, more or less regularly alternate except uppermost ones, the lower

Meconopsis exilis in habitat; Bijiang, W slope of Biluoxueshan, NW Yunnan, 3750 m. PHOTO: TOSHIO YOSHIDA

Above left: *Meconopsis exilis*; Bijiang, W slope of Biluoxueshan, NW Yunnan, 3750 m. PHOTO: TOSHIO YOSHIDA

Above right: *Meconopsis exilis*; Bijiang, Fugong Xian, W slope Biluoxueshan, NW Yunnan, 3700–3800 m. PHOTO: TOSHIO YOSHIDA

rather smaller than the upper, petiolate, the petiole 2–6 cm long, ascending, often subtending a shoot with a flower bud and 1–2 leaves (bracteoles), the lamina ovate to ovate-elliptic or oblong-spathulate, the base rounded to attenuate, the apex obtuse or acute, the margin entire or shallowly 1–6-sinuate; upper cauline leaves (bracts) sessile and subtending a flower bud, the uppermost 2–3 often aggregated into a false whorl, the lamina ovate-lanceolate, 2–6 cm long, 1–2.8 cm wide, the base semi-auriculate (half-clasping), the apex obtuse or acute. *Pedicels* 2–11.4 cm long (extending to 16 cm long in fruit) in the terminal flower, erect but abruptly curved 1–2 cm from the top. *Inflorescence* racemose with a terminal flower; flowers 1–5 per individual including immature flower buds, nodding, lantern to saucer-shaped, 2.5–3.4 cm across. *Sepals* 7–10 mm long, sparsely hairy. *Petals* 4 (–5), pale blue-purple, usually obovate to elliptic-obovate, 15–22 × 8–16 mm, the apex rounded to subobtuse, the upper margin minutely and irregularly erose. *Stamens* 12–24, with filiform filaments, 7–9 mm long, similar to petals in colour; anthers 1.75–2 mm long, curved inward, orange-yellow. *Ovary* ellipsoid-oblong, glabrous with a 2–4 mm long style that terminates in a c. 1 mm capitate stigma. *Fruit capsule* narrowly oblong or cylindrical, 3–5 cm long, 3–4 mm diameter, glabrous, 3–4-valved.

DISTRIBUTION. China, Yunnan (Fugong Xian, on the western slopes of Biluo Xueshan); Myanmar (western slopes of Gaoligongshan, around the Chimili Pass); 3600–3900 m (Map 24).

HABITAT. Seasonally wet slopes with a dense assemblage of herbs, open shrubberies: in the Bijiang type locality the plants are found scattered on south-facing steep, wet slopes densely covered with tall herbs such as *Anemone*,

Angelica, Epilobium, Geranium, Ligularia, Nomocharis, Phlomis, Potentilla and *Salvia*, the slopes often shrouded in dense fog in summer and deep snow in winter that can lie until early May.

FLOWERING. June–July.

SPECIMENS EXAMINED. *Farrer* 1123 (E); *Forrest* 18276 (BM, E), 18279 (E), 25047 (BM, E, K), 25530 (BM), 25991 (BM, K), 26939 (BM, E, K), 26940 (K), 26942 (BM, E), 27289 (BM, E, K), 29921(BM, E), 30098 (BM), 30350 (BM, E); *Kunming Edinburgh Exped. Sichuan* 271 (E); *Yoshida* K2 (E, KUN, TI).

Meconopsis exilis is similar to *M. compta*, *M. polygonoides* and *M. lyrata*, but differs from *M. compta* in the entire or crenate, not lobed, leaf-lamina and in the upper leaves being sessile with half-clasping bases; it differs from *M. polygonoides* primarily in the inflorescence with the presence of axillary flowers in addition to the terminal one, and also in the obovate non-acuminate petals; it differs from *M. lyrata* in the sessile, half-clasping upper cauline leaves and in the thicker, erect, non-decumbent, stem base.

NOTE. Although this plant has been known for almost 100 years its relationship to *Meconopsis lyrata* has not been recognised until recently. In his 1934 monograph of *Meconopsis* George Taylor makes no specific mention of specimens from north-western Yunnan nor the neighbouring part of Myanmar (then Burma). In fact Taylor took a very broad view of the species, sinking all the variants known at the time into *M. lyrata* without subdivision and with only scant reference to the geographical separation of the various entities. Since the publication of his monograph, more material has been collected and, more importantly, field observations along with photographs have greatly added to our knowledge. As a result it is now possible to present the *M. lyrata* aggregate as a series of species. They are, although very closely related, nonetheless, distinct in various botanical details as well as in their geography and habitat preferences.

In summation the key features of *M. exilis* are as follows:

The stem is relatively rigid and erect straight from the base and bears 4–7 cauline leaves, that are alternate except for the uppermost 2–3 that are aggregated into a false whorl. While the lower leaves are petiolate, the upper are sessile with a semi-amplexicaul base. 1–5 flowers (terminal and axillary) are produced, the lower flowers, when present, borne on short lateral shoots with 1–2 accompanying bracteoles. *M. exilis* is found in just a handful of scattered localities in the extreme north-west of Yunnan reaching across the border into Myanmar.

73. MECONOPSIS LAMJUNGENSIS

Meconopsis lamjungensis Tosh. Yoshida, H. Sun & Grey-Wilson, *Curtis's Bot. Mag.* 29 (2): 14 (2012). Type: central Nepal, Lamjung Himal, near Rambrong, 4050 m elevation, 10 July 1954, *Stainton, Sykes & Williams* 6254 (BM, holotype; E, KATH, TI, isotypes).

DESCRIPTION. *Plant monocarpic*, 18–40 cm tall in flower, to 50 cm tall in fruit, with a napiform to subcylindrical taproot, 10–25 mm long, 4–7 mm diameter, the lowermost leaves generally withered by flowering time, the entire plant sparsely white-hairy or subglabrous, the hairs weak but somewhat crisped, to 1.5 mm long. *Stem* stout and erect, 2–3 mm diameter near the base, bearing a few vestiges of old petioles bases. *Leaves* few, the basal with long petioles, to 22 cm long, the lamina ovate-elliptic to ovate-oblong, 4–10 × 1.5–4 cm, with a shallowly cordate or truncate base and an obtuse apex, the margin coarsely crenate or subentire; cauline leaves 2–4, often 3, the lower and middle ovate, ovate-oblong, spathulate, or lyrate, 2.5–6 cm long, 1–2.5 cm wide, often partly parallel-sided, with a shallowly cordate or truncated base and a subobtuse apex, the margin somewhat sinuate to more or less entire, the petioles 0.5–5 cm long, linear with narrow wings; uppermost cauline leaves subsessile, ovate-lanceolate, to 3.3 cm long and 1.5 cm wide, with a cuneate base and a triangularly acute apex. *Pedicels* stout and erect, 4–10 cm long, extending to 22 cm long in fruit in the terminal flower, similar to the stem in diameter. *Flowers* often solitary and terminal, occasionally accompanied by a flower with or without a bracteole at the axil of a lower stem leaf, nodding or half-nodding, lantern to saucer-shaped, 2.2–3 cm across. *Petals* 4, pale blue, or white flushed with violet, ovate or ovate-elliptic, 15–18 × 7–11 mm, with a rounded to subacute apex. *Stamens* 12–24, two-thirds the length of the gynoecium, the filaments white or pale violet, filiform, 5–8 mm long, the anthers yellow, c. 1.5 mm long. *Ovary* ellipsoid to oblong, 5–7 mm long, c. 2 mm diameter, glabrous, the style 2–4 mm long, terminating in a capitate, 0.7–1 mm diameter. *Fruit capsule* narrow oblong-cylindrical, 3.3–5.8 cm long, 3–4.5 (–6) mm diameter, 4-valved, somewhat curved at the base into the pedicel, green marked with red, glabrous, the persistent style 2–4.5 mm long.

DISTRIBUTION. Central Nepal, Lamjung Himal, near Rambrong (Lamjung Himal), S Tibet (Jilong); 3500–4050 m (Map 24).

HABITAT. On open south-facing slopes amongst coarse herbs or in open scrub.

FLOWERING. June–July.

SPECIMENS SEEN. *Stainton, Sykes & Williams* 1954 (BM), 6254 (BM, E, KATH, TI); 6599 (BM, KATH), 8583 (BM, E); *K. N. Sharma* 40 (BM); *C. Y. Wu et al.* 75-636 (KUN).

Meconopsis lamjungensis is closely related to *M. exilis*, but differs from the latter in the non-aggregation of the uppermost leaves, in the lower leaves having a shallowly cordate or truncated base to the lamina, and in the subsessile, non-clasping uppermost cauline leaves.

Meconopsis lamjungensis is a slender plant with a rather small flower and, as a consequence, is easily overlooked in the field, especially when it is in fruit. It comes from a relatively well-botanised part of Nepal, yet there are very few specimens of it in herbaria. Once more material has come to light it may well be possible to expand the description. The species finds its closest ally in *M. exilis*, also recently described, and although the two look superficially very similar, there are significant differences on close inspection. Notably, in the former the leaves are always strictly alternate, the uppermost not aggregated into a false whorl towards the top of the plant. In addition, the leaf-lamina base of *M. lamjungensis* is noticeably subcordate or truncated, rather than rounded to attenuate, while the uppermost leaf is subsessile (in *M. exilis* the uppermost leaves are sessile with clasping subauriculate bases). Another significant difference can be observed in flowering plants: while *M. exilis* has a subracemose inflorescence with up to 5 flowers (the uppermost terminal), the others arising in an upper leaf-axil, *M. lamjungensis* bears a solitary terminal flower, while towards the stem base there is often, in addition, a single flower bud, borne on a short 'pedicel' with or without a bracteole. The two species are located at a considerable distance from each other, *M. exilis* in north-eastern Myanmar and north-western Yunnan, while *M. lamjungensis* is restricted to a small region of central Nepal and the neighbouring part of Tibet, as far as our present knowledge goes. Within the *M. lyrata* aggregate *M. lamjungensis* has the westernmost distribution while *M. wumungensis* has the easternmost. The Tibetan locality for *M. lamjungensis* lies a short distance due north-east of the Nepalese location.

74. MECONOPSIS POLYGONOIDES

Meconopsis polygonoides (Prain) Prain, *Bull. Misc. Inform., Kew* 1915: 143 (1915).
[syn. *Cathcartia polygonoides* Prain, *J. Asiat. Soc. Bengal, Pt. 2, Nat. Hist.* 64 (2): 326 (1896)]. Type: S Tibet (Xizang), Chumbi, *Dr King's Collector* 13 July 1884 (BM, holotype; K, isotype); *M. lyrata* sensu Taylor, *The Genus Meconopsis*: 73 (1934), pro parte]

DESCRIPTION. A more substantial plant either *Meconopsis lyrata* or *M. compta*, being 15–40 cm tall in flower, with a napiform taproot, 10–40 mm long and 4–6 mm diameter. *Stem* erect but thin, to 27 cm long, 2–3 mm diameter near the base, hairy or subglabrous, the hairs weak, to 2 mm long; bulbils absent. *Leaves* often 4, evenly scattered, although the upper two sometimes subopposite at flowering time; lowermost leaves long-petiolate, the lamina ovate, triangular-ovate or oblong-spathulate (resembling a species of *Bistorta*), 2–7 × 0.6–2.2 cm, the petiole 1–5.2 cm long, slender, linear but broadening towards the base; upper leaves sessile, elliptic to lanceolate-ovate, ovate-elliptic, usually with a subamplexicaul base, the margin often coarsely crenate and undulate. *Flowers* solitary, terminal, nodding. *Pedicels* slender, 3–10 cm (to 19.5 cm long in fruit). *Petals* 4 (–6), pale blue to lavender-blue, narrow-obovate to narrow-elliptic, 15–22 × 11–15 mm, the apex acute to somewhat acuminate, the margin minutely erose. *Stamens* 16–24, the filaments the same colour as the petals, anthers orange-yellow. *Ovary* ellipsoid-oblong, glabrous, with a 1–3 mm long style. *Fruit capsule* linear-cylindrical, 22–45 mm long, 2.5–4 mm diameter, the persistent style greenish, 2–4 mm long.

DISTRIBUTION. S Tibet (Chumbi); W & NW Bhutan; 3500–4100 m (Map 23, as for *Meconopsis primulina*).

HABITAT. Among tall herbs, ferns and shrubs on the wet, shady floor of forests, edge of shrubberies.

Meconopsis polygonoides; Chomolhari trail, W Bhutan, 3700 m.
PHOTO: TOSHIO YOSHIDA

Above: *Meconopsis polygonoides*; Thongbu La, Thimphu district, W Bhutan, 4300 m. PHOTOS: TIM LEVER

SPECIMENS SEEN. *Dungboo* s.n. Chumbi (K); *B. J. Gould* 1118 (K); *Dr King's Collector* 331 (K), s.n. 13 July, 1884 (BM, K), s.n. 17 July 1884 (BM, K).

In recent years this interesting, yet little-collected species, has been observed and photographed by various travellers in western Bhutan, especially on the Chomolhari Trail. These photos show the unique habit of the species that are clearly different from those of the closely related *Meconopsis lyrata*: plants are free-standing and support themselves without the aid of other tall herbs and surrounding shrubs. Because of collecting restrictions in Bhutan, few additional herbarium specimens have been made and yet this species seems quite distinct within the *M. lyrata* aggregate. The pale, relatively small flowers are campanulate but, when fully open, often resemble a Dutch-cap in shape.

Meconopsis polygonoides probably finds its closest allies in *M. lamjungensis* and *M. exilis*, differing from both in its solitary flowers and unbranched stem, in always having semi-amplexicaul upper leaves and in the more pointed petals. It is a demure species scarcely likely to attract a horticultural audience but, at the same time, it would be an important addition to any collection of the genus, especially those of botanic gardens.

75. MECONOPSIS WUMUNGENSIS

Meconopsis wumungensis K. M. Feng ex C. Y. Wu & H. Zhuang, *Fl. Yunnanica* 2: 33 (1979), Pl. 11: 3. Type: N Yunnan, Luquan, Wumungshan (Jiaozishan), 3600 m, 1 June 1952, *D. I. Mao* 1081 (KUN, holotype).

DESCRIPTION. *A small monocarpic herb* to 25 cm tall in flower (taller in fruit), but often less than 10 cm tall in scapose plants, with a narrow-napiform to subcylindrical taproot, 3–6 cm long, 0.3–0.7 cm diameter, usually elongate with a gradually narrowed extension. *Leaves* rather bright green, 5–10, often forming a lax basal rosette, dying away in the autumn to a resting bud below ground level, the lamina broadly ovate to oblong, or lanceolate, 1–6 × 1–2 cm, coarsely crenate, pinnatifid, pinnately lobed, or entire, sometimes with one or two pairs of lobes separated from the rest of the lamina by a slender rachis, often rounded with somewhat contracted bases, the base shallowly cordate, truncate or attenuate, somewhat decurrent onto the petiole, the apex obtuse, acute or acuminate, glabrous overall or with scattered patent whitish hairs above; petiole slender, 2.5–6 cm long, slightly expanded towards the base into a 2–4 mm wide sheath; stem leaves, when present, 2–4, with a short petiole or subsessile, the lamina similar to, or larger than, the basal leaves, the upper 2–3 aggregated into a false whorl. *Flowers* solitary on basal scapes, or terminal and 1–2 axillary additional flower(s) in caulous plants. *Pedicels* (scapes) ascending, rather slender, 4.5–11 cm long, arched over at the top in flower, glabrous or patent to somewhat deflexed white-hirsute. *Buds* nodding, narrow ovoid, 12–15 mm long, 5–6 mm wide, pale green. *Sepals* oval, glabrous to sparsely patent-hirsute. *Petals* 4, rarely 5, pale blue or lilac-blue with a deeper blue staining outside towards the base, obovate to elliptic, 17–24 × 8–12 mm, with an obtuse to acute apex. *Stamens* 12–24, with whitish narrow-linear filaments, 8–12 mm long, and orange-yellow anthers 1.5–2 mm long. *Ovary* narrow-ellipsoidal, sparingly fulvous-hirsute; style 1.5–3 mm long with a 3–4-lobed capitate stigma, 1–1.5 mm diameter. *Fruit capsule* narrowly oblong-cylindrical, 3–5 cm long, 2–3.5 mm diameter.

DISTRIBUTION. N Yunnan, Luquan Xian; known only from Wumengshan (Jioazishan); 3600–3800 m (Map 24).
HABITAT. Rocky, often moss-covered, banks in damp semi-shaded places, rocky crags, rock ledges, rock fissures.
FLOWERING. June–August.
SPECIMENS SEEN. *P. Y. Mao* 1081 (KUN); *R. Z. Fang & Z. W. Lu* 44 (KUN), 113 (KUN); *H. Peng et al.* 9284 (K, KUN); *T. Yoshida* K75 (KUN).

This interesting species has been associated with the small scapose species such as *Meconopsis concinna* and *M. delavayi*. However, although the species was based on a scapose specimen, recent collections, along with ample photographic evidence, show that it can have a distinct stem. This evidence clearly places the species in the *M. lyrata* aggregate. Within this aggregate, *M. wumungensis* closely matches *M. compta* in various details. Both are found in Yunnan; however, whereas *M. compta* is distributed in NW Yunnan, N Myanmar and SE Tibet, *M. wumungensis* appears to be confined to Wumengshan (Wumungshan), north-eastern Yunnan. The mountain where *M. wumungensis* was found is called Jiaozishan which is located near the village of Wumeng at the western end of Wumengshan range: the range itself extends east into Guizhou Province. Because the area is relatively easily accessed from the Yunnan capital Kunming, just to the south, and because it has long been a famed area for the abundance of medicinal herbs, Jiaozishan has been botanised several times by Chinese botanists. As a result several new endemic taxa, including species of *Cypripedium*, *Rhododendron* and *Corydalis*, as well as *Meconopsis*, have been described from there in recent years. Besides *M. wumungensis*, the far larger and more obvious *M. wilsonii* subsp. *orientalis* was recorded

Meconopsis wumungensis; Wumengshan (Jioazishan), NE Yunnan, 3700 m.
PHOTO: TOSHIO YOSHIDA

Meconopsis wumungensis; Wumengshan, NE Yunnan, ca. 3900 m. PHOTO: PAM EVELEIGH

Meconopsis wumungensis showing distinct stem; Wumengshan, NE Yunnan, c. 3900 m. PHOTO: PAM EVELEIGH

recently from the same mountain. Although the flora of northern and north-eastern Yunnan has yet to be fully recorded, the Jiaozishan does appear to represent an outlier from other *Meconopsis* populations in Yunnan.

Meconopsis wumungensis in its dwarf habit and in the transitional nature of its leaves (from entire at the base of the plant to pinnatifid higher up) looks very similar to *M. compta*, yet the leaves are generally larger and paler and the leaf lobes less well defined and more rounded. In addition, scapose plants are often found with all the leaves aggregated into a basal rosette. On the other hand, it should be noted that there is in general a remarkable consistency in the shape and colour of the flowers in the *M. lyrata* aggregate. The unique habitat of *M. wumungensis* is also interesting: plants grow characteristically in rock crevices and pockets filled with moist soil into which the subcylindrical taproots delve.

76. MECONOPSIS COMPTA

Meconopsis compta Prain, *Bull. Misc. Inform., Kew* 1918: 212 (1918). Type: SE Tibet (Xizang), Sarong (= Chawalong, Tsawarong), 12–13,000 ft, *G. Forrest* 14306 (K, holotype; BM, E, isotypes).

[syn. *M. lyrata* sensu Taylor, *The Genus Meconopsis*: 73 (1934), pro parte]

DESCRIPTION. *Plant* monocarpic, rarely more than 25 cm tall in flower, sometimes as little as 6 cm, with a napiform taproot, 6–15 mm long, 4–6 mm diameter, with a caudal extension. *Stem* erect, to 15 cm long, generally 1–7 cm, occasionally more or less obsolete, 2–3 mm diameter at ground level, patent-bristly, especially in the lower half, glabrescent or glabrous above; basal underground portion of stem, short, curved up from the taproot. *Leaves* generally 5–7, crowded towards the base of the plant, reduced in size up the stem, the lamina elliptic-lanceolate to ovate, linear-lanceolate or oblong-spathulate, occasionally lyre-shaped, 10–40 × 7–16 mm, the margin entire to slightly lobed or pinnately-lobed, usually with 2–4 pairs of lateral lobes; the leaves generally transitional on the plant, the lowermost being entire, the uppermost lobed, sometimes much reduced in size; all leaves petiolate, the petiole 0.5–4.8 cm long, somewhat expanded, submembranous and sheathing at the base, the petiole of uppermost leaves short. *Pedicels* 3.5–11.4 cm long (to 23 cm long in fruit), glabrous or with a few scattered hairs. *Flowers* solitary, half-nodding, terminal, rarely with 1–2 smaller axillary flowers borne from the uppermost leaf axils. *Petals* 4, pale blue or lavender-blue, narrowly obovate-elliptic, 12–18 × 7–9 mm, more or less entire, with a rounded to obtuse apex. *Stamens* 10–24, the filaments the same colour as the petals, the anthers yellow or golden. *Ovary* green often with purplish ribs, glabrous, the style 1–2 mm long, the stigma capitate, c. 1 mm diameter; stigma and style green. *Fruit capsule* narrowly oblong or cylindrical, 3–4.7 cm long, 3–4 mm diameter, glabrous, 3 (–4)-valved. (Fig. 31B1–3).

DISTRIBUTION. N Myanmar (Adung Valley, Seinghku Valley), SE Tibet (Chawalong (Sarong), north-western side of Meilixueshan (Ka-gwr-pu), Doshong La); SW China; NW Yunnan (Meilixueshan (Ka-gwr-pu, Doker La), northern Gaoligongshan (Salwin-Kiu-chiang divide), northern Biluoxueshan (Sewalongba, Si La, Mekong-Salwin divide), Baimashan (Beimashan), Mekong-Yangtse divide, Lijiang range, Cangshan (Tali or Dali range); 3658–4268 m (Map 24).

HABITAT. Wet grassy and stony slopes, open shrubberies, at the lower edge of rhododendrons and other scrub on mountain slopes.

FLOWERING. June–August.

FIG. 31. Lower cauline leaves, lower petioles and pubescence not shown: **A1–3**, *Meconopsis ludlowii*; **B1–3**, *M. compta*; **C**, *M. sinuata*. All × 1¹/₃.
DRAWN BY CHRISTOPHER GREY-WILSON

SPECIMENS SEEN. *K. M. Feng* 5137 (KUN), 6213 (KUN); *G. Forrest* 14306 (BM, E, K), 14463 (E, K), 15491 (BM, E, K), 16762 (BM, E, K), 18276 (E, K), 18869 (BM, E, K), 19727 (BM, E), 19729 (E), 25530 (BM, E, K), 25991 (BM, E), 29921 (E), 30098 (BM, E), 30350 (E); *Handel-Mazzetti* 9066 (E); *Kingdon Ward* 7003 (BM, K), 7031 (K), 9775 (BM); *Ludlow, Sherriff & Taylor* 5226 (E), 5228 (BM, E), 5228A (BM), 6559 (E); *McLaren* 161 (BM, E); *Rock* 22328 (BM, E, K), 22419 (E), 22994 (E, KUN); *T. T. Yü* 22519 (E, KUN), *T. Yoshida* K74 (KUN).

Meconopsis compta is perhaps the most widely distributed species of the *M. lyrata* aggregate wih the exception of *M. sinuata*, with a centre of distribution in the border region between Yunnan and Tibet. It is a small plant with neat, generally lobed leaves that are clustered together in the lower part of the plant. As in some other species of *Meconopsis* (notably *M. impedita* Prain) the leaves on individual plants can be transitional in form with the lowermost undivided, while those above are increasingly lobed. In *M. compta* all the leaves are petiolate, the lowermost with the longest petioles. The flowers, which are closely similar to those of *M. exilis* and *M. polygonoides*, are generally solitary and terminal but rarely they can be accompanied by 1–2 axillary, usually rather smaller, flowers issued from the uppermost leaf-axils. Such plants are generally found growing at the lower elevations recorded for the species.

Meconopsis compta; W of Si La, Biluoxueshan, NW Yunnan, 3950 m. PHOTO: TOSHO YOSHIDA

Meconopsis compta; Lansey La, S of Si La, Biluoxueshan, NW Yunnan, 3750 m. PHOTO: TOSHO YOSHIDA

Kingdon Ward came across *Meconopsis compta* (recorded as *M. lyrata*!) in the Burmese Himalaya Seinghku valley, commenting that:

"*M. lyrata* [*M. compta* sic] looks like an annual. It is a frail, slender creature, six or eight inches high, bearing a solitary wan blue flower, so pale and delicate that it looks unable to defy the rude winds; yet it too grows at 12,000 feet, widely scattered on steep meadow slopes, or on the fringes of the ultimate forest, flowering in June and swamped presently amongst the coarse herbage. Very difficult is it to find the plants for seed later, nor do the thin capsules contain much ….."

Plant Hunting on the edge of the World: 263 (1930).

77. MECONOPSIS LUDLOWII

Meconopsis ludlowii Grey-Wilson **sp. nov.** very closely related to *M. sinuata*, differing in its short-stemmed or subacaulous habit, in the long-pedicelled scapose flowers, in the less elaborately lobed leaves and in the glabrous ovary and fruit capsules. Type: E Bhutan, Orka La, Sakden, 13,900 ft, 10 July 1934, *Ludlow & Sherriff* 642 (BM, holotype).

[syn. *M. concinna* sensu D. G. Long non Prain, *Fl. Bhutan* 1 (2): 408 (1984)]

DESCRIPTION. *Small monocarpic plant* to 18 cm tall in flower, often only 10–15 cm, with a narrow, napiform root 2.4–7.5 cm long, to 0.9 cm diameter, simple or sometimes with one or two branches. *Stem* to 7.5 cm long, but generally shorter and plants often appearing to be acaulous, glabrous. *Leaves* petiolate, pale green, few, generally 3–8, the lamina elliptic-oblong to lanceolate-elliptic, 1.7–5.6 × 1.1–2 cm, the margin subentire to pinnately-lobed on the same plant, or all pinnately lobed, with 2–4 shallow, obtuse lobes on each side, the basal pair often rather distant from the others, the terminal lobe the largest, ovate to oblong, 12–18 × 6–12 mm, obtuse, glabrous or with a few hairs beneath; petiole narrow, 10–40 mm long, glabrous. *Flowers* 2–8, issued from the axils of the leaves but often appearing to be scapose because of the reduced nature of the stem, nodding, broad-campanulate, 2–3.5 cm across (more when flattened). *Pedicels* (scapes) ascending to erect, slender, 5.5–14 cm long (to 33 cm long in fruit), somewhat pubescent to subglabrous, occasionally glabrous or glabrescent. *Petals* 4–5, lilac-blue to dark purplish violet or violet-blue, often with a deeper purplish blue zone or ring at the base within, obovate to almost rounded, 14–24 × 13–22 mm, the margin finely erose. *Stamens* c. 36, with linear filaments and pale yellow anthers. *Ovary* ellipsoid, glabrous, the style 2–4 mm long. *Immature fruit capsule* erect, cylindric-ellipsoid, 22–33 × 4–5 mm, green with purplish sutures, 5–7-valved, glabrous, the persistent style rather thick, purple, 5–7 mm long, to 1.5 mm diameter at the base. (Fig. 31A1–3).

DISTRIBUTION. Eastern Bhutan (Tawang Chu, Mago, Milakatong La, Orka La), adjacent parts of NE India (Arunachal Pradesh); 4115–4740 m (Map 24).

HABITAT. Grassy hillsides, low open scrub, often of juniper and dwarf rhododendrons, rocky places, grassy cliff ledges.

FLOWERING. Late June–July.

SPECIMENS SEEN. *Ludlow & Sherriff* 642 (BM), 728 (BM), 1080 (BM).

Ludlow & Sherriff 642 (BM), collected on the Orka La, eastern Bhutan, was wrongly recorded as *Meconopsis concinna* in the *Flora of Bhutan* and was filed by me as an 'undescribed species' related to *M. compta*. I asked those travelling to eastern Bhutan to keep an eye open for it. Because of travel restrictions in that part of the country this has not been possible up until the present time, a pity because, apart from this particular collection, Ludlow & Sherriff made some very interesting collections of *Meconopsis* from there, most notably some very fine forms of *M. grandis*. There matters remained until recently when a team including Martin Walsh, Tim Lever, David and Margaret Thorne ventured on the other side of the eastern Bhutan border in Arunachal Pradesh (Assam in the old days). There they made a number of exciting *Meconopsis* finds, their observations and photographs being of great value to this monographic work. Among these was a small blue-flowered species which clearly matches L&S 642 in detail. This information leaves me in no doubt that this is another new species in series *Cumminsia*, which in detail of flower, leaf and fruit capsule appears to come closest not to *M. compta* as I originally surmised, but to *M. sinuata*. Like *M. sinuata*, and indeed most of the members of *Cumminsia*, it is a slight plant but, unlike that species, the leaves are smaller and with fewer lobes, while the stem is much reduced or more or less absent and the flowers are borne on long basal or sub-basal scapes rather than in a well-defined raceme. In the reduction of the stem it clearly parallels *M. primulina* but that species has entire or somewhat sinuate leaves and larger, flatter rather than campanulate flowers with 5 or more, rather than 4, petals, while the fruit capsule is only 4-valved.

I have no hesitation in naming this charming little species after Frank Ludlow, one half of a formidable collecting partnership (with George Sherriff), primarily in Bhutan and south-eastern Tibet. Apart from rhododendrons and primulas, they paid special attention to *Meconopsis*, making many collections over the years they were in the field, a period of some 20 years from 1932. While the distinctive, bold, pink-flowered *Meconopsis sherriffii* commemorates the latter half of the partnership no *Meconopsis* commemorates the former and it seems only right that this is now addressed. Frank Ludlow (1885–1972) was an English naturalist, botanist and plant hunter. With George Sherriff he collected large numbers of herbarium specimens and made numerous seed collections. Their main herbarium collections reside at the British Museum (Natural History) where there is also a fine collection of half-tone and colour photographs.

"In addition to the Nyuksang La Sherriff and Danon paid a fleeting visit east to the Orka La (13,900 feet). 'I found some good flowers ….. and one could easily spend a month or so in the area. A new meconopsis and some primulas and a beautiful little corydalis and androsace were found. The best spot seems to be on the big rounded cliff to the south of the pass; on the north side of this there were many flowers.' The meconopsis was the low-growing dark purplish-violet-flowered *M. lancifolia* var. *concinna* (642)……".

(Harold Fletcher, *A Quest of Flowers*, 1975).

Meconopsis ludlowii in habitat; Kari La, W Arunachal Pradesh, NE India, 4460 m.
PHOTO: DAVID & MARGARET THORNE

77. MECONOPSIS LUDLOWII

Above left: *Meconopsis ludlowii* leaf-rosette; Kari La, W Arunachal Pradesh, NE India, 4460 m. PHOTO: DAVID & MARGARET THORNE
Above right: *Meconopsis ludlowii*; Lenang, W Arunachal Pradesh, NE India, 4500 m. PHOTO: TIM LEVER

Ludlow & Sherriff made additional collections of this plant in eastern Bhutan during 1934. Harold Fletcher recalls (op. cit.) that "They marched for 5 miles up the Gorjo Chu valley to Lap[40] (14,200 feet) and apart from the dwarf deep grape-purple *Lilium nanum* ((726), a slaty-blue-flowered ally of the gentians *Swertia kingii* (727), *Meconopsis lancifolia* var. *concinna* (728) .... " and later in October Danon, a Lepcha from a quinine estate near Darjeeling who had collected with L&S returned to the region as Fletcher relates: "Danon returned from Tawang with a good haul of seeds including those of several primulas and rhododendrons, of *Meconopsis lancifolia* var. *concinna* (1080) .... Danon paid another hasty visit to the Orka La and returned with seeds of *Meconopsis lancifolia* (1095), of *Primula atrodentata* (1094) and of the new species gathered in July, *P. occlusa* (1093)."

It is perhaps surprising that L&S should have equated the Orka La plant with *Meconopsis lancifolia* or *M. concinna*, two primarily Chinese species, and not with the members of the section *Primulinae*. What happened to the seed collected under the number 1080 is unclear and I cannot find any record of its history in cultivation, otherwise perhaps this dainty species would have been recognised long before now.

---

[40] This locality lies a relatively short distance to the north of the Orka la.

Meconopsis ludlowii; W Arunachal Pradesh, NE India, 4450 m.
PHOTO: DAVID & MARGARET THORNE

Meconopsis ludlowii; Lenang, W Arunachal Pradesh, NE India, 4400 m.
PHOTO: MARTIN WALSH

The following observations have been emailed to me from the group that explored in Arunachal Pradesh in the summer of 2012:

"The small blue poppy [*M. ludlowii* sic] was also very common in the areas that it occurred often associating with *M. bella* in open grassy hillsides amongst low scrub of *Rhododendron lepidotum* and *R. anthopogon*, juniper etc and sometimes with *M. paniculata*."   Martin Walsh.

"There was a lot of *M. bella* and for several days walking it was common. However, it grew with two other blue-flowered species from which it was quite distinct — *M. sinuata* and what Ludlow & Sherriff called *M. lancifolia* var. *concinna* [*M. ludlowii* sic]. I believe this latter plant is a new species. It is most like *M. primulina* with the long thin seed capsule (which *M. sinuata* also has), but it has completely different leaves (pinnately cut but with a broad rounded end lobe) …… This new species has a ring of slightly darker purplish blue towards the base of the corolla and the flower is noticeably less cupped than *M. bella*."

Margaret Thorne, extract from an email.

Meconopsis ludlowii; Lenang, W Arunachal Pradesh, NE India, 4500 m.   PHOTO: TIM LEVER

78. MECONOPSIS SINUATA

Meconopsis sinuata Prain, *J. Asiat. Soc. Bengal, Pt. 2, Nat. Hist.* 64 (2): 314 (1896). Type: NE India, Sikkim, 1877, *Dr King's collector* 4194 (K, holotype).
[syn. *M. sinuata* var. *typica* Prain, *J. Asiat. Soc. Bengal, Pt. 2, Nat. Hist.* op. cit.].

DESCRIPTION. *Monocarpic plant* with a narrow napiform taproot up to 30 cm long, generally far less, to 9 mm diameter near the top, simple or somewhat branched, the stem erect, variable in height from 14 to 80 cm tall in flower, densely patent-setose, the setae mostly 4–7 mm long, red-brown. *Leaves* green above, sometimes with brown blotches, paler beneath, the basal leaves in a lax rosette, generally withered by flowering time, the stem leaves scattered, often rather distant, the lower with an oblong to obovate-elliptic or oblanceolate lamina, 6.5–18 × 0.8–3.5 cm, the base cuneate to subtruncate, the apex obtuse, margin distantly sinuate-crenate, the upper leaves subsessile to sessile, decreasing in size and often bract-like, mostly 2.6–4.4 × 0.9–1.5 cm, more or less auriculate at the base; all leaves with a sparse to moderately dense indumentum of brown or reddish brown, occasionally orange, hairs on both surfaces; petiole of lower and basal leaves narrow, flattened, 3.8–6.5 cm long, slightly expanded at the base. *Inflorescence* a short raceme with 3–11 (–16) flowers, often only one or two opening at a time, mostly bracteate, the bracts of the uppermost flowers sometimes displaced; flowers nodding or half-nodding, cupped to saucer-shaped, 3.6–7 cm across. *Pedicels* 2.2–5.6 cm long, those of the terminal flower longer, to 11.5 cm, elongating in fruit, then 6.8–20.5 cm long, greenish or, more often, dull reddish purple, beset with patent reddish or brownish setae like the stem. *Buds* nodding, ovoid to oblong, 14–18 × 11–16 mm, patent-setose. *Petals* 4, pale to mid-blue, grey-blue, lilac-blue, pale mauve-blue, violet-blue or purple-blue, occasionally white, obovate to subrounded, (1.5–)2–35 × (14–)19–29 mm, the apex obtuse to subobtuse, the margin somewhat erose. *Stamens* numerous with filiform filaments generally the same colour as the petals but darker, about half the length of the petals, the anthers yellow or golden-orange. *Ovary* ellipsoid, green, beset with dense ascending, reddish bristles; style 4–5.5 mm long, with a capitate stigma. *Fruit capsules* narrow-obovoid to cylindric-ellipsoidal, 33–43 × 7–8 mm, broadest above the middle, narrowed at the base towards the pedicel, patent-setose, prominently 3–4-ribbed, 3–4-valved, splitting for about a third of its length from the top at maturity, the persistent style 5–7 mm long. (Fig. 31C).

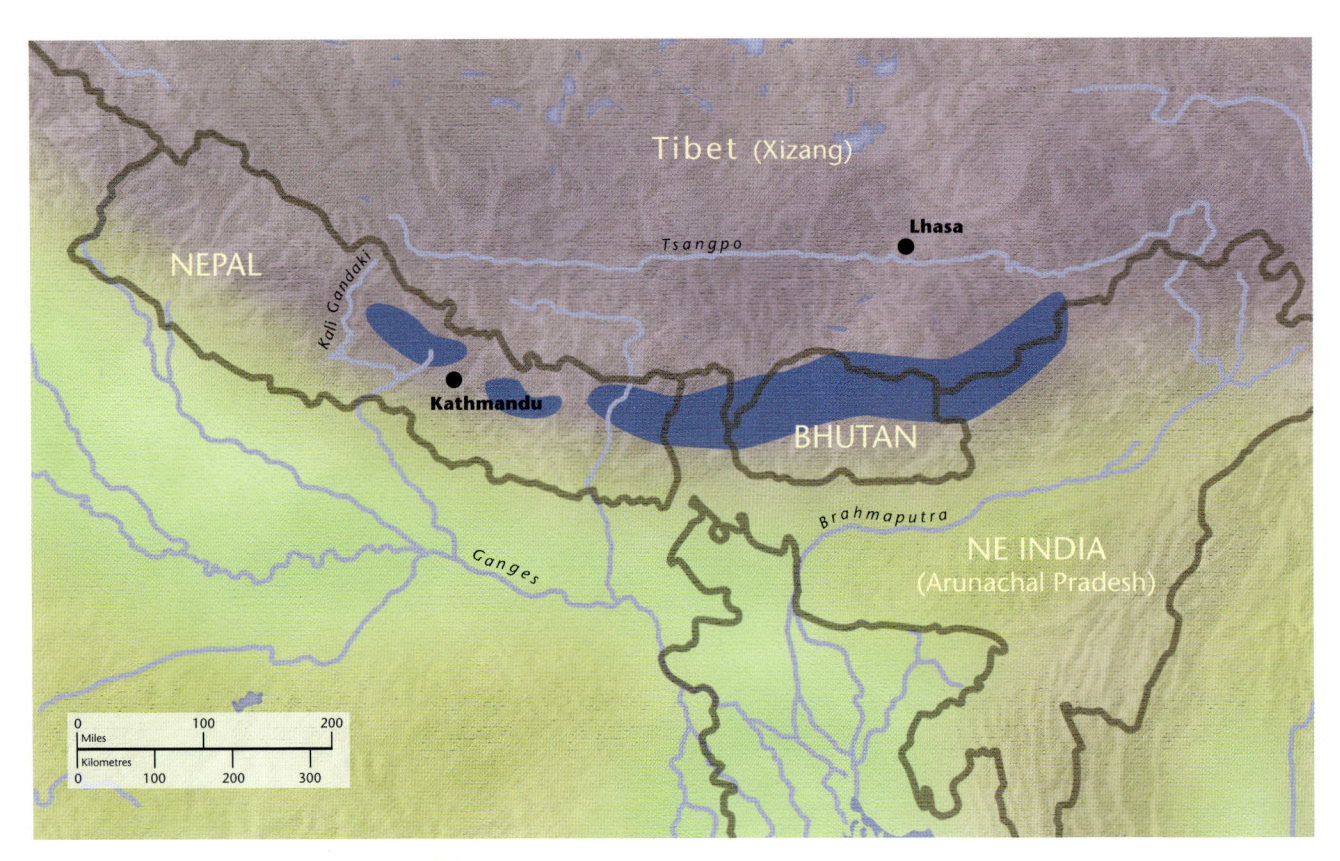

MAP 25. Distribution of *Meconopsis sinuata* ■.

Meconopsis sinuata, basal leaves; Padima Tsho, Tongsa district, C Bhutan, 4000 m. PHOTO: TIM LEVER

Meconopsis sinuata; Se La camp track, W Arunachal Pradesh, NE India. PHOTO: DAVID & MARGARET THORNE

DISTRIBUTION. C & E Nepal (Dhading, Kaski, Lamjung, Nuwakot, Solukhumbu & Taplejung districts; Ganesh and Lamjung Himal eastwards), NE India (Sikkim, W & N Arunachal Pradesh), C & N Bhutan, SE Tibet (Kashong (Chang) La, Cho La); (3200–)3355–4575 (–5335[41]) m (Map 25).

HABITAT. Rocky places on grassy slopes, stony banks, boulder fields, peaty pockets amongst dwarf shrubs, rhododendron scrub, rocky streamsides, by waterfalls, fir and bamboo forests, often in partial shade, sometimes deep shade.

FLOWERING. June–early September.

SPECIMENS SEEN. *R. Bedi* 245 (K); *L. W. Beer* 8260 (BM), 10014 (BM), 25429 (BM), 25445 (BM); *Bowes Lyon* 13091 (E), 13128 (E), 15088 (BM); *Cave* 9 (E), 1012 (E), 7110 (E), s.n. 24 Oct. 1914 (E), s.n. 2.09.1916 (BM), s.n. 24 Oct. 1916 (BM), s.n. 6 Nov. 1917 (BM), s.n. 10 Nov. 1917 (E); *R. E. Cooper* 86 (E), 573 (K); *Dr King's Collector* 4194 (K); *Kingdon Ward* 8420 (K), 12409 (BM), 13657 (BM), 14321 (BM); *Long, McBeath, Noltie & Watson* 451(E), 508 (E), 683 (E); *Ludlow & Sherriff* 394 (BM), 886 (BM), 2384 (BM, E), 3286 (BM, E) 1033 (BM); *Ludlow, Sherriff & Hicks* 16889 (BM, E), 17299 (BM), 19052 (BM, E), 19166 (BM), 19745 (BM, E), 20712 (BM, E), 21131 (BM), 21194 (BM); *D. M. McCosh* 409 (BM); *K. N. Sharma* 32/94 (BM), E369 (BM), E445 (E); *G. F. Smith* 35 (BM); *Stainton* 1703 (BM), 3975 (BM, E); *Stainton, Sykes & Williams* 6103 (E).

Meconopsis sinuata was described by David Prain at the end of the nineteenth century from a collection made by Dr King in Sikkim in 1877. In the original description Prain recognised two varieties, var. *typica* (equates to var. *sinuata*) and var. *prattii*. The latter, based on a Pratt specimen collected in western Sichuan close to Tatsienlu (now Kanding), was later accorded specific status but was sunk into synonymy under *M. horridula* Hook. f. & Thomson in George Taylor's 1934 monograph of the genus, although in this account it is recognised as a species in its own right. There is no doubt about the distinctiveness of *M. sinuata*, a central and eastern Himalayan species. In his monograph Taylor allies *M. sinuata* with both *M. aculeata* Royle and *M. latifolia* (Prain) Prain, but this seems to be wrong: the membranous undulate and often markedly sinuate leaves, the weak hairs on stems and leaves, the 4-petalled flowers borne in a bracteate raceme and the narrow ellipsoidal fruits clearly ally it to the *M. lyrata* aggregate. As with the other members of the aggregate the small, nodding flowers are

---

[41] K. N. Sharma records an altitude of 17,500 ft for his collections E369 (Dangrey, Nepal) but this is almost certainly erroneous.

78. MECONOPSIS SINUATA

Above, from left:

Meconopsis sinuata; Jule La, C Bhutan, 4450 m. PHOTO: DAVID & MARGARET THORNE

Meconopsis sinuata, white-flowered form; Sele La, Haa, W Bhutan, 3700 m. PHOTO: DAVID & MARGARET THORNE

Meconopsis sinuata; Se La camp track, W Arunachal Pradesh, NE India. PHOTO: DAVID & MARGARET THORNE

fleeting and generally open just one or two at a time. The position of *M. sinuata* has been strengthened by the work of Kadereit *et al.* (2011) based on a phylogenetic analysis and molecular dating which places *M. sinuata* in a close grouping with *M. lyrata* in particular but also with *M. primulina* and *M. wumungensis*, all members of the *M. lyrata* aggregate (section *Primulinae*). The other members of the *M. lyrata* aggregate were not included in the study which was basically focused on the position and affinities of *M. cambrica* (see p. 41).

Meconopsis sinuata is a slender plant often with just one flower opening at a time; however in the more robust forms the flowers can be reasonably large and a good rich colour and a plant may occasionally bear up to three open flowers at once. The basal leaves are often withered or partly withered by anthesis and the lower leaves often show browning by this time, especially in the sinuses along the margin. Although the inflorescence is bracteate the upper bracts are often displaced so that the flowers appear to be ebracteate.

Meconopsis sinuata probably finds its closest allies in the taller caulous species such as *M. lamjungensis* and *M. exilis* but is readily distinguished by its racemose inflorescence in which the flowers are evenly spaced, and in the more or less narrow oblong leaves with their rather neat sinuate-crenate margins. However, the species can vary greatly in vigour, even within the same colony, with the largest plants somewhat over 60 cm tall and bearing up to 8 flowers, while depauperate plants may only reach about 15–20 cm and carry just one or two flowers, specimens with a solitary flower being not infrequent. One collection, *Ludlow & Sherriff* 3286 (BM sheet), bears eight individual plants on the same sheet, all in flower (1–3 flowers open) and ranging in height from 8–24 cm. The specimens collected in central Bhutan (Dungshinggang or Black Mountain) clearly show that this is an underrated species horticulturally, being sturdy and compact specimens with good-sized flowers recorded as being purplish blue. Another collection, *Ludlow, Sherriff & Hicks* 19502 (BM specimen) is accompanied on a separate sheet

THE GENUS MECONOPSIS | 357
78. MECONOPSIS SINUATA

by an excellent black and white photograph of several plants in the wild, reminiscent of a rather exotic harebell. Although many recent photographs of the species taken in the wild (mostly in Bhutan) in recent years show that the lower foliage is dying back as the plants come into flower this is not always the case looking at older photographs and herbarium specimens and may be a result of seasonal factors such as unusually dry springs or delay to the onset of the summer monsoon.

Despite being recorded widely from central Nepal through the eastern Himalaya and the adjacent part of Tibet, *Meconopsis sinuata* is represented in herbarium by relatively few collections, presumably because it is easily overlooked in the field. A lack of records suggest that the species has rarely been in cultivation and then only transitory. In the *Quarterly Bulletin of the Alpine Garden Society* in 1955 Marjorie Brough records having grown *M. sinuata* in her Hertfordshire garden and that it "kindly set a few seeds this year".

Meconopsis sinuata is a slight plant, easily overlooked in the wild, especially in the fruiting stage. Although a single plant can bear as many as 8 flowers these are nearly always borne in succession, one at a time, so the plant does not make any great impact. For this reason it has never been much sought after as a garden plant, although aficionados of *Meconopsis* will certainly want to have it in their collections.

Left, from top:

Meconopsis sinuata; Jule La, C Bhutan, 4450 m.   PHOTO: DAVID & MARGARET THORNE
Meconopsis sinuata; Jule La, C Bhutan, 4450 m.   PHOTO: DAVID & MARGARET THORNE

Below, from left:

Meconopsis sinuata; Tshochenchen, Bumthang district, C Bhutan, 3750 m.   PHOTO: TIM LEVER
Meconopsis sinuata, immature fruit capsule; Tshochenchen, Bumthang district, C Bhutan, 3750 m.   PHOTO: TIM LEVER

section BELLAE

Section *Bellae* contains a single species, an alpine gem, that is primarily a denizen of cliffs and cliff ledges, although venturing onto moist banks and sometimes into open dwarf rhododendron scrub. It is a very distinctive little plant, which, along with *Meconopsis delavayi*, is one of only two polycarpic members of subgenus *Cumminsia*.

79. MECONOPSIS BELLA

Meconopsis bella Prain, *J. Asiat. Soc. Bengal, Pt. 2, Nat. Hist.* 63 (2): 82 (1894). Type: Sikkim, Nepal frontier, Nyegu 14,000ft, July 1887, *Dr King's Collector* s.n. (syntypes BM, G, K, P); Nepal frontier, Isyking La, October 1887 (syntypes BM, G, K, P)[42].

DESCRIPTION. *Small tufted evergreen perennial* to 20 cm tall in flower but often as little as 8–10 cm, with a stout dauciform to elongated taproot, to 50 cm long, 0.5–3.2 cm diameter at the widest, supporting a short condensed stem at the top which is congested with the remains of old persistent, ascending, closely overlapping petioles. *Leaves* all borne in a basal rosette, one or several congested together, deep green above, paler, sometimes glaucous, beneath, the leaves wide-spreading to ascending, the lamina elliptic-oblong, elliptic-lanceolate or spathulate to ovate in general outline, entire to regularly or somewhat irregularly pinnately or bipinnately lobed, (1.2–)1.8–11 × 0.7–4(5.6) cm, sometimes dimorphic on the same plant, with the outermost leaves entire, the innermost increasingly dissected (but plants with wholly entire leaves or bipinnately divided ones are not uncommon), the lobes elliptic to oblong, ovate-lanceolate or obovate, the apex rounded or subobtuse, the ultimate lobe frequently trifid, to 7 × 2 cm, glabrous or sparsely weak-bristly; petioles slender, 1.2–12.5 (–18) cm long, expanded at the base, glabrous or with a few scattered bristles. *Flowers* solitary, scapose, up to 20 per plant, occasionally more, flowering in succession, cup-shaped, nodding or half-nodding, 4–8.3 cm across when flattened, fragrant. *Pedicels* (scapes) slender, 5.2–18 cm long (up to 28 cm long in fruit), ascending, often arching, sparsely bristly, more densely so just below the flower, occasionally subglabrous. *Buds* nodding, ovoid to subglobose, the sepals bristly, sometimes sparsely so. *Petals* 4 (rarely 5–6), pale sky to deep blue, sometimes flushed with pink or purple, pink to deep violet-blue, subrounded to ovate to obovate, 2–4.6 × 1.3–4 cm, slightly pleated or undulate towards the margin or somewhat erose. *Stamens* numerous, one third the length of the petals, with filiform filaments the same colour as the petals, although generally darker, 8–10 mm long, the anthers oblong, cream, yellow or golden, 1.25–1.5 mm long. *Ovary* ovoid to ellipsoid or pear-shaped, sometimes constricted towards the base, greenish, adorned with appressed bristles and a capitate stigma, the style short 1–2.5 mm long. *Fruit capsule* obovoid to pyriform or ellipsoid, 13–28 mm long, 4–7 mm diameter, 4–7-ribbed, the short persistent style becoming swollen and conical, capping the top of the fruit, glabrous to moderately ascending bristly, the persistent style 2–5 mm long, 4–7-valved, splitting for a short distance from the top at maturity, borne on a curved to recurved or somewhat contorted pedicel. *Seeds* ellipsoid-falcate, 1.2–1.4 mm long, with a finely reticulated, somewhat ribbed testa.

DISTRIBUTION. Widely scattered in the central and eastern Himalaya from central Nepal to Sikkim, Bhutan, Arunachal Pradesh (NE India; Upper Assam); 3658–4725 (Map 26).

Meconopsis bella subsp. *subintegrifolia* leaf-rosette; Tsho Jumep, C Bhutan, 4450 m.
PHOTO: DAVID & MARGARET THORNE

---

[42] Although the type material is fragmentary, enough details of this distinct species can be ascertained. Both collections are presented on the same herbarium sheets.

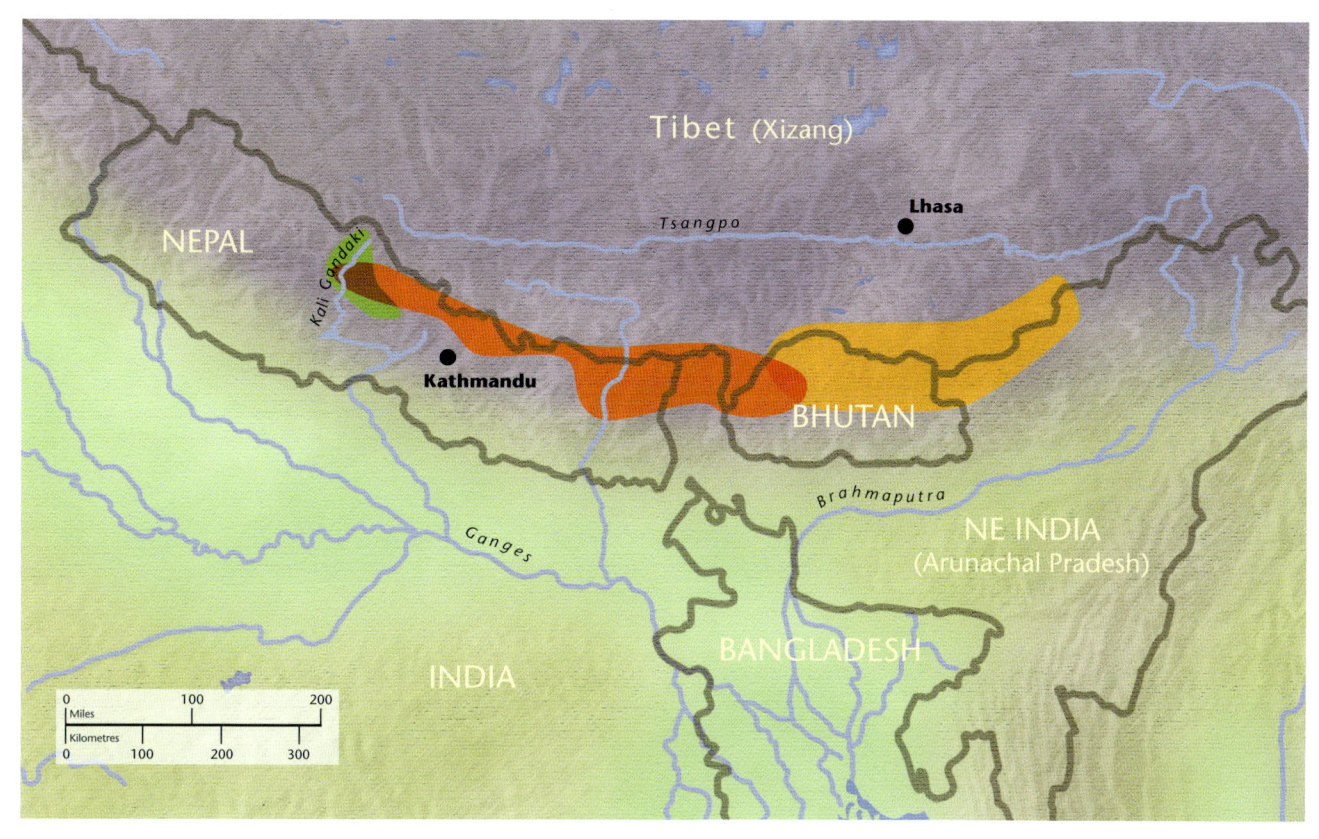

MAP 26. Distribution of *Meconopsis bella*: ■ *M. b.* subsp. *grandifolia*, ■ *M. bella* subsp. *bella*, ■ *M. b.* subsp. *subintegrifolia*.

HABITAT. Cliff crevices and ledges, sloping rock banks, sometimes in dwarf *Rhododendron* scrub, often of north- or east-facing aspect.

FLOWERING. June–July (–early August).

Meconopsis bella is a very distinctive small perennial species, perhaps the loveliest of all the small alpine gems within the genus. The flowers are beautifully cupped and relatively large for the size of the plant. The foliage, often a rather dense basal tuft, is very variable from entire to coarsely or finely divided, in some instances partly transitional on the same plant. One of the most characteristic features of *M. bella* is to be found in the pedicels (scapes). At flowering time these are generally ascending to erect but as the fruit begins to develop they become curved or recurved, sometimes markedly contorted. In the wild this has the effect of bringing the maturing fruit capsules close to the banks or cliffs on which it grows, often hiding the mature capsules in the surrounding growth. This feature is unique in *Meconopsis*, as is the swollen base of the style that forms a cap-like process over the top of the capsule. These various characters give *M. bella* a rather unique position in the genus and for that reason it is accorded a section of its own.

Meconopsis bella is a true alpine gem, one of the smallest and neatest species in the genus with a wide west-east range that encompasses much of the Nepalese Himalaya from the region of Mustang and the upper Kali Gandaki valley eastwards to Sikkim and Bhutan as far as NE India (Himachal Pradesh) and the corresponding regions of southern Tibet. In his monograph of 1934 George Taylor comments rightly on the variability found within the species but does not make any attempt at subdivision; however, since his monograph was written further and more complete material has come to light, especially from the eastern end of the range of the species and this has led to a change in concept. Over its range *M. bella* reveals quite a range of variability, especially in vigour and the size shape and substance of the flowers, but most obviously in the character of the leaves. All those who have written about *M. bella* in the past have commented on the leaf-lamina that can range from entire to pinnately or bipinnately lobed. In some plants all the leaves are entire while in others they are all dissected. At the same time, some plants bear a transition of leaves, the basal one or two being entire, the following leaves being increasingly dissected; such plants are generally to be found in the west of the range of the species, while those bearing only entire leaves are primarily found to the east. This geographical transition can be summarised in the following key:

79. MECONOPSIS BELLA

KEY TO SUBSPECIES OF MECONOPSIS BELLA

1. Leaves mostly simple, lanceolate-elliptic to ovate or paddle-shaped, occasionally with a few uneven lobes or trilobed, often on the same plant; fruit capsules 15–27 mm long, 4–5-valved, moderately to densely bristly . . . . subsp. **subintegrifolia** (W Bhutan to NE India, S Tibet)
   Leaves pinnately divided, generally with 2–4 pairs of primary divisions, these simple or pinnatisect; fruit capsules 13–15 mm long, 5–7 (–8)-valved, glabrous to sparsely bristly. 2

2. Leaf-lamina with 1–3 (–4) pairs of simple lobes, the end lobe (leaflet) large, at least 2.8 × 1.1 cm, lower (outermost) leaves occasionally subentire on the same plant . . . . subsp. **grandifolia** (C Nepal)
   Leaf-lamina with finely dissected pinnatisect or bipinnatisect divisions, the end lobe not more than 1.2 cm long . . . . subsp. **bella** (C Nepal to W Bhutan)

Seed of *Meconopsis bella* has been introduced from the wild on many occasions over the years, especially from Nepal, but the species has failed to establish in cultivation despite the fact that several have been brought to flowering size. Even expert alpine growers consider this to be a difficult and challenging species which is a pity because, being truly perennial, it would make an excellent addition to any collection. George Taylor (1934) relates that "The history of the species in cultivation is a melancholy one. Seeds were sent to this country and other parts of Europe by King as long ago as 1888, but, although several attempts were subsequently made to introduce the plant, its horticultural merit remained unknown until 1906, when seed received early in 1904 from the Calcutta Botanic Garden produced flowering plants at the Royal Botanic Garden, Edinburgh. The species has since disappeared from gardens, but it is hoped that the recent acquisition from Nepal may be the means of reintroducing this charming garden plant".

James Cobb (1989: 48–52) writes in full about the trials and tribulations of trying to grow *Meconopsis bella* and there is no need for me to repeat this here; however, this may inspire others to attempt to grow and perhaps succeed with this species when seed becomes available.

In many ways *Meconopsis bella* parallels *Paraquilegia microphylla* in character and habitat requirements. Both are predominantly chasmophytes forming neat tufted plants and bearing solitary cupped flowers. The two overlap in distribution over much of their range in the wild, although the *Paraquilegia* is far more widespread, being present throughout the Himalaya and much of mountainous western China. Although tricky in cultivation, it can flourish in the skilled hands of specialist alpine plant growers, both outdoors in troughs or on raised beds or tufa walls, but also within the confines of an alpine house. At one time thought to be almost impossible to grow in cultivation it has now been mastered, so it must be hoped that someone will unlock the key to success and succeed with the cultivation of *Meconopsis bella*.

FIG. 32. *Meconopsis bella* subsp. *bella*, leaf-laminas, lower petioles not shown. All × 1¹/₃. DRAWN BY CHRISTOPHER GREY-WILSON

Above: *Meconopsis bella* subsp. *bella*; above Manang, Marsyandi valley, C Nepal, 4350 m. PHOTOS: CHRISTOPHER GREY-WILSON

subsp. bella

DESCRIPTION. *Leaves* pinnatisect to bipinnatisect, generally with 2–4 pairs of prime divisions, these often further lobed (bipinnatisect) the terminal lobe not more than 1.2 × 0.6 mm, no larger than the supporting lobes, the lobes primarily obtuse, the basal leaf or two on the same plant sometimes smaller, entire or subentire. *Fruit capsule* 13–15 mm long, 5–7 (–8)-valved, glabrous to sparsely bristly. (Figs 32, 33).

DISTRIBUTION. Nepal (Gorkha, Manang, Myagdi, Nuwakot, Rasuwa, Sankhuwasabha, Sindhupalckok, Solukhumba & Taplejung districts), N Sikkim, W Bhutan (probably not east of the Paro Valley), adjacent parts of Tibet (Map 26).

HABITAT. As above; 3500–4270 m.

SPECIMENS SEEN. *F. M. Bailey* s.n. 12 July 1937 (E); *F. M. Bailey's Collector* s.n. 24 June 1935 (BM); *Cave* s.n. 24 July 1916 (E), s.n. 1 Sept. 1919 (E); *R. E. Cooper* 327 (E), 898 (E), 1674 (E), 1978 (BM, E, K), 2965 (E), s.n. Parshong, Bhutan (BM); *Dhwoj* 0367 (BM); *Durham University Exped.* D192 (BM); *B. J. Gould* 468 (K),1129 (K); *de Haas* 2745A (BM); *Herb. Hort. Bot. Calcutta* s.n. 1896 (BM); *Dr King's Collector*, s.n., July 1887 (BM, G, K, P), s.n. Oct 1887 (BM, G, K, P), s.n. Sikkim (K); *Rohmoo Lepcha* 6443 (E); *Lowndes* 1138 (E); *McBeath* 1496 (E); *Ohba et al.* 8330740 (E); *K. N. Sharma* E84 (BM); *W. W. Smith* 3962 (BM); *Stainton* 627 (BM), 754 (E), 8546 (E); *Stainton, Sykes & Williams* 6203 (E).

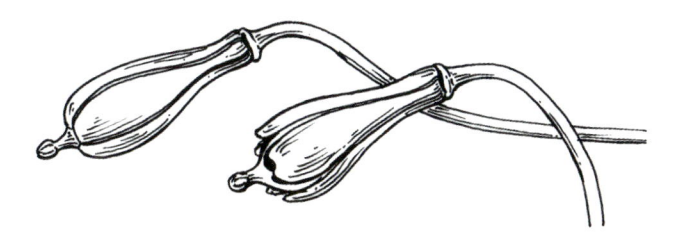

Subsp. *bella* is the most widely distributed in the wild spanning the Himalaya from central Nepal to Sikkim and western Bhutan. In leaf, plants look very like those of a *Paraquilegia* (Ranunculaceae) and both sometimes occupy the same cliffs in the wild. Apart from its finely dissected, rather small leaves, this subspecies is distinguished by its relatively small fruit capsules. The plant depicted in *Curtis's Botanical Magazine* in 1907 (tab. 8130) is of subsp. *bella*, seed having been sent to the Royal Botanic Garden Edinburgh in 1904 from Calcutta Botanic Garden, and presumably collected in Nepal or perhaps Sikkim.

George Taylor comments that:

"From the field notes available it appears that *M. bella* prefers a northern exposure on moist, sheltered, mossy rock ledges or crevices and it is never found in any great abundance in its native state. The rosette of leaves is closely applied to the surface of the rock, while the long tap-root delves into the crevices."

Genus Meconopsis: 106 (1934).

FIG. 33. *Meconopsis bella* subsp. *bella*, fruit capsules × 1¹/₃.
DRAWN BY CHRISTOPHER GREY-WILSON

79. MECONOPSIS BELLA

FIG. 34. *Meconopsis bella* subsp. *grandifolia*, leaf-laminas, lower petioles not shown. All × 1¹/₃. DRAWN BY CHRISTOPHER GREY-WILSON

Meconopsis bella subsp. grandifolia Grey-Wilson, **subsp. nov.** Differs from the other subspecies in its large, coarsely pinnatisect leaves with 1–3 (–4) pairs of lanceolate-elliptic lobes, the terminal lobe at least twice as large as in the other subspecies, and twice the size of the supporting lobes. Type: C Nepal, Annapurna Himal, Seti Khola, 13,500 ft, 30 July 1954, *Stainton, Sykes & Williams* 6551 (holotype, BM; isotype, E).

DESCRIPTION. Leaves mostly pinnatisect with 1–3 coarse elliptic to lanceolate-elliptic, primarily subobtuse, lobes (occasionally ovate to ovate-lanceolate and entire on the same plant), the end segment the largest, 2.8–7 × 1.1–2 cm, twice the size of the supporting lobes. *Fruit capsule* 18–24 mm, 6–7-valved. (Fig. 34).

DISTRIBUTION. Central Nepal (Kaski, Lamjung, Mustang, Myagdi districts); 3965–4420 m (Map 26).

HABITAT. Grassy and rocky banks, amongst dwarf shrubs, especially rhododendrons, cliff ledges.

FLOWERING. June–August.

SPECIMENS SEEN. *G. Miehe* 287 (BM); *Stainton, Sykes & Williams* 1322 (BM), 1795 (BM), 3604 (BM, E), 4559 (BM), 6551 (BM, E), 8546 (E), 8562 (BM).

Subsp. *grandifolia* is the most restricted of the three subspecies, limited to a relatively small region of central Nepal, being most common in the upper Kali Gandaki valley, the Mardi and Seti Kholas, the upper Marsyandi and the adjacent Lamjung Himal. In general appearance it looks a coarse version of subsp. *bella* with fewer, larger leaf divisions, the basal leaves in the rosette sometimes entire or subentire.

Meconopsis bella subsp. subintegrifolia Grey-Wilson, **subsp. nov.** Distinguished from the other two subspecies by having mostly simple leaves, occasionally with a few uneven lobes or trilobed, often on the same plant; fruit capsules larger, 4–5-valved, moderately to densely bristly. Type: Bhutan, Namdating, Upper Trongsa Chu, 13,500–14,500 ft, 17 July 1949, *Ludlow, Sherriff & Hicks* 19437 (BM, holotype; E, isotype).

[syn. *M. zangnanensis* L. H. Zhou, *Acta Phytotax. Sin.* 17 (4): 112–113 (1979), F1, 1–3. Type: Tibet, Cuona, Lebu, c. 4000 m, *Guo Benzhao* 22993 (PE, holotype).

Meconopsis bella subsp. *subintegrifolia*; Jule Tsho, Thimphu district, C Bhutan, 4500 m.
PHOTO: TIM LEVER

79. MECONOPSIS BELLA

Meconopsis bella subsp. *subintegrifolia* leaf-rosette; Bangajang track, W Arunachal Pradesh, NE India. PHOTO: DAVID & MARGARET THORNE

DESCRIPTION. *Leaves* entire to occasionally pinnatisect (often on the same plant), with up to 1–2 pairs of prime divisions, the leaf-tip or lobes subacute to obtuse. *Fruit capsule* ellipsoid, 15–27 mm long, 4-valved, moderately to densely bristly. (Fig. 35).

DISTRIBUTION. West, central and north-east Bhutan (particularly Chesha La, Chojo Dzong, Dhur Chu, Me La, Ritang, Thampe La, Worthong La); NE India (Upper Assam; Ze La); S Tibet (Bimbi La (Tsari), Chayul-Charme, Cuona (Tsona), Kashong La, Le La, Mago, Pö La, Tulung La) (Map 26).

HABITAT. As for subsp. *bella*; 3658–4725 m.

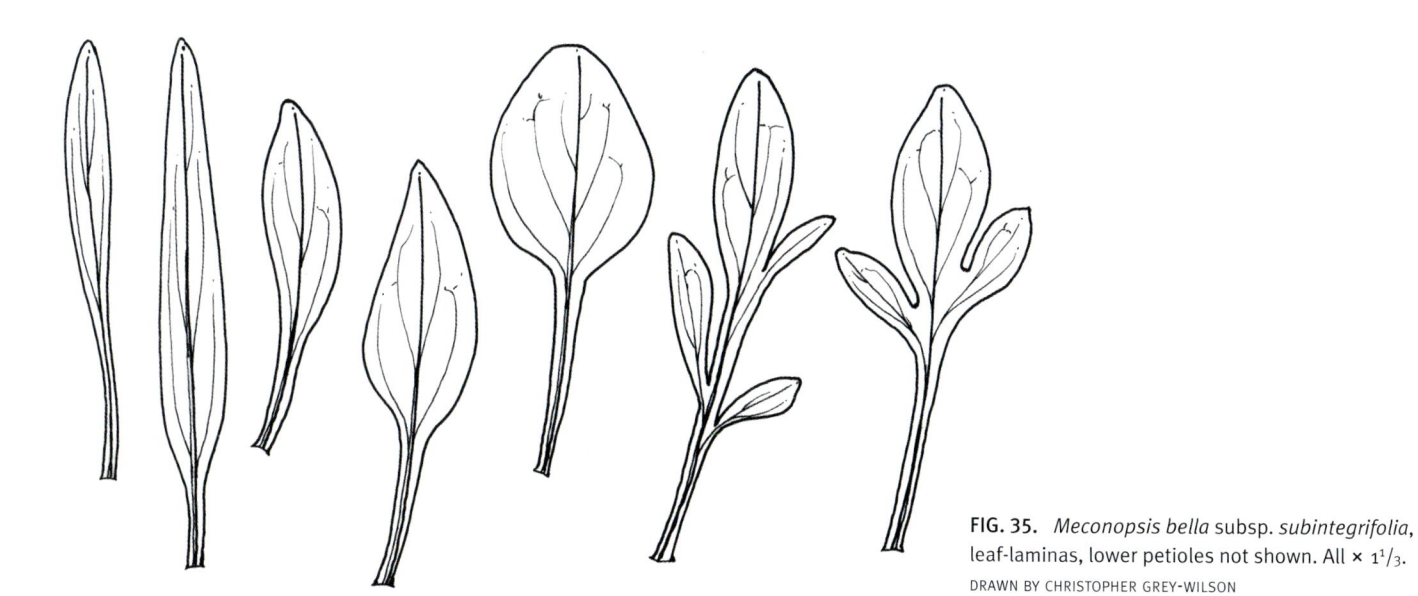

Meconopsis bella subsp. *subintegrifolia*; Tempe Tsho, Tongsa district, C Bhutan, 4600 m. PHOTO: TIM LEVER

FIG. 35. *Meconopsis bella* subsp. *subintegrifolia*, leaf-laminas, lower petioles not shown. All × $1^{1}/_{3}$.
DRAWN BY CHRISTOPHER GREY-WILSON

364 | THE GENUS MECONOPSIS
79. MECONOPSIS BELLA

Above: *Meconopsis bella* subsp. *subintegrifolia* in habitat; Tampe Tso, Wangdi-Phodrang, C Bhutan, 4300 m. PHOTO: TOSHIO YOSHIDA

Right from top:

Meconopsis bella subsp. *subintegrifolia*; Lenang, W Arunachal Pradesh, NE India, 4400 m. PHOTO: DAVID & MARGARET THORNE

Meconopsis bella subsp. *subintegrifolia*; Jule Tsho, Bumthang district, C Bhutan, 4500 m. PHOTO: MARTIN WALSH

Above left: *Meconopsis bella* subsp. *subintegrifolia*; Se La camp track, W Arunachal Pradesh, NE India. PHOTO: DAVID & MARGARET THORNE

Above right: *Meconopsis bella* subsp. *subintegrifolia* growing on a damp rock outcrop; Jule Tsho, Bumthang district, C Bhutan. PHOTO: MARTIN WALSH

SPECIMENS SEEN. *B. J. Gould* 468 (K), 1118 (K); *Kingdon Ward* 11821 (BM), 14151 (BM); *Ludlow & Sherriff* 708 (BM), 806 (BM), 2208 (BM), 2283 (BM), 2358 (BM), 2905 (BM), 3241 (BM), 3361 (BM), 3377 (BM, E), 3395 (BM), 3995 (BM), *Ludlow, Sherriff & Hicks* 16660 (BM), 16703 (BM, E), 17312 (BM), 19437 (BM, E), 19488 (BM, E), 19793 (BM), 20452 (BM, E), 21124 (BM); *Ludlow, Sherriff & Taylor* 6344 (BM), 6402 (BM), 6413 (BM).

This subspecies appears to overlap in distribution in western Bhutan with subsp. *bella* particularly in the region of Parshong (Barshong), near Thimphu; however, it is not clear whether the two subspecies can be found growing in the same localities or in intermixed colonies.

Meconopsis zangnanensis L. H. Zhou was described from the vicinity of Cona (= Cuona, Tsona) a remote corner of SE Tibet to the immediate north-east of Bhutan, some 200 km south-east of Lhasa, but fits well within the circumscription and distribution of *M. bella* subsp. *subintegrifolia*. Indeed several specimens confirmed as *M. bella* subsp. *subintegrifolia* were collected not far from Cona (e.g. *Ludlow & Sherriff* 708 from the Tulung La, for instance).

EXCLUDED SPECIES

(a full synonymy is not presented)

Meconopsis cambrica (L.) Vig. = *Papaver cambricum* L. = *Parameconopsis cambrica* (L.) Grey-Wilson (p. 367)
Meconopsis chelidonifolia Bureau & Franch. = *Cathcartia chelidonifolia* (Bureau & Franch.) Grey-Wilson (p. 374)
Meconopsis crassifolia Benth. = *Stylophorum diphyllum* (Michx.) Nutt.
Meconopsis diphylla DC. = *Stylophorum diphyllum* (Michx.) Nutt.
Meconopsis glabra Hook. = *Stylophorum diphyllum* (Michx.) Nutt.
Meconopsis heterophylla Benth. = *Stylomecon heterophylla* (Benth.) G. Taylor
Meconopsis oliveriana Franch. = *Cathcartia oliveriana* (Franch.) Grey-Wilson (p. 376)
Meconopsis petiolata DC. = *Stylophorum diphyllum* (Michx.) Nutt.
Meconopsis smithiana (Hand.-Mazz.) G. Taylor = *Cathcartia smithiana* Hand.-Mazz. (p. 373)
Meconopsis villosa (Hook. f.) G. Taylor = *Cathcartia villosa* Hook. f. ex Hook. (p. 370)

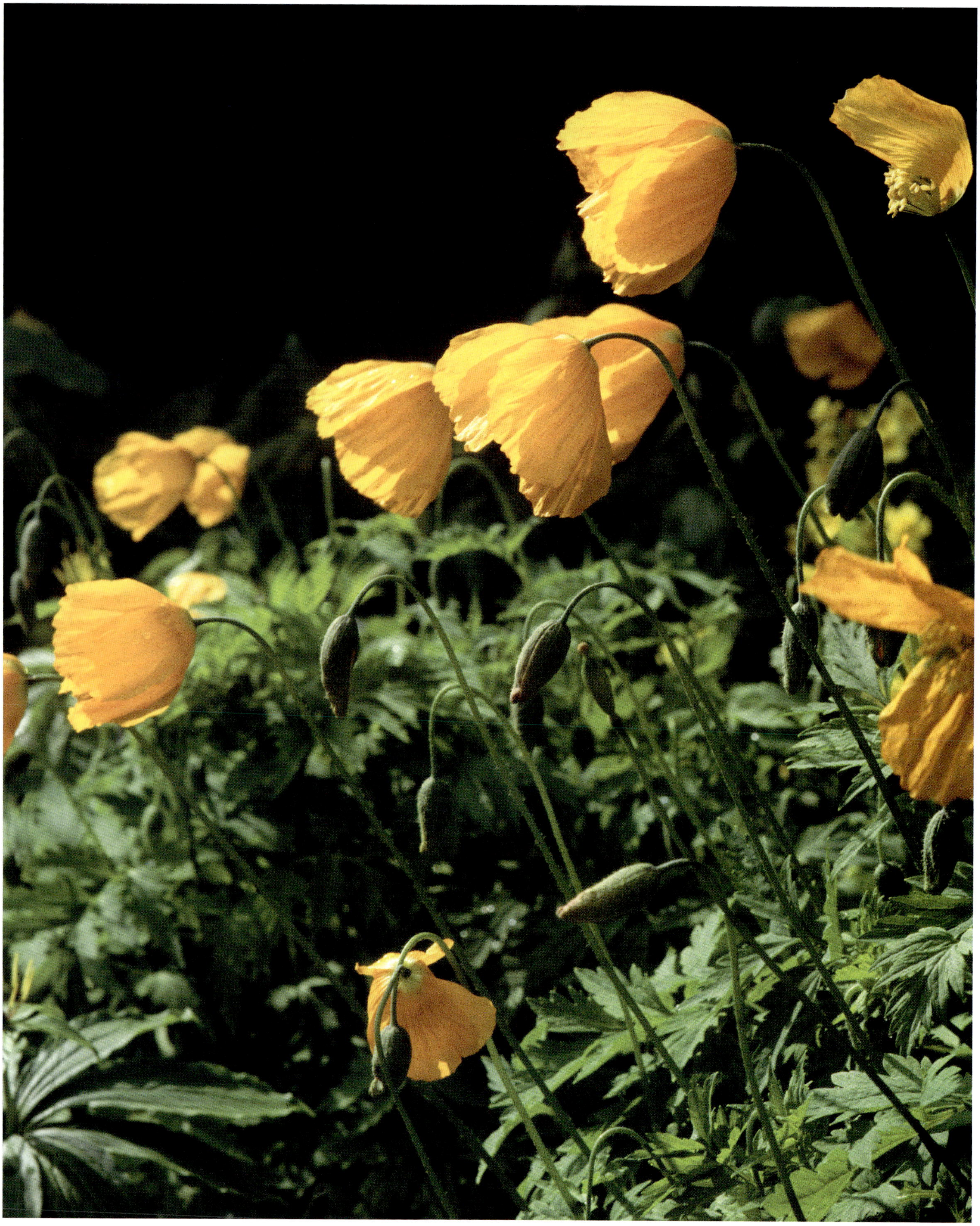

7. THE GENUS PARAMECONOPSIS

PARAMECONOPSIS Grey-Wilson
genus nova

Distinguished from *Papaver* by the presence of a well-developed style and by the absence of a cartilaginous stigmatic disk, this sometimes reduced to stigmatic ray tissue only or by deep incisions between the stigmatic rays (*Papaver* sect. *Meconella*), and from *Meconopsis* by the presence of pronounced pseudo-dorsal traces to the gynoecium. Type: *Papaver cambricum* L., *Sp. Pl.* (1753) based on a specimen in the *Hort. Cliff.*

[Syn. *Meconopsis* subgenus *Eumeconopsis* (Prain) Fedde section *Cambricae* (Prain) Fedde in Engl., *Pflanzenr.* 1v, 104: 251 (1909); *M.* section *Eumeconopsis* Prain pro parte; *Papaver* L., *Sp. Pl.* 1: 506 (1753) & *Gen. Pl.*, Ed. 5: 224 (1754) pro parte, quoad sp. 7; *Stylophorum* Nutt., *Gen. N. Amer. Pl.* ii: 7 (1818) pro parte, quoad *Papaver cambricum*; *Cerastites* Gray, *Nat. Arrang. Brit. Pl.* ii: 703 (1821) pro parte, quoad sp. 1; *Meconopsis* ser. *Chelidonifoliae* Prain, *J. Asiat. Soc. Bengal, Pt. 2, Nat. Hist.* 64, 2: 311 & 313 in obs; *M.* ser. *Cambricae* Prain, *Ann. Bot.* 20: 343 (1906)]

A monotypic genus native to western Europe: western Ireland, Wales, south-western England, western and southern France (especially the Pyrenees), northern Spain, growing primarily in damp shaded or part-shaded damp rocks, often in woodland, occasionally in meadows, to 2000 m; widely naturalised and cultivated in western Europe both within and without the area of natural distribution.

For a general discussion on the affinities and status of this taxon see p. 41.

PARAMECONOPSIS CAMBRICA

Parameconopsis cambrica (L.) Grey-Wilson, **comb. nov.**
[Syn. *Papaver cambricum* L., *Sp. Pl.* 1: 508 (1753); *Argemone cambrica* (L.) Desf., *Dict. Sci. Nat.* II: 481 (1804); *Meconopsis cambrica* (L.) Vig., *Hist. Pavots Argémon.* 48, fig. 3 (1814); *Cerastites cambricus* (L.) Gray, *Nat. Arrang. Brit. Pl.* II: 704 (1821); *Stylophorum cambricum* (L.) Spreng. in L., *Syst. Veg.* Ed. 16 (2): 570 (1825).]

Opposite:

Parameconopsis cambrica 'Aurantiacaa' in a Suffolk garden.
PHOTO: CHRISTOPHER GREY-WILSON

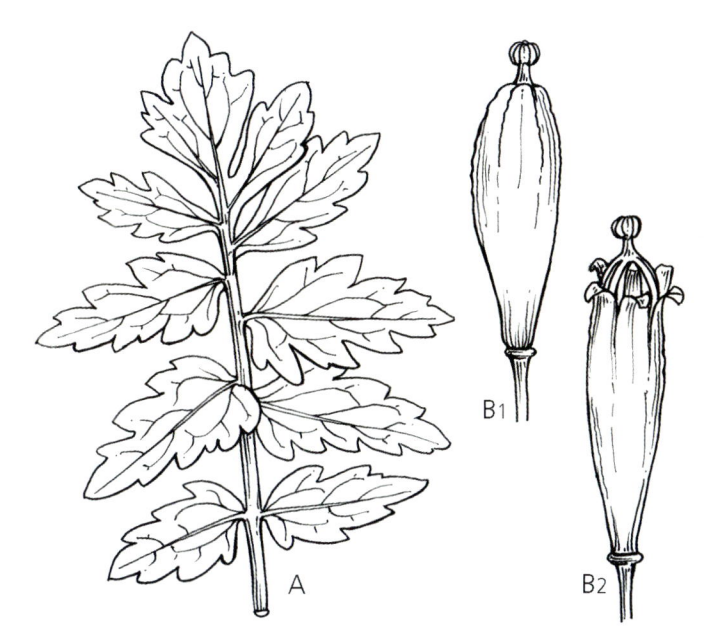

FIG. 36. *Parameconopsis cambrica*: **A**, basal leaf, lower petiole not shown; **B1**, fruit capsule undehisced; **B2**, fruit capsule dehisced. All × 1^1/$_3$. DRAWN BY CHRISTOPHER GREY-WILSON

Tufted perennial with a branched, rather fleshy rootstock, beset at the base with persistent scale-like leaf bases, bearing in mature plants a number of growing points. *Stems* erect, to 60 cm tall, often less, sparingly branched, leafy, glabrous to sparsely pubescent, producing a watery latex when cut. *Leaves* bluish or yellowish green, the basal long-petiolate, the lamina ovate-oblong in outline, pinnately lobed, pinnatisect in the lower part but increasingly pinnatifid towards the apex, the segments ovate-elliptic to oblong, variously lobed and dentate, glabrous to sparsely hairy, the petiole to 20 cm long, slender, somewhat expanded and part-sheathing at the base; cauline leaves spirally arranged, similar to the basal but increasingly smaller up the stem, the lower short-petiolate, the uppermost sometimes subsessile. *Inflorescences* ascending 1–several per leaf rosette, arising from the axils of the basal rosette leaves. *Flowers* 1–5, borne singly, terminal or from the axils of the upper leaves, cupped to saucer-shaped, 4.5–7.8 cm across, ascending to upright. *Pedicels* erect, slender, 16–25 cm long, glabrous to sparsely hairy. *Petals* 4, rarely 5, obovate to subrounded, 20–38 mm long, somewhat undulate,

Parameconopsis cambrica naturalised in a lane in SW Devon. PHOTO: CHRISTOPHER GREY-WILSON

yellow, occasionally orange. *Stamens* numerous, with linear, whitish filaments and yellow anthers. *Ovary* pale green, ovoid to ellipsoid, glabrous, narrowed above into a short but distinct, slender style, 1.5–3 mm long, the stigma capitate, 4–7-lobed. *Fruit capsule* ellipsoid-oblong to subobovoid, 20–40 mm long, 4–7-ribbed, glabrous, splitting at the top at maturity by 4–7 valves for about a quarter of the capsule length. 2n=22. (Fig. 36).

DISTRIBUTION. SE England, Wales and SW Ireland, SW France, N Spain.

HABITAT. Moist rocky places, stream margins, woodland and other shady places, but widely naturalised from gardens in parts of northern Britain, western and northern Europe, as well as Switzerland; May–September (–October)

Prior to 1753 the species, which had been long-cultivated as both a medicinal and ornamental herb, had acquired quite a few synonyms of which the following are prominent:

Papaver erraticum Pyrenaicum luteo flore C. Bauhin, *Prodr. Theatr. Bot.*: 92 (1620) & *flore flavo* C. Bauhin, *Pinnax Theatr. Bot.*: 171 (1623).

Argemone lutea Cambro-Britanica John Parkinson, *Theatr. Bot.*: 369, fig. dextr. (1640).

Argemone Cambro-Britanica lutea John Parkinson, op. cit.: 370 (1640).

Papaver luteum perenne, laciniato folio Cambo-britannicum Ray, *Syn. Meth. Stirp. Brit.*: 122 (1690).

Papaver cambricum perenne, flore sulphureo Dill., *Hort. Eltham.* 2: 300, tab. 223 (1732).

Parameconopsis cambrica is widely cultivated in gardens today, often sowing itself around freely in damper situations and frequently escaping into the countryside, especially in the cooler, wetter places of western and south-western Europe. Various forms are recognised in cultivation including those with orange flowers (var. *aurantiaca* Hort. ex Wehrh.), or crimson flowers and those with double or semi-double flowers in which smaller, narrower petals replace some of the outer stamens.

The Welsh poppy has been a popular garden plant for hundreds of years and can be considered an easy plant in all but very dry soils, enduring sun or part-shaded situations. Indeed it can be prolific in some gardens, seeding around freely. While some gardeners regard it as a weed, others cherish its cheerful poppy flowers borne over a long season from early summer to late autumn. It is often treated as a wilding in gardens, seeding around into those damper, shadier corners where its only competition are ferns and other shade-lovers. Plants can form substantial multi-stemmed clumps in time, dying down to a series of crowns in the autumn and reappearing during the early days of spring. While yellow (var. *cambrica*) and occasionally orange (var. *aurantiaca* = 'Aurantiaca') are the colours seen in the wild, others have been selected in cultivation, including those with double or semi-double flowers of which the following are best known: 'Flore Pleno Aurantiaca' (semi-double orange), 'Flore Pleno' (semi-double yellow), 'Francis Perry' = 'Rubra' (single, deep orange-crimson) and 'Muriel Brown' (semi-double red).

8. THE GENUS CATHCARTIA

CATHCARTIA Hook. f. ex Hook.

Cathcartia Hook. f. ex Hook., *Curtis's Bot. Mag.* 77: t. 4596 (1851).
[Syn. *Cathcartia* section *Eucathcartia* Prain, *Ann. Bot.* 20: 368 (1906); *Meconopsis* section *Eumeconopsis* Prain, *Ann. Bot.* 20: 343 (1906) pro parte; *M.* subgenus *Eumeconopsis* (Prain) Fedde section *Eucathcartia* (Prain) G. Taylor, *The Genus Meconopsis*: 24 (1934)]

The genus *Cathcartia* was established in 1851, based on a single species *Meconopsis villosa*, a species restricted to the central Himalaya. Additional species were referred to *Cathcartia* in the following years: most notably *C. delavayi*, *C. integrifolia*, *C. lancifolia*, *C. lyrata*, *C. polygonoides*, *C. punicea* and *C. smithiana*, all but the last named were subsequently transferred to *Meconopsis*. However, in 1934 George Taylor included all these species within *Meconopsis* upon publication of his monograph. Taylor placed *M. villosa* (Hook. f.) G. Taylor and *M. smithiana* (Hand.-Mazz.) G. Taylor ex Hand.-Mazz. in section *Eucathcartia* (Prain) G. Taylor adding two further species, *M. chelidonifolia* Bureau & Franch. described in 1891 and *M. oliveriana* Franch. ex Prain in 1896. Taylor in fact recognised two series within *Eucathcartia* as follows:

1. Stem unbranched, with flowers borne singly in the axils of the uppermost leaves (series *Chelidonifoliae*): *M. smithiana*, *M. villosa*

2. Stem branched, the branches leafy and generally bearing several flowers (series *Villosae*): *M. chelidonifolia*, *M. oliveriana*

These four species are undoubtedly closely related, being polycarpic plants with branching root systems, the tufted bases of the plants beset with numerous rufous, barbellate hairs, which are also present on the branched or unbranched stems and leaves. The flowers are consistently yellow, rather small and 4-petalled, being borne singly in the axils of the upper leaves/bracts, or several on leafy side shoots. The leaf-laminas are very distinctive, being basically trilobed or trifoliolate, the lobes often pinnatisect, sometimes pinnately lobed. Although very similar in flower it is in the character of the fruit capsule that most divergence is to be found: in both *Cathcartia chelidonifolia* and *C. oliveriana* the capsule is oblong or ellipsoid, splitting by 4–6 valves for a short distance from the top; *C. villosa* has subcylindric capsules splitting for a half to three-quarters their length into 4–7 valves; *C. smithiana* has small subglobose, 5-valved capsules heavily beset with appressed bristles.

Cathcartia has a distribution from the central Himalaya to western and south-western China with one species, *C. oliveriana* found further east in S Shaanxi, Henan, E Sichuan and W Hubei. All the species are typically mesic forest floor inhabitants, particularly favouring moist forest glades, ravines and streamsides, occasionally venturing into moist meadows or shrubberies. In contrast, *Meconopsis* are species of open habitats, the majority being found above the treeline, although one or two species such as *M. betonicifolia* and *M. wallichii* are generally found in monsoon-saturated open forests.

The relationship of *Cathcartia* to *Meconopsis* is an interesting one both from a morphological as well as a phylogenetic point of view. Kadereit *et al.* (1997, 2011) have clearly demonstrated that the Asian representatives of *Meconopsis* consist of two sister groups with a small basal-most clade represented by *M. chelidonifolia* and *M. villosa* (along with *M. oliveriana* and *M. smithiana* which were not used in their studies) and the remainder of Asian *Meconopsis*, some 79 species as currently recognised, as a distinct group. It would be inappropriate to detail all the various findings and arguments in the various papers from Carolan *et al.* and Kadereit *et al.* (see bibliography at the end of this work), however, Kadereit *et al.* (2011) have summed up the position very precisely in a paper which set out to examine the position of the Welsh poppy, *Meconopsis cambrica*, in relation to Asian *Meconopsis* and *Papaver* in general and as these have a bearing on the treatment of these various taxa in this monograph, I quote directly from their work:

1. The Arctic-montane species of *Papaver* sect. *Meconella* need to be placed either in a new genus by themselves or together with the larger of the two clades of Asian *Meconopsis*. Morphology would suggest treatment as two separate genera. The morphological distinctness of sect.

Meconella and arguments in support of a possible close relationship to Asian *Meconopsis* have been discussed by Kadereit *et al.* (1997). To our knowledge, no generic name is available for the treatment of sect. *Meconella* at generic rank.

2. *Papaver* sect. *Argemonidium* should be included in *Roemeria*, and there indeed exists strong evidence from pollen and sepal morphology (discussed by Kadereit *et al.* 1997) for a close relationship between these two groups. This approach differs from that recently adopted by Stace (2010) who included *Roemeria* in *Papaver*. In *Roemeria*, a combination at least for *P. argemone* is available, i.e. *Roemeria argemone* (L.) C. Morales, R. Mend. & Romero García.

3. The basal lineage of Asian *Meconopsis* needs to be treated as an independent genus, for which the name *Cathcartia* Hook. f.[43] is available.

4. The large clade of Asian *Meconopsis* also needs to be treated as an independent genus, for which to our knowledge no generic name is available. In order to avoid an inordinate number of nomenclatural changes, the conservation of the generic name *Meconopsis* Vig. should be proposed[44], with a different, conserved type taken from this large Asian clade.

---

[43] Although many authors have included *Cathcartia* within *Meconopsis* over the years (for instance George Taylor 1934) there has been no overall agreement. Based on their general morphology both Grey-Wilson in *Poppies* 1993 and David Long *Flora of Bhutan* 1984, opted for recognition of *Cathcartia* as separate from *Meconopsis* and subsequent molecular evidence supports this view.

[44] Such a proposal was presented by the current author during the preparation of this monograph and the proposal for the conservation of the generic name *Meconopsis* for the Asian representatives of the genus can be found in full in Appendix One as it appeared in *Taxon*, April 2012.

KEY TO SPECIES OF GENUS CATHCARTIA

1a. Stems unbranched, bearing flowers singly in the axils of the uppermost leaves (Section *Villosae*). 2
 + Stems branched, the branches leafy and bearing several flowers in the leaf-axils (Section *Chelidonifoliae*) . . 3

2a. Stems bearing small vegetative buds in the axils of the upper leaves; fruit capsule ellipsoid, 5–6-valved, glabrous or slightly hairy . . . . **C. chelidonifolia** (W, NW & SW Sichuan, NE Yunnan)
 + Stem lacking small vegetative buds; fruit capsule narrow oblong-cylindric, 4–5-valved glabrous. . . . . **C. oliveriana** (Chongqing, Henan, W Hubei, S Shaanxi, E Sichuan)

3a. Leaf-lamina palmately lobed; fruit capsules subcylindric, glabrous, 4–7-valved. . . . . **C. villosa** (C Himalaya, E Nepal to Bhutan)
 + Leaf-lamina trifoliolate; fruit capsule subglobose, bristly, 5-valved. . . . .**C. smithiana** (NE Myanmar, NW Yunnan)

section VILLOSAE (G. Taylor) Grey-Wilson comb. nov.

[Syn. *Meconopsis* series *Villosae* G. Taylor, *Genus Meconopsis*: 27 (1934)]
Type species: *Cathcartia villosa* Hook. f. ex Hook.

CATHCARTIA VILLOSA

Cathcartia villosa Hook. f. ex Hook., *Curtis's Bot. Mag.* 77: tab. 4596 (1851)]. Type: ? E Sikkim, *Hooker & Thomson* s.n. 1849 (K).
[syn. *Meconopsis villosa* (Hook. f. ex Hook.) G. Taylor, *Genus Meconopsis*: 28 (1934)].

DESCRIPTION. *Polycarpic plant* with a rather short, branching, stoutish rootstock, invested at the top with persistent petiole remains and rufous, finely barbellate bristles. *Stems* erect, generally several per plant, 30–60 cm tall, occasionally somewhat taller, generally unbranched, rufous or greyish villous throughout. *Leaves* mid grey-green above, somewhat glaucous beneath, the *basal* few, clustered, long-petioled, the lamina broad-ovate to suborbicular, 6–12.5 × 8–13.5 cm, 3–5-lobed, the base truncated to cordate, the lobes coarsely and obtusely to subacutely toothed, densely rufous villous on both surfaces, the petiole 15–25 cm long, villous; cauline leaves spirally arranged, well-spaced, similar to the basal but with increasingly shorter petioles, the uppermost subsessile, the lamina similar in shape and size to the basal leaves, sometimes somewhat larger. *Flowers* up to 5 per stem, solitary in the axils of the uppermost leaves, nodding to half-nodding, open bowl-shaped, 4.5–6.5 cm across. *Pedicels* rather slender, 3–13.5 cm long, rufous villous, arched towards the top. *Buds* pendent, narrow ellipsoid, patent rufous villous. *Petals* 4, yellow, obovate to suborbicular, 2.5–3.5 × 1.5–3.8 cm long, often somewhat broader, the apex rounded to obtuse. *Stamens* numerous with filiform yellow filaments and yellow anthers that brown on ageing. *Ovary* oblong-cylindric, glabrous; stigma sessile, rarely with a very short style, with 4–7 radiating lobes. *Fruit capsule* erect, subcylindric, 4–8.2 cm long, 0.5–0.7 cm wide, somewhat ribbed, glabrous, at maturity splitting from the top for more than half its length, sometimes almost to the base, into 4–7 narrow valves. *Seeds* subfalcate, the testa with longitudinal rows of narrow pits. (Fig. 37A1–2, B1–2).

DISTRIBUTION. Eastern Nepal (Sankhuwasabha & Taplejung districts; apparently not found west of the Arun River) to Sikkim and N & C Bhutan (Sakden, Thimphu, Tongsa and Upper Kulong Chu districts), SE Tibet (Tawang); 2700–4000 m (Map 27).

HABITAT. Rocky *Tsuga* and *Abies* forests, rocky streamsides, and rock clefts, in partial shade.

FLOWERING. May–July.

SPECIMENS SEEN. *East Nepal Exped.* (1992) 9240918 (E); *Grierson & Long* 405 (E); *Ludlow & Sherriff* 649 (BM), 3251 (E); *Ludlow, Sherriff & Hicks* 20817 (E), 21446 (E); *Sinclair & Long* 5574 (E); *Sixth Botanical Exped. Himalaya* (1977) 772695 (E).

Cathcartia villosa, the only Himalayan representative of the genus, is the most attractive and garden-worthy of the species, a handsome plant with neat, rounded, scalloped foliage and attractive semi-nodding flowers of good size. It is unsurprising therefore that it has been

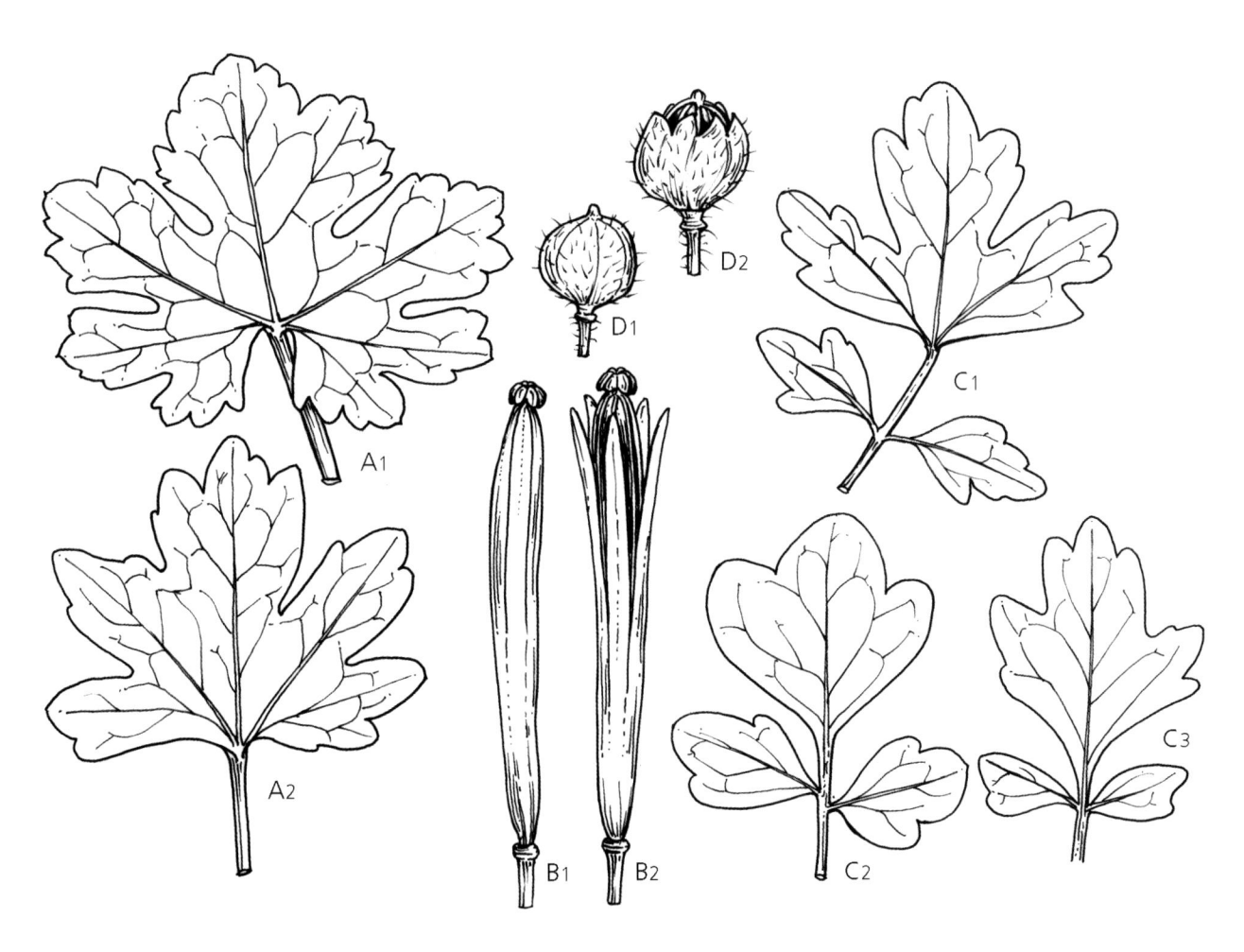

FIG. 37. *Cathcartia villosa*: **A1**, basal leaf; **A2**, cauline leaf; **B1**, undehisced fruit capsule; **B2**, dehisced fruit capsule. *C. smithiana*: **C1**, basal leaf; **C2–3**, cauline leaves; **D1**, undehisced fruit capsule; **D2**, dehisced fruit capsule. Pubescence and lower petioles not shown. All × 1¹/₂.

DRAWN BY CHRISTOPHER GREY-WILSON

Above left: *Cathcartia villosa*; Maroothang, C Bhutan, 3500 m. PHOTO: MARTIN WALSH

Above right: *Cathcartia villosa*; S of Maroothang, W Bhutan, 3600 m. PHOTO: TOSHIO YOSHIDA

a favourite woodlander in favoured gardens for many years. George Taylor (1934) remarks that "it is a plant of considerable horticultural merit. The perennial habit, rich yellow flowers and altogether attractive appearance are sufficient, one would have thought, to ensure to a wider reputation than it enjoys at present". The species was introduced into Britain by its discoverer Joseph Dalton Hooker from Sikkim, and first flowered in Britain in 1851. It is a plant primarily of forest glades, although Hooker found it growing in rock crevices within the forest zone. While plants can be propagated by division, this method gives rather unsatisfactory results. Far better, and indeed easier, is to grow new plants from seed; seed is produced in quantity in most seasons and sown fresh gives very good results. While most *Meconopsis* species thrive best in cooler more northerly or westerly gardens, *C. villosa* is more tolerant of the warmer south provided that is it is grown in a moist humus-rich soil in partial shade, in fact in similar places to other Papaveraceae, *Stylophorum diphyllum* and *Chelidonium majus* for instance.

The fruit capsule of *Cathcartia villosa* is very distinct and cannot be confused with any other *Cathcartia* or indeed any species of *Meconopsis*: they are long and cylindrical, splitting when ripe for at least half of their length from close to the apex, sometimes splitting almost to the base with up to seven valves.

CATHCARTIA SMITHIANA

Cathcartia smithiana Hand.-Mazz., *Anz. Akad. Wiss. Wien, Math.-Naturwiss. Kl.* 60: 182 (1923). Type: SW China, NW Yunnan, 1916, *Handel-Mazzetti*.
[syn. *Meconopsis smithiana* (Hand.-Mazz.) G. Taylor ex Hand.-Mazz., *Symb. Sin.* VII: 337 (1931)]

DESCRIPTION. *Herbaceous perennial* 30–90 cm tall in flower, with a short, thickened, stout stock, bearing at its apex a dense tuft of persistent leaf-bases with accompanying brown barbellate hairs, with a fibrous root system below; resting buds globose, scaly, to 20 mm diameter, at ground level. *Stems* simple, erect to ascending, somewhat deflexed barbellate-hirsute, becoming more or less glabrescent ultimately, with withered leaf remains at the base at flowering. *Cauline leaves* distant, petiolate or the uppermost subsessile, the lamina trifoliate, 8–11.2 × 5.5–8 cm, sparsely appressed barbellate-hirsute on both surfaces, the terminal leaflet broad-ovate to rounded, pinnately-lobed with often 5–7, rounded to subacute, lobes, the lateral leaflets somewhat smaller, usually 3-lobed, separated from the terminal leaflets by a 1–3 cm long rachis; upper leaves increasingly smaller, ovate, pinnately- or bipinnately-lobed; petioles of lower and middle leaves 1–10 cm long, appressed to slightly deflexed hirsute; petiolules up to 10 mm long, sometimes absent, pubescent as the petioles. *Inflorescence* a bracteate raceme with generally 3–4 flowers borne from the uppermost leaves (bracts), the flowers saucer-shaped, 3.4–3.8 cm across. *Buds ellipsoid*, the sepals hirsute. *Petals* 4, yellow, broad-obovate, c. 18 × 18 mm, slightly emarginate at the obtuse apex. *Ovary* subspherical, densely appressed brown barbellate-hirsute, the style very short of obsolete, bearing a 5–7-lobed stigma. *Fruit capsule* obconic, c. 15 mm long with a 10 mm long apex, 5–7-valved, sparsely covered in brown barbellate bristles, splitting for about a third of its length from the apex when mature. *Seeds* ellipsoid, with longitudinal grooves. (Fig. 37C1–3, D1–2).

DISTRIBUTION. NW Yunnan (Dongshan, Lahwangtum, Ludjiang, Djioudjiang), NE Myanmar (Meku-ji Pass, Nam Tamai Valley); 3100–3600 m (Map 27).

HABITAT. Moist habitats in forests and along forest margins, as well as grassy slopes, stony alpine meadows, gulleys amongst coarse herbs, open shrubberies, margin of bamboo thickets.

FLOWERING. June–August.

SPECIMENS SEEN. *Bartholomew et al.* 12945 (E); *Farrer* 1780 (E); *Gaoligong Shan Expedition 1999*, 9592 (E); *Handel-Mazzetti* 1755 (E), 9254 (E); *T. T. Yü* 20663 (E); *Kingdon Ward* 13200 (K).

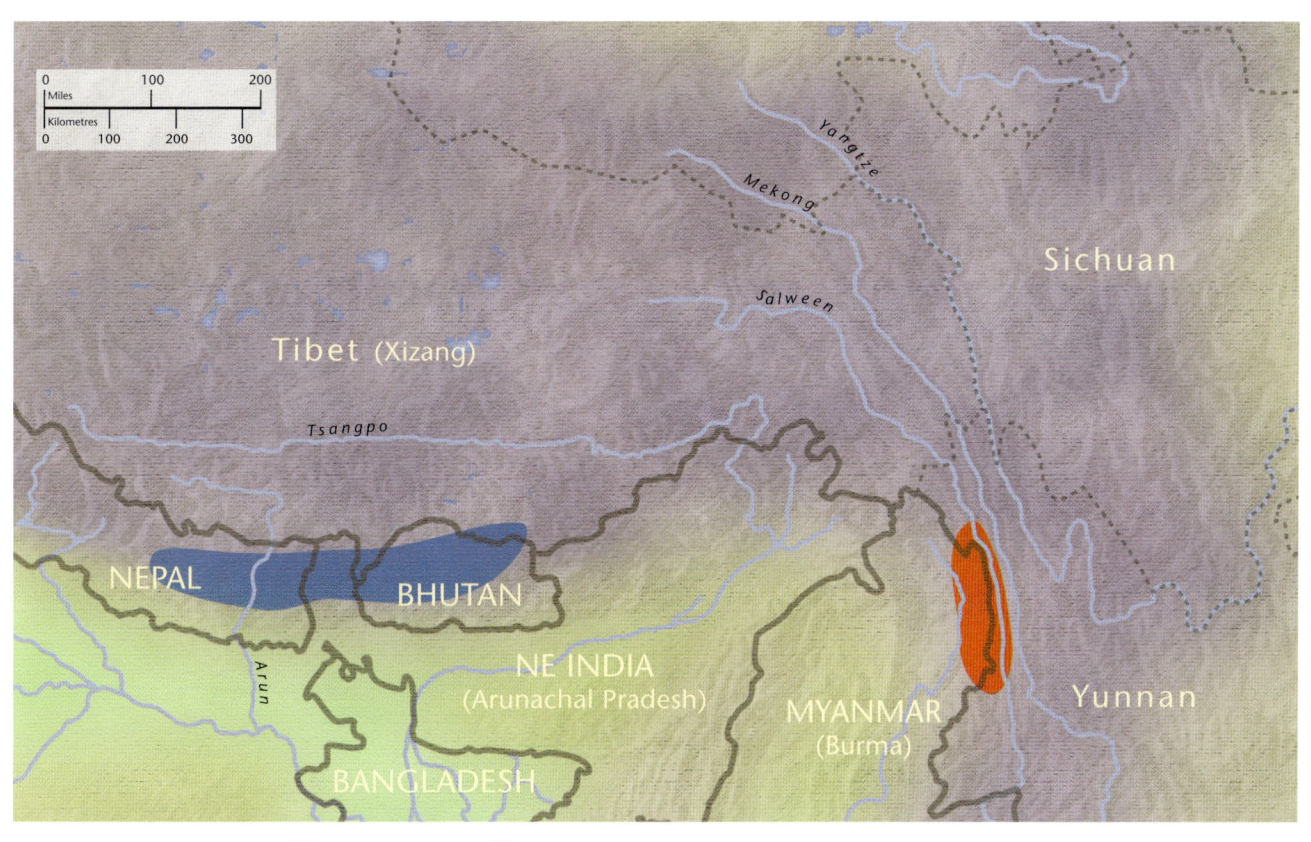

MAP 27. Species distribution: ■ *Cathcartia villosa*, ■ *C. smithiana*.

In many respects *Cathcartia smithiana* is the Chinese equivalent of the Himalayan *C. villosa*, being similar in habit and stature, although *C. smithiana* has smaller flowers. However, the two species differ in several important respects: most notable are the leaves which are trifoliate in *C. smithiana* but palmately lobed in *C. villosa*, but perhaps most notably in the characters of the fruit capsules, short, obconic and appressed-bristly in *C. smithiana* and splitting for about a third of its length from the apex, but long, cylindric and glabrous in *C. villosa* and splitting for at least half its length. Unfortunately, only *C. villosa* is in cultivation, while *C. smithiana* has never been in cultivation to my knowledge which is a pity for it would be good to make comparisons of living material.

Cathcartia smithiana was discovered by Handel-Mazzetti in north-western Yunnan in 1916 but not described by him until some 15 years later. The species is named in honour of Sir William Wright Smith, a former Regius Keeper of the Royal Botanic Garden, Edinburgh. Since its initial discovery this interesting and clearly localised species has been collected just a handful of times, mostly in north-western Yunnan but also just across the border in Myanmar (Burma) by Reginald Farrer. The only recent collection that I have been able to examine was made by the Gaoligong Shan Expedition of 1999, NW Yunnan very close to the Myanmar frontier. Although the original description states that *C. smithiana* is monocarpic, perennating basal buds on several of the herbarium collections show that it is in fact a sound perennial. Despite the fact that this interesting species has been seldom collected it was stated to be locally abundant; however, in the modern era few collectors have ventured close to Yunnan-Myanmar frontier area which appears to be its stronghold.

section CHELIDONIFOLIAE (Prain) Grey-Wilson, comb. nov.

[Syn. *Meconopsis* series *Chelidonifoliae* Prain, *J. Asiat. Soc. Bengal*, Pt. 2, *Nat. Hist*. 64 (2): 311 & 313 in obs. (1896), nom. nud., pro parte, excl. *Meconopsis cambrica*, & *Ann. Bot*. 20: 364 (1906); *M*. sect. *Polychaetia* Prain op. cit.: 352 (1906) pro parte; *M*. subgen. *Polychaetia* (Prain) Fedde in Engl., *Pflanzenr*. 4, 104: 262 (1909) pro parte, quoad sect. 9; *M*. sect. *Chelidonifoliae* (Prain) Fedde, op. cit.: 207 (1909)]. Type species: *M. chelidonifolia* Bureau & Franch.

CATHCARTIA CHELIDONIFOLIA

Cathcartia chelidonifolia (Bureau & Franch.) Grey-Wilson, **comb. nov.**

[syn. *Meconopsis chelidonifolia* Bureau & Franch., *J. Bot. (Morot)* V:19 (1891). Type: W Sichuan, Tatsienlu [Kangding], 1890, *Bonvalot & Henri d'Orléans* (syntypes P)]

DESCRIPTION. Herbaceous perennial 50–150 cm tall in flower, with a much-branched slender rootstock terminating in short branching rhizomes, the stem bases beset with persistent fibrous petiole remains and dense rufous, barbellate setae; over-wintering buds ovoid, to 2.5 cm diameter, with overlapping ovate scales, invested with dense rufous bristles, borne at ground level. *Stems* erect, rather slender, 5–8 mm diameter close to the base, longitudinally striate, green, often with a purplish flush, appressed rufous-setose in the lower part, especially close to the base, otherwise glabrous; vegetative buds (bulbils) present in some of the upper leaf-axils. *Leaves* mid green above, glaucous beneath, sparsely barbellate-hirsute, the basal withered by flowering time; basal and lower cauline leaves, ovate to ovate-oblong, 7–8 × 6.5–7 cm, pinnately-lobed with 3–5 rather distant lobes, these further pinnately-lobed or pinnatisect, the lobules ovate with an obtuse apex; upper cauline leaves sessile or almost so, the petiole, if present, 2–3 mm long, smaller than the lower leaves, the lamina broad-ovate, weakly 3-pinnatisect or 3-pinnately-lobed, the lobes pinnately-lobed; petiole of lower leaves 2–10 cm long, those of the middle stem to 2.7 cm long, subappressed pubescent to subglabrous. *Inflorescence* a very lax cymose panicle, 1 (–2) flowers borne from the uppermost leaf- (bract-) axils, the flowers half-nodding to ascending, saucer-shaped, 3–4.7 cm across; bracteoles present. *Pedicels* very slender, 5.2–7.2 cm long, glabrous. *Buds* subglobose, the sepals suborbicular, 8–12 mm long, glabrous, with one margin membranous. *Petals* 4, yellow, obovate to suborbicular, 15–23 × 15–25 mm. *Stamens* numerous, with filiform filaments, 4–5 mm long, the same colour as the petals, the anthers yellow, narrow-oblong, 3.5–5 mm long. *Ovary* ellipsoid, 5–6 mm long, glabrous or slightly hispid towards the base; style very short, c. 1 mm long or more or less obsolete, with a capitate stigma. *Fruit capsule* ellipsoid, 10–15 × 7–10 mm, 4 (–6)-valved, glabrous or somewhat bristly towards the base, splitting for a short distance from the apex when mature. *Seeds* falcate-oblong, c. 0.8 mm long, adorned with slightly impressed longitudinal pitted stripes. (Fig. 38A1–2, B).

DISTRIBUTION. W & S Sichuan (particularly Emeishan and Minyakonka-Kangding), NE Yunnan (Daguan, Yongshan); 1400–2700 m (Map 28).

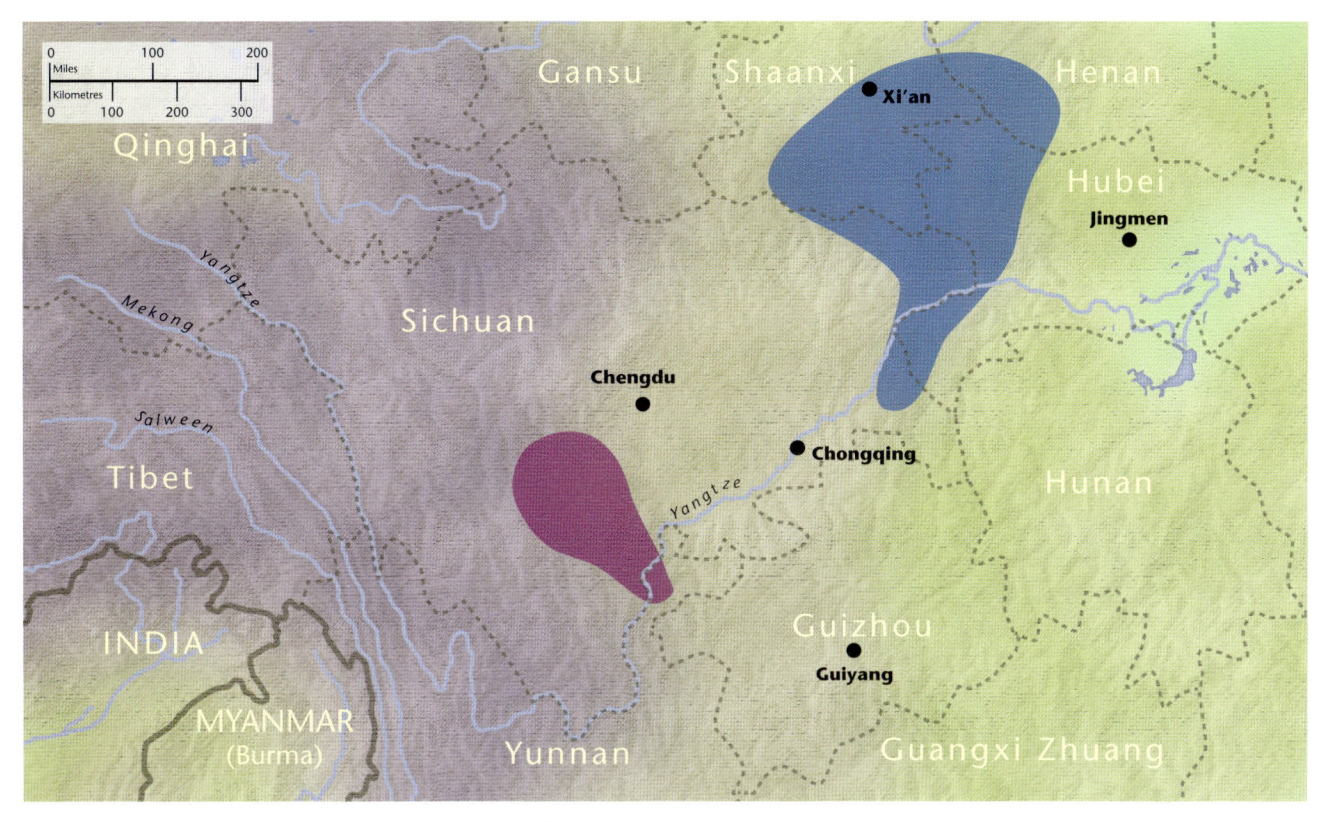

MAP 28. Species distribution: ■ *Cathcartia oliveriana*, ■ *C. chelidonifolia*.

HABITAT. Forests, ravines, stream and roadsides.
FLOWERING. May–August.
SPECIMENS SEEN. *Botanical & Forestry Department Hong Kong* 1410 (K); *Faber* 713 (K), 1387 (K); *Henry* 8793 (K); *A. E. Pratt* 134 (BM, K), 135 (BM, K); *H. Smith* 10121 (BM); *E. H. Wilson* 945 (K), 3161 (BM), 4717 (K).

Cathcartia chelidonifolia was first collected by Faber on Mt Omei (Emeishan) in Sichuan and sent to Britain in 1887; however, it was three years later that the species was named (as a species of *Meconopsis*) by Bureau and Franchet based upon specimens obtained by Bonvalot and Prince Henry d-Orléans who had collected it

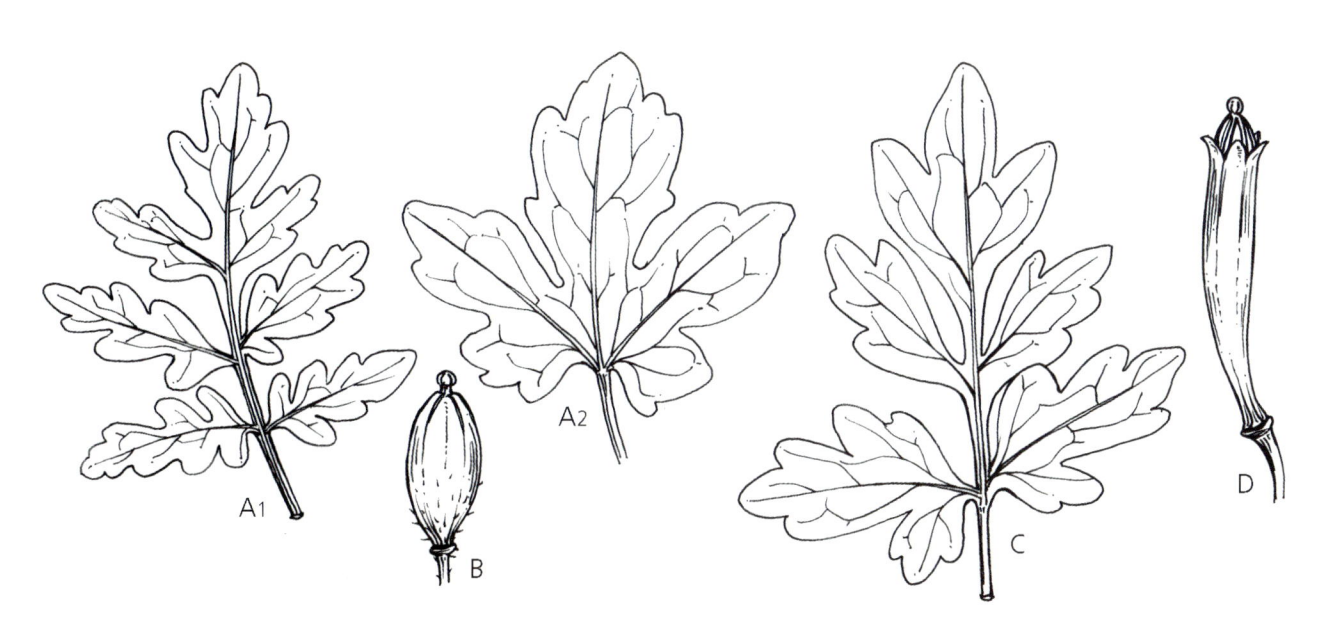

FIG. 38. *Cathcartia chelidonifolia*: **A1**, basal leaf; **A2**, middle cauline leaf; **B**, undehisced fruit capsule. *C. oliveriana*: **C**, lower cauline leaf; **D**, dehisced fruit capsule. Pubescence and lower petioles not shown. All × 1¹⁄₂. DRAWN BY CHRISTOPHER GREY-WILSON

Cathcartia chelidonifolia in cultivation. PHOTO: GRAHAM CATLOW

close to Tatsienlu, now Kangding. It appears to be especially common in the Kangding area and on Mt Omei (Emeishan) where most of the specimens have been gathered in the past. Ernest Henry Wilson also collected it close to the type locality, commenting that it grew "about 3 ft in height with clear yellow flowers, saucer-shaped and 2 1/2 inches across". Both Henry and Wilson described it as being common in thickets in the Kangding area. This is perhaps surprising as it has been little seen or photographed in more recent times.

Wilson was apparently responsible for its introduction into cultivation in Britain in 1904, but although it presents the grower with little difficulty to cultivate and is not tricky to propagate, it has never been popular in gardens, often being dismissed as a quiet woodlander. Yet the species has a refined charm and, in gardens at least, has an extended flowering season, that often reaches on into the autumn. Its superficial resemblance to a large-flowered greater celandine, *Chelidonium majus*, considered a weed in many gardens, has scarcely helped its cause in our gardens. *C. chelidonifolia* has the unusual faculty of producing vegetative buds in the axils of the upper leaves and these provide a ready means of propagation. Among the species of *Meconopsis* only *M. bulbilifera* has such a feature.

Cathcartia chelidonifolia is very closely related to *C. oliveriana* and the two could easily be mistaken for one another at a glance, especially in details of their general habit and foliage. However, the latter lacks stem bulbils (vegetative buds) and the fruit capsule is more cylindric, narrower and at least twice as long, and consistently glabrous. Both species have a very short or more or less obsolete style. While *C. chelidonifolia* is a western and south-western Chinese species, *M. oliveriana* is a central Chinese species, in fact it is the most easterly distributed of any *Cathcartia* or *Meconopsis*.

CATHCARTIA OLIVERIANA

Cathcartia oliveriana (Franch. ex Prain) Grey-Wilson, comb. nov.

[syn. *Meconopsis oliveriana* Franch. ex Prain, *J. Asiat. Soc. Bengal*, Pt. 2, *Nat. Hist.* 64: 312 (1896). Type: Hubei, *Henry* 6863 (syntypes, BM, K); Sichuan, *Farges* 390, "Sutchuen oriental, Tchen-keov-tin", China 1895–7 (syntypes, BM, P). Lectotype chosen here: Hubei, *Henry* 6863 (K; isolectotypes BM, K).

DESCRIPTION. *Herbaceous perennial* to 100 cm tall in flower, but as little as 40 cm, the rootstock rather stout, bearing perennating buds on short stolons, with numerous slender roots. *Stem* leafy, erect to ascending, branched, prominently grooved, to 6 mm diameter near the base glabrous or sparsely soft-bristly in the lower part, the base, beset with the persistent remains of withered leaves and their accompanying dense, rufous, barbellate bristles. *Leaves* dark green above, glaucous beneath, sparsely rufous-hirsute above and below, but the uppermost leaves and bracts often glabrous or subglabrous, the basal and lower leaves generally withered by flowering time, petiolate; middle and upper cauline leaves increasingly smaller, sessile or subsessile, the uppermost semi-amplexicaul; lamina ovate to oblong-ovate, 5–10 × 3–5 cm, pinnatisect towards the base, pinnatifid towards the apex, the lobes 3–5, rather distant, petiolulate to subsessile, ovate to obovate, pinnatifid, the lobules ovate to obovate, the terminal lobe to 6.5 × 4.8 cm, slightly heart-shaped and semi-amplexicaul at the base, the apex obtuse; upper leaves trilobed, to unlobed, the uppermost merging with the sessile, unlobed leaf-like bracts; petiole 0.4–4.5 (–6.8) cm long, sometimes more or less obsolete; petiolules 4–15 mm long, sparsely to moderately pubescent as are the petioles and rachises. *Flowers* borne in lax cymose panicles, solitary or paired, occasionally 3, at the upper leaf-axils, generally accompanied by small leaf-like bracteoles, ascending to semi-nutant, saucer-shaped, 2.2–4.5 cm across. *Pedicels* very slender, 3.6–8.5 cm long (to 14.5 cm long in fruit), glabrous. *Buds* spherical or ovoid, 9–13 mm long, semi-nutant to patent, the sepals elliptic, glabrous. *Petals* 4, yellow, rounded to ovate, 10–18 (–20) × 8–18 (–20) mm. *Stamens* numerous, the filaments filiform, 4–7 mm long, the anthers yellow, c. 1 mm long. *Ovary* narrow-oblong to oblong-cylindrical, glabrous, costate, with a very short (c. 1 mm long) style, sometimes obsolete, terminating in a 4–5-lobed, slightly decurrent stigma. *Fruit capsule* narrow-oblong to subcylindric, 24–45 × 3.5–6 mm, straight or slightly curved, 4–5-valved, with a very short or obsolete style, glabrous, splitting for a short distance from the apex at maturity. *Seeds* numerous, shiny-brown, elliptic-ovoid, c. 1 mm long, longitudinally striate and with an impressed lattice. (Fig. 38C–D).

DISTRIBUTION. C China: Chongqing, Henan, W Hubei (Badong, Shennongjia), S Shaanxi, E Sichuan; 1500–2400 m (Map 28).

HABITAT. Forests and forest glades, moist slopes within the forest, shrubberies.

FLOWERING. May–August.

SPECIMENS SEEN. *Farges* s.n. (BM), 390 (BM, K); *Henry* 6863 (BM, K); *E. H. Wilson* 2290[45] (E, K).

Meconopsis oliveriana was based upon collections of Abbé Farges from eastern Sichuan and those of Augustine Henry from W Hubei. It is clearly very closely related to *Cathcartia chelidonifolia* (see under that species for the prime differences), yet in the character of the fruit capsules it comes closest to the Himalayan *C. villosa*, although unlike those of the latter, the capsule split only for a short distance from the apex. The species is not, nor has it ever been, in cultivation. It is so similar in general appearance to *C. chelidonifolia* that its introduction is hardly like to have great impact, especially as the latter is not considered of great garden value; however to aficionados of *Meconopsis* and related genera its introduction into cultivation would certainly be worthwhile. From a taxonomic point of view it would be very interesting to compare living plants of these two species side by side.

The original description of *Meconopsis oliveriana* states that the flower colour was not yellow, however, subsequent collectors, including E. H. Wilson, state clearly that the flowers were yellow.

---

[45] Misquoted as *E. H. Wilson* 2390 in Engler, *Das Pflanzenreich*.

Overleaf:

Meconopsis rudis forma; W Haba Shan, NW Yunnan, 4120 m. PHOTO: ALAN DUNKLEY

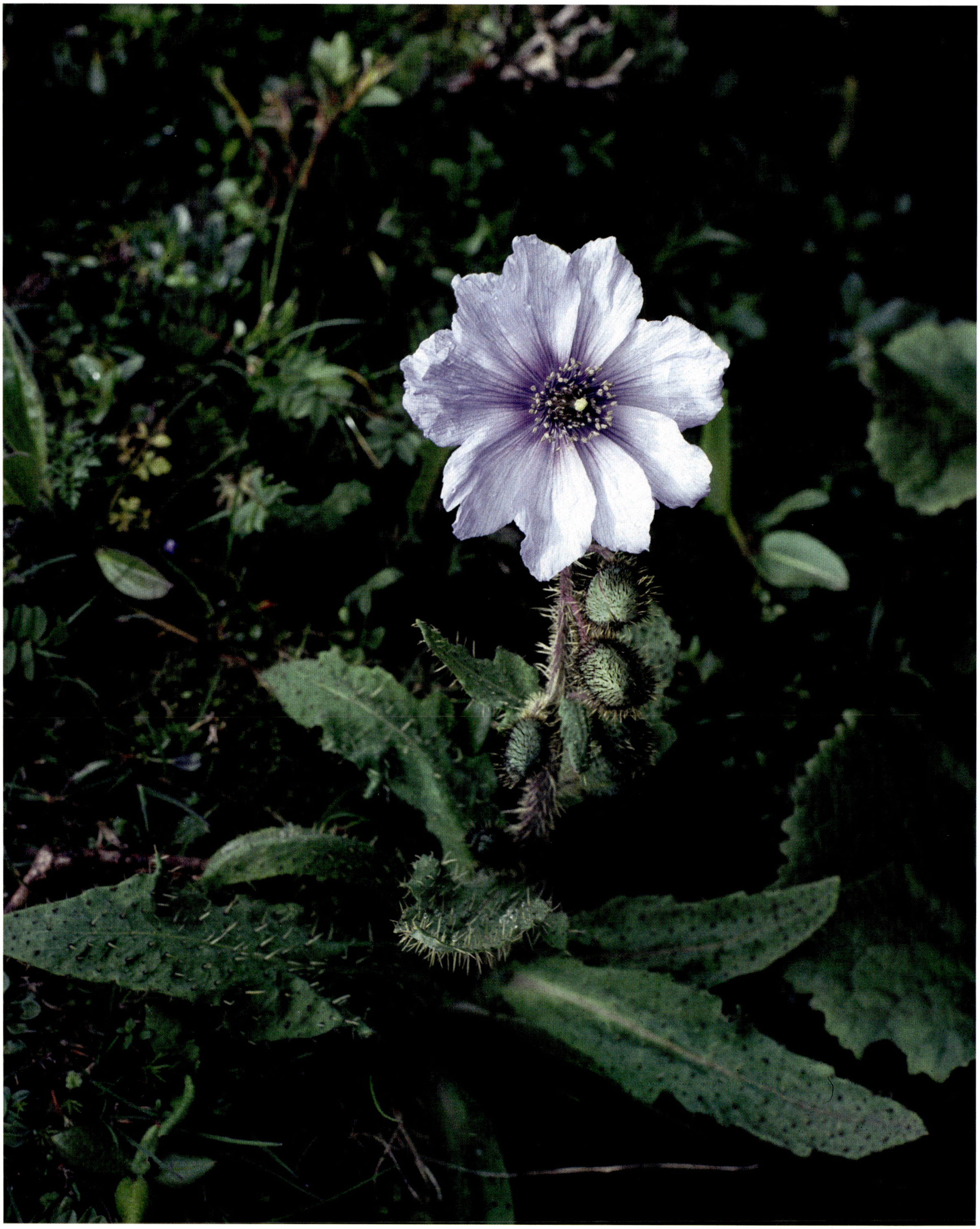

GLOSSARY

acuminate: bearing a gradually diminishing apex

agglutinate (agglutinated, agglutinosed): glued together, closely adhering

allopatric: species or taxa occupying different and disjunct geographical areas

amplexicaul: clasping the stem

aneuploid: having a chromosome number that is more or less than, but not an exact multiple of the basic chromosome number

anther: the part of the stamen containing the pollen

anthesis: flowering time

aperture: an opening

apiculate: bearing a short sharp point

appressed: pressed close to, e.g. hairs that are pressed close to the stem or leaf surface

auct. (*auctorum*): of authors

auricles: small ear-like flaps

axil: the angle between a leaf or bract and the stem, or between a lateral branch and the main stem

barbellate: armed with small barbs

biennial: a plant that germinates in one year and flowers in the next, then dies

bract: organ, often leaf-like, subtending a flower

bristle: a stiff hair; a stiff pointed hair

bulbil: a small, swollen, vegetative bud capable of becoming an independent plant; bulbils are located as offsets of a parent bulb or in the axils of leaves

capsule: a dry dehiscent seed vessel

cauline: with a stem; cauline leaves are those attached to the stem

chasmophyte: plant growing on cliffs, usually rooting into crevices or fissures

ciliate: bearing hairs along the margin

clade: a monophyletic group; a group of taxa sharing a closer common ancestry with each other than with any other clades

cladogram: a branching diagram of taxa showing their relationships

clavate: club-shaped

complex: a group of closely related taxa within which the circumscription of the various elements is not satisfactorily resolved taxonomically

conspecific: belonging to the same species

cordate: heart-shaped

corolla: collective term for the petals

crenulate: with small rounded teeth

cultivar: a cultivated variety; a horticultural taxonomic rank that is clearly distinct, uniform and stable, and which, when propagated, maintains those characteristics

cuneate: wedge-shaped

cylindrical: cylinder-like

cyme: an inflorescence in which each axis is terminated by a flower and with subsequent flowers developing on branches arising from the axis

cymule: a lateral branch of an inflorescence terminating in a flower

decurved: curved downwards

dilated: swollen, expanded

dimorphic: with two types, e.g. leaves

diploid: an organism or cell with twice the haploid number of chromosomes in its nuclei

disjunct: a distribution that is fragmented into two or more potentially interbreeding populations by such a distance as to prohibit gene flow

ellipsoid: shaped like an ellipse

emarginate: notched

epithet: the second part of a species binomial (genus name being first) or combination of words, which denotes an individual taxon

erose: having a gnawed-looking margin

exserted: protruding beyond

f., fil. (*filius*): son of

f., forma: form, a secondary category in hierarchy below that of variety

fertilisation: the fusion of male and female gametes

filament: the stalk of a stamen

filiform: thread-like

fimbriate: bordered with slender processes

fusiform: spindle-shaped

glabrescent: becoming glabrous

glabrous: lacking hairs

haploid: a nucleus or individual containing only one representative of each pair of chromosomes

holotype: a designated specimen (more rarely a painting or photograph) forming the basis for the original description of a species or other taxon

homologous: similar in structure and origin

homonym: a name rejected because it has previously been applied to another taxon

hybrid: a plant arising as a result of crossing between two distinct taxa (species, subspecies, varieties or other hybrids)
ICBN: International Code of Botanical Nomenclature
ICNCP: International Code of Nomenclature of Cultivated Plants
indumentum: hair covering; can also apply to scales
inflorescence: the floral axis bearing a few to numerous flowers
isolectotype: a duplicate of a lectotype
isotype: a duplicate of a holotype; it is always a specimen
lamina: the leaf-blade; can also refer to a petal
lanate: woolly; long, dense curled hairs
lanceolate: lance-shaped, broadest about one-third from the base
lectotype: a specimen or illustration selected to represent the type of a taxon, particularly when a number of syntypes were cited in the original description, or when no type was designated at all, or if the holotype is missing or lost
ligulate: tongue-shaped
lithophyte: plant growing on rocks
monad: a single cell, e.g. pollen (see also tetrad)
monocarpic: a plant that takes three or more years from germination until reaching its flowering phase, then dies having set seed in the process
mss.: manuscript
napiform: turnip-shaped
nom. ambig. (*nomen ambiguum*): an ambiguous name; a name consistently used by different authors for distinct taxa
nom. nud. (*nomen nudum*): a name unaccompanied by a description or reference to a published description
nom. superfl. (*nomen superfluum*): a name incorrectly applied to a taxon
obcordate: inversely heart-shaped
obovate: inversely ovate
orthographic error: an unintentional spelling error
orthographic variant: an alternative spelling for a word
ovary: the flower part containing the ovules
ovate: egg-shaped (two dimensional)
palynology: the study of pollen
paniculate: a branched raceme
pannose: felt-like
patent: spreading
pedicel: the flower stalk
pedicellate: with a pedicel
peduncle: the inflorescence axis (main or lateral) below the lowest flower
perianth: the calyx and corolla

pers. comm.: a personal observation by oneself
persistent: remaining attached
petal: one segment of the corolla
petiolate: with a petiole
petiole: the leaf stalk
pilose: softly hairy; generally short, weak, thin hairs
pinnate: a leaf in which a number of leaflets are arranged along a central rachis and terminated by a leaflet
pinnatifid: a leaf-lamina that is pinnately lobed but not divided to the base (for instance, in some *Meconopsis* species the lamina is pinnatisect in the lower part whilst being pinnatifid in the upper)
pinnatisect: a leaf-lamina that is lobed to the rachis in a pinnate manner
plicate: pleated
pollen: the male reproductive part of the flower
pollination: the act of transferring pollen (male) to the stigma (female) on the same or different flowers, usually of the same species
polycarpic: perennial, a plant that, having reach its flowering phase, continues to flower for a number, sometimes many, years thereafter
polyploid: an organism or cells with three or more full sets of chromosomes, i.e. having two or more sets of homologous chromosomes
pro parte: in part; used in conjunction with a plant name to indicate that it now only partly covers the former concept of a particular taxon (usually a species)
protologue: all the material associated with the original publication of a particular taxon
puberulous: minutely hairy
pubescence: hairiness
pungent: spiny
pyriform: pear-shaped
raceme: a simple (unbranched) inflorescence bearing stalked (pedicellate) flowers
rank: any category in the nomenclatural hierarchy
reniform: kidney-shaped
reticulate: a network pattern or ridges and hollows, e.g. seed coats, pollen grains
retrorse: bent or curved backwards
rhachis: the part of the inflorescence axis bearing the flowers
rhizome: a horizontal stem, above or below ground
scape: a leafless floral axis bearing a solitary or several flowers; always solitary in *Meconopsis*
sepal: one segment of the calyx
sericeus: silky; covered with fine close-pressed hairs
series: a term used (in the context of this work) to denote an assemblage of species within a section that share certain characteristics in common

serrulate: bearing small sharp teeth, saw-like
sessile: stalkless
setacaeus: bristle-like
sic: in this manner, thus
sinus: a recess e.g. the gap between leaf lobes
s.n. (*sine numero*): without number
spathulate (spatulate): spatula-shaped
species: the basic category in nomenclatural hierarchy
spheroidal: shaped like a sphere
spinule: tiny spine-like structures
stigma: the receptive tip of the style which receives the pollen
subspecies, subsp.: the nomenclatural category between species and variety (in the context of this work subspecies differ geographically from one another and also in several morphological characters, although sharing the defining characteristics of the species)
sulcate: grooved
sympatric: species or taxa occupying the same geographical areas
synonym: a different, alternative, name or epithet for a taxon that is not the accepted name or epithet
synonymy: a list of synonyms
syntype: one or two or more specimens cited with the description (protologue) of a new taxon when no holotype is designated

taxon (taxa pl.): a unit of classification at any rank
tectum: roof
terete: circular in cross-section
testa: the seed coat (outer covering of a seed)
tetrad: formations of four united pollen grains
tetraploid: organism or cells with four times the haploid number of chromosomes
tomentose: a dense covering of rather rigid short hairs
triploid: organism or cells with three times the haploid number of chromosomes
truncate: cut off
type collection: the collection or collections upon which the original description of a taxon is based
valvate: with valves (in an ovary or fruit the number of valves indicates the number of compartments in which the seeds are housed)
variety: a secondary category in nomenclatural hierachy below that of species and subspecies (in the context of this work a variety denotes those plants within the geographical range of a species or subspecies, that differ in one or two characters, although sharing the defining characteristics of the species)
vascular tissue: the cells (xylem and phloem) which transport water and nutrients around the plant
villous: shaggy hairs; generally long weak hairs

REFERENCES

An, L. Z., Chen, S. Y. & Lian, Y. S. (2009). A new species of *Meconopsis* (Papaveraceae) from Gansu, China. *Novon* **19**: 286–288.

Baker, W. J. (1994). Three men and an orchid. *Bull. Alpine Gard. Soc. Gr. Brit.* **62**: 99–114; **62**: 181–199.

Baldwin, B. G., Sanderson, M. J., Porter, J. M., Wojciechowski, M. F., Campbell, C. S. & Donoghue, M. J. (1995). The ITS region of nuclear ribosomal DNA: a valuable source of evidence on angiosperm phylogeny. *Ann. Missouri Bot. Gard.* **82**: 247–277.

Bhuju, U. R., Shakya, P. R., Basnet, T. B. & Shrestha, S. (2007). *Nepal Biodiversity Resource Book. Protected Areas, Ramar Sites, and World Heritage Sites*. International Centre of Integrated Mountain Development, Ministry of Environment, Science and Technology, in cooperation with United Nations Environment Programme, Regional Office for Asia and the Pacific, Kathmandu.

Bittkau, C. & Kadereit, J. W. (2002). Phylogenetic and geographical relationships in *Papaver alpinum* L. (Papaveraceae) based on RAPD data. *Bot. Jahrb. Syst.* **123**: 463–479.

Carolan, J. C. (2004). *Phylogenetic analysis of Papaver L.* PhD thesis, University of Dublin, Trinity College.

——, Hook, I. L. I., Chase, M. W., Kadereit, J. W. & Hodkinson, T. R. (2003). Phylogenetics of *Papaver* and related genera based on DNA sequences from ITS nuclear ribosomal DNA and plastid trnL intron and trnL-F intergenic spacers. *Ann. Bot.* **98**: 141–155.

——, ——, Walsh, J. J. & Hodkinson, T. R. (2002). Using AFLP markers for species differentiation and assessment of genetic variability of in vitro cultured *Papaver bracteatum* (section *Oxytona*). In vitro cellular and developmental biology. *Plant* **38**: 300–307.

——, ——, Chase, M. W., Kadereit, J. W. & Hodkinson, T. R. (2006). Phylogenetics of *Papaver* and Related Genera Based on DNA Sequences from ITS Nuclear Ribosomal DNA and Plastid *trnL* Intron and *trnL-F* Intergeneric Spacers. *Ann. Bot.* **98**: 141–155.

Chase, M. W., de Bruijn, A. Y., Reeves, G., Cox, A. V., Rudall, P. J., Johnson, M. A. T. & Eguiarte, L. E. (2000). Phylogenetics of Asphodelaceae (Asparagales): an analysis of plastid rbcL and trnL–F DNA sequences. *Ann. Bot.* **86**: 935–956.

Chuang, H. (1981). The systematic evolution and the geographical distribution of *Meconopsis* Viguier. *Acta Bot. Yunnan.* **3**: 139–146.

Cobb, J. L. S. (1984). *Meconopsis* Hybrids. *Bull. Alp. Gard. Soc. Gr. Brit.* **52**: 63–73.

—— (1989). *Meconopsis*. Timber Press in association with Christopher Helm.

Cordell, G. A. (1981). Alkaloids derived from phenylalanine and tyrosine. In: G. A. Cordell (ed.), *Introduction to alkaloids: a biogenetic approach*. Pp. 490–512. John Wiley and Sons, New York.

Cox, E. H. M. (1986). *Plant Hunting in China*. Reprint Oxford University Press.

Cui, Z. & Lian, Y. (2005). A new variety of *Meconopsis* from Gansu. *Guihaia* **25**: 106.

Cullen, J. (1965). Papaveraceae. In: P. H. Davis (ed.), *Flora of Turkey and the east Aegean Islands*, Vol. I: 213–247. Edinburgh University Press.

De Candolle, A. P. (1824). Papaveraceae. In: *Prodromus Systematis Naturalis Regni Vegetabilis*, Vol. **1**: 117–124.

Debnath, H. S. & Nayar, M. P. (1986). *The Poppies of Indiana Region (Papaveraceae)*. Botanical Survey of India, Calcutta.

—— & —— (1987). Short communication. Neotype of *Meconopsis longipetiolata* G. Tayl. ex Hay (Papaveraceae) and some additional note. *Bull. Bot. Surv. India* **27**: 244–245.

Doyle, J. J. & Doyle, J. L. (1987). A rapid DNA isolation procedure for small quantities of fresh leaf tissue. *Phytochemical Bull.* **19**: 11–15.

Egan, P. (2011). *Meconopsis autumnalis* and *M. manasluensis* (Papaveraceae), two new species of Himalayan poppy endemic to central Nepal with sympatric congeners. *Phytotaxa* **20**: 47–56.

——, Pendry, C. A. & Shrestha, S. (2011). Papaveraceae. In: M. F. Watson (ed.), *Flora of Nepal* **3**: 78–95. Royal Botanic Garden, Edinburgh.

Ernst, W. R. (1962). *A comparative morphology of the Papaveraceae*. Ph.D. thesis, Stanford University, Stanford.

—— (1965). *Meconopsis* in Documented chromosome numbers of plants. *Madroño* **18**: 123–126.

Farrer, R. (1921). *Rainbow Bridge*. Edward Arnold & Co., London.

—— (1928). *The English Rock Garden* **1**: 474–483, pl.47–50; **2**: 477–480.

Fedde, F. (1909). Papaveraceae–Hypecoideae et Papaveraceae–Papaveroideae. In: A. Engler (ed.), *Das Pflanzenreich*, p. 228. Englemann, Leipzig.

Felsenstein, J. (1985). Confidence limits on phylogenies: an approach using the bootstrap. *Evolution* **39**: 783–791.

Fletcher, H. R. (1975). *A Quest for Flowers*. Edinburgh University Press.

Gielly, L. P. & Taberlet, P. (1996). A phylogeny of the European gentians inferred from chloroplast trnL (UAA) intron sequences. *Bot. J. Linn. Soc.* **120**: 57–75.

——, Yuan, Y. M., Kupfer, P. & Taberlet, P. (1994). Phylogenetic use of noncoding regions in the genus *Gentiana* L.: chloroplast trnL (UAA) intron versus nuclear ribosomal internal transcribed spacer sequences. *Molec. Phylogenet. Evol.* **5**: 460–466.

Goldblatt, P. (1974). Biosystematic studies in *Papaver* section *Oxytona*. *Ann. Missouri Bot. Gard.* **61**: 264–296.

Grierson, A. J. C. & Long, D. G. (1984). Papaveraceae. In: *Flora of Bhutan*, **1 (2)**: 404–409. Royal Botanic Garden, Edinburgh.

Grey-Wilson, C. (1996). *Meconopsis integrifolia*, the yellow poppywort and its allies. *New Plantsman* **3 (1)**: 22–39.

—— (2000 revised edition). *Poppies*. B. T. Batsford, London.

—— (2002). A New Blue Poppy from Western China. *New Plantsman* **1 (4)**: 221–227.

—— (2006a). A New *Meconopsis* from Tibet. *Alpine Gardener* **74**: 212–225.

—— (2006b). The true identity of *Meconopsis napaulensis* DC. *Curtis's Bot. Mag.* **23 (2)**: 176–209.

—— (2009). Bailey's Blue Poppy Restored. *Alpine Gardener* **77**: 217–226.

—— (2012). Proposal to conserve the name *Meconopsis* (Papaveraceae) with a conserved type. *Taxon* **61 (2)**, 2061: 473–4.

—— & Cribb, P. J. (2011). *Flowers of Western China*. Royal Botanic Gardens Kew.

—— & Mitchell, J. (2007). Establishing and maintaining collections of monocarpic meconopsis. *Sibbaldia* **5**: 115–128.

—— & Mitchell, J. (2010). *Meconopsis grandis* — the true Himalayan blue poppy. *Sibbaldia* **8**: 75–96.

——, Rankin, D. W. H. & Zhikun, Wu (2011). *Meconopsis wilsonii* subsp. *orientalis*. *Curtis's Bot. Mag.* **28 (1)**: 32–46, t. 700.

Günther, K. F. (1975). Beitrage zur Morphologie und verbreitung der Papaveraceae 2. Teil: Die Wuchsformen der Papaverae, Escoscholzieae un Platystemonoideae. *Flora* **164**: 393–436.

Hammer, K. & Fritsch, R. (1977). Zur Frage nach der Ursprungsart des Kulturmohns (*Papaver somniferum* L.). *Kulturpflanze* **25**: 113–124.

Haw, S. G. (1980). *Meconopsis* in Western China. *Quart. Bull. Alp. Gard. Soc.* **48**: 236.

Henderson, D. M. (1965). The Pollen Morphology of *Meconopsis*. *Grana Palynol.* **6**: 191–209.

Hodkinson, T. R., Chase, M. W., Takahashi, C., Leitch, I. J., Bennett, M. D. & Renvoize, S. A. (2002). The use of DNA sequencing (ITS and trnL–F), AFLP and fluorescent in-situ hybridisation to study allopolyploid *Miscanthus* (Poaceae). *Amer. J. Bot.* **89**: 279–286.

Hooker, J. D. (1874–75). *Flora of British India*. L. Reeve, London.

Hopper, S. D., Fay, M. F., Rossetto, M. & Chase, M. W. (1999). A molecular phylogenetic analysis of the bloodroot and kangaroo paw family, Haemodoraceae: taxonomic, biogeographic and conservation implications. *Bot. J. Linn. Soc.* **131**: 285–299.

Huang, R. F., Shen, S. D. & Lu, X. F. (1996). Studies on the chromosome number and polyploidy for a number of plants in north-east Qinghai-Xizang Plateau. *Acta Bot. Boreal.-Occid. Sin.* **16**: 310–318.

Jee, V., Dhar, U. & Kachroo, P. (1989). Cytogeography of some endemic taxa of Kashmir. *Himalay. Proc. Ind. Nat. Sc. Acad.* Part B, Biological Sciences **55**: 177–184.

Jingwei, Zhang ed. (1982). *The Alpine Plants of China*. Gordon & Breach, New York.

Jørk, K. B, & Kadereit, J. W. (1995). Molecular phylogeny of the Old World representatives of Papaveraceae subf. Papaveroideae with special emphasis on the genus *Meconopsis* Vig. In: U. Jensen & J. W. Kadereit (eds), Systematics and evolution of the Ranunculiflorae. *Plant Syst. & Evol.* **9** (Suppl.): 171–180.

Kadereit, J. W. (1986a). A revision of *Papaver* sect. *Argemonidium*. *Notes Roy. Bot. Gard. Edinburgh* **44**: 25–43.

—— (1986b). A revision of *Papaver* L. sect. *Papaver* (Papaveraceae). *Bot. Jahrb. Syst.* **108**: 1–16.

—— (1987). A revision of *Papaver* sect. *Carinatae* (Papaveraceae). *Nordic J. Bot.* **7**: 501–504.

—— (1988a). Sectional affinities and geographical distribution in the genus *Papaver* L. (Papaveraceae). *Beitr. Biol. Pflanzen* **63**: 139–156.

—— (1988b). *Papaver* L. sect. *Californicum* Kadereit, a new section of the genus *Papaver*. *Rhodora* **90**: 7–13.

—— (1988c). The affinities of the south-hemispherical *Papaver aculeatum* Thunb. (Papaveraceae). *Bot. Jahrb. Syst.* **109**: 335–341.

—— (1989). A revision of *Papaver* section *Rhoeadium* Spach. *Notes Roy. Bot. Gard. Edinburgh* **45**: 225–286.

—— (1990). Some suggestions on the geographical origin of the central, west and north European synantropic species of *Papaver* L. *Bot. J. Linn. Soc.* **103**: 221–231.

—— (1993a). Papaveraceae. In: K. Kubitzki, J. G. Rohwer & V. Bittrich (eds), *The families and genera of vascular plants*, Vol. II: 20–33. Springer-Verlag, Berlin.

—— (1993b). A revision of *Papaver* sect. *Meconidium*. *Edinburgh J. Bot.* **50**: 125–148.

—— (1996). A revision of *Papaver* L. sects. *Pilosa* Prantl and *Pseudopilosa* M. Popov ex Gunther (Papaveraceae). *Edinburgh J. Bot.* **53**: 285–309.

—— & Erbar, C. (2011). Evolution of gynoecium morphology in Old World Papaveroideae: a combined phylogenetic/ ontogenetic approach. *Amer. J. Bot.* **98 (8)**: 1243–1251.

——, Preston, C. D. & Valtueña, F. J. (2011). Is Welsh Poppy, *Meconopsis cambrica* (L.) Vig. (Papaveraceae), truly a Meconopsis? *New J. Bot.* **1**: 80–88.

——, & Sytsma K. J. (1992). Disassembling *Papaver*: a restriction site analysis of chloroplast DNA. *Nordic J. Bot.* **12**: 205–217.

——, Schwarzbach, A. E. & Jørk, K. B. (1997). The phylogeny of *Papaver* s.l. (Papaveraceae): polyphyly or monophyly? *Plant Syst. Evol.* **204**: 75–98.

Kelchner, S. A. (2000). The evolution of non-coding chloroplast DNA and its application in plant systematics. *Ann. Missouri Bot. Gard.* **87**: 182–198.

Kingdon Ward, F. (1913). *The Land of the Blue Poppy*. Cambridge University Press.

—— (1923). *Mystery Rivers of Tibet*. Seeley Service & Co. Ltd, London.

—— (1924). *The Romance of Plant Hunting*. Edward Arnold & Co., London.

—— (1926). The Genus *Meconopsis*. *Gard. Chron.* Ser. III, **79**: 252–53, 306–08, 438–93, 459–60.

—— (1926). Notes on the genus *Meconopsis*, with some additional species from Tibet. *Ann. Bot.* **40**: 535–541.

—— (1927). Burmese species of *Meconopsis*. *Gard. Chron.* Ser. III **82**: 151.

—— (1928). Burmese species of *Meconopsis*. *Ann. Bot.* **42, no. 168**: 855–863.

—— (1930). *Plant Hunting on the edge of the World*. Victor Gollancz Ltd, London.

—— (1934). *A Plant Hunter in Tibet*. Jonathan Cape Ltd, London.

—— (1937). *Plant Hunter's Paradise*. Jonathan Cape Ltd, London.

—— (1941). *Assam Adventure*. Jonathan Cape Ltd, London.

—— (1949). *Burma's Icy Mountains*. Jonathan Cape Ltd, London.

—— (1960). *Pilgrimage for Plants*. George G. Harrap & Co. Ltd, London.

Lancaster, R. (1995). *A Plantsman in Nepal*, revised edition. The Antique Collectors' Club, Suffolk.

Levan, A., Fredga, K. & Sandberg, A. A. (1964). Nomenclature for centrometric position on Chromosomes. *Hereditas* **52**: 201–220.

Li, R., Yang, J. & Dao, Z. (2012). *Meconopsis xiangchienensis* (Papaveraceae), a New Species from Sichuan, China. *Novon* **22**: 180–182.

Markgraf, F. (1958). Familie Papaveraceae. In: P. Hegi (ed.), *Illustrierte Flora von Mitteleuropa* Band IV, 2nd edn. Teil 1.: 15–49. Hanser, Munich.

McAllister, H. (1999). The importance of living collections for taxonomy. In: S. Andrews, A. Leslie & C. Alexander (eds), *Taxonomy of cultivated plants*, pp. 3–10. Third International Symposium. Royal Botanic Gardens, Kew.

Morales Torres, C., Mendoza Castellon, R. & Romero Garcia, A. T. (1988). La posicion sistematica de *Papaver argemone* L.: interes evolutivo del orden Papaverales: 1. *Lagascalia* **15**: 181–189.

Novak, J. & Preininger, V. (1980). Sect., Glauca-nova sekce rodu *Papaver*. *Preslia* **52**: 97–101.

O'Donnel, S. & Lawrence, M. J. (1984). The population genetics of the self-incompatibilty polymorphism in *Papaver rhoeas*. IV. The estimated number of alleles in a population. *Heredity* **53**: 495–508.

Ohba, H. & Akiyama, S. (1992). *The Alpine Flora of the Jaljale Himal, East Nepal*. The University Museum, The University of Tokyo.

——, Yoshida, T. & Sun, H. (2009). Two New Species of *Meconopsis* (Papaveraceae) from Southern Biluo Xueshan, Yunnan, China [J]. *J. Jap. Bot.* **84 (5)**: 294–302.

Proctor, M., Yeo, P. & Lack, A. (1996). *The Natural History of Pollination*. HarperCollins, London.

Podlech, D. & Dieterle, A. (1969). Chromosomenstudien an Afghanischen Pflanzen. *Candollea* **24**: 185–243.

Polunin, O. & Stainton, A. (1984). *Flowers of the Himalaya*. Oxford University Press.

Popov, M. G. (1937). Papaveraceae. In: V. L. Komarov (ed.), *Flora of the USSR*: 470–474. Nauka, Moscow-Leningrad.

Prain, D. (1906). A review of the genera *Meconopsis* and *Cathcartia*. *Ann. Bot.* **20**: 323–370.

—— (1914). *Meconopsis rudis*. *Curtis's Bot. Mag.* **140**, t. 8568.

—— (1915a). Some additional species of *Meconopsis*. *Bull. Misc. Inform., Kew* **1915 (4)**: 129–177.

—— (1915b). *Meconopsis lyrata* (Cummins & Prain) in Fedde ex Prain. *Bull. Misc. Inform., Kew* **1915**: 142

—— (1915c). *Meconopsis polygonoides* (Prain) Prain. *Bull. Misc. Inform., Kew* **1915**: 143.

—— (1918). *Meconopsis compta* Prain. *Bull. Misc. Inform., Kew* **1918**: 212.

Preininger V. (1986). Chemotaxonomy of Papaveraceae and Fumariaceae. *The alkaloids*: 2–98. Academic Press, London.

——, Novak, J. & Santavy, F. (1981). Isolierung und Chemie der Alkaloide aus Pflanzen der Papaveraceae LXXXXI. Glauca, eine neue Sektion der Gattung *Papaver*. *Pl. Med.* **41**: 119–123.

Proctor, M., Yeo, P. & Lack, A. (1996). *The Natural History of Pollination*. HarperCollins.

Rändel, U. (1974). Beitrage zur Kenntnis der Sippenstruktuir der Gattung *Papaver* L. sectio *Scapiflora* Reihenb. Im vergleich mit *P. alpinum* L. (Papaveraceae). *Feddes Repert.* **86**: 19–37.

—— (1977). Uber die grondlandischen Vertreter der section *Lasiotrachyphylla* Bernh. (Papaveraceae). *Feddes Repert.* **88**: 421–450.

Ratter, J. A. (1968). Cytological studies in *Meconopsis*. *Notes Roy. Bot. Gard. Edinburgh* **28**: 191–200.

Reckin, J. (1973). A contribution to the cytology of *Papaver gracile* Auch. Including proposals for a revision of the section *Mecones*. *Caryologia* **26**: 245–251.

Safonova, I. N. (1991). Chromosome numbers in some species of the family Papaveraceae. *Bot. Žhurn. (Moscow & Leningrad)* **76**: 904–905.

Salamin, N., Chase, M. W., Hodkinson, T. R. & Savolainen, V. (2003). Assessing internal support with large phylogenetic DNA matrices. *Molec. Phylogenet. Evol.* **27**: 528–539.

Seelanan, T., Schnabel, A. & Wendel, J. F. (1997). Congruence and consensus in the cotton tribe (Malvaceae). *Syst. Bot.* **22**: 259–290.

Savolainen, V. & Chase, M. W. (2003). A decade of progress in plant molecular phylogenetics. *Trends Genet.* **12**: 717–724.

Schwarzbach, A. E. & Kadereit, J. W. (1999). Phylogeny of prickly poppies, *Argemone* (Papaveraceae), and the evolution of morphological and alkaloid characters based on ITS nrDNA sequence variation. *Pl. Syst. Evol.* **218**: 257–279.

Stace, C. A. (2010). *New flora of the British Isles*. 3rd edn. Cambridge University Press, Cambridge.

Sun, Y., Skinner, D. Z., Liang, G. H. & Hulbert, S. H.(1994). Phylogenetic analysis of *Sorghum* and related taxa using internal transcribed spacers of nuclear ribosomal DNA. *Theor. Appl. Genet.* **89**: 26–32.

Stebbins, G. L. (1971). *Chromosomal Evolution in Higher Plants*. Edward Arnold Ltd, London.

Sugiura, T. (1937). A list of chromosome numbers in angiosperm plants. III. *Bot. Mag. (Tokyo)* **51**: 425–426.

Sugiura, T. (1940). Chromosome studies on Papaveraceae with special reference to phylogeny. *Cytologia* **10**: 558–576.

Swofford, D. L. (2003). PAUP*: Phylogenetic analysis using parsimony (* and other methods), version 4.0b 10. Sinauer Associates, Sunderland, MA.

——, Olsen, G. J., Waddell, P. J. & Hillis, D. M. (1996). Phylogenetic inference. In: D. M. Hillis, C. Moritz & B. K. Mable (eds), *Molecular systematics*, 2nd edn, pp. 407–514. Sinauer Associates, Sunderland, MA.

Taberlet, P. L., Gielly, L., Pautou, G. & Bouvet, J. (1991). Universal primers for amplification of three non-coding regions of chloroplast DNA. *Pl. Molec. Biol.* **17**: 1105–1109.

Tanaka, R. (1971). *Types of resting nuclei in Orchidaceae. Bot. Mag. (Tokyo)* **84**: 118–122.

—— (1977). *Recent Karyotype Studies*. In: Y. Ogawa *et al.* (eds), *Plant Cytology*. Asakura Shoten, Tokyo.

Taylor, G. (1930). *Stylomecon*: a new genus of Papaveraceae. *J. Bot.* **68**: 138–140.

—— (1934). *An Account of the Genus Meconopsis*. New Flora and Silva Ltd., London.

—— (1951). Two New Meconopsis Hybrids. *J. Roy. Hort. Soc.* **76**: 231.

Terry, B. (2009). *Blue Heaven (encounters with the blue poppy)*. TouchWood Edition, British Columbia.

Thompson, D. (1986). Germination responses in *Meconopsis*. *J. Roy. Hort. Soc.* **93**: 336.

Webb, C. J. & Lloyd, D. G. (1986). The avoidance of interference between the presentation of pollen and stigmas in angiosperms. II. Herkogamy. *New Zealand J. Bot.* **24**: 163–178.

Wendel, J. F. & Doyle, J. J. (1998). Phylogenetic incongruence: window into genome history and molecular evolution. In: D. E. Soltis, P. S. Soltis & J. J. Doyle (eds), *Molecular systematics of plants II*: 265–296. Kluwer Academic Publishers, London.

——, Schnabel, A. & Seelanan, T. (1995). An unusual ribosomal DNA sequence from *Gossypium gossypioides* reveals ancient, cryptic, intergenomic introgression. *Molec. Phylogenet. Evol.* **4**: 298–313.

Williams, L. H. J. ((1972). *Meconopsis taylorii*, a new Species from Nepal. *Trans. Bot. Soc. Edinburgh* **41**(3): 347–349.

Wilson, E. H. (1913). *A Naturalist in Western China*. Methuen & Co. Ltd.

Wu, Z. Y. (1999). *Flora of China* (in Chinese) **32**: 7–51. Science Press, Beijing.

—— & Zhuang, X. (1979). *Meconopsis* Vig. In: *Flora Yunnanica* **2**: 24–42. Science Press, Beijing.

—— & —— (1980). A study on the taxonomic system of the genus *Meconopsis*. *Acta Bot. Yunnan.* **2** (4): 371–381.

——, Lu, A. M., Tang, Y. C., Chen, Z. D. & Li, D. Z. (2003). *The families and genera of Angiosperms in China: a comprehensive analysis*. Science Press, Beijing.

——, Zhuang, X. & Su, Z. (1999). Papaveraceae, *Flora Reipublicae Popularis Sinicae* [in Chinese]: **32**: 7–51. Science Press Beijing.

Xie, H.Y. (1999). *Ethnobotanical investigation of Meconopsis Vig. in NW Yunnan*; master degree dissertation, Kunming Institute of Botany, Chinese Academy of Science.

Yang, F-S., Qin, A-L., Li, Y-F. & Wang, X-Q. (2012). Great Genetic Differentiation amng Populations of *Meconopsis integrifolia* and its implications for Plant Speciation in the Qinghai-Tibetan Plateau. *PloS ONE* **7** (5): e37196. doi: 10-1371/journal.pone.0037196.

Ying, M., Xie, H-Y., Nie, Z-L., Gu, Z-J. & Yang, Y-P. (2006). A karyomorphological study on four species of *Meconopsis* Vig. (Papaveraceae) from the Hengduan Mountains, SW China. *Caryologia* **59**(1): 1–6.

Yoder, A. D., Irwin, J. A. & Payseur, B. A. (2001). Failure of the ILD to determine data combinability for slow Loris phylogeny. *Syst. Biol.* **50**: 108–121.

Yoshida, T. (2005). *Himalayan Plants Illustrated*. Yama-kei Publishers, Tokyo.

—— (2006). Geobotany of the Himalaya. *Newslett. Himal. Bot.* **37**: 1–24, **38**: 1–30. The Society of Himalayan Botany, Tokyo.

—— (2011). Endemism in the Sino-Himalaya. *Alpine Gardener* **79**: 181–190.

—— & Grey-Wilson, C. (2012). A new species of blue poppy (A new *Meconopsis* from Bhutan). *New Plantsman* **11** (2): 96–101.

——, Sun, H. & Boufford, D. E. (2007). *Meconopsis wilsonii* subsp. *wilsonii* (Papaveraceae) rediscovered [J]. *Acta Bot. Yunnan.* **29** (3): 286–288

——, —— & —— (2010). New species of *Meconopsis* (Papaveraceae) from Mianning, southwestern Sichuan, China [J]. *Acta Bot. Yunnan.* **32** (6): 503–507.

——, —— & —— (2011). New species of *Meconopsis* (Papaveraceae) from Balangshan, western Sichuan, China [J]. *Acta Bot. Yunnan.* (*Plant Diversity and Resources*) **33** (4): 409–413.

——, —— & —— (2012). New species of *Meconopsis* (Papaveraceae) from Laojun Shan and Yao Shan, northern Yunnan, China. *Acta Bot. Yunnan.* (Plant Diversity & Resources) **34** (2): 145–149.

——, —— & Grey-Wilson, C. (2012). A Revision of *Meconopsis lyrata* (Cummins & Prain) Fedde ex Prain and its allies. *Curtis's Bot. Mag.* **29** (2): 1–23.

Zhang, M. & Grey-Wilson, C. (2008). *Meconopsis* (Papaveraceae). In: C. Y. Wu, P. H. Raven & D. Y. Hong (eds), *Flora of China* **7**: 264–269. Science Press, Beijing/Missouri Botanical Garden Press, St. Louis.

Zhou, L-H. (1979). New Taxa of *Meconopsis* from Qinghai-Tibet Plateau. *Acta. Phytotax. Sin.* **17** (4): 112–115.

—— (1980). Study on the *Meconopsis* of the Qinghai-Tibet Plateau. *Bull. Bot. Lab. N.E. Forest Inst.*, Harbin **8**: 91–101.

In addition The Meconopsis Group has produced a series of newsletters for its members. Many of these have a great deal of interesting and factual information on the genus. Copies of all the newsletters are housed in the Group's archive. For further information and membership details **refer to:** www.meconopsis.org

APPENDIX 1

(2061) PROPOSAL TO CONSERVE THE NAME MECONOPSIS (PAPAVERACEAE) WITH A CONSERVED TYPE

CHRISTOPHER GREY-WILSON

Reproduced from
TAXON 61 (2) • April 2012: 473–474

Meconopsis Vig., *Hist. Nat. Pavots*: 48. Jan. 1814 (*Papaver*), nom. cons. prop. Type: *M. regia* G. Taylor, typ. cons. prop.

As presently circumscribed, *Meconopsis* includes a single European species and about 65 species in the Sino-Himalayan region. The genus is currently under revision by the present author, and the number of species is likely to increase to about 70 (Yoshida, pers. comm.). It is an iconic genus in Himalayan botany and very well known in horticulture, with about one third of the species in cultivation. The Asian species are a conspicuous element of the Sino-Himalayan herbaceous flora, the largest reaching 2 m high or more in flower; they are found in montane habitats from western Pakistan to western China. The European species, *M. cambrica* (L.) Vig., was originally described in *Papaver* by Linnaeus. It is native to western parts of the British Isles (Wales, Ireland, S.W. England), France and Spain, but is widely naturalised in other parts of western Europe.

At present the genus is distinguished from *Papaver* mainly by a distinct style on top of the ovary in contrast to the stigma being sessile with radiating lobes on a disk surmounting the ovary. In a very few species of *Meconopsis* the style is obsolete or almost so (sometimes within the same taxon), but then the stigmatic rays are prominently decurrent over a fluted stigmatic column and there is no hint of a disk as such. Both comparative morphological (Ernst, Comp. Morphol. Papaveraceae, Ph.D. thesis, Stanford University, 1962) and molecular research (Jørk & Kadereit in *Pl. Syst. Evol.* 9 (Suppl.): 171–180. 1995; Carolan & al. in *Ann. Bot. (Oxford)* 98: 141–155. 2006; Kadereit & al. in *New J. Bot.* 1: 80–87. 2011) indicate that *M. cambrica* is more closely related to other genera, especially *Papaver* and *Stylomecon* G. Tayl., than it is to the Asian species of *Meconopsis*.

The genus *Meconopsis* cannot be defined in such a way that it includes both the European and Asiatic taxa; if retained in its entirety it would have to be included in *Papaver*. The European *M. cambrica* differs from the Asiatic representatives by the lack of true dorsal traces to the gynoecium but has instead pseudo-dorsal traces that originate from the placental bundles (Ernst, l.c.). From an external morphological viewpoint, the branched fleshy rootstock of *M. cambrica*, along with the tufted habit and multiple, few-flowered, inflorescences are features not found in Asiatic *Meconopsis*, allowing that is for the exclusion of species that have been transferred to the genus *Cathcartia* Hook.

While the taxonomy seems clear enough, the nomenclatural consequences need to be addressed. The generic name *Meconopsis* was published in 1814 and based on the single European species, *M. cambrica*, which thus provides the type of the name. However, restricting the generic name to this species would be widely deplored in the botanical and horticultural worlds. No other names seem to be available for either the single European species or the 65 or so Asian species. *Cathcartia* Hook. f. has been applied to some of the Asian species, but is now considered to be a separate genus of two or three species (Grierson & Long, *Fl. Bhutan* 1: 409–410. 1984; Kadereit & al., l.c. 2011). *Cerastites* Gray, published in 1821, included *P. cambricum* L. among several other Linnaean European species of *Papaver*, and so is superfluous and illegitimate. It is therefore proposed here that *Meconopsis* be conserved with an Asian species providing the type. This will keep the name in use for the many species well known as Himalayan blue poppies. As for the European species, known in anglophone circles as the Welsh poppy, recent research by Kadereit & al. (l.c. 2011) and Carolan & al. (l.c.) strongly suggests that *M. cambrica* is embedded in *Papaver* and that it should be returned to that genus along with the Californian genus *Stylomecon*. Any other option at this stage would require extensive splitting of *Papaver* itself.

The choice of the type proposed for conservation requires consideration. The second species to be described within the genus as now defined was *M. napaulensis* DC., described in 1824. However, this species has been long misinterpreted (Grey-Wilson in *Curtis's Bot. Mag.* 23: 176–211. 2006) due primarily to the poor and rather fragmentary nature of the type material (based on Nathaniel Wallich collections from central Nepal). The species that most characterises the genus and around which there is no nomenclatural doubt is *M. regia* G. Taylor based on two collections made by Lall Dhwoj in central Nepal (the field notes state W Nepal but the localities are certainly central). The holotype is "Western Nepal: Barpak, 12–15,000 ft. …, *Lall Dhwoj* 18", in flower and young fruit (BM; with isotypes at E, K). The paratype collection, *Lall Dhwoj* 195, represented at BM & K, is from Michet and is in fruit.

Virtually all literature relating to the Asian species has used the name *Meconopsis* for nearly 200 years, and a great many horticultural works, including the monograph 'An Account of the Genus *Meconopsis*' by George Taylor (1934), have done the same. *Meconopsis* has countless listings on the Internet, the vast majority of these being referable to the Asian species. At the same time, the genus *Meconopsis* has a high profile in horticultural trade, particularly the famous Himalayan blue poppies (*M. baileyi* Prain and *M. grandis* Prain and their numerous spectacular hybrids and cultivars). In addition, the *Meconopsis* Group, based in Scotland, which is one of the prime areas for *Meconopsis* cultivation, has an international following. Any change in the generic name for the Asian species would have profound consequences both botanically and economically. It seems unthinkable not to keep the name in use in this way.

APPENDIX 2

COLLECTORS AND RESEARCHERS' BIOGRAPHIES

It is very difficult in a series of potted histories to give full credit to the plant collectors who have so enriched our knowledge of the plant world. Without them we would be much the poorer. I have presented below those who have, through their diligence and hard work, often under extremely difficult, indeed sometimes dangerous, conditions, collected *Meconopsis* for science and horticulture. In particular, the rich collections of dried herbarium specimens of *Meconopsis* have been fundamental to this monograph. To the uninitiated, dried specimens might seem dull and uninteresting, but each one holds valuable information of a plant collected at a particular place on a particular date. Perhaps more importantly they reveal the range of variability within and between species. While the field notes record details of the plant that are often lost in the drying process, they also contain, in many instances, details of habitat, location, altitude and flowering time, as well as the name of the expedition or individual collector. As valuable are the specimens themselves from which many details of the plant in question can be deduced. Perhaps most importantly, series of specimens of the same taxon give the researcher an overview of the variation found in the wild. This information may not be complete but it is a very good indication. Variation is what fascinates most botanists and it is often the most difficult aspect to analyse. Over the years I have found that seeing and studying plants in the wild makes it far easier to interpret herbarium specimens at a later date. This is especially true of the genus *Meconopsis*.

Below in alphabetical order are potted biographies of some of the principal plant collectors who collected *Meconopsis* amongst other genera, or who specifically targeted the genus in the wild.

Frederick Marshman (F. M.) Bailey (1882–1967)

Born in Lahore, India, F. M. Bailey became a captain in the British Army, Royal Engineers, undertaking clandestine operations in the Himalaya and Tibet. This gave him the opportunity to pursue his hobbies, particularly trophy hunting (he presented the British Museum, Natural History with more than 2000 bird skins and countless butterflies from the region). In 1913 he made an unauthorised expedition to the Tsangpo valley in SE Tibet in the company of Capt. Henry Morshead (who in 1921 joined George Mallory on the ill-fated expedition to conquer Mt Everest). In the Rong Chu he came across the now famous blue poppy that was to bear his name, *Meconopsis baileyi*. Although his herbarium specimens are poor and often fragemented, Bailey collected many species of *Meconopsis* including *M. bella*, *M. dhwojii*, *M. gracilipes*, *M. horridula*, *M. napaulensis*, *M. robusta*, *M. simplicifolia*, discovering apart from *M. baileyi*, *M. argemonantha* and *M. impedita* subsp. *morsheadii*, while he also collected specimens of two recently described species, *M. bhutanica* and *M. chankheliensis*. He died at Stiffkey in Norfolk, U.K.

Jean Marie Delavay (1834–1895)

Abbé Delavay was a French missionary and botanist who first ventured to China in 1867 to Guangdong Province but later to Yunnan. He was borne in Abondance in the French Savoie Alps and became an indefatigable collector, primarily of dried specimens which, by the end of his life, amounted to in excess of 200,000 numbers, the prime collection being located in the Paris Natural History Museum. Delavay spent a great deal of time collecting in north-western Yunnan finding there numerous new species (estimated to be some 1500), as well as a number of new genera. George Forrest retraced much of Delavay's ground and was able to introduce seed of many of Delavay's species. Among Delavay's specimens are many *Meconopsis*, including several new species, *M. betonicifolia*, *M. rudis* and the species named in his honour, *M. delavayi*. In addition, Delavay was probably the first person to collect the plant described in this work under the name *M. zhongdianensis*.

Paul Guillaume Farges (1844–1912)

A French missionary and plant collector, Farges was born at Monclar-de-Quercy in France. In 1867 he was sent to NE Sichuan where he worked for some 35 years. During this time he collected some 4000 dried specimens, including many ornamental species of trees and shrubs as well as *Meconopsis*, which were sent to the Museum in Paris. Several plants were named in his honour including the shrub *Decaisnea fargesii*. He died in Chongqing.

Reginald John Farrer (1880–1920)

A Yorkshire man, Farrer was a writer, painter, traveller and plant hunter. Born with a cleft palette and a speech impediment, he became one of the foremost writers on alpine plants as a result of travels in Europe and Asia. His flowery prose and exuberant enthusiasm for certain plants made him popular amongst Victorian readers and his books such as *Among the Hills*, *On the Eaves of the World* and *The Rainbow Bridge* (published posthumously) are classics of their genre, while *The English Rock Garden*, his single greatest achievement, is still much sought after today. Farrer was also an accomplished watercolour artist. In 1907 he visited Ceylon and became a Buddhist. His journey in 1914 (in conjunction with William Purdom) to NE Tibet and Gansu produced a wealth of plant material, including many *Primula* and *Meconopsis* species, including two new taxa, *M. psilonomma* and *M. lepida* (now a subspecies of *M. lancifolia*). His last expedition took him to northern Burma (now Myanmar) where he made a number of collections of *Meconopsis* but, unfortunately, where he died.

George Forrest (1873–1932)

Born in Falkirk in Scotland, Forrest became one of the most famous and successful plant collectors of all time, making seven important expeditions to south-western China and the adjacent parts of south-eastern Tibet and northern Burma (Myanmar) from 1904 until 1932, sometimes under extremely perilous conditions. He made extensive herbarium collections (primarily located at Edinburgh Botanic Garden as well as Kew and the British Museum) and innumerable collections of seeds and live plants, with a particular interest in *Rhododendron* and *Primula* including many new species, perhaps the most famous being the exquisite autumn-flowering gentian, *Gentiana sino-ornata*. His expeditions, like those of Kingdon Ward and others, were sponsored by various bodies or individuals including A. K. Bulley of Cheshire. He was also renowned for training a team of native collectors to enhance his collecting range and ultimately the collections that were sent to the West. He died at Tengyueh in Yunnan, although some of his native collectors continued on after his demise. Forrest made many collections of *Meconopsis*, discovering more new species than any other collector: both *M. forrestii* and *M. georgei* were named in his honour (by David Prain and George Taylor respectively), while his other discoveries included *M. compta*, *M. concinna*, *M. eximia* (now *M. lancifolia* subsp. *eximia*), *M. pseudovenusta*, *M. speciosa* and *M. venusta*. In addition, two recently described subspecies, *M. pseudointegrifolia* subsp. *daliensis* and *M. wilsonii* subsp. *australis*, are based upon Forrest collections.

Joseph Dalton Hooker (1817–1911)

J. D. Hooker was without question one of the greatest British botanists. Both botanist and explorer, he was a close friend of Charles Darwin and for some twenty years he was the Director of the Royal Botanic Gardens Kew, succeeding his father William Jackson Hooker. J. D. travelled widely around the world but one of his most famous journeys was to the Himalaya from 1847 until 1851. Based initially on Darjeeling, he explored the Sikkim Himalaya and the neighbouring part of Nepal making substantial collections including numerous new species, particularly of *Impatiens* and *Rhododendron*, but many other plants besides. Three species of *Meconopsis* were described as a result of this expedition, the widespread *M. horridula*, as well as *M. wallichii* and *M. villosa* (the latter now transferred back to the genus *Cathcartia* where it was first described). Later J. D. teamed up with Thomas Thomson, venturing further east to collect in the Sylhet and Khasia Hills region of NE India (Assam). He then collaborated with Thomson on two important works *Rhododendrons of the Sikkim Himalaya* (1849–1851) and *Flora Indica* (1851). However, his most famous and detailed work was the *Flora of British India* published in 7 volumes from 1872 onwards.

Frank (Francis) Kingdon Ward (1885–1958)

Born in Manchester, Kingdon Ward (sometimes Kingdon-Ward) was one of the most thorough and prolific plant explorers, mounting a series of important expeditions to south-western China, south-eastern Tibet, Burma (now Myanmar) and Arunachal Pradesh (Assam) in north-east India in the period 1909 until 1956. His expeditions were recorded in a series of fact-filled and adventurous books such as *The Land of the Blue Poppy*, *The Riddle of the Tsangpo Gorges* and *Plant Hunting on the Edge of the World*. Besides plant hunting, Kingdon Ward was also a skilled geographer, these endeavours being recognised by the Royal Geographical Society in London. His collecting span covered a longer period than almost any other collector, enriching herbarium collections as well as botanic and private gardens: trees and shrubs (particularly *Rhododendron*), and herbs (he had a particular interest in *Lilium*, *Meconopsis* and *Primula*, besides many other genera). He made numerous collections of *Meconopsis* and is credited for introducing the famous blue poppy, *M. baileyi* (*M. betonicifolia* var. *baileyi*) into general cultivation in western gardens. His *Meconopsis* discoveries include *M. baileyi* subsp. *pratensis*, *M. florindae*, *M. prainiana* and *M. violacea*, as well as the hybrid *M. ×kongboensis* (formerly equated with *M. ×harleyana*) found on the Sang La in south-eastern Tibet. In addition, he made fine collections (particularly herbarium specimens) of a number of already well-established species such as *M. bella*, *M. compta*, *M. horridula* sensu lato, *M. impedita*, *M. paniculata*, *M. simplicifolia*, *M. speciosa* and *M. sulphurea*. Kingdon Ward wrote a series of important and thought-provoking articles on *Meconopsis* (see References, pp. 383–384).

Frank Ludlow (1885–1972)

An English naturalist, botanist and plant explorer born in London (Chelsea) who, in partnership with George Sherriff (see below), made large and detailed collections mainly in Bhutan and south-eastern Tibet, primarily between 1932 and 1949, although Ludlow, who had been employed by the Tibetan Government in the 1920s, first made collections on the Tibetan Plateau during the period 1923–26. Between them they made extensive fine, detailed herbarium collections as well as live plants and seeds for introduction to botanic gardens and British horticulture. They made expeditions to Bhutan in 1933 and 1934, later often linking up with other collectors: in 1936 they were accompanied by Dr Kenneth Lumsden, in 1938 with George Taylor, in 1947 they were accompanied by George Sherriff's wife Betty and Colonel Henry Elliot, and in 1949 again by Betty Sherriff with the addition of Dr J. H. Hicks. Their collections were wide ranging but *Meconopsis*, *Primula* and *Rhododendron* were high on their priority list. Their fine *Meconopsis* collections included *M. argemononatha*, *M. baileyi* (as *betonicifolia*), *M. bella*, *M. discigera* (now *M. bhutanica*), *M. florindae*, *M. grandis* (subsp. *orientalis*), *M. horridula* sensu lato, *M. impedita*, *M. pseudointegrifolia* (as *M. integrifolia*), *M. paniculata*, *M. primulina*, *M. sherriffii*, *M. simplicifolia* (primarily subsp. *grandiflora*), *M. sinuata*, *M. speciosa*, *M. superba*, *M. torquata*; *M. sherriffii* was discovered by Ludlow and Sherriff during their 1936 expedition to northern Bhutan. In addition, they first discovered on the Shagma La, Tsari, in south-eastern Tibet the yellow form of *M. prainiana*, var. *lutea* (first assigned to *M. horridula* by George Taylor). In this work I have described a new species from eastern Bhutan in honour of Frank Ludlow (*Meconopsis ludlowii*) from a specimen first collected by the Ludlow-Sherriff partnership on the Orka La in 1934. It seems only fitting that Ludlow should be honoured, especially as his collecting partner has long been recognised in *Meconopsis sherriffii*, described in 1937.

Oleg Vladimirovitch Polunin (1914–1985)

One of three English brothers (Nicholas 1909–1997, was an environmentalist and Ian 1920–2010, a photographer and ethnographist) Oleg Polunin was a teacher, botanist and traveller, who taught at Charterhouse School, Godalming, for more the 30 years. He became best known for his series of popular and authoritative books on the European and Himalayan floras, the last, *Flowers of the Himalaya* written in collaboration with Adam Stainton. He travelled widely in the Himalaya collecting in conjunction with others such as Sykes & Williams from the British Museum, adding greatly to our knowledge on *Meconopsis* in the process, but many other plants as well; the first set of their collections are to be found at the British Museum (Natural History).

David Prain (1857–1944)

Born in Fettercairn in Kincardineshire, Scotland, Prain was a graduate of Edinburgh University, becoming a demonstrator at the College of Surgeons in the same city. Following this, he travelled to India as a physician/botanist for the Indian Medical Service. In 1887 he was appointed Curator of the Calcutta Herbarium and in 1898 Director of the Royal Botanic Garden in Calcutta and of the Botanical Survey of India. During this period he wrote numerous scientific papers including important ones on *Meconopsis*, describing new species such as *M. bella*, *M. grandis*, *M. primulina* and *M. sinuata*. In 1905 he returned to Britain taking up the important post of Director of the Royal Botanic Gardens, Kew. For his services to science he was awarded a knighthood in 1912. There is no doubt that David Prain contributed very significantly to our knowledge and understanding of the genus *Meconopsis*. Other species that bear his authorship include *M. argemonantha*, *M. discigera*, *M. forrestii*, *M. impedita*, *M. latifolia*, *M. speciosa*, *M. torquata* and *M. venusta*.

Nicolai Mikhailovich Przewalski (1839–1888)

Przewalski was an extraordinary man, soldier, explorer, geographer and naturalist, who was borne near Smolensk in Russia. His first great mission was to the Ussuri region of eastern Siberia from 1867 to 1869, but from 1870 until 1885 he made extensive journeys in central Asia, particularly northern China, collecting innumerable zoological and botanical (herbarium) specimens, the latter primarily for the St Petersburg Botanic Garden. He also gathered many seeds of plants for the botanic garden and other places in Europe. He died in Karakol at the commencement of another journey. He is credited with the discovery of *Meconopsis integrifolia*, *M. punicea*, *M. quintuplinervia* and *M. racemosa*.

William Purdom (1880–1921)

As a young man Purdom, who had trained at Kew, was sent by the firm of Veitch & Sons and the Arnold Arboretum to explore the Yellow River Valley in northern China. As a result of his enthusiasm many plants and seeds were dispatched to Britain and America. Later he was appointed Inspector of Forests by the Chinese Government. In 1914 he joined forces with Reginald Farrer on an ambitious two year expedition to Gansu and the Tibetan borderlands. As a result many new plants were collected and discovered and these are vividly described in Farrer's book *On the Eaves of the World*, published in 1917.

George Sherriff (1898–1967)

He was born in Larbert, Stirlingshire, Scotland and became a soldier (attaining major in the British army) and later a plant hunter. His famous collecting partnership with Frank Ludlow (see above), one of the best known and most scrupulously recorded partnerships in twentieth century plant exploration, led to extensive herbarium collections (housed primarily at the British Museum, Natural History) and the introduction of innumerable plants and seed from the eastern Himalaya and adjacent regions of Tibet. On occasions they were accompanied in the field by Sherriff's wife Betty. The Sherriff's established a renowned garden at Ascreavie in the Scottish highlands. The striking and distinctive pink-flowered *Meconopsis sherriffii*, described by George Taylor in 1937, was discovered by Sherriff on the Drichung La in southern Tibet and found some 12 years later by Ludlow & Sherriff in far greater abundance in northern Bhutan, although to be fair it was found there by two of their native helpers by the names Danon and Ramzana.

Jean André Soulié (1858–1905)

Another French missionary and plant collector, Soulié was born in Saint Juéry, Aveyron. He was stationed in China's western provinces for many years centred on Tatsienlu (now Kangding). He collected numerous dried plant specimens in the rugged mountain terrain along the Tibet frontier, his collection of some 7000 numbers being sent to the Paris Natural History Museum. Tragically he was murdered in Taregong by Tibetan monks during one of numerous uprisings in what was, in those days, extremely hostile border territory.

Adam Stainton (1921–1991)

A traveller, amateur botanist and plant collector, Stainton undertook many journeys in the Himalaya, particularly in Nepal. He co-authored the widely acclaimed book *Flowers of the Himalaya* published in 1984, and was the sole author to the supplement that followed in 1988. He made a number of gatherings of *Meconopsis*, one proving to be new to science, *M. ganeshensis* from central Nepal. Adam Stainton, William Russell (Bill) Sykes & Leonard Howard John (L. H. J.) Williams (1915–1991) made a formidable collecting team when they joined forces on a British Museum (Natural History) expedition to western and central Nepal in 1954. Their collections, some of the finest and best annotated of any Himalayan expedition, included many *Meconopsis*, especially the large evergreen monocarpic species such as *M. paniculata* and *M. regia*, as well as a new species from the Annapurna region, *M. taylori*. More recently several additional new species have been based upon their 1954 collections: *M. bulbilifera*, *M. lamjungensis*, *M. simikotensis* and *M. staintonii*. Two years previously Sykes and Williams had teamed up with Oleg Polunin (see above) in western Nepal, during which time two new taxa were discovered but only recently recognised and described, *M. chankheliensis* and the western populations of *M. grandis*, subsp. *jumlaensis*.

George Taylor (1904–1993)

A Scotsman, George Taylor was one of the most significant and influential botanists of his era with a profound knowledge of plants and a deep and lasting interest in garden design and gardening. During his early career he partook in plant collecting expeditions to South Africa and Rhodesia (1927–28) and the Ruwenzori Mountains (1934). From 1940–45 he was Deputy Keeper of Botany at the British Museum (Natural History), and later promoted to Keeper of Botany (1950–56). From 1956 until 1971 he was Director of the Royal Botanic Gardens, Kew, being knighted in recognition of his considerable promotion of the gardens during this period. Perhaps his most significant expedition was in 1938 when he joined forces with Frank Ludlow and George Sherriff on a long collecting trip to Bhutan and south-eastern Tibet: how he must have enjoyed seeing some of his cherished *Meconopsis* growing in the wild. His monograph of the genus *Meconopsis*, published in 1934 when he was a young botanist at the British Museum, has proved the standard work on the subject for many years. He is commemorated in *Meconopsis taylorii*, a species published in 1972.

Te-tsun Yú (1908–1986)

Born in Gansu province, he became a distinguished botanist and eventually Senior Professor of the Institute of Botany, Academia Sinica in Beijing (Peking). He was an authority on Chinese Rosaceae in particular and collected widely in China, particularly in Yunnan and Sichuan. He was a strong and wise collaborator with western scientists and in 1937 and 1940 led two British- and Irish-backed Chinese expeditions in these two provinces. As a result of his efforts many interesting plants were introduced into cultivation in western gardens. He made, amongst many other gatherings, some fine collections of *Meconopsis*, particularly of *M. integrifolia*, *M. pseudointegrifolia* and the fairly recently described *M. wilsonii*, which in China harboured for many years under the name *M. napaulensis*.

Nathaniel Wallich (1786–1854)

Of Danish origin, Wallich was both a surgeon and botanist working in India, first on a Danish settlement near Calcutta but later for the East India Company. Wallich developed the world renowned Calcutta Botanic Garden and assisted many plant collectors in India and the Himalaya, describing at the same time numerous newly discovered species. He built up a substantial Catalogue of dried specimens both of his and other collectors, duplicates of which were distributed in various herbaria in Europe. The Wallich Collection itself is housed at the Royal Botanic Gardens Kew and kept as a separate collection from the main collections. It contains some important type specimens of *Meconopsis* collected and described by Wallich from central Nepal: *M. napaulensis*, *M. paniculata* and *M. simplicifolia*. In addition, *M. robusta* was described from a collection which he made in NW India's Kumaon region. *Meconopsis wallichii* was named by J. D. Hooker in his honour.

Ernest Henry Wilson (1876–1930)

Known as 'Chinese' Wilson, he was born in Chipping Camden in Gloucestershire and was widely travelled, becoming both a plant hunter and author and eventually the Keeper of the Arnold Arboretum in Massachusetts, USA, who sponsored most of his important expeditions. He was best known for his explorations of the mountains of western China, especially Sichuan and Hubei, during four expeditions between 1899 until 1911. He collected a large amount of living material, both plants and seeds, as well as innumerable herbarium specimens. Although his chief interests appear to have been in the woody flora he also collected many herbs, his most famous and best known introduction being *Lilium regale* which he discovered in the Min Valley in northern Sichuan. He made many dried collections of *Meconopsis* during his travels, particularly the fairly recently described *M. wilsonii* which he first found near Baoxing (then Moupin) in western Sichuan. He also made fine gatherings of *M. henrici*, *M. integrifolia*, *M. prattii*, *M. psilonomma*, *M. punicea* and *M. racemosa*. Much of his Chinese travels are recorded in his book *A Naturalist in Western China*, published in two volumes in 1913. Tragically, both he and his wife were killed in a motorcar accident in Massachusetts in 1930.

Toshio Yoshida (b. 1949–)

A botanical photographer and amateur botanist, born in Japan, is an associate researcher at the University of Tokyo and a member of the Society of Himalayan Botany. In 2005 his seminal book *Himalayan Plants Illustrated*, the most comprehensive book of photographs of Himalayan plants ever published, received widespread critical acclaim. Yoshida has studied many plants in the field, especially *Meconopsis*, *Nomocharis* and *Primula*. He has been keen to explore parts of western and south-western China new or little visited and as a consequence of his painstaking studies he has, in association with others (David E. Boufford, Harvard University Herbarium, Cambridge, USA and Hang Sun, Key Laboratory of Biodiversity and Biogeography, Kunming Institute of Botany, Chinese Academy of Sciences, Kunming, Yunnan, in particular) discovered and named several new species: *M. balangensis*, *M. heterandra*, *M. muscicola*, *M. pulchella* and *M. yaoshanensis*. In addition, the present author has worked with Toshio Yoshida naming several new species based on field and herbarium studies: *M. bhutanica*, *M. bulbilifera*, *M. exilis* and *M. lamjungensis*.

APPENDIX 3

CHINESE PLACE NAMES

It is often very difficult to trace Chinese place names, especially those used in the older literature. In 1958 the People's Republic of China (PRC) replaced the Wade-Giles Mandarin Romanisation system by the Pinyin system, which was adopted by the International Organisation for Standardisation in 1982. Some important and classic localities in western China, particularly those known to the earlier European and American explorers, plant-hunters and naturalists, have changed. In respect to the Chinese and to modern collectors the Pinyin equivalents are presented below. I have, however, kept to the familiar (pre Pinyin) names for rivers and some other geographical locations as these are those that will be familiar to the reader and are still widely used in China today; for instance the Mekong, Salween and Yangtse rivers. Besides this, these names are widely used in literature and in collectors' field notes. Some other equivalents are also included in the list below:

Atendse = Deqin
Atuntsi = Deqin
Beima Shan = Baimashan
Big Snow Mountain = Daxueshan
Bijiang (a county name now included in Fugong county, NW Yunnan)
Burma = Myanmar
Chienchuan = Jian Chuan
Chungtien = Zhongdian (sometimes referred inappropriately to as Shangri-La)
Dadjienlou = Kangding
Dali Range = Cangshan
Dalifu = Erhai
Dardo = Kangding
Dechen - Deqin
Diqin = Deqin
Domggrergo = Huanglong
Dschöngdu = Chengdu
Dschungdien = Zhongdian
Erhlang Shan = Erlangshan
Four Maidens Mountain = Siguniangshan
Four Sisters Mountain = Siguniangshan
Ganghoba = Gangheba
Gaocheng = Litang
Genyen Shan = Genyanshan
Gompa La = Kawagebo (the highest mountain in the Gaoligongshan, see also under Karkapo which refers to a different mountain in NW Yunnan)
Guenhenshan = Genyashan
Ha-la-ma = Hongyuan
Heng Tuan = Hengduanshan
Hianglong = Huanglongsi
Hsiagwan = Xiaguan

Hsia-Kuan = Xiaguan
Hupeh = Hubei
Hurama = Hongyuan
Ichang = Yichang
Jade Dragon Snow mountain = Yulongxueshan
Jone = Zhuoni
Kagurpo = Kawagebo
Kansu = Gansu
Kao-li-kung = Gaoligongshan
Karkapo = Kawagebo = Meilixueshan
Kawagabu = Kawagebo
Kyirong = Jilong
Leishan = Lechan
Lichiang = Lijiang
Lidjiang = Lijiang
Likiang = lijiang
Litiping = Lidiping
Little Snow Mountain = Xiaoxueshan
Little Zhongdian = Xiaozhongdian
Lu-ting = Luding
Mekong River = Lancangjiang
Min River = Minjiang
Minya Konka = Gonggashan
Moupin = Baoxing
Mt Emei = Emeishan
Mt Genyuen = Genyanshan
Mt Omei = Emeishan
Omei Shan = Emeishan
Pei-mas Shan = Baimashan
Salween River = Nujiang
Schigu = Shigu
Shangri-La = Zhongdian (the name Shangri-La was proposed in 2002 to promote tourism)

Shan-hsi = Shaanxi (also Shanxi)
Shansi = Shaanxi (also Shanxi)
Shensi = Shaanxi
Shi-ch'ang = Xichang
Shih-kou = Shigu
Shikou = Shigu
Shiku = Shigu
Shweli = Lungchuanchiang
Siakwan = Xiaguan
Sinkiang = Xinjiang
Sung-p'an = Songpan
Szechuan = Sichuan
Tachienlu = Kangding
Ta-chien-lu = Kangding
Tali = Dali
Tali Range = Cangshan
Talifu = Erhai
Tasienlu, Tatsienlu = Kangding
Teng-yueh = Tengchong

Tien Bao = Tianbao
Tiger Leaping Gorge = Hutiao Xia
Tsangpo River = Yarlung Zangbojiang
Tulung = Dulong
Upper Yangtse River = Jinshajiang
Varma-la = Nyima La
Weishi - Weixi
Wengsui = Wengshui
Western Hills = Xishan
Woolong = Wolong
Ya-chou-fu = Ya'an
Yangtse River = Jinshajiang (in upper part), Changjiang (lower part)
Tsarong = Cawarong
Yulong Shan = Yulongxueshan
Yunnanfu = Kunming
Yungning = Yongning
Zitsa Degu = Juizhaigou

INDEX TO SCIENTIFIC NAMES

Page numbers for illustrations are in **bold**.

Abies 72, 79, 114, 293, 371
   spectabilis 125
Ajuga lupulina 278
Allium kansuense 309
Androsace 2, 351
   lehmannii 130
   robusta subsp. *purpurea* 130
Anemone 343
   coronaria 310
Angelica 344
Argemone 5, 9
   cambrica (L.) Desf. 5, 367
     Cambro-Britanica lutea John Parkinson 368
     lutea Cambro-Britanica John Parkinson 368
Asplenium scolopendrium 326

Berberis 194
   temolaica 333
Betula 210
Bistorta 345

Campanula 212
Cardamine macrophylla 278
Carex incurva 130
Cassiope fastigiata 128
   wardii 333
Cathcartia Hook. f. ex Hook. 5–8, 10, 11, 15, 17, 41, 42, 177, 190, 206, 369, 370, 372, 376, 386, 389
   section *Chelidonifoliae* (Prain) Grey-Wilson 370, 374
   section *Cumminsia* Prain 45, 47
   section *Eucathcartia* Prain 369
   section *Villosae* (G. Taylor) Grey-Wilson 370
   betonicifolia (Franch.) Prain 137
   chelidonifolia (Bureau & Franch.) Grey-Wilson 5, 22, 365, 369, 370, 374, **375**, **376**, 377
   delavayi Franch. 5, 315, 369
   integrifolia Maxim. 176, 177, 187, 369
   lancifolia Franch. 5, 318, 369
   lyrata H. A. Cummins & Prain ex Prain 339, 369

   oliveriana (Franch.) Grey-Wilson 369, 370, **375**, 376
   polygonoides Prain 345, 369
   punicea Maxim. 204, 369
   smithiana Hand.-Mazz. 365, 369, 370, **371**, 373, 374
   villosa Hook. f. ex Hook. 9, 22, 365, 369, 370, **371**, **372**, 373, 374, 377
Cerastites Gray 5, 8, 367, 386
   cambricus (L.) Gray 367
   laciniatus 8
   macrocephalus 8
Chelidonium majus 25, 372, 376
Cistaceae 25
Cistus 25
Codonopsis bhutanica 130
   thalictrifolia 128
Corydalis 347, 351
   calycosa 278
Cremanthodium oblongatum 130
Cypripedium 347
   tibeticum forma 278

Decaisnei fargesii 388
Diapensia himalaica 333

Epilobium 344

Fabaceae 25
Festuca tibetica 130

Gentiana 28, 256, 309
   sino-ornata 389
Geranium 344
   pogonanthum 278

Helianthemum 25
Hippophae 139
Hyacinthus orientalis 15

Impatiens 389
Incarvillea himalayensis Grey-Wilson 80, 158

Larix 210
Ligularia 344
Liliaceae 25

Lilium 28, 389
   nanum 352
   regale 100, 391
Lloydia serotina 130

Meconopsis Vig. emend. Grey-Wilson 42
   group *Grandes* 186
   group *Primulina* 310
   group *Primulinae* 310
   section *Aculeatae* (Prain) Fedde 45, 49, 216
   section *Bellae* (Prain) Fedde 22, 47, 49, 358
   section *Cambricae* (Prain) Fedde in Engl. 6, 8, 43, 367
   section *Chelidonifoliae* (Prain) Fedde 374
   section *Cumminsia* (Prain) Grey-Wilson 22, 47, 232, 329, 338
   section *Eucathcartia* (Prain) G. Taylor 6, 8, 43, 369
   section *Eumeconopsis* Prain 6, 367, 369
   section *Forrestianae* C. Y. Wu & H. Chuang 22, 46, 49, 318
   section *Forrestii* C. Y. Wu & H. Chuang 6, 46
   section *Grandes* (Prain) Fedde 14, 44, 48, 136, 193
   section *Horridulae* 300
   section *Impeditae* Grey-Wilson 46, 49, 281, 300, 317
   section *Meconopsis* 43, 48, 56, 57, 60, 62, 70
   section *Nyingchienses* L. H. Zhou 7, 45
   section *Pinnatifoliae* C. Y. Wu & N. Zhang 7, 44
   section *Polychaetia* Prain 1, 6–8, 31, 43–48, 56, 70, 75, 78, 79, 87, 93, 100, 107, 116, 118, 374
   section *Primulinae* (Prain) Fedde 15, 46–48, 52, 334, 352, 365
   section *Racemosae* C. Y. Chuang & H. Chuang 6, 44, 45, 49, 232
   section *Robustae* (Prain) Fedde 12, 43, 60
   section *Simplicifoliae* (G. Taylor) C. Y. Wu & H. Chuang 6, 7, 44–48, 136, 193

THE GENUS MECONOPSIS
INDEX TO SCIENTIFIC NAMES

section *Superbae* 31
section *Torquatae* (Prain) Fedde 44, 118
series *Aculeatae* Prain 6, 8, 45, 46, 48, 219
series *Bellae* Prain 6, 8, 47
series *Cambricae* Prain 367
series *Chelidonifoliae* Prain 6, 367, 369, 374
series *Cumminsia* (Prain) Prain 6, 47, 337, 351
series *Decorae* 6
series *Delavayanae* G. Taylor 6, 8, 44–47, 49, 281, 315, 318
series *Grandes* Prain 6, 8, 44, 45, 48, 136, 137, 144, 152, 153, 158, 338
series *Henricanae* C. Y. Wu & H. Chuang 44, 46, 49, 281, 304
series *Heterandrae* Grey-Wilson 7, 46, 49, 232, 237, 275
series *Impeditae* 46, 49, 281, 295, 303, 315, 318
series *Integrifoliae* Grey-Wilson 6, 44, 48, 136, 172, 175
series *Lancifoliae* 318
series *Lyratae* 17, 48, 329
series *Polychaetia* 43, 48, 70, 80, 81, 88, 117, 338
series *Primulinae* Prain 6, 8, 46–48, 329, 331, 334, 337
series *Puniceae* Grey-Wilson 45, 48, 193, 204
series *Racemosae* 45, 49, 232, 237
series *Robustae* Prain 6, 8, 43, 48, 70, 338
series *Simplicifoliae* G. Taylor 6, 8, 44, 45, 48, 152, 193
series *Superbae* 6, 8, 12, 43
series '*Surperbae*' 43
series *Torquatae* Prain 44, 118
series *Villosae* G. Taylor 6, 369, 370
subgenus *Aculeatae* 14
subgenus *Cumminsia* (Prain) Grey-Wilson 1, 7, 8, 12, 13, 22, 33, 45, 48, 119, 232, 337, 358
subgenus *Discogyne* G. Taylor 6–8, 12, 19, 20, 22, 33, 44, 48, 118–120, 125, 129, 131, 134, 135
subgenus *Eumeconopsis* (Prain) Fedde 6, 8, 43–47, 367, 369
subgenus *Grandes* (Prain) Grey-Wilson 1, 7, 8, 12, 14, 22, 33, 44, 48, 119, 136, 152
subgenus *Meconopsis* (Prain) Fedde 1, 2, 7, 8, 11, 12, 14, 22, 25, 27, 31, 33, 43, 48, 56, 59, 72, 76, 82, 99, 105, 110, 119, 152

subgenus *Polychaetia* (Prain) Fedde in Engl. 44, 118, 374
subsection *Cumminsia* (Prain) G. Taylor 6, 8, 43–47, 216
subsection *Eupolychaetia* 6, 7, 43, 82
aculeata Royle 1, 5, 6, 22, 18, 45, 54, 216, **217**, 218, **219–222**, 224, 225, 228, 243, 244, 262, 318, 355
  f. *acutiloba* Prain 217
  f. *normalis* Prain 217
  var. *nana* Prain 217
  var. *typica* Prain 217
aculeata T. Smith non Royle 223
argemonantha Prain 6, 18, 47, 53, 262, 318, 329, **330**, **331**, 332–334, 388–390
  var. *argemonantha* 332, 334
  var. *genuina* G. Taylor 332
  var. *lutea* G. Taylor 262, 330, 332, 334
×*auriculata* Stapf 33
autumnalis P. A. Egan 14, 43, 51, **74**, 80, 82, 113, **114–116**
baileyi Prain 1, 9, 11, 12, 14, 18, 22–25, 27, 28, 30–33, 37, 40, 44, 52, 118, 135–137, 139–142, 144, **145**, 146, 148, 149, 151, 153, 158–166, 202, 331, 333, 387–389
  'Hensol Violet' **148**
  'Inverewe' **150**
  subsp. *baileyi* **138**, 139, **142**, 146, **147**, 150
  subsp. *multidentata* Grey-Wilson **138**, 146, 150, **151**
  subsp. *pratensis* Kingdon Ward ex Grey-Wilson **138**, 139, **142**, 146, 148, **149**, **150**, 389
  var. *alba* 148
  var. *pratensis* Kingdon Ward 148, 149
balangensis Tosh. Yoshida, H. Sun & Boufford 12, 14, 17, 35, 46, 54, 180, 237, 257, 267, **275–278**, **280**, 297, 306, 391
  var. *atrata* Tosh. Yoshida, H. Sun & Boufford **279**
  var. *balangensis* 278, 279
barbiseta C. Y. Wu & H. Chuang ex L. H. Zhou 289, 308, 309, 311
'Barney's Blue' 162
×*beamishii* Prain 30, 33, 163
bella Prain 6, 12, 14, 15, 18, 20, 37, **40**, 41, 47, 55, 216, 288, 291, 292, 316–318, 353, 358–361, 388–390
  subsp. *bella* 359, **360**, **361**, 362, 363, 365
  subsp. *grandifolia* Grey-Wilson 359, 360, **362**

subsp. *subintegrifolia* Grey-Wilson 318, **358**, 359, 360, **362–365**
betonicaefolia 141
betonicifolia Franch. 5, 6, 11, 15, **16**, 18, **21**, 22, 23, 44, 52, 118, 137, **138**, 139, **140–144**, 145, 146, 151–153, 160, 161, 163, 166, 214, 331, 369, 388, 389
  baileyi 140, 142, 148
  f. *baileyi* (Prain) Cotton 140, 141, 145, 151
  f. *betonicifolia* 141
  f. *franchetii* Stapf 137, 141
  var. *baileyi* (Prain) Edwards 140, 145, 149, 151, 389
  var. *franchetii* (Stapf) L. H. & E. Z. Bailey 137
  var. *pratensis* 149
betonicifolia hort. 33, 159, 163, 164
betonicifolius 141
bhutanica Tosh. Yoshida & Grey-Wilson 17, 44, 50, 119, 120, 122, **123–126**, 127, 131, 264, 388, 389, 391
bijiangensis H. Ohba, Tosh. Yoshida & H. Sun 45, 54, 237, 254, 268–270, 274, 276, 326
  subsp. *bijiangensis* **269–271**, 276
  subsp. *chimiliensis* Grey-Wilson 271, 276
biloba L. Z. An, Shu-Y. Chen & Y. S. Lian 209, 214
'Bobby Masteron' 164
brevistyla Hort. ex Prain 172, 184, 189
'Bryan Conway' 164
bulbilifera Tosh. Yoshida, H. Sun & Grey-Wilson 11, 47, 53, 334, **338**, **340**, 341, 342, 376, 390, 391
calciphila Kingdon Ward 233, 234, 257, 258
cambrica (L.) Vig. 5–10, 17, 41, 42, 221, 222, 233, 356, 365, 367, 369, 374, 386
castanea H. Ohba, Tosh. Yoshida & H. Sun 273, 274, 326
cawdoriana Kingdon Ward 231, 232
chankheliensis Grey-Wilson 43, 50, 70, 75, 79, 80, 82, 388, 390
chelidonifolia Bureau & Franch. 5–7, 9, 11, 41, 365, 369, 374
compta Prain 15, 20, 47, 53, 118, 334, 339, 344, 345, 347, 348, **349**, **350**, 351, 389
concinna sensu D. G. Long non Prain 350

INDEX TO SCIENTIFIC NAMES

concinna Prain 11, 12, 14, 15, 46, 55, 56, 281–283, **284**, **290–292**, 295, 297, 303, 334, 336, 347, 351, 352, 389
×*cookei* G. Taylor 22, 26, 32, **39**, 64, 210, 213, **214**, **215**, 216
×*coxiana* G. Taylor 31, 33
crassifolia Benth. 5, 9, 365
cyanochlora Farrer in mss 271
×*decora* Prain 33
delavayi (Franch.) Franch. ex Prain 1, 6, 9, 18, **20**, 22, 30, 33, 41, 46, 55, 216, 310, **315–317**, 318, 320, 325, 328, 347, 358, 388
dhwojii G. Taylor ex Hay 1, 6, **13**, 18–20, 22, 28, 33, 43, 50, 70, **71**, 72, **73**, 75, **76–79**, 80, 388
diphylla DC. 365
discigera Prain 6, 9, 11, 17, 20, 44, 50, 118–120, **121**, **122**, 123–127, 129, 131–135, 264, 318, 389, 390
duriuscula Prain 238
eucharis Farrer ex Irving 324
exilis Tosh. Yoshida, H. Sun & Grey-Wilson 47, 53, 334, 339, **342**, **343**, 344–346, 349, 356, 391
eximia Prain 235, 320, 322, 389
×*finlayorum* G. Taylor 33
florindae Kingdon-Ward 6, 18, 47, 53, 329, 331, **332**, 333, 334, 389
forrestii Prain 6, 11, 14, 18, **21**, 47, 56, 318, 326, 327, **328**, 329, 389, 390
ganeshensis Grey-Wilson 43, 51, **74**, 81, 82, 87, **88**, 89, 390
georgei G. Taylor 6, 9, 16, 18, **21**, 45, 55, 237, 254, 264, 270, 272–274, 276, 318, 319, 324, 326, 329, 389
   f. *castanea* (H. Ohba, Tosh. Yoshida & H. Sun) Grey-Wilson **272–274**, 276
   f. *georgei* 274
George Sherriff group 165, 166
glabra Hook. 365
gracilipes G. Taylor 6, 18–20, 22, 40, 43, 50, 70, **71**, 72, **73**, 74, 75, **76**, 78, 80, 87, 388
grandis Prain 1, 6, 9, 11, 14, 15, 18, 20, 22, 23, 25, 27–33, 39, 44, 52, 80, 136, 137, 140, 144, 151–155, 157–166, 169, 170, 172, 186, 194, 200, 313, 351, 387, 390
   GS600 160–162, 165
   'Prain's Variety' 164
   subsp. *grandis* 18, 23, 32, 33, 151, 152, **153**, **154**, 155, **156**, 157, **160–162**, 164, 166

subsp. *jumlaensis* Grey-Wilson 18, 23, 152, **153**, 155, 157, 158, **159**, 165, 390
subsp. *occidentalis* Grey-Wilson 155
subsp. *orientalis* Grey-Wilson 18, 23, 127, 151, 152, **153**, **154**, 155, **157**, **158**, 160, 162, **163–165**, 166, 389
guilelmi-waldemarii Klotzsch & Garcke 217
×*harleyana* G. Taylor 30, 33, 163, 202, 203, 389
henrici Bureau & Franch. 5, 6, 11, 17, 18, 46, 55, 207, **276**, 304, **305**, 306, **307**, 308–314, 318, 391
   var. α *genuina* G. Taylor 305
   var. *genuina* 308
   var. *henrici* 308
   var. *psilonomma* (Farrer) G. Taylor 308, 314
heterandra Tosh. Yoshida, H. Sun & Boufford 14, 17, 46, 54, 237, 275, **276**, 277, 279, **280**, 306, 391
heterophylla Benth. 5, 9, 365
himalayensis (*Himalayense*) Hort. ex Hay 104
horridula aggregate 232, 236, 237, 242, 254, 267
'*Horridula*' complex 172, 224, 233, 234, 235, 237, 252, 256, 261, 300
horridula Hook. f. & Thomson 2, 5, 6, 9, 12, **13**, 14–16, 18, 19, 22, 27, 29, 35, 37, 39, **40**, 45, 54, 78, 105, 129, 133, 134, 169, 170, 173, 216, 225, 228, 230, 232–240, **241**, 242–244, 248–250, 253–255, 257, 260, 262, 264, 267, 288, 294, 300, 313, 318, 320, 355, 388, 389
   forma **241**, **244**, 245
   subsp. *drukyulensis* Grey-Wilson 240, **245–247**
   subsp. *horridula* 16, **21**, **233**, **238**, **239**, **241–243**, 245, 247
   subsp. *tibetica* K. B. Jørk mss 261
   var. *abnormis* Fedde in Engl. 238
   var. *lutea* G. Taylor 262, 263, 274
   var. *prattii* Prain 233
   var. *racemosa* (Maxim.) Prain 233, 249, 257
   var. *rudis* Prain 233, 265
   var. *spinulifera* L. H. Zhou 17, 244, 245, 254
   var. *typica* Prain 238
horridula Hort. 1, 242–244, 26
horridula Kingdon Ward 238
×*hybrida* Puddle ex *Gard. Chron.* 33, 200

impedita Prain 6, 11, 12, 14, 15, 18, 40, 46, 56, 118, 230, 231, 262, 281–284, **285**, 286–289, 294, 295, 297, 300, 302–304, 317, 334, 349, 389, 390
   subsp. *impedita* 20, 21, 283, **284**, **285**, **286–288**
   subsp. *morsheadii* 286, 388
   subsp. *rubra* (Kingdon Ward) Grey-Wilson 257, 283, 284, **287**
   var. *morsheadii* Prain 231, 282
   var. *rubra* Kingdon Ward 286
integrifolia (Maxim.) Franch. 1, 5, 6, 9, 14, 15, 18–20, 22, 27, 28, 30, 33, 35–37, **39**, 40, 44, 45, 51, 136, 144, 159, 163, 171–178, 180–184, 186–190, 192, 202, 203, 207, 210, 212, 214, 262, 389–391
   subsp. *integrifolia* 36, 38, 39, **172**, 174, **176**, 177, 178, **179**, **180**, 181, 311, 313
   subsp. *lijiangensis* Grey-Wilson 182, 183
   subsp. *souliei* (Fedde) Grey-Wilson **16**, **38**, **173**, 174, 177, 178, **179**, 180, **181**, 278
   var. *microstigma* Prain ex Kingdon Ward 184, 187
   var. *souliei* Fedde 180
   var. *uniflora* C. Y. Wu & H. Chuang 184
'Ivory Poppy' Kingdon Ward 202
'Kingsbarns' 165
×*kongboensis* Grey-Wilson 26, 32, 64, **203**, 214, 389
lamjungensis Tosh. Yoshida, H. Sun & Grey-Wilson 47, 53, 334, 339, 344–346, 356, 390, 391
lancifolia Franch. ex Prain 6, 9, 11, 13, 14, 18, 47, 56, 219, 273, 274, 291–295, 297, 315, 317, 318, **319**, 320, 324–326, 328, 329, 352, 389
   subsp. *eximia* (Prain) Grey-Wilson **13**, **39**, 273, 320, 321, **322–324**, 325, 389
   subsp. *lancifolia* 320, **321**, 325
   subsp. *lepida* (Prain) Grey-Wilson 320, 324, 325
   var. *concinna* (Prain) G. Taylor 290, 336, 351–353
   var. *georgei* (G. Taylor) G. Taylor mss 272
   var. *limprichtii* Fedde ex Limpr. f. 321
   var. *solitariifolia* Fedde 321

THE GENUS MECONOPSIS
INDEX TO SCIENTIFIC NAMES

latifolia (Prain) Prain 6, 18, 22, 33, 45, 54, 216, 218–220, **223**, **224**, 225, 228, 257, 262, 355, 390
latifolia sensu Silva Tarouca & Schneider 217
leonticifolia Hand.-Mazz. 298
lepida Prain 310, 324, 389
lhasaensis Grey-Wilson 45, 54, **234**, 236, 237, **241**, **248**, **249**
lijiangensis (Grey-Wilson) Grey-Wilson 39, 44, 52, **174**, 175, 177, **179**, **182–184**, 187
'Lingholm' 23, 28, 159, 164
longipetiolata G. Taylor ex Hay 6, 22, 81, 87
ludlowii Grey-Wilson 47, 53, 263, 292, 334, 339, **349**, 350, **351–353**, 389
lyrata aggregate 329, 336, 338, 344–349, 355, 356
lyrata (H. A. Cummins & Prain) Fedde ex Prain 6, 9, 15, 18, 19, 45, 47, 53, 118, 334–336, **337**, 338, 339, **340**, 341, 342, 344–346, 350, 356
lyrata sensu Taylor 339, 340, 342, 345, 348
'Maggie Sharp' 164
manasluensis P. A. Egan 44, 49, 119, 120, 124, **134**, **135**
morsheadiana 286
morsheadii (Prain) Kingdon Ward 282, 286–288
morshediana 231
×*musgravei* 31, 33
muscicola Tosh. Yoshida, H. Sun & Boufford 12, 46, 56, 281, 282, **284**, **293–295**, 391
napaulensis DC. 5, 6, 9, 10, 12, 14, 18, 19, 22, 31, 33, 40, 43, 51, 60, 61, 69, 74, 78–81, **82**, **83**, 84, **85**, **86**, 87, 89–91, 93, 98, 99, 106, 107, 110, 116, 318, 387, 388, 391
napaulensis (hort.) 33, 84
napaulensis Walp. non DC. 104
neglecta G. Taylor 6, 45, 46, 54, 216, 218, 219, 222
nepalensis var. *elata* B. 104
nipalensis Hook. f. & Thomson 81, 104, 106
nyingchiensis L. H. Zhou 45, 52, 193, 198, 200, **201**, 202, 232
oliveriana Franch. ex Prain 6, 7, 9, 41, 365, 369, 376, 377
'Old Rose' 216
'Ormswell' 164
ouvrardiana Hand.-Mazz. 225

paniculata (D. Don) Prain 1, 5, 6, 10–12, **13**, 14–16, 18–20, **21**, 22, 24, 25, 28, 30, 31, 33, 35, 40, 43, 51, 58–62, 68, 72, 79, 80, 82–84, 87, 88, 91, 93, **104**, 105, **106–110**, 111, 114–118, 151, 152, 158, 169, 170, 263, 264, 273, 331, 353, 389–391
  Ghunsa form 106, **110**
  subsp. *paniculata* **58**, **74**, 105, 107–109, 111, 112
  subsp. *pseudoregia* Grey-Wilson **58**, 105, 107, **112**, **113**, 115
  var. *elata* Prain 104
  var. *paniculata* 110, 111
  var. *rubra* Grey-Wilson 109, 110, **111**
  var. *typica* 104
petiolata DC. 365
pinnatifolia C. Y. Wu & X. Zhuang 44, 50, 119, 120, 124, 132
polygonoides (Prain) Prain 47, 53, 339, 344, **345**, 346, 349
prainiana Kingdon Ward 15, 16, **21**, 45, 54, 233, 234, 236, 237, 240, **241**, 254, 258, 260–264, 273, 300, 313, 389
  var. *lutea* (G. Taylor) Grey-Wilson **236**, **263**, **264**, 389
  var. *prainiana* **262**, 263, 264
prattii Prain 1, 18, 45, 55, 225, 235–237, **241**, 242–244, 250–254, **255–258**, 260, 261, 300, 318, 391
primulina Prain 6, 18–20, 47, 52, 329, 331–334, **335**, **336**, 345, 351, 353, 356, 389, 390
principis Bulley 305
pseudohorridula C. Y. Wu & H. Chuang 231, 232
pseudointegrifolia Hort. 1, 189
pseudointegrifolia Prain 15, **40**, 44, 51, 136, 144, 172–175, 178, **179**, 183, 184, **185**, **186**, 187, 189, 191, 202, 203, 262, 389, 391
  subsp. *daliensis* Grey-Wilson 177, 186–188, 389
  subsp. *pseudointegrifolia* 177, 186, 188
  subsp. *robusta* Grey-Wilson 186, 188
  var. *brevistyla* Prain ex Kingdon Ward 183, 184, 187
pseudovenusta G. Taylor **viii**, 6, 11, 12, 14, 18, 22, 35, 37, 46, 55, 281, 283, **284**, 295, 298, 300, **301–304**, 318, 389
psilonomma Farrer 17, 27, **37**, 39, 46, 55, 207, 289, 304, 306, **308–312**, 314, 389, 391

var. *sinomaculata* (Grey-Wilson) Ohba 312
psilonomma Wehrh. non Farrer 223
pulchella Tosh. Yoshida, H. Sun & Boufford 46, 56, 282, **284**, 295, **296**, **297**, 391
punicea Maxim. 1, 5, 6, 9, 11, 15, **16**, 17–19, 26, 27–29, 32, **36**, **37**, **39**, 45, 52, 136, 180, 193, **204–208**, 210–216, 309–311, 313, 390, 391
  f. *albiflora* L. H. Zhou 208, 211
  var. *elliptica* Z. J. Cui & Y. S. Lian 204
  var. *glabra* M. Z. Lu & Y. S. Lian 204
  var. *limprichtii* Fedde 209
quintuplinervia Regel 1, 5, 6, 9, 11, 14, 18, 19, 22, 26, 27, 29, 32, 33, **39**, 40, 45, 52, 136, 143, 193, **204–206**, 208, **209–212**, 213–216, 253, 309–311, 313, 318, 390
  'Kaye's Compact' form 33, **34**, **213**
  var. *glabra* M. C. Wang & P. H. Yang 210
  var. *quintuplinervia* 210
racemosa Maxim. 1, 5, 17, 18, 22, 36, 40, 45, 55, 233–237, **241**, 242–245, 249, 250, **251–254**, 255–261, 265, 324, 390, 391
  forma *horridula* Farrer 238
racemosa sensu Silva Tarouca 217
×*ramsdeniorum* G. Taylor 33, 79
rebeccae Debnath & M. P. Nayar 74
regia G. Taylor 1, 6, 9–12, **13**, 14, 18, 20, 22, 25, 28, 30, 31, 33, 42, 43, 50, 56, 57, **58–63**, 64–66, 69, 84, 91, 110, 112, 113, 264, 273, 386, 387, 390
rigidiuscula Kingdon Ward 238
robusta Hook. f. & Thomson 5, 6, 18, 40, 43, 50, 56, 70, **71–74**, 75, 78, 81, 82, 100, 186, 188, 388, 391
  var. *gracilis* Kingdon Ward
robusta Prain 104
rubra (Kingdon Ward) Kingdon Ward 118, 257, 286, 288
rudis (Prain) Prain 12, 14, **20**, 22, 45, 54, 181, 224, 234–237, **241**, 242–244, 254, 255, 257, 261, **265–268**, 274, 277, 278, 294, 300, 324, 388
  forma 268, **378**
  var. *prattii* mss? 254
×*sarsonsii* Sarsons ex *Gard. Chron.* 30, 33, 163
×*sheldonii* 30, 32, 164, 165
'Sherriff 600' 160, 161

sherriffii G. Taylor **4**, 6, 14, 18, 22, 33, 44, 52, 136, 137, 144, 152, 153, **167–171**, 190, 351, 389, 390
'Sikkim Form' 166
simikotensis Grey-Wilson 44, 49, 119, 120, 124, **130**, **131**, 390
simplicicaulis Wood 193
simplicifolia (D. Don) Walp. 5, 6, 10, 11, 14–16, 18, 19, 22, 23, 26, 30, 32, 33, 44, 45, 52, 93, 128, 129, 136, **142**, 151, **156**, 159, 163, 164, 166, 169, 170, 186, 193–195, 199, 200, 202, 203, 212, 262, 331, 388, 389, 391
  baileyi Farrer 193
  'Bailey's Form' 199, 202
  Bailey's var. 202
  subsp. *grandiflora* **194**, 195, 197, **198–200**, 389
  subsp. *simplicifolia* 195, **196**, **197**, 199, **203**
  var. *baileyi* Kingdon Ward 193
  var. *eburnea* Hort. ex R. L. Harley 202
'Single-headed Blue' 166
sinomaculata Grey-Wilson 16, 17, 27, 39, 46, 55, **276**, 304, 306, 309, 311, 312, **313**, **314**
sinuata Irving non Prain 223
sinuata Prain 6, 18, 19, 47, 53, 224, 255, 262, 335, 338, **349**, 350, 351, 353, 354, **355–357**, 389, 390
  var. *latifolia* Prain 223
  var. *prattii* Prain 233, 254, 355
  var. *sinuata* 355
  var. *typica* Prain 354, 355
'Slieve Donard' 23, 28, 163, 164
smithiana (Hand.-Mazz.) G. Taylor ex Hand.-Mazz. 6, 7, 41, 365, 369, 373
souliei (Fedde) Farrer 180
speciosa Prain 6, 11, 14–16, 18, 22, 45, 54, 202, 216, 219, 225, 226, **227**, 228, 230, 232, 273, 318, 333, 389, 390
  subsp. *cawdoriana* (Kingdon-Ward) Grey-Wilson **21**, 226, **227**, 228, **231**, 232, 248
  subsp. *speciosa* **21**, **225**, 226, **227**, 228, **229**, **230**, 231, 232
  subsp. *yulongxueshanensis* 226, 228, 231
staintonii Grey-Wilson 1, 12, 14, 16, 22, 25, 28, 30, 31, 33, 43, 51, 59–62, 64, 65, 69, **74**, 79–82, 84, 85, 88, **89–91**, 106, 110, 390

sulphurea Grey-Wilson 1, 22, 26, 28, 32, 44, 45, 52, 136, 174, **175**, 178, 184, 186, **188–192**, 202, 203, 313, 389
  subsp. *gracilifolia* Grey-Wilson 177, 190, **192**
  subsp. *sulphurea* 16, **20**, 177, **179**, 190, 192
  var. *gracilis* 191, 192
superba King ex Prain 1, 6, 12, **16**, 18, **20**, 31, 33, 35, 43, 50, 56, 57, **58**, **60**, **65–67**, 68, 69, 110, 139, 152, 389
taylorii L. H. J. Williams 6, 18, 20, 43, 56, 57, **58**, **60**, 62, 64, 68, 69, 390, 391
  aff. *taylorii* 68
tibetica Grey-Wilson 16, 44, 49, 50, 119, 120, 122, **124**, **127–129**, 130, 131
torquata Prain 6, 9, 17, 19, 27, 35, 44, 49, 118, 120, **124**, 125, 131, 132, **133**, 134, 245, 389, 390
uniflora Gumbl. 193
venusta Prain 6, 14, 18, **21**, 35, 37, 41, 46, 55, 281, 283, **284**, 295, **298–300**, 301–304, 318, 389, 390
villosa (Hook. f. ex Hook.) G. Taylor 6–9, 17, 41, 42, 365, 369, 370, 389
violacea Kingdon-Ward 6, **21**, 31, 33, 43, 50, 51, 56, **74**, 81, 87, 99, 100, 117, **118**, 150, 257, 389
wallichii Hook. 1, 5, 14–16, **21**, 24, 28, 31, 33, 40, 43, 51, 60, **74**, 81, 82, 84, 91–95, 100, 106, 110, 116, 117, 152, 369, 389, 391
  forma *fusco-purpurea* Bulley 95
  forma *purpurea* Bulley 95
  var. *fusco-purpurea* Hook. f. 81, 82, 92, 95, **96**, **97**, 109, 110
  var. *rubro-fusca* Prain 95
  var. *typica* Prain 91
  var. *wallichii* **34**, **92–95**
wardii Kingdon Ward 305, 310
wilsonii Grey-Wilson 14, 18, 24, 28, 43, 51, 81, 82, 98–100, 102, 117, 391
  subsp. *australis* Grey-Wilson 19, **74**, 99, 100, **101**, 102, 389
  subsp. *orientalis* Grey-Wilson, D. W. H. Rankin & Z. K. Wu 99–101, **102**, **103**, 347
  subsp. *wilsonii* **98–101**
wollastoni Regel 104
'Wolong' 181

wumungensis K. M. Feng ex C. Y. Wu & H. Zhuang 46, 47, 53, 99, 332, 334, 335, 339, 345, 346, **347**, **348**, 356
xiangchengensis R. Li & Z. L. Dao 46, 56, 281, **289**, 290
yaoshanensis Tosh. Yoshida, H. Sun & Boufford 20, 47, 56, 318, 325, **326**, **327**, **391**
zangnanensis L. H. Zhou 362, 365
zhongdianensis Grey-Wilson 18, 22, 40, 45, 55, 225, **235**, 236, 237, 240, **241**, 242–244, 254, 257, 258, **259**, **260**, 261, 388

Nomocharis 27, 334, 391

Orobanche coerulescens 130

Paeonia 25
Papaver L. 1, 5, 7–10, 15, 18, 20, 24, 25, 30, 41, 42, 118, 194, 367, 369, 370, 386
  section *Argemonidium* 10, 41, 370
  section *Californicum* 10
  section *Carinatae* 10
  section *Horrida* 10
  section *Meconella* 10, 41, 42, 367, 369, 370
  section *Meconidium* 10
  section *Oxytona* 10
  section *Papaver* 10
  section *Pilosa* 10
  section *Pseudopilosa* 10
  section *Rhoeadium* 10
  section *Scapiflora* 9
  alpinum aggregate 41
  anomalum 41
  argemone 370
  californicum 7, 9
  cambrica 10
  cambricum L. 8, 221, 365, 367, 389
  cambricum perenne, flore sulphureo Dill. 368
  croceum 41
  dubium 8
  erraticum Pyrenaicum luteo flore C. Bauhin 368
  hybridum 8
  involucratum 41
  luteum perenne, laciniato folio Cambro-britannicum Ray 368
  miyabeanum 41
  orientalis 30
  paniculatum D. Don 5, 9, 81, 82, 104, 106
  radicatum 41
  rhoeas 24, 25

simplicifolium D. Don 5, 9, 193, 194
Papaveraceae 1, 6, 7, 9–11, 15, 18, 19, 24, 25, 41, 66, 119, 372, 386
Papaveroideae 9, 335
Parameconopsis Grey-Wilson 8, 10, 367
  cambrica (L.) Grey-Wilson 17, 22, 365, **366–368**
    'Aurantiaca' **366**, 368
    'Flore Pleno' 368
    'Flore Pleno Aurantiaca' 368
    'Francis Perry' 368
    'Muriel Brown' 368
    'Rubra' 368
    var. *aurantiaca* Hort. ex Wehrh. 368
    var. *cambrica* 368
Paraquilegia 360, 361
  microphylla **265**, 268, 360
Parrya forrestii 267
Pedicularis bella 130
  nana 130
Phlomis 344
Picea 152, 250
Polychaetia paniculata Wall. ex Prain 104
  scapigera Wall. 193
Potentilla 203, 250, 344
  fruticosa 128
    var. *grandiflora* 202
  microphylla 128
Primula 27, 28, 144, 149, 160, 221, 222, 309, 389, 391

atrodentata 352
cawdoriana 333
longipetiolata **38**, **181**
occlusa 352
primulina 130
pulchella 297
tangutica 309
tenella 130
walshii 130
waltonii 160
Quercus 79

Ranunculaceae 1, 361
Ranunculus 25
Rheum nobile 333
Rhodiola himalensis 128, 130
  sp. 128
Rhododendron 27, 28, 119, 128, 129, 144, 146, 148, 194, 230, 269, 270, 347, 359, 389
  anthopogon 128, 353
  'Lapponicum' 192
  laudandum 201
  lepidotum 128, 353
  rupicola 270
  setosum 128
Roemeria 7, 9, 10, 370
  argemone (L.) C. Morales, R. Mend. & Romero García 370

Romneya 9
Rosaceae 391
Roscoea purpurea 89

Salix 194, 205
  lindleyana 130
Salvia 344
Saussurea 129
Saxifraga calcicola 267
  signata 333
Solanaceae 25
Sphagnum 28
Spiraea 129
Stylomecon G. Taylor 5, 7, 9, 10, 41, 386
  heterophylla G. Taylor 5, 7, 9, 365
Stylophorum Nutt. 5, 367
  cambricum (L.) Spreng. in L. 367
  diphyllum (Michx.) Nutt. 365, 372
  nepalense (DC.) Spreng. in L. 104
  simplicifolium (D. Don) Spreng. in L. 193
Swertia kingii 352
Syrphidae 24

Tiliaceae 1
Trifolium 25
Trollius europaeus 173, 177
Tsuga 371

INDEX TO COMMON NAMES

American celandine 5
aster 178, 207

Bailey's blue poppy 140, 160
bamboo 355
barberry 139
berberid 178
'Betty's dream poppy' 161
Betty Sherriff's dream poppy 160
big blue poppies 158, 159, 163, 164
big perennial blue poppies 137, 154, 159, 161, 163, 164, 166
birch 129, 210
blood poppy 207
blue poppy 1, 2, 27, 118, 127, 146, 221, 338
broomrape 130
bumblebee 24, 270, 278
buttercup 25
butterfly 24

Cambridge blue poppy 229
Cambridge blue poppywort 229
celestial poppy 255
clover 25
corn poppy 25

dainty poppy 325
delphinium 28

evening primrose 25

Farrer's lampshade poppy 172
fir 90, 355
fritillary 325

gentian 352, 389
globeflower 173, 177
greater celandine 25, 376

harebell 357
harebell poppy 1, 136, 212
hart's-tongue fern 326
Himalayan blue poppy 9, 28, 140, 153, 199, 221, 386, 387
honey bee 23
honeysuckle 168
horridulous blue poppy 235
hoverfly 24
hyacinth 15, 226

ivory poppy 26, 202, 203

juniper 129, 168, 330, 350, 353

lampshade poppy 178
larch 210
large yellow meconopsis 172
legume 25
lettuce 78
lily 25, 222
lousewort 130

monkey-nut 22
mountain ash 129

nightshade 25

oak 90

peony 25, 207, 212
perennial big blue poppies 163
perennial blue poppies 158
poppywort 172, 207, 229
prickly poppy 251, 253, 257
primula 351, 352

rhododendron 158, 168, 202, 203, 205, 212, 262, 291, 293, 310, 334, 348, 350–352, 355, 358, 362
rockrose 25
rose 139
roseroot 130
ruby poppy 288

sapphire poppy 252
saxifrage 212
Shirley poppy 149
slipper orchid 278
snowdrop 212
sunrose 25

Tibetan blue poppy 140
Tibetan poppy 212

Welsh poppy 5, 8, 233, 368, 369, 386
white poppy 264
willow 129, 130, 205
woodland blue poppy 139

yellow poppy 173, 262
yellow poppywort 172, 173, 186

Meconopsis
Distribution of the genus